TRANSITION TO
COLLEGE
MATHEMATICS
AND STATISTICS

Christian R. Hirsch • Eric W. Hart

Ann E. Watkins • Beth E. Ritsema

with

James T. Fey • Brin A. Keller

Rebecca K. Walker • James K. Laser

McGraw Hill Education

Bothell, WA • Chicago, IL • Columbus, OH • New York, NY

mheonline.com

 This material is based upon work supported, in part, by the National Science Foundation under grant no. DRL-1020312. Opinions expressed are those of the authors and not necessarily those of the Foundation.

Send all inquiries to:
McGraw-Hill Education
STEM Learning Solutions Center
8787 Orion Place
Columbus, OH 43240

ISBN: 978-0-07-662626-7
MHID: 0-07-662626-1

Transition to College Mathematics and Statistics
Student Edition

Printed in the United States of America.

1 2 3 4 5 6 7 8 9 DOW 18 17 16 15 14

McGraw-Hill is committed to providing instructional materials in Science, Technology, Engineering, and Mathematics (STEM) that give all students a solid foundation, one that prepares them for college and careers in the 21st century.

Senior Curriculum Developers

Christian R. Hirsch (Director)
Western Michigan University

Eric W. Hart
American University in Dubai

Ann E. Watkins
California State University, Northridge

Beth E. Ritsema
Western Michigan University

Contributing Curriculum Developers

James T. Fey (Emeritus)
University of Maryland

Brin A. Keller
Michigan State University

Rebecca K. Walker
Guttman Community College

James K. Laser
Western Michigan University

Principal Evaluator

Steven W. Ziebarth
Western Michigan University

Advisory Board

Diane Briars (formerly)
Pittsburgh Public Schools

Jeremy Kilpatrick
University of Georgia

Robert E. Megginson
University of Michigan

Kenneth Ruthven
University of Cambridge

David A. Smith
Duke University

Mathematics Consultants

Christine Franklin
University of Georgia

Bernard L. Madison
University of Arkansas

Steven B. Maurer
Swarthmore College

Doris Schattschneider
Moravian College

Instructional Technology Consultant

James P. Hirsch
Plano Independent School District

Evaluation Consultant

Robert E. Reys (Emeritus)
University of Missouri – Columbia

Technical and Production Coordinator

James K. Laser
Western Michigan University

Support Staff

Hope Smith
Teresa Ziebarth
Western Michigan University

Graduate Assistants

AJ Edson
Nicole L. Fonger
Josh Goss
Western Michigan University

Transition to College Mathematics and Statistics is a fourth-year college preparatory mathematics course developed for students whose intended college program of studies does not require calculus. It was designed to help ensure college readiness for students who had completed a single-subject sequence of algebra, geometry, and advanced algebra or a three-year integrated mathematics program.

Consistent with the design of this research-based course, the student and teacher support materials were field-tested in three high schools in which students had completed courses in algebra, geometry, and advanced algebra (each school using texts from a different publisher) and in three high schools in which students had completed Courses 1–3 of an integrated mathematics program—*Core-Plus Mathematics*.

Special thanks are extended to the following teachers and their students who participated in the testing and evaluation of *Transition to College Mathematics and Statistics*.

Brighton High School
Rochester, New York
Tracy Pearce
Melissa Staloff

Cooper High School
Union, Kentucky
Eric Van Laningham

Holland Christian High School
Holland, Michigan
Brian Lemmen
Mike Verkaik

Legacy High School
Broomfield, Colorado
Daniel Hummel
Mike Meyer

Plano Senior High School
Plano, Texas
Cassandra Dailey
Elizabeth Doyle

Royal Oak High School
Royal Oak, Michigan
Jay Vought

Design, development, and evaluation of the student text materials, teacher support materials, assessments, and computer software for *Transition to College Mathematics and Statistics* (*TCMS*) was funded through a grant from the National Science Foundation. We express our appreciation to NSF and, in particular, to our program officers Patricia Wilson and Karen King for their support and thoughtful input as the project evolved.

As seen on page iii, *TCMS* has been a collaborative effort among mathematics educators at several institutions. The diversity of experiences and ideas has been a particular strength of the project. Special thanks is owed to the exceptionally capable support staff, Hope Smith and Teresa Ziebarth, at Western Michigan University.

We are grateful to our Advisory Board, Diane Briars (formerly Pittsburgh Public Schools), Jeremy Kilpatrick (University of Georgia), Robert E. Megginson (University of Michigan), Kenneth Ruthven (University of Cambridge), and David A. Smith (Duke University) for their guidance and advice.

Special thanks are owed to the following statistician: Christine Franklin (University of Georgia), and to mathematicians: Bernard L. Madison (University of Arkansas), Steven Maurer (Swarthmore College), and Doris Schattschneider (Moravian College) who reviewed and commented on units as they were being conceptualized, developed, tested, and refined.

We were also indebted to James Hirsch who helped us think hard about the present and future affordances of instructional technologies.

Our gratitude is expressed to the teachers and students at the evaluation sites listed on page iv. Their experiences using the *TCMS* units provided constructive feedback and suggested improvements that supported use of the materials with students possessing a wide range of backgrounds and college aspirations.

Finally, we want to acknowledge Catherine Donaldson, Angela Wimberly, Michael Kaple, Justin Moyer, Karen Corliss, and their colleagues at McGraw-Hill Education who contributed to the design, editing, and publication of this program.

Transition to College Mathematics and Statistics (TCMS) is a Common Core State Standards (CCSS)-grounded, problem-based, inquiry-oriented, and technology-rich fourth-year high school mathematics course. It was developed to help ensure student success in college and careers in an increasingly technological, information-laden, and data-driven global society. *TCMS* was specifically designed for the large numbers of students whose undergraduate programs of study do *not* require calculus—such as business; management; economics; the information, life, health, and social sciences; and many teacher preparation programs.

According to the most recently available reports from the Conference Board of Mathematical Sciences (2010) and the National Center for Educational Statistics (2011), it is estimated that about 5.8% of undergraduate students nationwide are enrolled in calculus or advanced courses with calculus as a prerequisite. At community colleges, that number drops to 1.8%.

All too often, college-bound noncalculus-intending students are enrolled in an inappropriate precalculus course or opt out of mathematics their senior year. But research has repeatedly shown that students who are not enrolled in an appropriate mathematics course their senior year are much more likely to be placed in a remedial (non-credit bearing) course in college.

Overview

Transition to College Mathematics and Statistics is designed to be used as a fourth-year capstone course for students who have successfully completed a CCSS-aligned conventional single-subject sequence of algebra, geometry, and advanced algebra or a three-year integrated sequence. The course has been carefully field tested in high schools with students using conventional curricula and with students using an integrated program. (See page iv.)

TCMS builds upon the theme of mathematics as sense-making. Through investigations of real-life contexts and problems, students develop a rich understanding of important mathematics that makes sense to them and that, in turn, enables them to make sense out of new situations and problems. This theme of sense-making as well as the pervasive expectation that students reason about mathematics align well with both CCSS recommendations and the recent National Council of Teachers of Mathematics recommendations for high school mathematics. (*See Principles to Actions: Ensuring Mathematical Success for All* (2014).)

Key content and instructional features as outlined below have been informed by the latest research on student learning and recommendations from client disciplines on the focus of undergraduate non-calculus-based mathematics and statistics courses.

- **Balanced Content** *Transition to College Mathematics and Statistics* reviews and extends students' understanding of important and broadly useful concepts and methods from algebra and functions, statistics and probability, discrete mathematics, and geometry. These branches of mathematics are connected by CCSS mathematical practices and by mathematical habits of mind such as visual thinking, recursive thinking, searching for and explaining patterns, making and checking conjectures, exploiting use of multiple representations, providing convincing explanations, and a disposition towards strategic use of technological tools.

- **Flexibility** *TCMS* consists of eight focused and coherent units, each of which is generally self-contained with attention to content prerequisites provided by "Just-in-Time" review tasks in lesson homework sets. The course has been organized to be as flexible as possible. The organization permits teachers to tailor courses that best meet the needs and interests of their students. For example, some teachers may choose to use the unit on *Mathematics of Democratic Decision-Making* as the second or third unit of the course to parallel state or national elections.

- **Mathematical Modeling** *TCMS* emphasizes mathematical modeling including the processes of problem formulation, data collection, representation, interpretation, prediction, and simulation. The modeling perspective supports students in connecting mathematical content with important mathematical practices and habits of mind.

- **Technology** Numeric, graphic, and symbolic manipulation capabilities such as those found in *TCMS-Tools®* and on many graphing calculators are assumed and appropriately used throughout the course. *TCMS-Tools* is a suite of software tools that provide powerful aids to learning mathematics and solving mathematical problems. (See page xviii for further details.) This use of technology permits the curriculum and instruction to emphasize multiple linked representations (verbal, numerical, graphical, and symbolic) and to focus on goals in which mathematical thinking and problem solving are central.

- **Active Learning** The instructional materials promote active learning and teaching centered around collaborative investigations of problem situations followed by teacher-led whole-class summarizing activities that lead to analysis, abstraction, and further application of underlying mathematical ideas and principles. Students are actively engaged in exploring, conjecturing, verifying, generalizing, applying, proving, evaluating, and communicating mathematical ideas.

- **Multi-dimensional Assessment** Comprehensive assessment of student understanding and progress through both curriculum-embedded formative assessment opportunities and summative assessment tasks support instruction and enable monitoring and evaluation of each student's performance in terms of mathematical practices, content, and dispositions.

Content Focal Points

TCMS features a coherent and connected development of important ideas drawn from four major branches of the mathematical sciences.

The *Algebra and Functions* units review and extend student ability to recognize, represent, and solve problems involving relations among quantitative variables. Central to the development is the use of functions as mathematical models. The key families of functions that are focused on in the course are linear, exponential, power, polynomial, logarithmic, and the circular functions sine and cosine. Methods are developed for customizing these functions to model more complex quantitative relations and data patterns whose graphs are transformations of basic patterns.

Students extend their skills in *symbolic manipulation*—rewriting expressions in equivalent forms to reveal important information about the corresponding function and to solve equations and inequalities. They also extend their skills in *symbolic reasoning*—making inferences about symbolic relations and connections between symbolic representations and graphical, numerical, and contextual representations. These enhanced skills serve as the cornerstone for the development and application of the mathematics essential for financial decision-making as related to investments, loans, and leases. This latter work aligns with the recently released National Standards for Financial Literacy (2013).

The *Statistics and Probability* units develop students' ability to summarize, represent, and interpret categorical data, to recognize and measure variation, to review and apply basic probability concepts, and to understand binomial distributions and how they are used for statistical inference. The ultimate goal is for students to understand how inferences can be made about a population by surveying a random sample taken from that population.

The key focal points are a flexible understanding of the concepts of statistical inference: sampling distributions, P-values, statistical significance, and margin of error.

The *Discrete Mathematics* units develop students' quantitative literacy and problem-solving ability related to fundamental aspects of contemporary life: enumeration, information processing and the Internet, voting, political apportionment, and fair division of assets.

Major focal points of the units are discrete mathematical modeling, algorithmic problem-solving, and combinatorial reasoning.

The *Geometry* unit focuses on *mathematical visualization*, the process of forming images (mentally,

Berett Wilber

with paper and pencil, or with the aid of technology), in the context of two- and three-dimensional geometry. It extends student ability to visualize and represent three-dimensional shapes using contour diagrams, cross sections, and relief maps.

In this unit, geometry and algebra become increasingly intertwined in the use of coordinate methods for representing and analyzing three-dimensional shapes and their properties, in modeling with systems of linear equations in three variables, and in solving linear programming problems.

Active Learning and Teaching

The manner in which students encounter mathematical ideas can contribute significantly to the quality of their learning and the depth of their understanding. *Transition to College Mathematics and Statistics* units are designed around multi-day lessons centered on big ideas. Each lesson includes 2–5 mathematical investigations that engage students in a four-phase cycle of classroom activities, described in the following paragraph—*Launch, Explore, Share and Summarize*, and *Check Your Understanding*. This cycle is designed to engage students in investigating and making sense of problem situations, in constructing important mathematical concepts and methods, in generalizing and proving mathematical relationships, and in communicating, both orally and in writing, their thinking and the results of their work. Most classroom activities are designed to be completed by students working collaboratively in groups of two to four students.

The Launch phase of a lesson promotes a teacher-led class discussion of a problem situation and of related questions to think about, setting the context for the student work to follow and providing important information about students' prior knowledge. In the second or Explore phase, students investigate more focused problems and questions related to the launch situation. This investigative work is followed by a teacher-led class discussion in which students summarize mathematical ideas developed in their groups, providing an opportunity to construct a shared understanding of important concepts, methods, and supporting justifications. Finally, students complete a formative assessment task related to their work.

Each lesson includes homework tasks to engage students in applying, connecting, reflecting on, extending the concepts and methods of the lesson, and

reviewing previously learned mathematics and statistics and refining their skills in using that content. These *On Your Own* tasks are central to the learning goals of each lesson and are intended primarily for individual work outside of class. We recommend that opportunities be provided for students to exercise some choice of tasks to pursue, and at times be encouraged to pose their own problems and questions to investigate.

Formative and Summative Assessment

Assessing what students know and are able to do is an integral part of *Transition to College Mathematics and Statistics*. There are opportunities for formative assessment in each phase of the instructional cycle. Initially, as students pursue the investigations that comprise the course, the teacher is able to informally assess student understanding of mathematical processes and content and their disposition toward mathematics. At the end of each investigation, a class discussion to Summarize the Mathematics provides an opportunity for the teacher to assess levels of understanding that various groups of students and individuals have reached as they share, explain, and discuss their findings. Finally, the Check Your Understanding tasks and the tasks in the On Your Own sets provide further opportunities for formative assessment of what students know and are able to do. Quizzes, in-class tests, take-home assessment tasks, and extended projects are included in the teacher resource materials for summative assessments.

College Readiness and Course Placement

The primary goal of the Common Core State Standards is to help ensure students' college or career readiness. However, college placement tests and career screening tests continue to be important factors in determining what mathematics or statistics students will first study or the career and apprenticeship opportunities available. To prepare students to perform well on these placement/screening tests, *TCMS* Review sets provide carefully designed distributed practice of core competencies that are a focus of these assessments.

In addition, each lesson is accompanied by an online College Readiness Assessment (CRA) in multiple-choice format with performance reports available to both the student and his/her teacher. The CRAs are also available as blackline masters accompanied by a recording sheet that students can use to monitor areas needing further work to achieve proficiency.

UNIT 1 Interpreting Categorical Data

Interpreting Categorical Data develops student understanding of two-way frequency tables, conditional probability and independence, and using data from a randomized experiment to compare two treatments.

Topics include two-way tables, graphical representations, comparison of proportions including absolute risk reduction and relative risk, characteristics and terminology of well-designed experiments, expected frequency, chi-square test of homogeneity, statistical significance.

Lesson 1 Comparing the Risk

Lesson 2 A Test of Significance

Lesson 3 The Relationship Between Two Variables

Lesson 4 Looking Back

UNIT 2 Functions Modeling Change

Functions Modeling Change extends student understanding of linear, exponential, quadratic, power, circular, and logarithmic functions to model quantitative relationships and data patterns whose graphs are transformations of basic patterns.

Topics include linear, exponential, quadratic, power, circular, and base-10 logarithmic functions; mathematical modeling; translation, reflection, stretching, and compressing of graphs with connections to symbolic forms of corresponding function rules.

Lesson 1 Function Models Revisited

Lesson 2 Customizing Models by Translation and Reflection

Lesson 3 Customizing Models by Stretching and Compressing

Lesson 4 Looking Back

UNIT 3 Counting Methods

Counting Methods extends student ability to count systematically and solve enumeration problems using permutations and combinations.

Topics include systematic listing and counting, counting trees, the Multiplication Principle of Counting, Addition Principle of Counting, combinations, permutations, selections with repetition; the binomial theorem, Pascal's triangle, combinatorial reasoning; and the general multiplication rule for probability.

Lesson 1 Systematic Counting

Lesson 2 Order and Repetition

Lesson 3 Counting Throughout Mathematics

Lesson 4 Looking Back

UNIT 4 Mathematics of Financial Decision-Making

Mathematics of Financial Decision-Making extends student facility with the use of linear, exponential, and logarithmic functions, expressions, and equations in representing and reasoning about quantitative relationships, especially those involving financial mathematical models.

Topics include forms of investment, simple and compound interest, future value of an increasing annuity, comparing investment options, continuous compounding and natural logarithms; amortization of loans and mortgages, present value of a decreasing annuity, and comparing auto loan and lease options.

Lesson 1 Financial Decision-Making: Saving

Lesson 2 Financial Decision-Making: Borrowing

Lesson 3 Looking Back

UNIT 5 Binomial Distributions and Statistical Inference

Binomial Distributions and Statistical Inference develops student understanding of the rules of probability; binomial distributions; expected value; testing a model; simulation; making inferences about the population based on a random sample; margin of error; and comparison of sample surveys, experiments, and observational studies and how randomization relates to each.

Topics include review of basic rules and vocabulary of probability (addition and multiplication rules, independent events, mutually exclusive events); binomial probability formula; expected value; statistical significance and *P*-value; design of sample surveys including random sampling and stratified random sampling; response bias; sample selection bias; sampling distribution; variability in sampling and sampling error; margin of error; and confidence interval.

Lesson 1 Binomial Distributions

Lesson 2 Sample Surveys

Lesson 3 Margin of Error: From Sample to Population

Lesson 4 Looking Back

UNIT 6 Informatics

Informatics develops student understanding of the mathematical concepts and methods related to information processing, particularly on the Internet, focusing on the key issues of access, security, accuracy, and efficiency.

Topics include elementary set theory and logic; modular arithmetic and number theory; secret codes, symmetric-key and public-key cryptosystems; error-detecting codes (including ZIP, UPC, and ISBN) and error-correcting codes (including Hamming distance); and trees and Huffman coding.

Lesson 1 Access: Set Theory, Logic, and Searching

Lesson 2 Security: Cryptography

Lesson 3 Accuracy: Error-Detecting and -Correcting Codes

Lesson 4 Efficiency: Data Compression

Lesson 5 Looking Back

UNIT 7 Spatial Visualization and Representations

Spatial Visualization and Representations extends student ability to visualize and represent three-dimensional shapes using contour diagrams, cross sections, and relief maps; to use coordinate methods for representing and analyzing three-dimensional shapes and their properties; and to use graphical and algebraic reasoning to solve systems of linear equations and inequalities in three variables and linear programming problems.

Topics include using contours to represent three-dimensional surfaces and developing contour maps from data; sketching surfaces from sets of cross sections; three-dimensional rectangular coordinate system; sketching planes using traces, intercepts, and cross sections derived from algebraic representations; systems of linear equations and inequalities in three variables; and linear programming.

UNIT 8 Mathematics of Democratic Decision-Making

Mathematics of Democratic Decision-Making develops student understanding of the mathematical concepts and methods useful in making decisions in a democratic society, as related to voting and fair division.

Topics include preferential voting and associated vote-analysis methods such as majority, plurality, runoff, points-for-preferences (Borda method), pairwise-comparison (Condorcet method), and Arrow's theorem; weighted voting, including weight and power of a vote and the Banzhaf power index; and fair division techniques, including apportionment methods.

CONTENTS

CONTENTS

UNIT 3   Counting Methods

UNIT 4   Mathematics of Financial Decision-Making

CONTENTS

UNIT 5   Binomial Distributions and Statistical Inference

UNIT 6   Informatics

UNIT 7   Spatial Visualization and Representations

CONTENTS

Have you ever wondered …

- How you can assess the risks of behavior such as using tanning beds?

- How it is possible to show which treatment tends to cause faster recovery from an illness when the treatments are tested on only a few hundred people?

- How climate scientists are able to predict potential effects of global warming such as rise in sea levels and pollution in the atmosphere?

- How the results of a national election can be accurately predicted by polling only 1,500 voters?

- How many different Web site passwords or ATM PINs are possible, and how long it would take a hacker to try them all?

- How interest compounding rates and periods affect the total cost of a student college loan?

- How a credit card number is sent securely when you buy music online?

- How large files, like photos or videos, get compressed so that they fit on your phone?

- Why contour maps are so useful in hiking?

- Why the results of some elections seem unfair and how mathematics might be used to make them fairer?

The mathematics you will learn in *Transition to College Mathematics and Statistics (TCMS)* will help you answer questions like these. In answering such questions, you will often use the process of mathematical modeling as shown here.

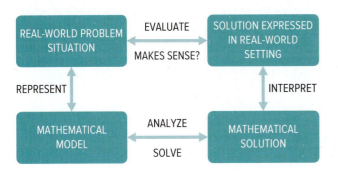

Because real-world situations often involve data, shape, quantity, change, or chance, you will review and extend your understanding of broadly useful ideas from statistics and probability, algebra and functions, discrete mathematics, and geometry. In the process, you will also see and use many connections among these branches of mathematics.

In this course, you will learn important mathematics as you investigate and solve interesting problems. You will develop the ability to reason and communicate about mathematics as you are actively engaged in understanding and applying mathematics. You will often be learning mathematics in the same way that many people work in their occupations—by working in teams and using technology to solve problems.

In the 21st century, anyone who faces the challenge of learning mathematics or using mathematics to solve problems can draw on the resources of powerful information technology tools.

Calculators and computers can help with modeling, calculations, drawing, and data analysis in mathematical explorations and solving mathematical problems.

Graphing calculators and computer software tools will be useful in your work on many of the investigations in *TCMS*. Set as one of your goals to learn how to make strategic decisions about the choice and use of technological tools in solving particular problems.

Jim Laser

The curriculum materials include computer software called *TCMS-Tools* that will be of great help in learning and applying the mathematical topics in this course. You can freely access the software at www.wmich.edu/tcms/TCMS-Tools/.

The software toolkit includes four families of programs:

- **Algebra**—The software for work on quantitative problems includes a spreadsheet and a computer algebra system (CAS) that produces tables and graphs of functions, manipulates algebraic expressions, solves equations and inequalities, and systems of equations.

- **Geometry**—The software for work on problems involving shape, size, and location includes an interactive drawing program for constructing, measuring, and manipulating geometric figures in a coordinate or coordinate-free environment. Also included is a geometric linear programming tool.

- **Statistics**—The software for work on problems involving data and chance provides tools for graphic display and analysis of data, testing statistical significance, simulation of probabilistic situations, and mathematical modeling of quantitative relationships involving variability.

- **Discrete Mathematics**—The software provides tools for creating and analyzing discrete mathematical models involving information processing and democratic decision-making.

In addition to these general-purpose tools, *TCMS-Tools* includes files of most data sets essential for work on problems in the text. When the opportunity to use computer tools in an investigation seems appropriate, select the *TCMS-Tools* menu corresponding to the content of, and your planned approach to, the problem. Then select the submenu items corresponding to the appropriate mathematical operations and required data set(s).

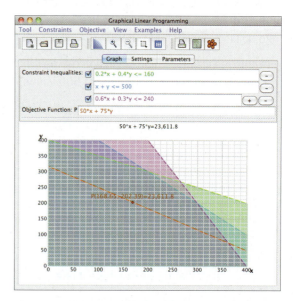

In *TCMS*, you will learn a lot of useful mathematics and it is going to make sense to you. You will also deepen your understanding of fundamental ideas and methods that support future coursework in college-level mathematics or statistics. You are going to strengthen your skills in working collaboratively on problems and communicating with others as well. You are also going to strengthen your skills in using technological tools strategically and effectively. Finally, you will continue to have plenty of opportunities to experience the creative side of mathematics. Enjoy!

Interpreting Categorical Data

Digital Resources at
ConnectED.mcgraw-hill.com

Watch · Tools · Audio · eBook

Categorical data involve variables like gender, intended college major, being a mountain climber (yes or no), or being a smoker (yes or no) that place an individual into one of two or more categories. Teen engagement or non-engagement in risky behaviors is a categorical variable that is so important that the U.S. government monitors the prevalence of such behavior. Statistical tools needed to calculate the possible consequences of engaging in risky behavior are developed in this unit.

At the age of 18, you become responsible for making your own medical decisions, albeit with possible advice of family members. It is likely that you will be asked to help make similar decisions for other members of your family.

In the three lessons of this unit, you will learn statistical methods to help you make those decisions.

LESSONS

1 Comparing the Risk

Use two-way frequency tables and bar graphs to compare the risk of some condition for two groups. Understand the limitations of anecdotal evidence and dangers of the placebo effect when evaluating the effectiveness of a treatment. Understand the characteristics of a well-designed experiment, especially the role of randomization, to compare the worth of different treatments.

2 A Test of Significance

Use proportional reasoning to calculate frequencies expected under homogeneity of two populations. Compute and use the chi-square statistic (χ^2) as a measure of difference between two samples. Conduct and interpret a chi-square test to make an inference about the homogeneity of two populations. Determine if the value of (χ^2) is statistically significant.

3 The Relationship Between Two Variables

Review conditional probability and independent events. Use the definition of independent events to calculate frequencies expected under independence of two categorical variables. Conduct and interpret a chi-square test to make an inference about the independence of two categorical variables in a population.

© Redlink/Corbis

Comparing the Risk

Some teens engage in very risky behavior. About 6,748,000 high school boys in the United States do not smoke cigarettes—but 1,645,000 do. The table below shows how many of these smokers and non-smokers can be expected to get lung cancer (if the smokers continue to smoke, based on Canadian data). (**Sources:** U.S. Census Bureau, Current Population Survey, October 2009, Internet release date: February 2011; P. J. Villeneuve and Y. Mao, "Lifetime Probability of Developing Lung Cancer, by Smoking Status, Canada," *Canadian Journal of Public Health*, Vol. 85, November 1994, pp. 385–388.)

Teenage Boys in the United States

	Smoke Cigarettes	Do Not Smoke Cigarettes
Get Lung Cancer	283,000	88,000
Do Not Get Lung Cancer	1,362,000	6,660,000
Total	1,645,000	6,748,000

THINK ABOUT THIS SITUATION

Use the data on the previous page to help you think about the question of whether or not smoking causes lung cancer.

a. Is a smoker or a non-smoker more likely to get lung cancer? Explain your reasoning using the data reported in the table.

b. Is a boy who gets lung cancer more likely to be a smoker than a boy who does not get lung cancer? Explain your reasoning.

c. Based on the reported data, complete this sentence: A male smoker is _____ times as likely to get lung cancer as is a male non-smoker.

d. How many cases of lung cancer could be prevented if none of these boys smoked?

e. What exactly do people mean when they say that smoking *causes* lung cancer? Use the information in the table to support their position. Are there any plausible reasons other than smoking why smokers get lung cancer at a greater rate than non-smokers do?

f. Have you ever heard someone doubt that smoking causes lung cancer because he or she knew a smoker who had not gotten lung cancer? Or perhaps they knew someone who did have lung cancer and did not smoke. Do you think observations like these help to answer the question of whether smoking causes lung cancer? Explain your thinking.

Soon you must be prepared to make your own medical decisions and monitor your own medical care. The situation described above is typical of those you should understand: How does the proportion of one group who have a condition compare to the proportion in another group who have the condition?

Summarizing and Displaying the Risk

The table on the previous page shows **categorical data**—the data collected are whether each boy falls into the category of smoker or non-smoker and into the category of lung cancer or not. People often find such a table difficult to interpret. Summary statistics and a thoughtfully made graph can make comparisons easier.

As you work on the problems in this investigation, look for answers to this question:

What summary statistics and graphs can make categorical data easier to comprehend?

1 **Absolute risk** is defined as the proportion or percentage of people in a group for whom an undesirable event occurs. In college classrooms, students typically can choose their own seats. Professors have noticed a difference in grades between students who choose to sit in the front and those who choose to sit in the back. For example, in one math class, 9 of the 20 students who sat in the back failed the class, but only 3 of the 20 students who sat in the front failed the class. What was the absolute risk of failing the class for students who sat in the back? For students who sat in the front? Give your answers as fractions, proportions, and percents.

2 According to breastcancer.org, "Researchers estimate that 1 in 8 women will be diagnosed with invasive breast cancer at some time in their lives."

What is the absolute risk, as a percent? Write your answer in a complete sentence, in context.

3 Medications to prevent loss of bone density may also suppress the body's ability to replace older bone tissue with new bone tissue. Among 52,595 women who had been treated with such medications, an unusual type of fracture occurred in 117 of them. What is the absolute risk, as a percent? Write your answer in a complete sentence, in context. (**Source:** L. Park-Wylie, *et al.* "Bisphosphonate Use and the Risk of Subtrochanteric or Femoral Shaft Fractures in Older Women," *Journal of the American Medical Association,* Vol. 305, 2011, pp. 783–89.)

4 Many teens have been vaccinated against the human papillomavirus (HPV), which can cause cancer in both men and women. According to the U.S. Centers for Disease Control and Prevention (CDC), this is a safe vaccine. Approximately 56 million doses have been distributed, and the CDC has received 21,194 reports of adverse events in females who received an HPV vaccine. Of these, 92.1% were classified as non-serious events such as fainting, nausea, or injection-site pain. The adverse events may or may not have been caused by the vaccine. (**Source:** www.cdc.gov/mmwr/preview/mmwrhtml/mm6229a4.htm)

a. What is the absolute risk of having an adverse event (whether caused by the vaccine or not), that is reported after receiving a dose of the HPV vaccine as a proportion? As a percent?

b. What is the absolute risk of a serious adverse event that is reported as a proportion? As a percent?

5 To display categorical data, you can create different types of *bar graphs.* Examine the three different types of bar graphs on the facing page that display the data on male teenage smoking and lung cancer provided on page 2. For these data, the two *groups* are the boys who smoke cigarettes and the boys who do not. Whether a boy smokes is called the *explanatory variable.* The *response variable* is whether the boy eventually gets lung cancer.

Stacked Bar Graph (Frequency)

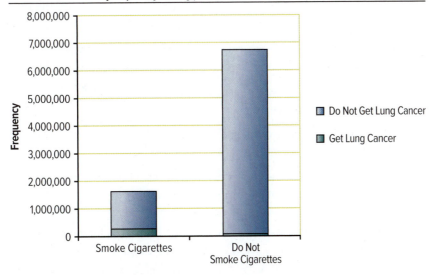

Stacked Bar Graph (Percent)

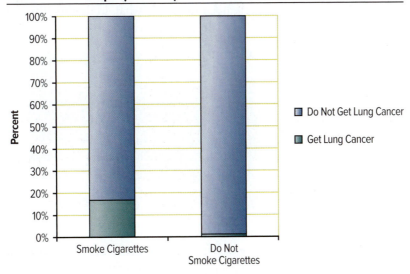

Grouped Bar Graph (Frequency)

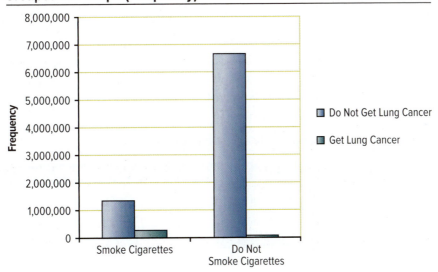

When answering the questions at the top of page 6, be prepared to explain your thinking.

a. In the **stacked bar graphs** (sometimes called *segmented* bar graphs), are different bars defined by the explanatory variable or by the response variable? Is each bar segmented according to the explanatory variable or the response variable?

b. For each bar graph, describe the meaning of the leftmost bar.

c. Which graph shows most clearly which group has the larger absolute risk of getting lung cancer?

d. Which graph shows most clearly that only a small percentage of high school boys smoke cigarettes?

e. Which graph makes it easiest to compare the number of boys in the two groups who do not get lung cancer?

6 The best of intentions sometimes results in unanticipated consequences. A program in the Junnar region of India relocated leopards away from human-dominated areas. The leopards were moved into forests, an average of 24 miles away. In the three years before the program began, there were 12 leopard attacks on humans, 2 of them lethal and 10 not lethal. In the first three years after the program began, there were 50 attacks, 18 of them lethal and 32 not lethal.

Clearly, there were more attacks after the relocation program, but did the seriousness of the attacks also change? (**Source:** Vidya Athreya *et al.* "Translocation as a Tool for Mitigating Conflict with Leopards in Human-Dominated Landscapes of India," *Conservation Biology*, Vol. 25, 2011, pp. 133–141.)

a. To answer the question, what two groups must be compared? What is the response variable?

b. The information may be summarized in a **two-way frequency table**, such as that below. When making such a table, *place the numbers for each group (explanatory variable) in different columns and place the levels of the response variable in different rows.*

Leopard Attacks on Humans

	Before Program	After Program
Lethal	2	18
Not Lethal	10	32
Total	12	50

Use the table to determine whether the absolute risk that an attack results in a death was higher before or after the relocation program. Explain your reasoning.

c. Use the data in the table to make a stacked bar graph (frequency on the vertical axis), a stacked bar graph (percent on the vertical axis), and a grouped bar graph (frequency on the vertical axis).

   d. For each bar graph, write a sentence that summarizes information shown best by that graph.

7 To investigate whether middle school students think that being overweight is under a child's control, students in 7th and 8th grade were given a questionnaire. One question asked was, "Is it the child's fault if they are fat?" Of the 84 boys who answered the question, 25 said yes, 24 said no, and 35 said they were not sure. Of the 138 girls, 23 said yes, 61 said no, and 54 said they were not sure.

(**Source:** Paul B. Rukavina and Weidong Li, "Adolescents' Perceptions of Controllability and Its Relationship to Explicit Obesity Bias," *Journal of School Health*, Vol. 81, January 2011, pp. 8–14.)

   a. What two groups were compared? What is the response variable?

   b. Summarize the information in a two-way frequency table.

   c. Compare two values to answer this question: Was a boy or a girl more likely to say yes?

   d. Make the bar graph that best compares the proportions of boys and girls who gave each answer.

   e. What can you conclude about any difference in how boys and girls respond to this question?

8 Sometimes you will see data such as that in Problem 7 displayed by a **pie chart**. In the pie chart below, the *central angle* for each of the three sectors is proportional to the percentage of boys who gave the indicated answer.

Boys' Answers

   a. Verify, by calculation, that the measure of the central angle for the "not sure" answer should be 150°.

b. Compute the measure of the central angle for the "yes" answer. For the "no" answer.

c. Compute the measure of the central angle for each of the girls' answers and make a pie chart.

d. Most statisticians recommend including stacked bar graphs rather than pie charts in reports, even if you have technology that easily makes both. Give a reason why you think stacked bar graphs are better than pie charts.

SUMMARIZE THE MATHEMATICS

In this investigation, you learned to use summary statistics and graphs to compare groups sorted by the same categorical variable.

a. What is absolute risk?

b. When comparing situations involving risk, how can you tell which is the explanatory variable and which is the response variable?

c. Describe three types of bar graphs and the characteristics of the data that are best shown by each one.

d. What steps are involved in making a pie chart?

Be prepared to share your ideas and reasoning with the class.

✔ CHECK YOUR UNDERSTANDING

A "food-secure" household is one that has access at all times to enough food for an active healthy life for all household members. In 2009 in the United States, 194,579,000 adults and 57,010,000 children lived in food-secure households; 20,741,000 adults and 16,209,000 children lived in households with low food security; and 12,223,000 adults and 988,000 children lived in households with very low food security. (**Source:** U.S. Census Bureau, *Statistical Abstract of the United States: 2012*, Table 214.)

a. Summarize the information in a two-way frequency table with *adult/child* as the explanatory variable.

b. In 2009, was an adult or a child more likely to live in a household with very low food security? What two values do you need to compare to answer this question?

c. Make a bar graph that best compares the proportions of adults and children who live in households with the three levels of food security. Describe what you can conclude from this graph.

Comparing Risk

In this investigation, you will examine how reports in the media use percentages to compare the prevalence of some condition in two different groups.

As you work on the following problems, look for answers to this question:

How is the risk associated with two different conditions best compared?

1 In a study of over 15,000 teens, 62% of boys and 55% of girls consumed sugar-sweetened soda on the previous day. (**Source:** Nalini Ranjit *et al.* "Dietary and Activity Correlates of Sugar-Sweetened Beverage Consumption Among Adolescents," *Pediatrics* published online Sep. 27, 2010.) When asked to compare the boys' percentage with the girls' percentage, two students gave the answers below. Show how each student computed the percentage.

Alma: The boys' percentage is 7% bigger than the girls'.

Bill: The boys' percentage is about 113% of the girls'.

2 Melanoma is the most dangerous type of skin cancer. The International Agency for Research on Cancer (IARC) reported that the absolute risk of melanoma over eight years for women who do not use tanning beds regularly is only about two-tenths of 1 percent. However, the risk of melanoma is increased by 75% when the use of tanning beds starts before age 30. That sounds serious, but what, exactly, do they mean? (**Sources:** Fatiha El Ghissassi *et al.* "A Review of Human Carcinogens—Part D: Radiation," *The Lancet Oncology*, Vol. 10, August 2009, pp. 751–752, www.delawareonline.com/article/20100304/NEWS/105070009, www.iarc.fr/en/publications/index.php)

   a. What could be the absolute risk of melanoma for those who *do* use tanning beds regularly? Give two answers to this question, based on the thinking of Alma and Bill in Problem 1.

   b. Which answer from Part a must be the absolute risk of melanoma for those who tan regularly? Why?

To try to prevent confusion, medical professionals typically give one of two standard statistics when helping patients compare risk:

absolute risk reduction = absolute risk for one group − absolute risk for another group

$$relative\ risk = \frac{absolute\ risk\ for\ one\ group}{absolute\ risk\ for\ another\ group}$$

Typically, the proportion for the group with the larger absolute risk is placed so that the absolute risk reduction is positive and the relative risk is greater than 1.

3 Refer back to Problem 1 involving the consumption of sugar-sweetened soda.

  a. Which student computed absolute risk reduction? Did either student compute something like relative risk?

  b. Read the descriptions below about Problem 1. Which statement is the clearest interpretation of absolute risk reduction? Which statement is the clearest interpretation of relative risk? Be prepared to explain your reasoning.

  I. The boys were 7% more likely than the girls to consume sugar-sweetened soda on the previous day.

  II. The difference in the percentage of the boys and the percentage of girls who consumed sugar-sweetened soda on the previous day is 7 percentage points.

  III. A boy was about 1.13 times more likely than a girl to have consumed sugar-sweetened soda on the previous day.

  IV. The boys were about 113% more likely than the girls to consume sugar-sweetened soda on the previous day.

4 Now refer to the information on the use of tanning beds and the development of melanoma in Problem 2.

  a. What is the absolute risk reduction in melanoma from regular tanning versus no regular tanning, expressed in percentage points? Use your answer in a sentence that describes its meaning.

  b. What is the relative risk of melanoma from regular tanning versus no regular tanning? Use your answer in a sentence that describes its meaning.

  c. The units for absolute risk reduction usually are *percentage points*. What can you say about the units for relative risk?

  d. In Part a, the absolute risk reduction may seem very small. However, it has been estimated that every year nearly 2.3 million American teens use tanning beds. How many cases of melanoma do you estimate could be prevented if they did not use tanning beds?

5 Baseball pitchers aged 9 to 14 were followed for 10 years. Four out of the 26 who had pitched at least 100 innings in one year had elbow surgery, shoulder surgery, or retirement due to a throwing injury. Of 128 who had not pitched this much, 6 had such problems. (**Source:** G. Fleisig, *et al.* "Risk of Serious Injury for Young Baseball Pitchers: A 10-Year Prospective Study," *American Journal of Sports Medicine*, Vol. 39, February 2011, pp. 253–257.)

  a. Compute the percentage of each group who have such problems. Compute the difference of the percentages. What is the name for this difference? Write an explanation, for parents of young pitchers, that explains its meaning.

  b. Compute the ratio of the percentages of each group who have such problems. What is the name for this difference? Then write an explanation, for parents of young pitchers, that explains its meaning.

6 Asthma is a chronic disease in which there are episodes of shortness of breath and wheezing due to inflammation and narrowing of small airways in the lungs. About 9.6% of teens and children under age 18 in the United States have asthma. One of the most-commonly prescribed medicines for asthma is Advair. Like all medications, Advair can have *side effects*—undesirable symptoms brought on by the medicine. A 28-week *clinical trial* was conducted to determine the side effects of one of the components of Advair, salmeterol. Available patients age 12 to 18 were randomly divided into two groups. One group of 1,653 patients received salmeterol. The other group of 1,622 patients received a *placebo*, which had no active ingredient in it. At the end of the study, researchers counted how many patients in each group had been hospitalized for any cause.

The sentence below summarizes the results of the Advair clinical trial.

  Hospitalization from all causes was increased in the salmeterol group $(2\%)\left(\dfrac{35}{1{,}653}\right)$ vs. the placebo group $(<1\%)\left(\dfrac{16}{1{,}622}\right)$ [relative risk 2.1].

(**Sources:** *National Health Statistics Reports*, Number 32, January 12, 2011; Advair Medication Guide, GlaxoSmithKline, 2010)

  a. What is the meaning of $\dfrac{35}{1{,}653}$ and $\dfrac{16}{1{,}622}$?

  b. Show how the 2% and <1% summaries were computed.

  c. Verify the computation of the relative risk.

  d. Compute the absolute risk reduction.

  e. Write a paragraph for patients explaining the risk.

7 Immunizations are not equally effective overall and are not effective for all people. For example, the polio vaccine is about 99% effective in children who receive the full series. That means that for every 100 immunized children who are exposed to polio, only 1 is expected to get the disease. In contrast, the influenza vaccine is only about 66% effective. (**Source:** www.cdc.gov) Here are three hypothetical diseases.

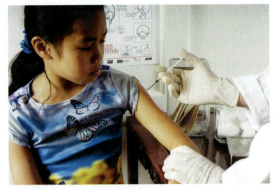

Disease A: The absolute risk of getting disease A for those who have been immunized is 1%. The absolute risk for those who have not been immunized is 2%.

Disease B: The absolute risk of getting disease B for those who have been immunized is 34%. The absolute risk for those who have not been immunized is 68%.

Disease C: The absolute risk of getting disease C for those who have been immunized is 98%. The absolute risk for those who have not been immunized is 99%.

a. For each disease, compute the absolute risk reduction for those who have been immunized.

b. Identify the two diseases with the same absolute risk reduction. Suppose that both diseases are similarly serious but you have the resources to provide vaccinations for only one of them. For which disease would you choose to vaccinate, or does it matter? Why?

c. For each disease, compute the relative risk.

d. Identify the two diseases with the same relative risk. Suppose that both diseases are similarly serious but you have the resources to provide vaccinations for only one of them. For which disease would you choose to vaccinate, or does it matter? Explain your reasoning.

SUMMARIZE THE MATHEMATICS

In this investigation, you learned two methods of comparing risk.

a. What is absolute risk reduction? Give an example of how it is interpreted.

b. What is relative risk? Give an example of how it is interpreted.

c. Sometimes a condition is extremely serious, but quite rare, such as melanoma. In such a case, discuss whether it is better to tell people about the absolute risk reduction from not engaging in the risky behavior, or tell them about the relative risk.

Be prepared to explain your ideas and reasoning to the class.

Measles is a highly contagious childhood disease that can cause complications including death. There was an outbreak of measles in Colorado in December 1994. Out of 609 children who had been vaccinated, 10 got measles (and those 10 children had been given only one of the two recommended doses). Of the 16 children who were not vaccinated, 7 got measles. (**Source:** C. R. Vitek *et al.* "Increased Protections During a Measles Outbreak of Children Previously Vaccinated with a Second Dose of Measles-Mumps-Rubella Vaccine," *Pediatric Infectious Diseases Journal*, Vol. 18, July 1999, pp. 620–623.)

a. What is the explanatory variable? What is the response variable?

b. Make a two-way frequency table that summarizes this information.

c. For each group, compute the absolute risk of getting measles.

d. Compute the absolute risk reduction.

e. Compute the relative risk of getting measles.

f. Write a paragraph for parents using your results in Parts c, d, and e.

INVESTIGATION 3

Design of Experiments

As you may have discussed in the Think About This Situation (page 3), it is impossible to prove that smoking causes lung cancer just by collecting data that show that smokers are more likely than non-smokers to get lung cancer. Such an *observational study* leaves open the possibility that the people who smoke tend to be the same people who do something else that is actually causing the increased rate of lung cancer.

As you work on the problems in this investigation, look for answers to this question:

How can you design an experiment that provides convincing evidence that one treatment causes a different response than another treatment?

Centers for Disease Control

1 Nosebleeds are common, and you may have heard one of many folk remedies about how to stop them. One such suggestion is to drop car keys down the back of the neck, inside the shirt. People have said things like, "I tried the car-key trick, and in no time at all, my nosebleed stopped." (**Source:** People's Pharmacy, *Los Angeles Times*, February 28, 2011.)

a. Evidence such as this ("it worked for me") is called **anecdotal evidence**. Why is anecdotal evidence not convincing that the car-key trick causes nosebleeds to stop?

b. Think about what is meant when a person says that a certain remedy causes nosebleeds to stop. Which statement below best conveys the intent of the person?

I. The person wants you to believe that it works in every case.

II. The person wants you to believe that a nosebleed will not stop unless this remedy is used.

III. The person wants you to believe that, for some people, nosebleeds usually stop sooner if they use the remedy than if they do not use the remedy.

Few treatments work in all cases. Even vaccines, as you saw in the previous investigation, are not effective for all people. To further complicate matters, not everyone who is exposed to a disease gets the disease even if he or she is unvaccinated. So, we say that a vaccine "works" when exposed people are less likely to get a disease if they are immunized than if they are not immunized. To show that this is the case, scientists must conduct an *experiment*.

In a typical **experiment**, two or more **treatments** (conditions you want to compare) are randomly assigned to an available group of people (or animals, plants, or products), called **subjects**. The purpose of an experiment is to establish cause and effect—does one treatment cause a different response than the other treatment? A well-designed experiment must have three characteristics:

- *Random assignment:* Treatments are assigned randomly to the subjects.

- *Comparison group or control group:* At least two groups get different treatments and then are compared. Alternatively, one group gets the treatment under study and then is compared to another group that does not get a real treatment (a **control group**).

- *Sufficient number of subjects:* Subjects will vary in their responses, even when they are treated alike. If there are not enough subjects, this variability of responses within each treatment group may obscure any difference between the effects of the treatments. Deciding how many subjects are sufficient is one of the more difficult tasks that statisticians perform.

After treatments are completed, the **response** of each subject is determined. Then the responses for each treatment are summarized and compared.

2 Injuries of the leg, knee, and foot are common in people who do a lot of running, such as military recruits undergoing basic training. Researchers wanted to know whether foot orthotics—shoe inserts custom-molded to the person's foot—might help prevent such injuries.

Four hundred military officer trainees were identified as medium or high risk for leg, knee, or foot injury. They were randomly split into two groups. One group got custom orthotics and the other did not. After seven weeks of basic training, 21 of the 200 who had received custom orthotics had a leg, knee, or foot injury that required that he or she stop physical training for two or more days. Sixty-one of the 200 who did not get orthotics had such injuries.

(**Source:** Andrew Franklyn-Miller, *et al.* "Foot Orthoses in the Prevention of Injury in Initial Military Training: A Randomized Controlled Trial," *American Journal of Sports Medicine*, Vol. 39, 2011, pp. 30–37.)

a. Who are the subjects in this experiment? What are the treatments? What is the response variable?

b. Does this experiment have the first two characteristics of a well-designed experiment? Explain.

c. What was the absolute risk of leg, knee, or foot injury for each group?

d. The researchers hypothesized that the injury rate would be lower among those who wear foot orthotics than among those who do not. Do you think that their experiment provides strong, weak, or no evidence for that hypothesis? Explain.

Many studies have shown that people tend to do better when they are given special attention or when they believe they are getting competent medical care. This is called the **placebo effect**. Even people with post-surgical pain report less discomfort if they are given a pill that they think is a painkiller but actually contains no medicine. One way to control for the placebo effect is to make the experiment **single blind**—the person receiving the treatment does not know which treatment he or she is getting. That is, subjects in all treatment groups appear to be treated exactly the same way. This often calls for the use of a **placebo**, a fake treatment that has no medical value, but looks like a real treatment to the person receiving it. In a **double-blind** experiment, neither the subject nor the people who administer the treatment and evaluate how well it works know which treatment the subject received.

3 What can you do if you have a toothache and cannot immediately get to the dentist? Some people recommend dabbing a small amount of clove oil near the tooth. (Clove oil is toxic, so only a very small amount must be used.) Does that work? Read the following report from the *New York Times*.

> In a study published in *The Journal of Dentistry* in 2006, for example, a team of dentists recruited 73 adult volunteers and randomly split them into groups that had one of four substances applied to the gums just above the maxillary canine teeth: a clove gel, benzocaine, a placebo resembling the clove gel, or a placebo resembling benzocaine. Then, after five minutes, they compared what happened when the subjects received two needle sticks in those areas. Not surprisingly, the placebos failed to numb the tissue against the pain, but the clove and benzocaine applications numbed the tissue equally well.

Source: well.blogs.nytimes.com/2011/02/17/remedies-clove-oil-for-tooth-pain/, Athbi Alqareer, Asma Alyahya, and Lars Andersson, "The effect of clove and benzocaine versus placebo as topical anesthetics," *Journal of Dentistry*, Vol. 34, November 2006, pp. 747–750.

a. In this experiment, what treatments were tested? What was the response variable?

b. Why was it important to split the volunteers randomly into the treatment groups rather than letting them decide which group they wanted to be in?

c. What were the placebos in this experiment?

d. This experiment was single blind. Why was this precaution necessary?

e. About how many adults would have been in each group? Could that be sufficient to convince you that the clove oil is as effective as the benzocaine? Create some hypothetical data from 73 adults in such an experiment that would be convincing.

4 A study of young baseball pitchers found that 7 of the 103 who had thrown curveballs before the age of 13 were injured seriously enough to have surgery or quit pitching. Of the 187 who had not thrown curveballs before age 13, 8 were injured seriously enough to have surgery or quit pitching. The researchers had hypothesized that throwing curveballs before age 13 put young pitchers at risk for injury. (**Source:** G. Fleisig, *et al.* "Risk of Serious Injury for Young Baseball Pitchers: A 10-year Prospective Study," *American Journal of Sports Medicine*, Vol. 39, February 2011, pp. 253–257.)

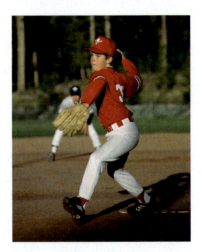

a. Display the data in a two-way frequency table.

b. What two conditions are being compared in this study?

c. What is the response variable?

d. What is the absolute risk of injury for those pitchers who had thrown curveballs before age 13? For those who had not?

e. Would you say that the researchers have strong, weak, or no evidence for their hypothesis? Explain.

f. Does this study satisfy the first two characteristics of a well-designed experiment?

5 A **lurking variable** helps to explain the association between the treatments and the response, but is not the explanation that the study was designed to test. Treatments are assigned randomly to subjects to equalize the effects of possible lurking variables among the treatment groups as much as possible. Refer to the measles study in the Investigation 2 Check Your Understanding (page 13). Can you think of possible lurking variables that might explain why those who were not immunized were more likely to get measles than those who were immunized?

6 Assume that you are the nurse at a large elementary school where children frequently come to you with nosebleeds.

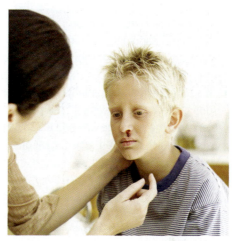

Design an experiment to test whether dropping keys down the back of the neck causes nosebleeds to stop. (Assume parents of all children in the school have given permission for their participation in the experiment and you have followed ethical guidelines for research with human subjects.)

One possible explanation for why keys down the back might actually be effective in stopping a nosebleed is that they are cold, causing blood vessels to constrict.

a. What are the treatments in your experiment? How might you use a placebo?

b. Describe the variable you will use as the response.

c. Treatments should be randomly assigned to the children. Describe how you could accomplish this. That is, when a child with a nosebleed comes to the nurse's office, how will you know which treatment he or she should get?

d. Write a *protocol*, or description of how to proceed, when a child with a nosebleed comes into the nurse's office.

e. Is your experiment single or double blind? Explain.

SUMMARIZE THE MATHEMATICS

In this investigation, you examined the characteristics of well-designed experiments.

a. What are the three characteristics of a well-designed experiment? Why is each necessary?

b. Why are single blinding and double blinding desirable in an experiment?

c. What is the placebo effect? How can you account for it when designing an experiment?

Be prepared to share your ideas and reasoning with the class.

✓ CHECK YOUR UNDERSTANDING

Parkinson's disease is a progressive disease that results in loss of motor function, among other problems. Currently, there is no cure and only partial treatment.

In a clinical trial to test a new gene therapy, 45 patients aged 30 to 75 years with typical, advanced Parkinson's disease had tubes implanted in their brains. Twenty-two were randomly selected to receive infusions through the tube of a saline solution containing a gene that should increase levels of a brain chemical that is missing in people with Parkinson's. The other 23 patients received only saline solution through the tube. Neither the patients nor the specialist who evaluated how well they were doing knew which treatment the patient received. After six months, the patients who received the gene therapy had a 23.1% improvement in their score on a test of motor ability (which included assessment of such things as problems walking, problems with speech, and severity of muscle tremors). The patients who received the sham treatment had a 12.7% improvement.

Source: Peter A. LeWitt, *et al.* "AAV*2-GAD* gene therapy for advanced Parkinson's disease: A double-blind, sham-surgery controlled, randomised trial," *The Lancet Neurology*, Vol. 10, April 2011, pp. 309–319

a. Who are the subjects in this study? What are the treatments?

b. Describe the response variable.

c. Does this study have the first two characteristics of a well-designed experiment? Explain.

d. Is this study single blind? Is it double blind? Explain.

APPLICATIONS

These tasks provide opportunities for you to use and strengthen your understanding of the ideas you have learned in the lesson.

1 College students in Korea and in the American Midwest participated in a study. They were told to pretend that they were sick and that a meeting with classmates to work on a group project was scheduled for tomorrow. Each student then composed an email to the members of their group asking that the meeting be rescheduled. Out of 127 Koreans, 105 included apologies in their message, while 51 out of 97 Americans included apologies. (Americans tended, instead, to write things like "I would really appreciate" or "Thank you," while Koreans rarely used such language.) (**Source:** Hye Eun Lee & Hee Sun Park, "Why Koreans Are More Likely to Favor 'Apology,' While Americans Are More Likely to Favor 'Thank You,'" *Human Communication Research*, Vol. 37, 2011, pp. 125–146.)

a. What two groups are being compared? What is the response variable?

b. Summarize the information in a two-way frequency table.

c. Make a bar graph that best compares the proportion of each group that included an apology.

d. Based on your graph, write a summary sentence or two describing how the two groups differ.

2 Artificial turf on athletic fields was first introduced in the 1960s. Its safety has been controversial since then. One issue that has been investigated is whether injuries of football players tend to be more serious on artificial turf than on grass. A study followed 24 NCAA Division 1A college football teams over three seasons. In the games played on FieldTurf (an artificial turf), there were 1,050 injuries, 875 of which were minor, 114 were substantial, and 61 were severe. In about the same number of games played on grass, there were 1,203 injuries, 938 of which were minor, 169 were substantial, and 96 were severe.

(**Source:** Michael C. Meyers, "Incidence, Mechanisms, and Severity of Game-Related College Football Injuries on FieldTurf Versus Natural Grass: A 3-Year Prospective Study," *American Journal of Sports Medicine*, Vol. 38, 2010, pp. 687–697.)

a. What two groups are being compared? What is the response variable?

b. Summarize the information in a two-way frequency table.

c. Make a stacked bar graph (frequency on the vertical axis) and a stacked bar graph (percent on the vertical axis). Check your bar graph against the stacked bar graph produced using *TCMS-Tools* data analysis.

d. Write a paragraph to answer the question of whether football injuries during games played on artificial turf and on grass tend to differ in severity.

3 People have become increasingly concerned that loud music, especially that played through headphones and MP3 players, is damaging the hearing of teens. In part to test whether that might be the case, hearing loss in teens in 1988–1994 (before MP3 players were available) was compared to hearing loss in teens in 2005–2006. Of 796 teens aged 18 or 19 in 1988–1994, 15.2% had some hearing loss. Of 413 teens aged 18 or 19 in 2005–2006, 20.1% had some hearing loss.

(**Source:** Josef Shargorodsky, *et al.* "Change in Prevalence of Hearing Loss in US Adolescents," *Journal of the American Medical Association*, Vol. 304, August 2010, pp. 772–778.)

a. What two groups are being compared? What is the response variable?

b. How many students in each group had hearing loss? Summarize the information in a two-way frequency table.

c. Compute and interpret the difference in absolute risk of hearing loss between the two groups.

d. In Part c, the absolute risk reduction may seem fairly small. But suppose that a high school has 4,000 students, all of whom use headphones and MP3 players, and it is true that these cause hearing loss. How many cases of hearing loss do you estimate could be prevented if the students did not use them?

e. Compute and interpret the relative risk of hearing loss for the two groups.

4 In another study of the safety of tanning beds, 375 women who got melanoma before age 40 were compared to 275 similar women who had not. Of the women who got melanoma, 103 had used tanning beds. Of those who had not gotten melanoma, 67 had used tanning beds. (**Source:** Anne E. Cust *et al.* "Sunbed Use During Adolescence and Early Adulthood Is Associated with Increased Risk of Early-Onset Melanoma," *International Journal of Cancer*, 2010.)

a. What two groups are being compared? What is the response variable?

b. Display the data in a two-way frequency table.

c. Was a woman who got melanoma more likely to have used a tanning bed than a woman who did not get melanoma? Explain your reasoning.

d. Complete this sentence: A woman who got melanoma before age 40 was _____ times more likely to have used a tanning bed than a woman who did not get melanoma.

5 Refer back to the information about the asthma medication study from Problem 6 in Investigation 2 (page 11).

a. Who are the subjects in this study? What are the treatments? What is the response variable?

b. Does this study have the first two characteristics of a well-designed experiment? Explain.

c. Is this study single blind? Is it double blind? Explain.

6 A group of college students wanted to see if other college students would obey instructions better when given by a professionally dressed person or a casually dressed person. Two videos of the same 22 year-old actor were made. She gave the same instructions in each, but she was wearing different clothing and hair style. Students were randomly assigned to watch one of the two videos. Of the 32 students who watched the professionally dressed version of the video, 41% correctly followed instructions to write their gender and grade level on a test they were given. Of the 35 students who watched the casually dressed video, 63% wrote this information. (**Source:** Anastacia E. Damon *et al.* "Dressed to Influence: The Effects of Experimenter Dress on Participant Compliance," *Undergraduate Research Journal for Human Sciences*, Vol. 9, 2010, www.kon.org/urc/v9/damon.html)

a. What are the treatments in this study? What is the response variable?

b. Does this study have the first two characteristics of a well-designed experiment?

c. Is this study single blind? Explain.

d. Identify a lurking (hidden) variable that could account for the results.

7 Look back at your work on Applications Task 4. Would you say that the researchers have strong, weak, or no evidence that melanoma is associated with the use of tanning beds? Explain your reasoning.

CONNECTIONS

These tasks will help you build links between mathematics you have studied in the lesson and to connect those topics with other mathematics you know.

8 The recommended level of physical activity for a 12th-grade boy or girl is at least 60 minutes per day at least five days a week of activity that increases heart rate and makes him or her breathe hard. Of 12th-grade girls, 20.6% meet this recommendation, 43.2% exercise at least 60 minutes but for only one to four days a week, and 36.2% do not exercise at least 60 minutes on any day of the week. Of 12th-grade boys, the respective percentages are 38.7%, 39.8%, and 21.5%. (**Source:** U.S. Census Bureau, *Statistical Abstract of the United States: 2011*, Table 209.)

  a. Construct pie charts for these data.

  b. Construct a stacked bar graph (percent on the vertical axis).

  c. Do boys or girls tend to do better meeting the recommendation? How much better? Which graphic better shows this?

9 Reports in the media often mention a percentage increase or percentage decrease. Carefully examine each of the following reports.

  a. Some people think that rhino horns have medicinal uses, so rhinos are often killed illegally so their horns can be ground into powder. A UPI headline read, "Rhino poaching in South Africa increased by 50 percent in 2013." The following article said that 1,004 rhinos were killed in South Africa in 2013, but did not give the number killed in 2012. (**Source:** www.upi.com/Science_News/Blog/2014/01/17/Rhino-poaching-in-South-Africa-increased-by-50-percent-in-2013/3981389975097/)

   i. About how many rhinos were killed in South Africa in 2012?

   ii. A CNN article published about the same time gave the number of rhinos killed in 2012 (the number you computed in part i) and the number of rhinos killed in 2013 (which is 1,004). It said that "The 2013 record is almost double the year before." Is CNN correct? (**Source:** www.cnn.com/2014/01/18/world/africa/south-africa-record-rhinos-poached/)

b. The Measles & Rubella Initiative reported:

> With intervention by the Measles & Rubella Initiative and commitment from governments around the world, global measles deaths worldwide fell by 74 percent between 2000 and 2010, from an estimated 535,000 to 139,300.

Source: www.measlesrubellainitiative.org

   i. Show how the percentage decrease of 74% was computed.

   ii. Is the 74% one of the statistics you studied in this lesson or something different? Explain.

   iii. Assuming that the world population was equal in 2000 and 2010, show how you can compute the relative risk of dying from measles in those two years.

   iv. Can you compute the absolute risk reduction? Explain.

c. The *Cary News* reported that 16,804 high school students dropped out of North Carolina's public schools in 2009–2010, a 12.4 percent decrease over the year before. The newspaper said the dropout rate in 2009–2010 was 3.75% but was 4.27% in 2008–2009. (**Source:** www.carynews.com/2011/03/16/30046/schools-retain-more-students.html, March 16, 2011)

   i. Compute the percentage decrease in number of high school dropouts based on the information given. Compare your answer with the reported answer.

   ii. How many students were there in North Carolina's public schools in 2009–10?

   iii. Assuming that the total number of students was the same as in 2009–10, how many students dropped out in 2008–09?

   iv. Compute the relative risk of dropping out for the two years. Then compute the absolute risk reduction.

d. A survey in 22 public high schools in Boston asked students about physical date violence. It found that physical date violence was associated with alcohol use. In a table giving the absolute risk of physical date violence perpetration by risk behaviors and sex, a relative risk of 2.05 was reported for boys who used alcohol in the past 30 days compared to those who had not. (**Source:** Emily F. Rothman *et al.* "Perpetration of Physical Assault Against Dating Partners, Peers, and Siblings Among a Locally Representative Sample of High School Students in Boston, Massachusetts," *Archives of Pediatric and Adolescent Medicine*, Vol. 164, December 2010, pp. 1118–1124.) Explain the meaning of the relative risk of 2.05, in the context of this situation.

10 In a rural area of Bangladesh, more than 10% of children die by 10 years of age. A study followed 144,858 Bangladeshi newborns. Among the 1,385 children whose mothers had died, only 24% survived to their 10th birthday. Of the 143,473 whose mothers lived, 89% survived to their 10th birthday. (**Source:** Carine Ronsmans et al. "Effect of Parent's Death on Child Survival in Rural Bangladesh: A Cohort Study," *The Lancet*, Vol. 375, June 2010, pp. 2024–2031.)

a. Make a two-way frequency table summarizing this situation.

b. Overall, what percentage of children died by age 10?

c. Of the children who died, what percentage had mothers who died?

d. Write a sentence or two for a newspaper report giving the absolute risk reduction of death for children whose mothers remained alive compared to those whose mothers died.

e. Write a sentence or two for a newspaper report giving the relative risk for children whose mothers died compared to those whose mothers remained alive.

11 Healthy adults need at least 7 hours of sleep a night. (Teens need at least 9 hours.) A survey of 74,571 Americans 18 and older found that 35.3% reported sleeping less than 7 hours per night on average. People who reported sleeping less than 7 hours were more likely to report unintentionally falling asleep during the day at least once in the preceding 30 days (46.2% versus 33.2%). They were more likely to report nodding off or falling asleep while driving in the preceding 30 days (7.3% versus 3.0%). They also were more likely to report snoring (51.4% versus 46.0%). (**Source:** *CDC Morbidity and Mortality Weekly Report*, March 4, 2011, www.cdc.gov/mmwr/preview/mmwrhtml/mm6008a2.htm?s_cid=mm6008a2_w)

a. Make a two-way frequency table, with one column for those who get at least 7 hours of sleep and one for those who do not. The rows will represent the number in each group who did and did not report snoring.

b. What percentage of all of the Americans surveyed snore?

c. What is the relative risk of nodding off or falling asleep while driving in the preceding 30 days for the two groups? Use this ratio in a sentence that explains what it means.

d. What percentage of all of the Americans surveyed nodded off or fell asleep while driving in the preceding 30 days?

e. Would you say that the increased risk of nodding off or falling asleep while driving is higher or lower than the increased risk of snoring for those who do not get 7 hours of sleep on average? Explain how you decided this.

12 People like to speculate about "curses" attached to various events that are otherwise positive for the person involved. For example, "Researchers compared actresses who won Best Actress statuettes from 1936 to 2010 to those who were nominated but didn't win, and found that winners were, indeed, 1.68 times as likely to divorce as non-winners. Of the 265 married nominees, 159 eventually divorced—a whopping 60 percent." In the last sentence, "nominees" means all of the actresses nominated, both those who won the Oscar and those who did not.

Your goal in this task is to compute the proportion of winners who were divorced and the proportion of non-winners who were divorced. (**Sources:** H. Colleen Stuart, *et al.* "The Oscar Curse: Status Dynamics and Gender Differences in Marital Survival," *Social Sciences Research Network*, January 27, 2011; www.huffingtonpost.com/2011/01/31/oscar-curse-study-researc_n_816295.html?ir=Entertainment)

a. Let x be the number of winners who eventually divorced. Let y be the number of non-winners who eventually divorced. Write an equation that gives the sum of x and y.

b. Typically, five women are nominated for Best Actress and one of them is the winner. What is a reasonable estimate of the number of the 265 married nominees who were winners? What is a reasonable estimate of the number who did not win?

c. Write the expression that was used to compute the relative risk of 1.68.

d. Parts a and c lead to two equations in two unknowns, say x and y.

   i. Write and solve the system of equations.

   ii. Then, find the proportion of winners who were divorced.

   iii. Find the proportion of non-winners who were divorced.

e. Do you think it is better to report that winners were 1.68 times more likely to divorce or would something else have been more clear to most people? Explain.

13 According to an article on magicvalley.com, "One in five Idahoans receive welfare benefits, almost double the level 10 years ago." The article goes on to say that nearly 321,700 Idahoans are enrolled in the state's food stamp, cash assistance, child care, or Medicaid program. But the number has dropped by 10% from the previous year. (**Source:** magicvalley.com/news/local/govt-and-politics/in-idahoans-receive-welfare-benefits-director-says/article_02ec414e-7cde-11e3-8a98-001a4bcf887a.html)

   a. What percentage of Idahoans receive "welfare benefits"? What percentage received welfare benefits 10 years ago?

   b. How many people live in Idaho? How many received "welfare benefits" in the previous year?

   c. The article also says that the median wage in Idaho is $11.15 an hour or $23,200 annually. Show how the yearly amount was computed from the hourly rate.

REFLECTIONS

These tasks provide opportunities for you to re-examine your thinking about ideas in the lesson and your use of mathematical practices and habits of mind that are useful throughout mathematics.

14 Refer to the bar graph below. Is it possible that more people in Group B than in Group A have the condition under study? If so, provide some sample data to illustrate this. If not, explain why not.

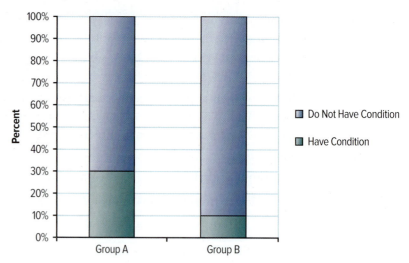

15 Sometimes the word *odds* is used when referring to risk. However, this word is not used in a consistent way, so it is important to pay close attention to what the person might mean. The mathematical definition of odds is illustrated by this example: If the odds that an event occurs are 3 to 5, then the probability the event occurs is $\frac{3}{3+5}$, or $\frac{3}{8}$. When outcomes are equally likely, the odds of an event are *number of favorable outcomes* to *number of unfavorable outcomes*.

Each statement below uses the word *odds*. For each, decide whether the person probably is using the word according to the mathematical definition or whether they mean something else. Justify your choice.

a. The odds against winning on an American roulette wheel are 37 to 1. On an American roulette wheel, there are 38 spaces and you win if your ball falls into the one you had selected.

b. An article in the *Chronicle of Higher Education* described the odds of getting an interview after submitting an application for a job as university professor as "somewhere in the neighborhood of one in 20 to one in 30." It then compared this to the odds of surviving the Hunger Games: one in 24. (In the original *Hunger Games*, there are two contestants from each of twelve districts and only one winner.)
(**Source:** chronicle.com/article/The-Odds-Are-Never-in-Your/144079/)

c. President Obama once said that the odds of completing final treaties in the Middle East "are less than fifty-fifty." (**Source:** www.newyorker.com/reporting/2014/01/27/140127fa_fact_remnick)

d. "Lotteries offer the worst odds in legal gambling—about 55 percent of what people pay for tickets is paid out in prizes. Yet we spend an average of $540 per household on lottery tickets every year …" (**Source:** opinionator.blogs.nytimes.com/2014/01/15/playing-the-odds-on-saving/)

e. The National Safety Council gives the "lifetime odds of death" by being bitten or struck by a dog as 1 in 122,216. It gives the odds of dying by any cause as 1 in 1. (**Source:** www.nsc.org/news_resources/injury_and_death_statistics/Documents/Injury_Facts_43.pdf)

16 When absolute risk is low, it is often reported as the number of cases per 1,000 people, per 10,000 people, or per 100,000 people. For example, Problem 4 (page 4) of Investigation 1 says that 54 people out of every 100,000 who get the HPV vaccine have an adverse event.

Read this statement about collisions of autos and deer in Pennsylvania: "The highest rate for 2009 was Fulton County, west of Adams County along the Maryland border, with 16.83 per 10,000 people." In 2009, Fulton County had 25 deer-related accidents. About how many people lived in Fulton County? (**Source:** www.dailylocal.com/article/20110310/NEWS/303109986&pager=full_story, March 10, 2011)

17 Write a short summary of what the research summarized below tells you about the placebo effect.

> Remifentanil is a very quick acting and effective pain-reliever. However, its effect doesn't last very long, with half of its usefulness gone in 3 to 4 minutes. In a test of the placebo effect, twenty-two healthy volunteer subjects were hooked up to an IV so that the Remifentanil could be administered. In the first step, before being given Remifentanil, the subjects were subjected to moderate pain (70 on a scale of 0 to 100) caused by a heat source attached to their calf and their reaction measured. In the second step, the subjects were given Remifentanil, but were led to believe that it hadn't been turned on yet. Their rating of their pain was significantly lower. In the third step, the subjects were told that the drug had been started, and their pain rating dropped about the same amount again as it had in the second step. In the final step, the subjects were told that the drug had been stopped, even though it hadn't. The subjects reported a considerable increase in pain intensity, almost to as high a level as in the first step when they were receiving no Remifentanil at all. The subject's brains were scanned during these tests and the activity in the areas of the brain that are involved with pain was consistent with the subjects' reports of their pain.

Source: Ulrike Bingel, et al. "The effect of treatment expectation on drug efficacy: Imaging the analgesic benefit of the opioid remifentanil," Science Transitional Medicine, Vol. 3, 16 February 2011.

18 Look back at your work on Applications Task 3. Does your analysis of these data convince you that loud music, especially played through headphones and MP3 players, is damaging the hearing of teens? Explain.

EXTENSIONS

These tasks provide opportunities for you to explore further or more deeply the mathematics you are learning.

19 Sometimes you will see risk given in the form "1 in n." For example, in Problem 2 (page 4) of Investigation 1, the risk of breast cancer for a woman was given as 1 in 8.

  a. Refer to the two-way frequency table analyzed in the Think About This Situation (page 2). Write the risk of getting lung cancer for the boys who smoke in the form "1 in n."

  b. Write the risk of getting lung cancer for the boys who do not smoke in the form "1 in n."

  c. Write a procedure that can be used to convert numbers like those in the table to the form "1 in n."

  d. In the Civil War, 77 of 425 Confederate generals were killed in action; 47 of 583 Union generals were killed. (**Source:** Jim Webb, *Born Fighting*, Broadway Books, 2004, page 221.) Use your procedure from Part c to convert these risks to the form "1 in n." Then compare them in a sentence.

20 Consider the risk of a 16–17-year-old driver being involved in a fatal crash in the next year. (**Source:** www.cdc.gov/mmwr/preview/mmwrhtml/mm5941a2.htm)

a. In a government report from the U.S. Centers for Disease Control (CDC), the number of 16–17-year-old drivers involved in fatal crashes in a year is given as 16.7 per 100,000. What is the absolute risk, of a 16–17-year-old driver being involved in a fatal crash in the next year?

b. The lowest absolute risk is in New Jersey, 0.000097. Convert this absolute risk to number per 100,000.

c. Write a procedure that can be used to convert absolute risk to number per 100,000.

d. The highest absolute risk is in Wyoming, 0.000596. Use your procedure from Part c to write this risk as number per 100,000.

e. Modify your procedure from Part c and use it to give the risk in Wyoming as number per 10,000.

f. The graph below gives the rates per 100,000 over all states over the years from 1990 to 2008. Write a one-paragraph summary of the most important information shown.

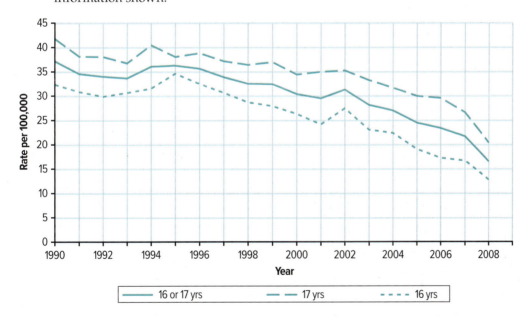

Image Source/Getty Images

21 Another statistic used in medical reports is the **number needed to treat**. This is the number of people who would have to be treated in order to prevent one adverse event. For example, suppose that there are 30 injuries per year for every 1,000 bicyclists who do not wear helmets and 10 injuries per year for every 1,000 who do. The absolute risk reduction is $\frac{30}{1,000} - \frac{10}{1,000} = \frac{20}{1,000}$. In other words, 20 injuries are prevented for every 1,000 riders who wear a helmet. To prevent 1 injury, $\frac{1,000}{20} = 50$ riders would have to wear a helmet. This is the number needed to treat.

a. The number needed to treat is the reciprocal of another statistic. Which one?

b. Millions of Americans take *statin* drugs to prevent heart attacks and strokes. In one experiment to see how effective atorvastatin was compared to a placebo, 5,168 patients with high blood pressure but normal cholesterol were randomly assigned to get atorvastatin and 5,137 to get a placebo. The numbers who had either coronary heart disease (CHD) or a stroke in three years were 175 for those who took atorvastatin and 258 for those who got a placebo. How many people need to be treated with atorvastatin in order to prevent one case of CHD or stroke in this type of patient? (**Source:** P.S. Sever *et al.* "Prevention of Coronary and Stroke Events with Atorvastatin in Hypertensive Patients Who Have Average or Lower-than-Average Cholesterol Concentrations, in the Anglo-Scandinavian Cardiac Outcomes Trial-Lipid-Lowering Arm (ASCOT-LLA): A Multicentre Randomised Trial," *Lancet*, Vol. 361, 2003, pp. 1149–1158. www.medicine.ox.ac.uk/bandolier/booth/cardiac/statascot.html)

c. Cryptococcal meningitis is a fungal infection that has a 55% death rate when treated with fluconzole alone. Each year, 957,000 people, mostly in sub-Saharan Africa or Southeast Asia, get this disease. A study found that when amphotericin B is added to the treatment, the death rate drops to 26%. (**Source:** Nancy Walsh, "Short Ampho Course Works in Cryptococcus, September 27, 2012, www.medpagetoday.com/InfectiousDisease/PublicHealth/34998)

   i. How many deaths from cryptococcal meningitis are expected among every 10,000 victims who are treated with fluconzole alone?

   ii. How many deaths are expected if all 10,000 receive amphotericin B as well as fluconzole?

   iii. What is the number needed to treat?

   iv. Use the number needed to treat from part iii in a sentence that explains what it means.

d. It has been reported that wearing a helmet while skiing/snowboarding reduces the probability of a head injury by 35%. This sounds impressive, but people have pointed out that the risk of a head injury even without a helmet is only about 0.09 per 1,000 outings. Compute and interpret the number needed to treat (wear a helmet) to avoid one head injury.

(**Source:** www.nhs.uk/news/2011/02February/Pages/head-injury-protection-ski-helmet.aspx)

22 A study was conducted to see how people respond to medical statistics. Each subject was given the following descriptions of three screening tests for cancer.

Test A: If you have this test every 2 years, it will reduce your chance of dying from cancer A by around one third over the next 10 years.

Test B: If you have this test every 2 years, it will reduce your chance of dying from cancer B from around 3 in 1,000 to around 2 in 1,000 over the next 10 years.

Test C: If around 1,000 people have this test every 2 years, 1 person will be saved from dying from cancer C every 10 years.

a. One of these describes the number needed to treat cancer (see Extensions Task 21). Which one?

b. One describes the absolute risk reduction. Which one?

c. One describes the relative risk. Which one?

d. The 306 subjects were asked if they would be likely to accept the test. The percentage who said yes was 80% for Test A, 53% for Test B, and 43% for Test C. Is this rational decision-making? Explain. (**Source:** Gerd Gigerenzer et al. "Helping Doctors and Patients Make Sense of Health Statistics," *Psychological Science in the Public Interest*, Vol. 8, 2008, pp. 53–96.)

ON YOUR OWN

These tasks provide opportunities for you to review previously learned mathematics and to refine your skills in using that mathematics.

23 Being able to work with and understand percentages and ratios is important in many different careers. Consider each of the following situations that require work with percentages or ratios.

a. Some government funding is related to the number of people living in poverty in a particular area. The 2010 population of Caswell County, NC, was 23,719 and 21.7% of the people were living in poverty. (**Source:** quickfacts. census.gov/qfd/states/37/37033.html) How many people were living in poverty in Caswell County, NC, in 2010?

b. Wisconsin farmers received an average price of $17.60 per 100 pounds for their milk in April 2012, down $1.60 from April 2011. By what percentage did the average price of milk decrease between April 2011 and April 2012? (**Source:** www.nass.usda.gov/Statistics_by_State/Wisconsin/Publications/Dairy/mkallpri.pdf)

c. The Centers for Disease Control estimates that 1 in 6 people in the United States suffer food poisoning in any given year. At the beginning of 2012, the population of the United States was approximately 312,781,000 people. How many people would you expect to have suffered from food poisoning during 2012? (**Sources:** www.cdc.gov/foodborneburden/, www.commerce.gov/blog/2011/12/30/census-bureau-projects-us-population-3128-million-new-years-day-2012)

d. Cystic Fibrosis (CF) is a genetic disease that causes mucus to build up and clog some of the organs in the body. The 2012 population of the United States was approximately 313 million people. Approximately 12 million people in the United States carried the defective gene and did not have CF. Suppose you want to say that 1 out of every x people was a carrier of the defective gene for CF without having CF. What is the correct value of x? (**Source:** www.iacfa.org)

24 Determine each number.

a. 36 is 75% of what number?

b. 90 is what percent of 150?

c. What number is $\frac{2}{10}$ of 1% of 4,100?

25 Solve each of the following equations for n.

a. $\frac{a}{b} = \frac{1}{n}$

b. $\frac{a}{b} = \frac{n}{1,000}$

c. $\frac{1}{a} = \frac{n+3}{k}$

d. $4n - 2 = an + 7$

e. $n(5 - k) = 11n + 8$

f. $\frac{8}{an} = a + b$

26 The amount of Tylenol (in mg) that a child should take depends on the child's weight. Suppose that $T(w)$ represents the maximum safe dosage for a child weighing w pounds.

a. $T(24) = 160$. Explain what this tells you.

b. The table below gives several weights and the associated dosages. Based on this table of values, would a linear function be a good model for this relationship? Explain.

w	T(w)
24	160
48	320
60	400

c. Based on the table, predict the maximum dosage for a child weighing 80 pounds. Explain your reasoning.

d. In the field of healthcare, functional relationships are conventionally called formulas. In the formula $T(w) = \frac{20}{3}w$, what does the $\frac{20}{3}$ tell you?

27 Recall that you can think of an angle as being formed by rotating a ray about its endpoint from an initial position to a terminal position. If the rotation is counterclockwise, the angle has positive measure. If the rotation is clockwise, the angle has negative measure.

The diagram at the right shows an angle with measure θ in **standard position** in a coordinate system. Its *initial side* $\overrightarrow{OA}$ coincides with the x-axis, its vertex is at the origin, and the *terminal side* of the angle contains the point $B(x, y)$. Recall that in this context,

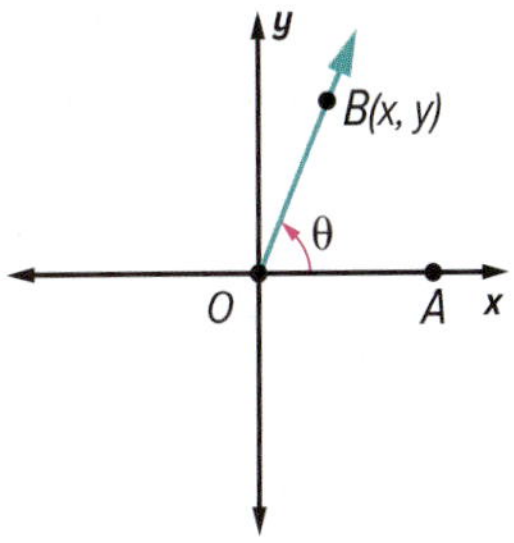

$$\cos \theta = \frac{x}{OB} \quad \text{and} \quad \sin \theta = \frac{y}{OB}.$$

a. For each point on the terminal side of an angle with measure θ in standard position, draw a sketch of the angle. Find cos θ. Find sin θ.

  i. $P(5, 12)$

  ii. $Q(-6, 4)$

b. How are these definitions for cosine and sine similar to, and different from, the corresponding right-triangle definitions?

c. Suppose the terminal side of an angle in standard position with measure θ is on the axis indicated below. Find cos θ and sin θ in each case.

  i. positive y-axis

  ii. negative x-axis

  iii. negative y-axis

  iv. positive x-axis

d. Copy the following table. Indicate whether the value of each function is positive or negative in the given quadrant.

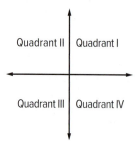

Quadrant	I	II	III	IV
cos θ				
sin θ				

e. Describe the pattern of *sign* change in cos θ as θ increases from 0 to 360°. In sin θ.

28 Recall that if $10^b = a$, then b is called the **base-10** (or **common**) **logarithm of** a. This is denoted $\log_{10} a = b$ or simply $\log a = b$.

a. Rewrite each equation using logarithmic notation.

  i. $10^0 = 1$

  ii. $10^2 = 100$

  iii. $10^5 = 100{,}000$

  iv. $10^{-2} = 0.01$

b. Rewrite each equation using exponential notation.

  i. $\log 10 = 1$

  ii. $\log 1{,}000 = 3$

  iii. $\log 0.1 = -1$

  iv. $\log 0.001 = -3$

c. Explain why $2 < \log 485 < 3$.

d. Complete a copy of the table below by making consecutive integer estimates (see Part c) of the following logarithms. Check your estimates using a graphing calculator or *TCMS-Tools*.

	Estimate	Calculation
log 16		
log 1.6		
log 1,600		
log 3		
log 0.3		
log 30		

e. Use the table in Part d to help determine how the values of log x and log 100x are related to each other. Justify why this will be true for any value of x.

29 Without using technology, determine the value of each expression if $a = -\frac{1}{3}$, $b = 4$, and $c = 3$. Then check your results using technology.

a. $\sqrt{a^2}$

b. $c^3 - ac$

c. a^{-2}

d. $|b - c|$

e. $-6|a|$

f. $|b^{-1} + a|$

30 Gabriel borrows $750 from his parents so that he can start a neighborhood lawn-mowing business. His parents will charge him 6% annual simple interest. Simple interest is calculated using only the actual amount of the loan (or deposit).

a. If Gabriel's parents add the interest to his loan balance on a monthly basis, how much will the interest be each month?

b. Suppose that for the first year, Gabriel does not make any payments to his parents. Make a table showing the loan balance at the end of each month for the first year.

c. Describe how the monthly balance of the loan is changing from one month to the next.

d. Suppose that $f(n)$ represents the loan balance after n months and $f(n + 1)$ represents the loan balance after $n + 1$ months, which of the following recursive formulas are correct?

Formula I: $f(n + 1) = 1.06f(n)$

Formula II: $f(n + 1) = 1.005f(n)$

Formula III: $f(n + 1) = 0.06f(n)$

Formula IV: $f(n + 1) = f(n) + 3.75$

e. Write a function rule that represents the amount B that Gabriel owes his parents as a function of the number of months n since he borrowed the money.

A Test of Significance

In some cases, it is obvious that there is a difference between two groups. In other cases, it is not so obvious. For example, the U.S. Centers for Disease Control and Prevention (CDC) periodically conducts a survey to track risky behavior by American youth. The two tables below give the responses of 12th-grade students to two questions on the most recent survey. The sample sizes differ somewhat because not every student answers every question. (**Source:** 2011 Youth Risk Behavior Survey, www.cdc.gov/healthyyouth/yrbs/)

Rode with a driver who had been drinking alcohol one or more times during the 30 days before the survey?

	12th-Grade Boys	12th-Grade Girls
Yes	488	503
No	1,292	1,295
Total	1,780	1,798

In a physical fight one or more times during the 12 months before the survey?

	12th-Grade Boys	12th-Grade Girls
Yes	616	356
No	1,191	1,480
Total	1,807	1,836

THINK ABOUT THIS SITUATION

Think about whether the evidence is comparable that the proportion of all 12th-grade boys who engaged in each risky behavior is different from the proportion of all 12th-grade girls who engaged in the behavior.

a. Compare the proportions of 12th-grade boys and girls who rode with a driver who had been drinking alcohol.

b. Compare the proportions of 12th-grade boys and girls who were in a physical fight.

c. Now suppose that the survey could have asked *every* 12th-grade boy and girl about these behaviors. For which behavior is there more convincing evidence that the proportion of all 12th-grade boys who engaged in the behavior would be different from the proportion of all 12th-grade girls who engaged in the behavior? Explain your reasoning.

d. If the survey had asked *every* 12th-grade boy and girl about these behaviors, for which question(s) is it *possible* that the proportions would be equal? For which question(s) does it seem *plausible* (*believable*) that the proportions would be equal?

In this lesson, you will learn how to decide, on the basis of random samples from two different populations, whether it is reasonable to believe that there is a difference in the proportion of all people in each population who have some characteristic. Most of the examples in this lesson will be from the fields of politics and education. You also will reconsider some of the medical situations from the previous lesson.

INVESTIGATION 1

Homogeneous Groups

In this investigation, you will learn how to decide if two groups you are comparing are *homogeneous*. As you work on the problems in this investigation, look for answers to this question:

What can help you decide whether two random samples were taken from populations with different proportions?

1 Two different groups are called **homogeneous** if, when they are sorted into the same categories, the proportion of people in the first group who fall into any given category is equal to the proportion of people in the second group who fall into the category. For example, the 2000 and 2010 U.S. Census asked Hispanic residents about their origin. The results are given in the table and graphs on page 38.

Origin of U.S. Hispanics	2000 Census	2010 Census
Mexican	20,640,711	31,798,258
Puerto Rican	3,406,178	4,623,716
Cuban	1,241,685	1,785,547
Other	10,017,244	12,270,073
Total	**35,305,818**	**50,477,594**

Source: *The Hispanic Population:* 2010, 2010 Census Brief, www.census.gov/prod/cen2010/briefs/c2010br-04.pdf

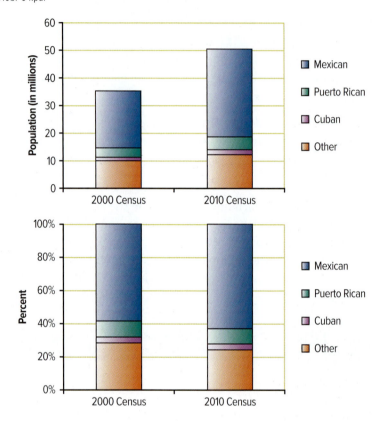

a. The top bar graph shows *frequency* of country of origin on the vertical axis. The bottom bar graph shows *percent* of country of origin on the vertical axis. Which bar graph best shows how much the total Hispanic population increased in the 10 years between the two censuses? Estimate this increase using the graph.

b. Are the two groups (2000 Hispanics and 2010 Hispanics) homogeneous with respect to origin or are they similar but not homogeneous? Which bar graph shows this best? How would you describe any change?

c. Suppose that the Hispanic population increases from the 2010 Census to the 2020 Census, but the percentage in each category of origin does not change. How would a stacked bar graph with percent on the vertical axis for the 2020 Census differ from the one for the 2010 Census?

You may recall computing an **expected number** in your previous studies. For example, if you roll a die 42 times, the expected number of times you roll a 4 is 7. That is because the probability of getting a 4 is $\frac{1}{6}$ and $\frac{1}{6} \cdot 42 = 7$. Similarly, the expected number of 4s in 33 rolls is $\frac{1}{6} \cdot 33 = 5.5$. (Expected number does not have to be a whole number.)

2 Suppose you plan to take a random sample of 1,000 Hispanics in the United States. Assume that the distribution of origin has not changed since 2010.

a. What proportion of Hispanics in the United States in 2010 were of Mexican origin? What is the expected number of Hispanics in your sample who will be of Mexican origin?

b. What proportion of Hispanics in the United States in 2010 were of Puerto Rican origin? What is the expected number of Hispanics in your sample who will be of Puerto Rican origin?

c. Now suppose that you take a random sample of 1,000 Hispanics from your state and find 521 Hispanics of Mexican origin. A friend takes a second random sample of 1,500 Hispanics from your state and finds 847 of Mexican origin. What is the best estimate of the proportion of Hispanics in your state who are of Mexican origin?

d. Suppose you plan to take yet another random sample from your state, this time of 500 Hispanics. Based on your work in Part c, what is the best estimate of the expected number of Hispanics of Mexican origin you will find in this sample?

3 The larger group from which a **sample** is taken is called the **population**. The proportions in small samples can vary quite a bit from those in the population from which the sample was taken and from each other.

The table at the top of the next page shows two random samples of size 50 taken from the national population of SAT Critical Reading scores.

a. Samples 1 and 2 were taken from the same population. Looking only at these two samples, what is your best estimate of the proportion of all Critical Reading scores that are in the category 500–590?

Score	Sample 1	Sample 2
700–800	2	2
600–690	8	6
500–590	13	11
400–490	22	26
300–390	4	3
200–290	1	2
Total	**50**	**50**

b. What is your best estimate of the expected number of scores in a random sample of size 50 that would fall in the category 300–390?

c. Results from a third random sample are given in the table below.

Score	Sample 1	Sample 2	Sample 3
700–800	2	2	1
600–690	8	6	3
500–590	13	11	17
400–490	22	26	11
300–390	4	3	14
200–290	1	2	4
Total	**50**	**50**	**50**

i. Match the following three stacked bar graphs (percent on the vertical axis) with the corresponding samples above.

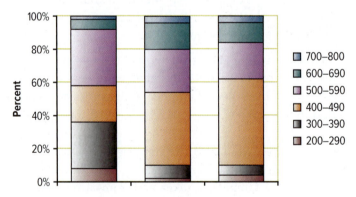

ii. Is it *possible* that this third sample also is a random sample from the population of all SAT Critical Reading scores? Do you think that is *plausible* or do you think it probably came from a different population? (**Plausible** means **reasonable to believe**.) Explain your reasoning using the data in the table or the stacked bar graphs.

SUMMARIZE THE MATHEMATICS

In this investigation, you learned how to decide whether two groups are homogeneous.

a. What does it mean to say that two groups are homogeneous?

b. How can you tell from a stacked bar graph (percent on the vertical axis) whether two groups are homogeneous? How can you tell from a stacked bar graph (frequency on the vertical axis) whether two groups are homogeneous?

c. How do you compute an expected number?

d. What is the difference between a sample and a population?

Be prepared to share your ideas and computation method with the class.

✓CHECK YOUR UNDERSTANDING

This table shows the age distribution of people in the United States in 2010.

a. Complete the last column of a copy of this table.

Age	Number of People	Proportion of Population
Under 18	74,181,467	0.2403
18–44	112,806,642	
45–64	81,489,445	0.2639
65 and Over	40,267,984	
Total	**308,745,538**	

Source: *Age and Sex Composition:* 2010, Census Briefs, www.census.gov/prod/cen2010/briefs/c2010br-03.pdf

b. The following table gives the age distribution for the state of Utah. Are the age distributions for the U.S. and Utah homogeneous? If so, show why. If not, describe the main way that they differ.

Age	Number of People
Under 18	871,027
18–44	1,096,191
45–64	547,205
65 and Over	249,462
Total	**2,763,885**

c. Suppose that you take a random sample of 5,000 people from the U.S. population. Make a table that shows the expected number of people in your sample who come from each of the four age groups.

The Chi-Square Statistic

In the last investigation, you learned that in homogeneous populations, the same proportion of people fall into each category. Frequently, however, you have only a sample from each population. In this investigation, you will learn a method of measuring how different the two samples are. In the next investigation, you will use this statistic to decide whether the samples are so different that you should conclude that the populations from which the samples came are different.

As you complete the following problems, look for an answer to this question:

If you have two samples in which the people are placed into various categories, how can you measure how different the two samples are?

1 Shown below is information on responses from another question on the U.S. Centers for Disease Control and Prevention (CDC) Youth Risk Behavior Survey. This time, only the (approximate) **marginal totals** are given.

Drank (non-diet) soda daily during the 7 days before the survey?

	12th-Grade Boys	12th-Grade Girls	Total
Yes			930
No			2,520
Total	1,725	1,725	3,450

a. Overall, what proportion of the 12th-grade students surveyed drank soda daily? What proportion did not?

b. Fill in the cells of a copy of the table so that the samples of boys and girls are homogeneous. Round your entries to the nearest tenth in this and similar problems.

2 Here is response information from another question on the CDC Youth Risk Behavior Survey. Again, only the marginal totals are given, but this time, the sample sizes are not equal because different numbers of boys and girls had ridden a bike during the previous 12 months.

Wore a bicycle helmet (among students who had ridden a bicycle during the previous 12 months)?

	12th-Grade Boys	12th-Grade Girls	Total
Rarely or Never			1,825
Yes			202
Total	1,160	867	2,027

a. Overall, what proportion of the 12th-grade students surveyed who had ridden a bicycle over the previous 12 months wore a helmet rarely or never? What proportion did wear a helmet?

b. Fill in the cells of a copy of the table so that the samples of boys and girls are homogeneous.

Brand X Pictures

3 The numbers you entered in the cells of the tables in Problems 1 and 2 are called **expected frequencies** or **expected counts**. They show what perfectly homogeneous samples would have looked like.

Reproduced below is the table of *observed frequencies* from Applications Task 4 (page 21) in Lesson 1.

Observed Frequencies

	Women with Melanoma	Women with No Melanoma	Total
Used Tanning Beds	103	67	170
Did Not Use Tanning Beds	272	208	480
Total	375	275	650

a. Overall, what proportion of the women had used tanning beds? What proportion of the women had not used tanning beds?

b. Fill in the cells of a copy of the table below of frequencies expected if the two samples of women were taken from homogeneous populations.

Expected Frequencies

	Women with Melanoma	Women with No Melanoma	Total
Used Tanning Beds			170
Did Not Use Tanning Beds			480
Total	375	275	650

c. Compare the table of observed frequencies with the table of expected frequencies. Do you think the evidence is strong, weak, or nonexistent that women with melanoma are more likely to have used tanning beds than those without melanoma?

4 In 2000 and 2010, the Gallup Poll asked national random samples of 1,000 parents of grades K–12 students,

> "Would you say that public education today in grades K through 12 is better, about the same, or worse than when you were a student?"

Here are the results.

Parent Perceptions of Public Education

	2000	2010
Better	340	278
About the Same	124	227
Worse	536	495
Total	1,000	1,000

Source: counts estimated from www.gallup.com/poll/1612/education.aspx

a. Make a table of the frequencies expected if the two samples of parents were taken from homogeneous populations.

b. Compare the table of observed frequencies with the table of expected frequencies. Does the evidence seem strong, weak, or nonexistent that if the 2000 and 2010 Gallup Polls asked all parents, the proportion giving at least one of the responses would have been different?

5 Suppose that you have two samples with marginal totals as given in the table below. For each cell of the table, use the totals and proportional reasoning to write a general formula that can be used to find the expected frequency for homogeneous samples. Compare your formula with that of others. Resolve any differences.

Expected Frequencies

	Sample 1	Sample 2	Row Total
Category 1			Row 1 Total
Category 2			Row 2 Total
Category 3			Row 3 Total
Column Total	Column 1 Total	Column 2 Total	Grand Total

6 Now that you can construct a table of expected frequencies that shows what the two samples would have looked like if they had been perfectly homogeneous, you can measure how different the actual samples are from the perfectly homogeneous ones. For this measurement, you can use the **chi-square statistic χ^2**. Follow these steps to compute χ^2 in Parts a–c.

Step 1. Make a table of expected frequencies.

Step 2. For each cell of the table, find the difference $O - E$ between the observed frequency O and the expected frequency E.

Step 3. Square each difference.

Step 4. Divide each squared difference by the expected frequency for that cell.

Step 5. Sum up all of the values from Step 4.

a. Compute χ^2 for these data from random samples of 9th- and 12th-graders.

Favor last-period pep rallies for home football games?

	9th-Graders	12th-Graders
Yes	40	42
No	60	58
Total	100	100

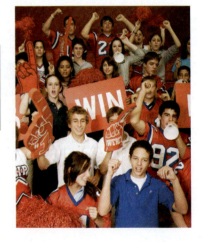

b. Compute χ^2 for these data from random samples of 9th- and 12th-graders.

Favor an off-campus site for junior-senior prom?

	9th-Graders	12th-Graders
Yes	39	43
No	61	57
Total	100	100

c. Compute χ^2 for these data from random samples of 9th- and 12th-graders.

Favor double-period math classes?

	9th-Graders	12th-Graders
Yes	34	48
No	66	52
Total	100	100

d. When sample sizes are the same, does χ^2 tend to be larger when the two samples are more different or when the two samples are more alike? Explain your reasoning.

7 Looking back at your work in Problem 6, which of the following gives a formula for the chi-square statistic χ^2? Recall that the symbol Σ indicates to add up all of the different values for the expression that follows.

I. $\quad \chi^2 = \sum \left(\dfrac{O - E}{E} \right)^2$

II. $\quad \chi^2 = \dfrac{\Sigma (O - E)^2}{E}$

III. $\quad \chi^2 = \sum \dfrac{(O - E)^2}{E}$

IV. $\quad \chi^2 = \dfrac{(O - E)^2}{E}$

8 Examine the formula for the chi-square statistic χ^2.

a. If two samples are identical, what is the value of χ^2? Justify your answer.

b. Why is it necessary to square the difference $O - E$? Justify your answer by giving an example of what happens if you do not.

9 Below are the results of two polls of random samples of high school seniors and juniors, who were asked if they approve of, disapprove of, or do not care about a proposed ban of energy drinks on high school campuses.

Poll I

	Seniors	Juniors
Approve	10	12
Disapprove	30	32
Do Not Care	60	56
Total	100	100

Poll II

	Seniors	Juniors
Approve	100	120
Disapprove	300	320
Do Not Care	600	560
Total	1,000	1,000

a. Compare the proportions of seniors and juniors who approve, disapprove, and do not care in the two polls.

b. For each poll, use your calculator or computer software to compute a table of expected counts and χ^2.

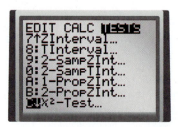 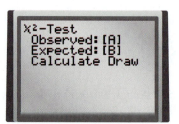

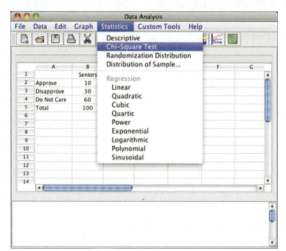

c. Explain why one value of χ^2 is much larger than the other.

SUMMARIZE THE MATHEMATICS

In this investigation, you learned the meaning of the chi-square statistic χ^2 and how to compute it.

a. How do you compute expected frequencies for a two-way frequency table? What is meant by "expected"? What is the sum of the expected frequencies for each group?

b. What does χ^2 measure? How do you compute it?

c. If two samples are homogeneous—exactly alike in the proportion that fall into each category—what can you say about χ^2?

d. If χ^2 for one pair of samples is larger than that for another pair of samples of the same sample sizes, what can you say about the two pairs of samples?

e. If all else is equal, what effect do larger sample sizes have on the size of the chi-square statistic?

Be prepared to explain your ideas and reasoning to the class.

✔ CHECK YOUR UNDERSTANDING

The Check Your Understanding (page 13) in the previous lesson provided the following information about an outbreak of measles in Colorado. Of 609 children who had been vaccinated, 10 got measles (and those 10 children had been given only one of the two recommended doses). Of the 16 children who were not vaccinated, 7 got measles.

a. Make tables of observed and expected frequencies for this situation. As usual, the explanatory variable defines the columns and the response variable defines the rows.

b. Compute χ^2.

c. Is χ^2 close to 0 or does it seem large? Explain what the value of χ^2 tells you about the two samples.

INVESTIGATION 3

Statistical Significance

You now know that a larger value of χ^2 indicates that the samples are more different from each other than they would be if the value of χ^2 were smaller. But how large does χ^2 have to be before you are convinced that the populations from which the samples were taken are different? After all, you cannot expect random samples, even those taken from the homogeneous populations, to be identical.

As you work on the problems in this investigation, look for answers to this question:

How large does χ^2 have to be before you can conclude that the populations from which the samples were taken are not homogeneous?

1 The survey in Problem 1 (page 42) of Investigation 2 asked 12th-grade boys and 12th-grade girls whether they drank soda daily. The table of observed frequencies from the actual survey is given below.

Drank soda daily during the 7 days before the survey?

	12th-Grade Boys	12th-Grade Girls
Yes	538	392
No	1,187	1,333
Total	1,725	1,725

a. Using the table of observed frequencies above and your table of expected frequencies from Problem 1 of the previous investigation, compute χ^2.

b. A value of χ^2 is called **statistically significant** if it is large enough to reasonably conclude that the populations from which the samples were taken are not homogeneous. Do you think that will turn out to be the case here?

c. One way to determine whether your value of χ^2 is statistically significant is to consider 200 pairs of random samples taken from homogeneous populations, each of which has two categories. Then calculate χ^2 for each of the 200 pairs of samples. Shown below is the distribution for 200 such values of χ^2. What is the largest value of χ^2 that occurred in any of these 200 pairs of samples?

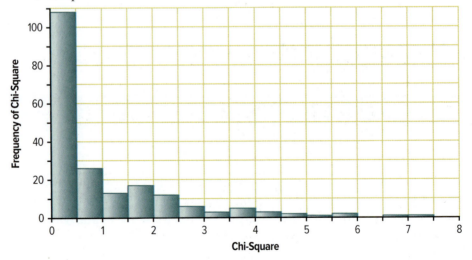

d. Where would the value of χ^2 calculated in Part a be located on this histogram?

e. How does the histogram help you conclude that the value of χ^2 from Part a is large enough to be statistically significant?

f. Select the best conclusion.

  A. If the survey had asked all 12th-grade boys and all 12th-grade girls, it is plausible (reasonable to believe) that the proportions who said they drink soda daily would be equal.

  B. If the survey had asked all 12th-grade boys and all 12th-grade girls, it is not plausible that the proportions who said they drink soda daily would be equal.

  C. Because the survey did not ask all 12th-grade boys and all 12th-grade girls, it is possible that anything could be the case in the two populations, so we cannot come to any reasonable conclusion.

2 Rather than comparing the value of χ^2 calculated from your samples to a histogram each time, statisticians have prepared a table that gives you a quick way to determine whether your χ^2 is statistically significant. Just compare your χ^2 to the appropriate value in the following table, called the *critical value*. If your χ^2 is larger than the critical value, the difference in the samples is statistically significant. In Problem 7, you will learn more about how the critical values were determined. For now, keep in mind that the larger your value of χ^2, the more evidence you have that the populations from which the two samples were taken are not homogeneous.

Chi-Square Critical Values

Number of Categories	2	3	4	5	6	7	8
Critical Value	3.84	5.99	7.81	9.49	11.07	12.59	14.07

The survey reported in Problem 2 (page 42) of the previous investigation asked 12th-graders whether they wore a helmet when riding a bicycle. The table of observed frequencies from the actual survey is given below.

Wore a bicycle helmet (among students who had ridden a bicycle during the previous 12 months)?

	12th-Grade Boys	12th-Grade Girls	Total
Rarely or Never	1,067	758	1,825
Yes	93	109	202
Total	1,160	867	2,027

a. Compute the proportion of each group who rarely or never wore a bicycle helmet. Do you think the difference will turn out to be statistically significant?

b. Using your table of expected frequencies from Problem 2 of the previous investigation and the table above, compute χ^2.

c. How many categories are there for each sample?

d. What critical value should be used for comparison?

e. Is there statistically significant evidence that the samples came from populations that are not homogeneous? Explain.

f. Is the result from Part e consistent with your judgment in Part a?

3 Refer to the two survey questions examined in the Think About This Situation on page 36.

  a. For which survey question do you think the results are most convincing that, if the survey had asked every 12th-grader in the country, the proportions for the two populations (boys and girls) would be different? Which is least convincing?

  b. For each survey question, compute χ^2.

  c. Is the difference in the two samples statistically significant for either survey question?

  d. Is the result from Part c consistent with your judgment in Part a about these polls?

4 Harris Interactive asked samples of adults (age 18 and over) in various countries,

"How important are brand names to you, if at all, when purchasing clothes and fashion accessories?"

The responses are given in the table below, as a percent.

	U.S.	China	India	Britain	France	Germany	Spain	Italy
Very Important	4	19	23	7	3	3	2	2
Important	21	53	51	17	22	19	26	26
Not That Important	48	24	21	41	49	49	50	51
Not At All Important	27	4	5	35	26	29	22	21
Column Total	100%	100%	100%	100%	100%	100%	100%	100%
Sample Size	2,309	500	500	1,293	1,179	1,058	1,019	1,064

Source: www.harrisinteractive.com

  a. Name two countries for which the difference in the percentages between those countries clearly is statistically significant. Explain your reasoning.

  b. Name two countries where you think that the difference in the percentages probably will not be statistically significant.

  c. For the countries you chose in Part b, convert each percentage to a frequency. Round to the nearest whole number.

  d. For the two countries you chose in Part b, compute χ^2.

  e. How many categories are there for each sample?

  f. What critical value should be used for comparison?

  g. Write a conclusion.

5 A chi-square test also can be used to analyze the result of an experiment. In the experiment reported in Problem 2 on page 15, after seven weeks of basic training, 21 of the 200 officer trainees who had received custom orthotics had leg, knee, or foot injury that required he or she stop physical training for two or more days. Sixty-one of the 200 who did not get orthotics had such injuries.

a. Make tables of observed and expected frequencies for this situation.

b. Compute χ^2.

c. Do you have statistically significant evidence that the custom orthotics caused a difference in the proportion of officer trainees who got such injuries? Explain.

6 In a study of conformity, a large group of volunteer Internet users was randomly divided into two groups, a control group and the treatment group. On the Internet, each group was asked the following multiple-choice question (among others):

In which city can you find Hollywood?

- New York
- Las Vegas
- Los Angeles
- San Francisco
- Bombay

Along with the multiple-choice question, the treatment group, but not the control group, was shown the following fabricated information about how others have answered the question.

Others from your community have given the following answers.

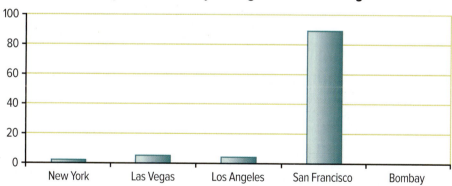

The responses to the multiple-choice question are given in the table below.

	Treatment Group	Control Group
Answered "San Francisco"	101	19
Gave Another Answer	376	430
Total	477	449

Source: Michael Rosander and Oskar Eriksson, "Conformity on the Internet—The role of task difficulty and gender differences," *Computers in Human Behavior, 28* (2012), pp. 1587–1595

a. Compute the value of χ^2 for this situation. Is it statistically significant?

b. Write a short conclusion for this study.

7 In Problem 1, critical values were determined by repeatedly taking pairs of random samples from two populations that are homogeneous. You then observed how large χ^2 tends to be when the populations from which the samples are taken are identical.

For this problem, 200 pairs of random samples were taken from homogeneous populations. In this case, each population had three categories. The chi-square statistic was calculated for each of the 200 pairs of samples. Shown below is the distribution of the 200 values of χ^2.

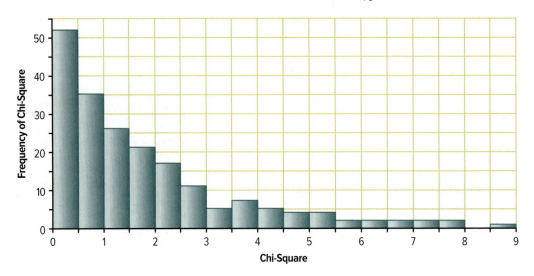

a. What is the largest value of χ^2 that occurred in any of these 200 pairs of samples?

b. What are the major differences between this distribution and the one in Problem 1 Part c?

c. Use the formula for the chi-square statistic to explain why there are no negative values of χ^2 in this distribution.

d. Using the table on page 49, find the critical value of χ^2 for the situation where there are three categories.

e. It is possible to get a value of χ^2 as large as the critical value even when taking two random samples from populations that are identical. But this is not very likely to happen. How many times out of the 200 different pairs of random samples was χ^2 larger than the critical value? What percentage of the time did this happen?

f. Critical values typically are located so that they cut off the upper 5% of the distribution generated using a method like that above. Thus, values higher than the critical value can happen when samples are taken from homogeneous populations, but this is not likely. The next histogram shows the distribution of the values of χ^2 from 200 different pairs of random samples taken from homogeneous populations. Where would you set the critical value of χ^2 for this distribution? How many categories do you think there were in this case?

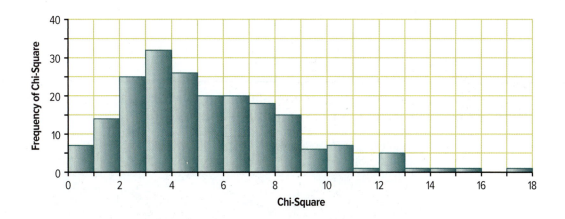

SUMMARIZE THE MATHEMATICS

In this investigation, you learned how to decide if a value of χ^2 is statistically significant.

a. How can you tell if the difference in the proportions in two random samples that fall into the various categories is statistically significant?

b. What does it mean when the difference is statistically significant?

c. How are critical values determined?

d. If a difference is not statistically significant, should you conclude that the two populations are homogeneous or should you conclude only that you do not have statistically significant evidence that they are different?

Be prepared to explain your ideas to the class.

✓ CHECK YOUR UNDERSTANDING

A survey collected information from 4,249 young adults age 18 to 28 from 30 countries who had reported crying in the last year. Of the 2,577 women, 1,344 said they felt better after their most recent crying episode, 964 said they felt the same, and 269 said they felt worse. Of the 1,672 men, 851 said they felt better, 674 said they felt the same, and 147 said they felt worse. (**Source:** Lauren M. Bylsma, *et al.* "When is Crying Cathartic? An International Study," *Journal of Social and Clinical Psychology,* Vol. 27, 2008, pp. 1165–1187. Numbers estimated from percents.)

a. Make tables of observed and expected counts so that you can compare the proportion of men and women who fall into the three categories. Round expected counts to the nearest tenth.

b. Compute χ^2.

c. What critical value should be used for comparison?

d. Is the difference in the proportion of men and women who fell into the different categories statistically significant? Write a sentence or two explaining what this means in the given context.

ON YOUR OWN

1 After a round of cut-backs at a manufacturing plant, 18 of 28 younger workers were laid off and 28 of 64 older workers were laid off.

a. Make a table that summarizes the information given.

b. Select the best answer to the question of whether there is evidence of possible age discrimination. Be prepared to describe weaknesses in the other answers.

   A. There may be discrimination against older workers because more older workers were laid off than younger workers.

   B. There may be discrimination against older workers because a larger proportion of older workers were laid off than younger workers.

   C. There may be discrimination against younger workers because more younger workers were laid off than older workers.

   D. There may be discrimination against younger workers because a larger proportion of younger workers were laid off.

   E. There is no evidence of discrimination because half of the workers were laid off and half were not.

c. Construct a bar graph to convince someone that your answer to Part b is correct. Which type of bar graph did you make? How does the bar graph help the person see that you are correct?

2 The table below shows the ages of the male and female actors who got regular roles in pilot shows for new television series, for those actors whose age could be determined (which was all but 8 males and 5 females).

Age	Number of Males	Number of Females
0–9	13	7
10–19	31	29
20–29	73	95
30–39	116	88
40–49	52	21
50–59	33	19
60 and over	13	10
Total	**331**	**269**

Source: James J. Jones, *The 2011 Pilot Report*, Premier Talent Group, f.cl.ly/items/3q1v2Z1W0Z3N19003839/PTG-2011-Pilot-Report.pdf

a. The following two bar graphs show the distribution of ages. Which graph better shows which group (males or females) got more regular roles in pilot shows? Describe this difference between the groups.

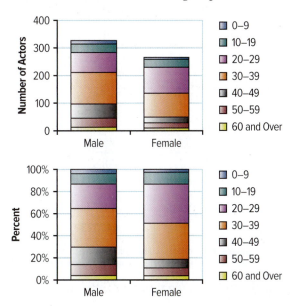

b. Are the two groups homogeneous? If so, explain why. If not, explain how they differ.

c. Suppose that over the next three-year period, 1,000 male actors and 800 female actors get regular roles in pilot shows. Assuming that the age distribution remains the same for males and females, what is the expected number of males aged 20–29 who get these jobs? What is the expected number of females in the same age range?

3 This table shows the distribution of scores on the SAT Critical Reading test for 2011 college-bound seniors.

Score	Number of Students	Proportion of Students
700–800	76,565	0.0465
600–690	256,676	0.1558
500–590	480,588	0.2918
400–490	531,429	
300–390	247,836	0.1505
200–290	54,029	
Total	1,647,123	

Source: *2011 College-Bound Seniors Total Group Report*, College Board, professionals.collegeboard.com/profdownload/cbs2011_total_group_report.pdf

a. What are the missing numbers in the third column of this table?

b. Suppose that you take a random sample of 1,000 scores from this population of SAT Critical Reading scores. What is the expected number of scores in the category 700–800? In the category 400–490?

4 The following table shows the distribution of scores on the SAT Mathematics test for 2011 college-bound seniors.

Score	Number of Students	Proportion of Students
700–800	111,893	0.0679
600–690	304,037	0.1846
500–590	481,170	0.2921
400–490	498,944	0.3029
300–390	210,645	
200–290	40,434	
Total	1,647,123	1.0000

Source: *2011 College-Bound Seniors Total Group Report*, College Board, professionals.collegeboard.com/profdownload/cbs2011_total_group_report.pdf

a. What are the two missing numbers in the third column of this table?

b. Suppose that you take a random sample of 1,000 scores from this population of SAT Mathematics scores. What is the expected number of scores in the category 700–800? In the category 400–490?

c. Make a stacked bar graph (percent on the vertical axis) to compare this distribution with the distribution of SAT Critical Reading scores in Applications Task 3.

d. Are the two populations homogeneous? Explain why or why not.

5 In Problem 4 (page 16) in the previous lesson, you read about a study of young baseball pitchers. It found that 7 of the 103 who had thrown curveballs before the age of 13 were injured seriously enough to have surgery or quit pitching. Of the 187 who had not thrown curveballs before age 13, only 8 were injured seriously enough to have surgery or quit pitching.

   a. Make tables of observed and expected frequencies for this situation.

   b. Compute χ^2.

   c. Is χ^2 close to 0 or does it seem large? Explain what this means in terms of the two samples.

6 Refer to Applications Task 2 (page 55) about the ages of male and female actors.

   a. Add the row totals to a copy of the table. Then, using only the marginal totals, make a table of expected frequencies that shows what perfectly homogeneous groups of males and females would have looked like. Round to the nearest tenth.

   b. Compute χ^2.

   c. Is χ^2 close to 0 or does it seem large? Explain what this means in terms of the two groups.

7 Refer to your table of observed frequencies from Applications Task 2 (page 19) in Lesson 1 about the safety of artificial turf.

   a. Make a table of expected frequencies for this situation.

   b. Compute χ^2.

   c. How many categories are there for each sample?

   d. What critical value should be used for comparison?

   e. Is the difference in the two samples statistically significant? Explain how you know.

   f. What can you conclude?

8 A worker over age 50 was laid off from a paper products company. He sued his company for age discrimination. Among other evidence presented was this: In one division of the company, 9 of 22 workers under age 50 were laid off while 19 out of 28 workers age 50 and older were laid off. (**Source:** Ann E. Watkins, Richard L. Scheaffer, and George Cobb, *Statistics: From Data to Decision*, Wiley, 2011.)

   a. Make tables of observed and expected frequencies for this situation.

   b. Compute χ^2.

   c. Is the difference in the proportion of younger and older workers laid off statistically significant? Explain how you know.

   d. Given just this evidence, how strong would you judge the worker's case to be?

9 The Harris Interactive Poll asked about 1,092 adult males and 1,092 adult females in the United States,

> "Overall, how satisfied are you with your life nowadays?"

The percentage giving various responses are given in the table below.

	Male (%)	Female (%)
Very Satisfied	25.5	24
Fairly Satisfied	53.5	59
Not Very Satisfied	15	13
Not At All Satisfied	6	4

Source: www.harrisinteractive.com

   a. Convert the percentages to frequencies, rounding to the nearest whole number.

   b. Compute χ^2.

   c. Is there statistically significant evidence that, if Harris had asked all adult males and all adult females in the United States, the distributions of responses would be different? Explain your reasoning.

10 In Investigation 2 Problem 6 (page 11) in Lesson 1, you read about an experiment to determine side effects caused by the use of salmeterol (an ingredient in Advair). Of 1,653 patients who were randomly assigned to receive salmeterol, 35 had been hospitalized. Of 1,622 patients who received the placebo, 16 were hospitalized. The value of χ^2 for this situation is 6.83.

Is there statistically significant evidence that salmeterol caused a change in the rate of hospitalization? Explain.

11 As you read in Extensions Task 21 Part b on page 30, millions of Americans take *statin* drugs to prevent heart attacks and strokes. In one experiment to see how effective atorvastatin was compared to a placebo, 5,186 patients with high blood pressure but normal cholesterol, were randomly assigned to get atorvastatin

and 5,137 to get a placebo. The number of patients who got either coronary heart disease (CHD) or stroke in three years was 175 and 258, respectively. The value of χ^2 for this situation is 17.44.

Is there statistically significant evidence that the use of atorvastatin caused a difference in the proportion of patients with high blood pressure, but normal cholesterol, who got either CHD or stroke in three years?

CONNECTIONS

12 The formula for the distance d between two points, (x_1, y_1) and (x_2, y_2), in the plane is given by:

$$d = \sqrt{(x_2 - x_1)^2 + (y_2 - y_1)^2}$$

a. Use this formula to find the distance between the points (3, 5.2) and (5, 3.2).

b. How is this formula similar to that for the chi-square statistic?

c. How is it different?

13 The chi-square statistic involves a sum of squared differences. You may have learned about other statistics that involve a sum of squared differences. One of these is the **standard deviation**, which measures how much the values in a set of data vary from their mean. An algorithm to calculate the standard deviation follows.

Step 1. Find the mean $\overline{x}$ of the n values:

$$\overline{x} = \frac{\Sigma x}{n}$$

Step 2. Find the *deviation from the mean* for each value:

$$deviation\ from\ the\ mean = value - mean = x - \overline{x}$$

Step 3. Square each deviation.

Step 4. Find the sum of the squared deviations.

Step 5. Calculate an average of the squared deviations by dividing the sum of the squared deviations by $n - 1$. (The reason for using $n - 1$ rather than n is technical.)

Step 6. Take the square root of the average from Step 5.

a. Use the above algorithm to compute the standard deviation for each set of grades on chapter tests.

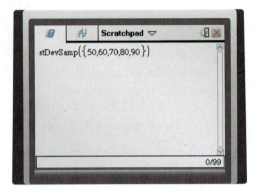

  i. 50, 60, 70, 80, 90

  ii. 60, 65, 70, 75, 80

b. Use the standard deviation feature of your calculator or computer software to check your answer.

c. Which set of grades in Part a has more variability? Does it have the larger standard deviation?

d. Look back at your calculations of the standard deviation. Which of the following gives a formula for the standard deviation s?

$$s = \sqrt{\frac{\Sigma\,(x - \bar{x}^2)}{n - 1}} \qquad s = \sqrt{\frac{\Sigma\,(x - \bar{x})^2}{n - 1}} \qquad s = \sqrt{\frac{(\Sigma\,x - \bar{x})^2}{n - 1}} \qquad s = \sqrt{\Sigma\left(\frac{x - \bar{x}}{n - 1}\right)^2}$$

e. If all values in a set of data are the same, what is the standard deviation? Use algebraic reasoning to show that this must be the case.

14 This table, from Problem 1 (page 38) of Investigation 1, gives the proportion of Hispanics of various origins.

Origin	2000 Census	2010 Census
Mexican	20,640,711	31,798,258
Puerto Rican	3,406,178	4,623,716
Cuban	1,241,685	1,785,547
Other Hispanic	10,017,244	12,270,073
Total	**35,305,818**	**50,477,594**

a. Use your calculator or computer software to compute a table of expected frequencies that shows how the results would have turned out if the distributions for the two years had been homogeneous. Round to the nearest whole number.

b. Use the matrix functions in *TCMS-Tools* or on your calculator to compute the values of $O - E$. Are the values of $O - E$ unusually large or small? Why is that the case?

c. Compute χ^2. Why is this so large?

15 Suppose that a polling organization takes a random sample of 100 older U.S. voters and a random sample of 100 younger U.S. voters. The voters are asked whether they approve of the job the President is doing. The results of Poll A are reported in the table at the right.

Poll A

	Older	Younger
Approve	35	30
Disapprove	65	70
Total	100	100

Another polling organization takes a random sample of 1,000 Republicans and a random sample of 1,000 Democrats and asks the same question. The results of Poll B are reported in the table at the right.

For Poll A, $\chi^2 \approx 0.57$. For Poll B, $\chi^2 \approx 5.70$.

Poll B

	Republicans	Democrats
Approve	350	300
Disapprove	650	700
Total	1,000	1,000

a. Is the result statistically significant in either case? What can you conclude about older and younger voters? About Republicans and Democrats?

b. Are the proportions any different in the two polls? In light of your answer, explain how the result could be statistically significant for one poll and not the other.

c. Compare the values of χ^2. By what factor is the one for Poll B larger than the one for Poll A? Use algebraic reasoning to explain why this must be the case.

REFLECTIONS

16 How many times would you have to flip a coin before convincing someone that it was fair? The plot below shows the number of flips of a fair quarter on the horizontal axis and the proportion of times, so far, that the coin came up heads.

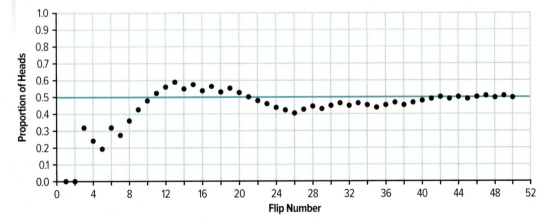

a. Was the first flip a head or a tail?

b. Was the 50th flip a head or a tail?

c. How many heads were in the longest run of consecutive heads?

d. Describe what is happening as the number of flips increases.

e. About how many flips did it take until the proportion of heads stays pretty close to 0.5?

f. Predict what would happen to the pattern in the plot if the person continued flipping the quarter. Explain why this should be the case.

17 Look back at Problem 3 (page 39) of Investigation 1. SAT Critical Reading score is a quantitative variable.

a. How were the data organized to permit SAT Critical Reading scores to be analyzed as a categorical variable using a frequency table?

b. Identify another problem in Investigation 1 for which a similar technique was used.

18 Suppose that the proportions in each category from two samples from Populations A and B differ more than the proportions from two other samples from Populations C and D.

a. Create an example to show that it is *not* necessarily the case that χ^2 for the samples from Populations A and B will be larger than χ^2 for the samples from Populations C and D.

b. Why is it reasonable that this could be the case?

19 To estimate the proportion of voters in favor of an issue, national polls typically use samples of 1,000 to 1,500 voters. Is that large enough? Some people say that such polls cannot possibly be accurate because no one asked them. Write an explanation to such a person illustrating that the proportion could not change very much whether the poll did ask them or did not.

EXTENSIONS

20 A chi-square statistic also may be used to compare how closely the proportions in a single random sample match specified proportions. This is called a *chi-square goodness of fit test*. For example, suppose that you have a die and want to test whether it is fair. You roll it 60 times and count the number of times each face lands on top.

Getty Images

a. Complete a copy of this table of expected frequencies for a fair die.

Result	1	2	3	4	5	6	Total
Frequency							60

b. Here are the observed frequencies of actual results from 60 rolls. Compute χ^2.

Result	1	2	3	4	5	6	Total
Frequency	12	9	10	6	11	12	60

c. If χ^2 is larger than the critical value in the Chi-Square Critical Values table reproduced below, the value of chi-square is statistically significant. Is your χ^2 larger than the critical value for 6 categories? Do you have statistically significant evidence that the die is unfair? Explain.

Chi-Square Critical Values

Number of Categories	2	3	4	5	6	7	8
Critical Value	3.84	5.99	7.81	9.49	11.07	12.59	14.07

21 Indian mythology holds that older snakes get energized from the full moon. This belief prompted researchers to study whether a full moon results in more snakebites of humans.

They looked at 125 consecutive snakebite deaths in Yavatmal, India. Thirty-seven of these deaths occurred during the nine days closest to and including the full moon; 33 occurred five through 10 days after the full moon; and the rest occurred more than 10 days after the full moon. (**Source:** Anil K. Batra and Ajay N. Keoliya, "Do Fatal Snakebites Occur More During a Full Moon? An Observational Analysis," *International Journal of Medical Toxicology & Legal Medicine*, Vol. 7, 2004, www.scribd.com/doc/23986204/Snake-Bites-full-Moon/)

a. How many snakebites occurred more than 10 days after the full moon? There are 29 days in the lunar cycle. How many days were in this period?

b. If the lunar cycle has nothing to do with a sample of 125 snakebites, what is the expected number of fatal snakebites in the nine days around and including the full moon? In days five through 10 days after the full moon? On days more than 10 days after the full moon?

c. Using the expected frequencies from Part a, complete a copy of this table.

Lunar Cycle	Closest to Full Moon	Five to Ten Days After Full Moon	More than Ten Days After
Observed Number of Snakebites			
Expected Number of Snakebites			

d. Compute χ^2 as in Extensions Task 20.

e. What is the critical value for this situation? Is the value of chi-square that you computed in Part d statistically significant?

f. What is your conclusion?

22 Learn to spin a penny on a flat surface by holding it on edge with one finger and then flicking it on one side with a finger of the other hand.

a. Spin your penny 50 times. Count the number of times it lands heads and tails.

b. Complete a copy of these tables, filling out the expected frequencies under the assumption that heads and tails are equally likely.

Result	Observed
Heads	
Tails	
Total	**50**

Result	Expected
Heads	
Tails	
Total	**50**

c. Compute χ^2 as in Extensions Task 20.

d. What is the critical value for this situation? Is the value of chi-square that you computed in Part c statistically significant?

e. What is your conclusion? If your result was not statistically significant, should you conclude that heads and tails are equally likely or can you conclude only that you do not have statistically significant evidence that they are different? Explain your reasoning.

f. If possible, combine results with the rest of the class and repeat Parts b through e.

23 When there are only two samples and two categories, another test for statistical significance, the two-proportion z-test, can be used instead of the chi-square test. For example, in the Check Your Understanding for Investigation 2 on page 47, there are two samples, the vaccinated children and the unvaccinated children. There are two categories, getting measles and not getting measles. Thus, you can use

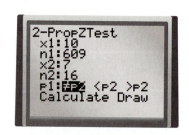

Jim Laser

either the chi-square test or the two proportion z-test. These two tests are equivalent. To find the two-proportion z-test on a TI-84 calculator, go to the STAT menu, arrow over to **TESTS**, and select **6:2-PropZTest**.... For **x1**, enter the number of people who fall into the first category in the first sample and for **n1**, enter the sample size for the first sample. Continue similarly for the second sample. Select **p1:≠p2**, then select **Calculate** and ENTER. On the resulting screen will be a value of z.

a. Use your calculator to find z for the data in the Check Your Understanding on page 47. Square this value of z and compare to the value of χ^2 that you computed in the Check Your Understanding.

b. Verify that $\chi^2 = z^2$ using the data in Applications Task 5.

c. The following formula for z looks impressive, but involves only the number of people in the first category in each sample and the sample sizes. (With a lot of algebra, you can prove that $\chi^2 = z^2$.)

$$z = \frac{\dfrac{x_1}{n_1} - \dfrac{x_2}{n_2}}{\sqrt{\left(\dfrac{x_1 + x_2}{n_1 + n_2}\right)\left(1 - \dfrac{x_1 + x_2}{n_1 + n_2}\right)\left(\dfrac{1}{n_1} + \dfrac{1}{n_2}\right)}}$$

Use this formula to verify the value of z that you found in Part b.

d. In the formula in Part c, what is given by the expression in the numerator?

REVIEW

24 The circle graph below represents the ethnic breakdown of the population of San Francisco County in 2010. Suppose that someone randomly selected a group of 500 people from San Francisco County in 2010.

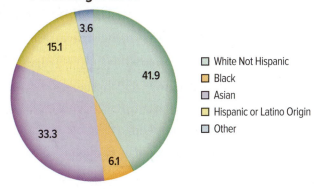

San Francisco County Population Percentages 2010

- White Not Hispanic
- Black
- Asian
- Hispanic or Latino Origin
- Other

Source: factfinder2.census.gov/faces/tableservices/jsf/pages/productview.xhtml

a. What is the expected number of Black people in the group?

b. What is the expected number of Hispanics/Latinos in the group?

c. What is the expected number of Asians in the group?

d. Suppose another person selected a group of people that was n times the size of the original sample. How would your answers to Parts a–c change? Support your answer with mathematical reasoning.

25 Darius has a savings account that earns 2.35% annual interest compounded annually. His current account balance is $5,278. Assume that Darius does not withdraw or deposit any money and that he leaves all earned interest in the account.

a. How much interest will Darius earn during the first year? What will his account balance be at the end of the first year?

b. During the second year, will Darius earn more interest or the same amount of interest than he did during the first year? Explain your reasoning.

c. What will his account balance be five years from now?

d. Explain how the balance in Darius's account changes from one year to the next.

e. Explain why the function rule $B(t) = 5{,}278(1.0235^t)$ can be used to determine the balance in Darius's account after t years.

26 In the diagram at the right $\angle AOB$ is in standard position; its initial side coincides with the positive x-axis, its vertex is the origin and its terminal side contains $\overline{OB}$. As you may recall, angles in standard position in a coordinate system can be measured in terms of revolutions of the initial side of the angle, in degrees, and in radians. Recall that a **radian** is the measure of a *central angle* that intercepts an arc equal in length to the radius of the circle. In the diagram at the right, the radian measure of $\angle AOB$ is 1.

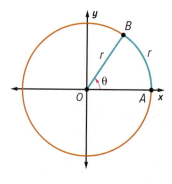

a. If the initial side of an angle is rotated one complete revolution, then the measure of the angle is 2π radians. Explain why this is the case.

b. Complete each of the following statements.

    i. ___?___ revolution = 90° = ___?___ radians

    ii. ___?___ revolution = ___?___ degrees = $\frac{\pi}{3}$ radians

c. Complete a copy of the following table to show equivalent revolution, degree, and radian measurements.

Revolution/Degree/Radian Equivalents

Revolutions	0							
Degrees	0	30			90		135	150	
Radians			$\frac{\pi}{4}$	$\frac{\pi}{3}$		$\frac{2\pi}{3}$			π

27 The histogram below displays the ages of the Green Bay Packers roster on July 5, 2011. On that day, the roster had 85 players.

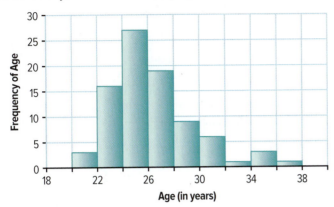

Source: www.packers.com/team/players.html

a. Three players were 20 or 21 years old. How can this be determined by looking at the histogram?

b. What percentage of the players were 28 or 29 years old?

c. What percentage of the players were at least 30 years old?

d. What percentage of the players were younger than 36 years of age?

e. When asked the median age of the players on the team, Dario said that it was 29 years. Do you agree or disagree with Dario? Explain your reasoning.

28 Consider the circle with radius 5 that is centered at the origin.

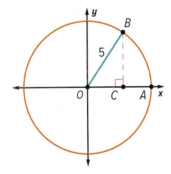

a. If $m\angle AOB = 60°$, find the coordinates of point B.

b. Describe the location of another point B_1 on the circle that has the same x-coordinate as point B. What is the relationship between the y-coordinates of the two points?

c. Describe the location of another point B_2 on the circle that has the same y-coordinate as point B. What is the relationship between the x-coordinates of the two points?

29 Shown at the right is a circle of radius 1, called a **unit circle**. $\angle POQ$ is an angle in standard position with radian measure θ.

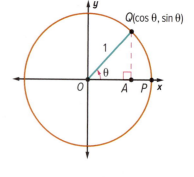

a. Why is $\cos \theta$ the x-coordinate of point Q?

b. Why is $\sin \theta$ the y-coordinate of point Q?

c. Suppose the terminal side of θ, $\overrightarrow{OQ}$, is in the quadrant given below. Determine, in terms of θ, the x- and y-coordinates of point Q.

   i. Quadrant II

   ii. Quadrant III

   iii. Quadrant IV

d. Suppose the radius of a circle is equal to the unit segment on a tape measure.

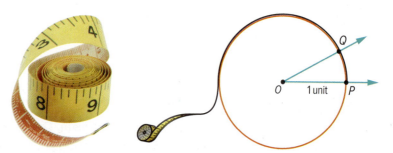

Imagine wrapping the tape measure counterclockwise around the circle, starting where $\overline{OP}$ meets the circle and noting the point on the tape measure where $\overline{OQ}$ meets the circle. The arc length measured by the tape measure will be the radian measure of the angle. Wrapping the tape measure around the circle (in both directions from point P) produces a correspondence between points on the tape measure and points on the circle. Consider the coordinates of the points on the circle under the above correspondence. How does this method construct cosine and sine functions with domain all real numbers?

30 The amount of force f (in pounds) needed to break a board (with fixed width and thickness) is a function of the length ℓ (in inches) of the board. For one type of board, the function rule is $f(\ell) = \dfrac{20}{\ell}$.

a. As the length of the board increases, how does the amount of force needed to break the board change? How is your answer reflected in a graph of $f(\ell)$?

b. Evaluate $f(12)$. What does your answer tell you about this situation?

c. Solve $f(\ell) = 0.5$. What does this solution tell you about this situation?

d. The length of board that Peter is trying to break is twice that of the one that his older brother just broke. How does the force needed by Peter compare to the force needed by his brother?

31 Mrs. Takemura surveyed the 440 juniors and seniors at Sayre High School and asked them whether they worked for pay in the previous week. The results are reported in the table below.

	Worked for Pay	Did Not Work for Pay	Total
Juniors	100	140	240
Seniors	90	110	200
Total	190	250	440

Suppose that you randomly pick one of these students to interview.

a. Compute this probability: *P(student worked for pay the previous week)*

b. Compute this probability: *P(student was a junior)*

c. Is a randomly selected senior or a randomly selected junior more likely to have worked for pay the previous week?

d. The notation for the probability that the student worked for pay the previous week, if the student you picked turned out to be a senior is *P(worked for pay the previous week | was a senior)* $= \frac{90}{200}$. The vertical line | is read "given that." Write the notation for the probability that the student worked for pay the previous week, if the student you picked turned out to be a junior.

e. Suppose that you pick a student who worked for pay the previous week. Compute the probability that the student was a junior, *P(was a junior | worked for pay the previous week)*.

f. Compute *P(was a senior | worked for pay the previous week)*.

g. Is a randomly selected student who worked for pay the previous week more likely to be a junior or a senior? Explain.

The Relationship Between Two Variables

Texting has increasingly become the way in which people, young and older, communicate with one another. About 77% of 12th-graders text, with about half of those sending at least 60 text messages per day. (**Source:** pewinternet.org/Reports/2010/Teens-and-Mobile-Phones/Summary-of-findings.aspx)

Suppose that you take a random sample of 12th-grade students who text and categorize them according to two variables,

* whether they send at least 60 texts per day and
* whether they text their parents every day.

The results are given in the table below.

Send At Least 60 Text Messages Daily

		Yes	No	Total
Text Parents Daily	**Yes**	275	65	340
	No	35	225	260
	Total	310	290	600

Allan Shoemake/Getty Images

THINK ABOUT THIS SITUATION

Think about whether there is an association between the number of texts sent by 12th-graders and whether they text their parents daily.

a. What proportion of 12th-graders who send at least 60 texts daily also text their parents daily? Is the proportion the same for 12th-graders who do not send at least 60 texts daily?

b. What proportion of 12th-graders who text their parents daily also send at least 60 texts daily? Is the proportion the same for 12th-graders who do not text their parents daily?

c. Are the two variables *send at least 60 texts daily* and *text parents daily* associated in this sample? Or are they independent variables; that is, does knowing whether a teen does one of these change the probability that he or she does the other? Explain your answer.

d. Does it seem plausible, just from looking at the sample, that the two variables *send at least 60 texts daily* and *text parents daily* would be independent in the population of *all* 12th-grade students who text?

e. Does it seem plausible, from knowing what you know about texting among 12th-graders, that the two variables *send at least 60 texts daily* and *text parents daily* would be independent in the population of all 12th-grade students? Explain.

In the previous two lessons, you analyzed situations where there were two groups, such as two samples taken at random from two different populations. In this lesson, you will examine single samples, classified on two different categorical variables. The previous table is an example of such a situation.

INVESTIGATION 1

Diagnostic Testing

If someone is suspected of having a medical problem, often he or she first is given an inexpensive and quick diagnostic test, sometimes called a *screening test*. These tests are not always very accurate, so if a positive result is obtained, a more accurate but usually more expensive, time-consuming, or invasive diagnostic test is given.

As you work on the problems in this investigation, look for answers to this question:

What are the statistical characteristics of a good diagnostic test?

1 When a pregnant woman receives a blow to the abdomen, doctors must check for pelvic free fluid (FF), which is associated with internal injury. A fast and inexpensive way to do this is with an ultrasound. But sometimes an ultrasound gives the wrong result. A more expensive and invasive test is available, which never gives the wrong result. To check the accuracy of the ultrasound test, 328 pregnant women who had received a blow to the abdomen were given both tests. The results are given in the table below. (**Source:** E. L. Ormsby, et al., "Pelvic Free Fluid: Clinical Importance for Reproductive Age Women with Blunt Abdominal Trauma," *Ultrasound in Obstetrics & Gynecology*, September 2005, pp. 271–278.)

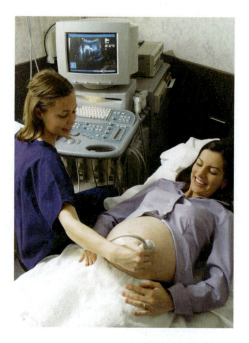

Result of Ultrasound

	Positive for FF	Negative for FF	Total
FF Actually Present	14	9	23
FF Actually Absent	15	290	305
Total	29	299	328

a. A **false positive** occurs when a test shows a positive result but is wrong. How many false positives were there? Explain the meaning of a false positive for the woman tested.

b. A **false negative** occurs when a test shows a negative result but is wrong. How many false negatives were there? Explain the meaning of a false negative for the woman tested.

c. Looking only at those who had FF, what proportion got a positive ultrasound? This is called the **sensitivity** of the test.

d. Looking only at those who did not have FF, what proportion got a negative ultrasound? This is called the **specificity** of the test.

e. Looking only at those who got a positive ultrasound, what proportion actually had FF? This is called the **positive predictive value (PPV)**.

f. Looking only at those who got a negative ultrasound, what proportion actually did not have FF? This is called the **negative predictive value (NPV)**.

g. What do you see as the major statistical strengths and weaknesses of the ultrasound test for FF? Share your ideas with your classmates. Resolve any differences.

2 The statistics defined in Problem 1 Parts c–f may be thought of as *conditional probabilities*. That is, they are all of the form:

$$\frac{cell\ count}{row\ (or\ column)\ total}$$

Conditional probability uses special notation. For example, the sensitivity of a test is the probability of a positive test result when the condition actually is present. This is written *P(positive test | condition present)*. Here, *P* stands for probability and the vertical line | is read, "given that" or "if it is true that."

Here is the complete definition of the sensitivity of a test:

$$sensitivity = P(positive\ test\ |\ condition\ present) = \frac{count\ with\ condition\ and\ positive\ test}{total\ with\ condition}$$

a. Write a definition of the specificity of a test using conditional probability.

b. Write a definition of the positive predictive value (PPV) using conditional probability.

c. Write a definition of the negative predictive value (NPV) using conditional probability.

d. If a diagnostic test is a good one, which of the four conditional probabilities defined above should be close to 0? Which should be close to 1?

3 Upon admission to college, students often must take placement tests in English and mathematics. Students who "fail," for example, the mathematics placement test must retake high school level mathematics (called remedial mathematics) before being able to take the mathematics required to graduate from college. Sometimes students fail the mathematics placement test because they did not bother to review basic ideas of algebra and geometry before taking the test. These students are placed into remedial mathematics classes even though they are capable of doing college level mathematics. The hypothetical table below shows the results for a random sample of 100 entering freshmen who did not bother to review for the placement test.

	Fail Placement Test (Test Positive for Remedial Math)	Pass Placement Test (Test Negative for Remedial Math)	Total
Need Remedial Math (Have Condition)	25	11	36
Do Not Need Remedial Math	26	38	64
Total	51	49	100

Use this table in completing the following questions and tasks.

a. Define a *positive test* as failing the placement test (testing positive for needing remedial mathematics) and define *condition present* as needing remedial mathematics.

    i. How many false positives were there? What is the meaning of a false positive for a student?

    ii. How many false negatives were there? What is the meaning of a false negative for a student?

b. Compute the sensitivity and specificity of this placement test for a student who does not review for it. Report each value in a sentence that explains its meaning.

c. Suppose a student fails the test. What is the probability that he or she actually needs to take remedial mathematics? What is the name for this value?

4 When the prevalence of a disease in the population is very low, such as HIV infection or certain cancers, there is controversy about the benefits of screening everyone for the disease. In this problem, you will see why this is the case.

The table below shows what is likely to happen if all roughly 300,000,000 Americans each were given an inexpensive enzyme immunoassay screening test for HIV infection.

	Test Positive	Test Negative	Total
Have HIV	1,339,370	4,030	1,343,400
Do Not Have HIV	4,479,849	294,176,751	298,656,600
Total	5,819,219	294,180,781	300,000,000

Sources: www.cdc.gov, R. Chou et al. (July 2005). "Screening for HIV: A Review of the Evidence for the U.S. Preventive Services Task Force," *Annals of Internal Medicine*, Vol. 143, July 2005, pp. 55–73, www.annals.org/content/143/1/55.full

a. How many false positives were there? Explain the consequence of a false positive for the person tested.

b. How many false negatives were there? Explain the consequence of a false negative for the person tested.

c. What is the sensitivity of this test? Explain the meaning of this statistic, in the context of this test.

d. What is the specificity of this test? Explain the meaning of this statistic, in the context of this test.

e. What is the positive predictive value (PPV)? Use this value in a sentence explaining to a person what his or her positive test might indicate.

f. What is the negative predictive value (NPV)? Use this value in a sentence explaining to a person what his or her negative test might indicate.

g. Based on your results, explain why people are reluctant to recommend universal screening for HIV.

SUMMARIZE THE MATHEMATICS

In this investigation, you learned how to evaluate a diagnostic test statistically.

a. What is a false positive? What is a false negative?

b. Write the sensitivity and specificity of a test as conditional probabilities. Describe what each of them tells you.

c. If you test positive for a disease, why is the positive predictive value (PPV) the single best statistic for you to know? Write the PPV as a conditional probability.

d. A screening test for prostate cancer, called a PSA, is available, but the proportion of positive test results that are false is around 75%.

   i. Which of the four statistics defined in Problem 1 can you compute from this information?

   ii. If the PSA test is positive, a biopsy must be performed, which often has harmful side effects. Why do many people believe that the test should not be given to all men?

Be prepared to share your responses and explain your ideas.

✓ CHECK YOUR UNDERSTANDING

High intracranial pressure (inside the skull) typically is a result of an injury to the head and can be very dangerous. A screening test for high intracranial pressure was proposed many years ago, based on the data in the following observations. This simple and non-invasive test involves observing the retinal vein to see if it is pulsating. Pulsation is normal and so would be considered a negative test result.

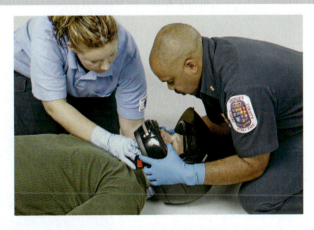

	Pulsation Absent (Positive Test)	Pulsation Present (Negative Test)	Total
High Intracranial Pressure (Condition Present)	43	0	43
Normal Intracranial Pressure (Condition Absent)	18	128	146
Total	61	128	189

Sources: B. E. Levin, "The Clinical Significance of Spontaneous Pulsations of the Retinal Vein." *Archives of Neurology*, Vol. 35, 1978, pp. 37–40; www.cebm.net/index.aspx?o=1042

a. How many false positives were there among these 189 people? Explain the meaning of a false positive for the person tested.

b. How many false negatives were there? Explain the meaning of a false negative for the person tested.

c. What is the sensitivity of this test?

d. What is the specificity of this test?

e. What is the positive predictive value (PPV)? Use this value in a sentence explaining to a person the meaning of his or her positive test.

f. What is the negative predictive value (NPV)? Use this value in a sentence explaining to a person the meaning of his or her negative test.

g. All in all, does this seem to be a good screening test? Explain your reasoning.

INVESTIGATION 2

Independence

In this investigation, you will review how to determine whether two events are independent and extend this idea to the independence of two categorical variables.

As you work on the problems in this investigation, look for answers to these questions:

What does it mean for two categorical variables to be independent?

How can you assess if two categorical variables are independent?

1 A graduate student in marine science observed bottlenose dolphins in Sarasota Bay, Florida that engaged in "depredation." That means that the dolphins were stealing or damaging bait or prey already captured by human fishing gear. She classified each of these

dolphins (that she could identify) by the categorical variables of age and sex, as shown in the table below.

	Young	Adult	Total
Male	5	10	15
Female	4	3	7
Total	9	13	22

Source: Jessica R. Powell and Randall S. Wells, "Recreational Fishing Depredation and Associated Behaviors Involving Common Bottlenose Dolphins (*Tursiops Truncatus*) in Sarasota Bay, Florida," *Marine Mammal Science*, Vol. 27, January 2011, pp. 111–129.

a. Suppose that you select one of the 22 dolphins at random. What is the probability that the randomly selected dolphin is young? That is, find $P(young)$.

b. What is the probability that the randomly selected dolphin is young, if you are told that it is female? Use the notation $P(young \mid female)$ in your answer.

c. What is the probability that the randomly selected dolphin is young, given that it is male? Write your answer using notation like in Part b.

d. Compare your probabilities in Parts b and c. What can you conclude?

e. Compare $P(female \mid young)$ and $P(male \mid young)$. What can you conclude?

2 In some situations, knowing which category a person falls into on one variable helps you better predict which category they fall into on a second variable. For example, suppose you want to estimate the probability that a teen has long hair. If you are told that the teen is female, then that fact helps you—the probability is higher than if you had been told the teen is male. The variables *gender* and *hair length* are *associated* (or, *dependent*). On the other hand, suppose that you want to estimate the probability that a teen is carrying a cell phone. Then knowing whether the teen is male or female probably does not help you at all—males and females are equally likely to be carrying cell phones. The variables of *gender* and *cell phone possession* are *independent*. (**Source:** www.pewinternet.org/Reports/2010/Social-Media-and-Young-Adults/ Part-2/ 1-Cell-phones.aspx)

a. In the situation in Problem 1, does knowing whether one of these dolphins is young or adult help you predict whether it is male? That is, are the variables independent or associated? Explain your answer.

b. If the table had been like the one below, would the variables be independent or associated? Explain your answer.

	Young	Adult	Total
Male	5	10	15
Female	3	6	9
Total	8	16	24

3 Complete a copy of the following tables using students in your class. First, agree on how to classify hair length and fingernail length into either long or short.

	Long Fingernails	Short Fingernails	Total
Long Hair			
Short Hair			
Total			

	Male	Female	Total
Likes Broccoli			
Does Not Like Broccoli			
Total			

a. Are the two variables in the first table independent or associated? In the second table?

b. Using the marginal totals for your class, fill in the cells of a copy of the first table to make the variables *hair length* and *fingernail length* independent.

c. Using the marginal totals for your class, fill in the cells of a copy of the second table to make the variables independent.

4 In Problem 3, you may have checked for independence using this rule:
Two events *A* and *B* are *independent* if $P(A) = P(A \mid B)$ (assuming $P(B) \neq 0$).

a. Suppose that you randomly select one of the dolphins from those in the table of Problem 1. Using the rule above with *male* as event *A* and *adult* as event *B*, determine whether the events *male* and *adult* are independent. Choose the correct word to interpret this result: *Adult dolphins are* [*more, less, equally*] *likely to be male than are dolphins overall.*

b. Use the above rule to determine whether the events *female* and *young* are independent events. Interpret your result.

An equivalent way to check whether events *A* and *B* are **independent** is to verify that $P(A \text{ and } B) = P(A) \cdot P(B)$. Again, suppose that you randomly select one of the dolphins from those in the table in Problem 1 reproduced below.

	Young	Adult	Total
Male	5	10	15
Female	4	3	7
Total	9	13	22

c. Using this rule with *male* as event *A* and *adult* as event *B*, determine whether the events *male* and *adult* are independent events.

d. Use this rule to determine whether the events *female* and *young* are independent events.

5 Two categorical *variables* are called **independent** if $P(A \text{ and } B) = P(A) \cdot P(B)$ for all categories *A* that make up the first variable and all categories *B* that make up the second variable.

a. Use this rule to determine whether the categorical variables of *sex* and *age* are independent in the table in Problem 1 reproduced above.

b. Use this rule to determine whether the categorical variables of *sex* and *age* are independent in the sample given in the following table from Problem 2.

	Young	Adult	Total
Male	5	10	15
Female	3	6	9
Total	8	16	24

c. When two categorical variables are independent, the columns are proportional. That is, each column is a multiple of the first column. (The multiplier does not have to be a whole number.) Is that the case for the table in Part b? Is that the case for the table in Problem 1?

d. When two categorical variables are independent, it also is true that each row is a multiple of the first row. Is that the case for the table in Part b? Is that the case for the table in Problem 1?

e. Make a bar graph for these data, using male and female as the two bars. How can you tell from this graph alone that the two variables of *sex* and *age* are independent?

6 The numbers given in the table below are the marginal totals.

	Wearing Jeans	Not Wearing Jeans	Total
Teen			35
Adult			65
Total	80	20	100

a. Fill in the four cells of a copy of the table so that the variables are independent.

b. Fill in the four cells of a copy of the table so that the variables are not independent.

c. Is there more than one possible answer for Part a? For Part b?

7 Write a formula that uses only the marginal totals and can be used to fill in the cells of a table to make the variables independent. Check your formula using this table.

	C	D	Total
A			40
B			60
Total	30	70	100

8 The table below shows only the marginal totals for the dolphin study in Problem 1. Use your formula from Problem 7 to complete a copy of the table with values that would make the two variables of *age* and *sex* independent. (The values will not be whole numbers.)

	Young	Adult	Total
Male			15
Female			7
Total	9	13	22

SUMMARIZE THE MATHEMATICS

In this investigation, you reviewed independent events and learned how to tell if two categorical variables are independent.

a. Describe two methods of deciding whether two events are independent.

b. How can you tell if two categorical variables are independent in a population? Give at least two ways.

c. If you have only the marginal totals, how can you determine frequencies for each cell that make the variables independent?

d. If two categorical variables are independent, does knowing which category a person belongs to on the first variable help you predict which category he or she belongs to on the second variable? Explain your reasoning.

Be prepared to explain your ideas to the class.

✓ CHECK YOUR UNDERSTANDING

In a study of 203 young male/female couples, each partner privately was asked whether they had ever cheated and whether they suspected or knew that their partner had cheated. The results are given in the tables below.

	Female Cheated	Female Had Not Cheated	Total
Male Right	29	162	191
Male Wrong	9	3	12
Total	38	165	203

	Male Cheated	Male Had Not Cheated	Total
Female Right	24	138	162
Female Wrong	35	6	41
Total	59	144	203

Source: Jane M. Watson, "Cheating Partners, Conditional Probability and Contingency Tables," *Teaching Statistics*, 33 (Autumn 2011), pp. 66–70.

a. If you select a couple at random, what is the probability that the male cheated? The female cheated?

b. Are males or females more likely to be wrong about whether their partner cheated?

c. Do males or females do a better job of detecting cheating? Justify your answer using conditional probability notation.

d. Are the events *male cheated* and *female right* independent? Show the computations needed to justify your answer.

e. Are the two variables independent or associated for the first table? For the second table?

f. In context, describe the association in the first table.

g. Fill in the cells of a copy of the following table so that the two variables are independent. Do the actual frequencies and those under the assumption of independence appear somewhat similar or quite different?

	Male Cheated	Male Had Not Cheated	Total
Female Right			162
Female Wrong			41
Total	59	144	203

The Chi-Square Test of Independence

As you probably found in Problem 3 (page 78) of the last investigation, in your small sample of males and females, the variables *likes/does not like broccoli* and *male/female* are not independent. However, the mathematical definition that you used there may seem too strict. Perhaps if you had a larger group, the proportions who liked broccoli would be more equal.

As you work on the problems in this investigation, look for answers to this question:

> *How can you tell whether it is plausible that two categorical variables are independent in the population from which a random sample was taken?*

1 Reproduced below is the text messaging table from the lesson Think About This Situation (page 70), which shows a random sample of 12th-grade students who text, categorized according to two variables, whether they send at least 60 texts per day and whether they text their parents everyday.

Send At Least 60 Text Messages Daily

		Yes	No	Total
Text Parents Daily	Yes	275	65	340
	No	35	225	260
	Total	310	290	600

a. Suppose that the variables *send at least 60 texts daily* and *text parents daily* had been independent in the sample. Fill in a copy of the following table, showing what the cell frequencies would have been. Round to the nearest tenth. As in Lesson 2, these values, which do not have to be whole numbers, are called expected frequencies.

Send At Least 60 Text Messages Daily

		Yes	No	Total
Text Parents Daily	Yes			340
	No			260
	Total	310	290	600

b. Compare the expected frequencies and the observed frequencies. Do the differences seem relatively large or small?

c. Compute the chi-square statistic χ^2 using the method in Lesson 2.

d. Reproduced below is the table on page 49 in Lesson 2. For the 2 by 2 tables (2 rows and 2 columns) in this lesson, you use the critical value for two categories. To what critical value should your value of χ^2 from Part c be compared?

Chi-Square Critical Values

Number of Categories	2	3	4	5	6	7	8
Critical Value	3.84	5.99	7.81	9.49	11.07	12.59	14.07

e. Is the value of χ^2 statistically significant?

f. Pick what you believe to be the best interpretation. Be prepared to defend your choice.

A. Conclude that the two variables *send at least 60 texts daily* and *text parents daily* are independent in the population of all 12th-grade students.

B. Conclude that it is plausible that the two variables are independent in the population.

C. Conclude that the variables are associated because it is not plausible that the two variables are independent in the population.

D. Conclude that the variables are associated because it is not possible that the two variables are independent in the population.

2 Baseball fans can be fanatic about statistics. In fact, baseball fans are interested in such things as whether there is any association between whether the game was played at night or in the day and whether the home team or visiting team won. The table below gives the outcomes from a random sample of 1,000 games.

	Day Game	Night Game	Total
Home Team Wins	242	316	558
Visiting Team Wins	183	259	442
Total	425	575	1,000

a. Using the marginal totals in the table above, complete a table of expected frequencies, giving the cell frequencies for the case that the variables *day/night game* and *home/visiting team wins* were independent in the sample. Round to the nearest whole number.

b. Compare the expected frequencies and the observed frequencies. Do the differences seem relatively large or small?

c. Compute the chi-square statistic χ^2.

d. To what critical value in the Chi-Square Critical Values table should your value of χ^2 be compared?

e. Is the value of χ^2 statistically significant?

f. Pick the best interpretation. Be prepared to defend your choice.

   A. Conclude that the two variables are independent in the population.

   B. Conclude that it is plausible that the two variables are independent in the population.

   C. Conclude that the variables are associated because it is not plausible that the two variables are independent in the population.

   D. Conclude that the variables are associated because it is not possible that the two variables are independent in the population.

3 Refer to Problem 3 (page 77) from Investigation 2, where you collected data from your class. Assume that you can think of your class as a random sample of all students your age. For each of the two tables, complete the following.

a. Compute the chi-square statistic χ^2.

b. Is the value of χ^2 statistically significant?

c. Write a conclusion in context describing the relationship between the variables.

SUMMARIZE THE MATHEMATICS

In this investigation, you learned how to conduct and interpret a chi-square test of independence.

a. When is it appropriate to use the chi-square test of independence?

b. Describe how to conduct this test, including how to make a table of expected frequencies.

c. When results are not statistically significant, why do you conclude that it is *plausible* that the two variables are independent rather than concluding that the variables *are* independent?

d. When results are statistically significant, what can you conclude about the two variables?

Be prepared to share your ideas and reasoning with the class.

✔ CHECK YOUR UNDERSTANDING

The following table is based on a random sample of people age 18 and older who said they were planning a vacation this summer. They were asked if they would bring a laptop computer with them.

	Male	Female	Total
Bringing a Laptop	354	307	661
Not Bringing a Laptop	356	320	676
Total	710	627	1,337

Source: frequencies estimated from The Harris Poll, Americans Work on Their Vacation, 2011

a. Compute a table of expected frequencies.

b. Compare the expected frequencies and the observed frequencies. Do the differences seem relatively large or small?

c. Compute the chi-square statistic χ^2.

d. Is the value of χ^2 statistically significant? Describe what that means for this situation.

1 Unwanted, unexpected, and often fraudulent email, called *spam*, typically is sent to try to get your money. Your email program most likely has a junk email filter, which places spam in a special folder or sends it directly to the trash. The filter decides which messages are spam based on the occurrence of certain words that are frequently used in spam. For example, the default settings on one widely used spam filter sends all messages that contain the phrase "hey bro," to the spam folder. (**Source:** spamassassin.apache.org/tests_3_3_x.html) Of course, sometimes "hey bro," or any other indicator of spam occurs in real email (called *ham*). Here are the results for a random sample of 100 messages, which were sorted using one spam filter.

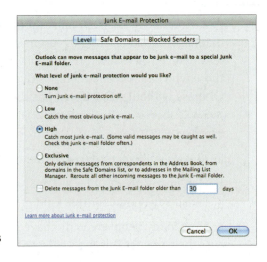

	Sent to Spam Folder	Sent to Inbox	Total
Actually Spam	36	9	45
Actually Ham	2	53	55
Total	38	62	100

a. Define a false positive as sending ham to the spam folder. How many false positives were there? Explain what a false positive means for the recipient of the message.

b. How many false negatives were there? Explain what a false negative means for the recipient.

c. What is the sensitivity of the spam filter? Explain the meaning of this statistic in this context.

d. What is the specificity of the spam filter? Explain the meaning of this statistic in this context.

e. What is the positive predictive value (PPV)? Use this value in a sentence explaining its meaning in this context.

f. What is the negative predictive value (NPV)? Use this value in a sentence explaining its meaning in this context.

g. All in all, does this seem to be a good spam filter?

2 Alzheimer's disease (AD) causes progressive cognitive impairment. A diagnostic blood test for AD recently was proposed. An evaluation of this blood test involved 50 known AD patients and 40 controls who did not have AD. Of those who had AD, 48 tested positive on the blood test and 2 tested negative. Of those without AD, 3 tested positive on the blood test and 37 tested negative.

(**Source:** Eric Nagele et al. "Diagnosis of Alzheimer's Disease Based on Disease-Specific Autoantibody Profiles in Human Sera," *PLOS ONE*, Vol. 6, 2011.)

a. Make a table that summarizes the information above.

b. How many people who did not have AD tested positive on the blood test? What is the name for this kind of test result?

c. How many people who did have AD tested negative on the blood test? What is the name for this kind of test result?

d. With a serious disease, it is important that a high proportion of those who have the disease be detected by the test. Was that the case here? What is the name for the proportion of cases that are detected by a test?

e. It also is important that a high proportion of those who do not have the disease should test negative. Was that the case here? What is the name for this proportion?

f. What is the positive predictive value (PPV)? Use this value in a sentence explaining to a person what his or her positive test suggests.

g. What is the negative predictive value (NPV)? Use this value in a sentence explaining to a person what his or her negative test suggests.

h. All in all, do you think this is a good test? Explain.

3 About 5.4 million people in the United States, almost all age 65 and over, have Alzheimer's disease (AD). In the Check Your Understanding on page 41, you saw that 40,267,984 people in the United States are age 65 and over. In Parts a–d, you will complete a copy of the following table, which reflects what would happen if all people age 65 and over were tested and the probabilities estimated from the data in Applications Task 2 are correct. (**Source:** www.alz.org/alzheimers_disease_facts_and_figures.asp)

Scott Bodell/Getty Images

	Tested Positive	Tested Negative	Total
Has AD			
Does Not Have AD			
Total			

a. About how many people in the United States age 65 and over do not have AD? Fill in the row marginal totals and the grand total in your table.

b. What *percentage* of the people who do have AD would you expect to test positive, based on the results in Applications Task 2? Fill in the *number* of the people who do have AD that you would expect to test positive. Then fill in the number that you would expect to test negative.

c. Repeat Part b for the people who do not have AD.

d. Finally, fill in the column totals.

e. What would be the negative consequences of universal screening for AD using this test?

4 In a classic psychology experiment, preschool children individually were put in a room with a marshmallow (or other treat they selected). They were told that the investigator had to leave for 15 minutes and if they did not eat the marshmallow until he returned, they would be given a second marshmallow and then they could eat both. Suppose that

this experiment is replicated with 4 year-olds to see if there is any relationship between ability to wait and whether the child can tie their shoes. The results are given below.

	Waited	Ate Treat	Total
Can Tie Shoes	8	20	28
Cannot Tie Shoes	14	35	49
Total	22	55	77

a. If you select one of these children at random, what is the probability that they can tie their shoes?

b. If you select one of the children at random from those who can tie their shoes, what is the probability that they were able to wait until the investigator returned?

c. If you select one of the children at random from those who waited, what is the probability that they can tie their shoes?

d. Are the events *can tie shoes* and *waited* independent? Explain how you know.

e. Are the categories *can/cannot tie shoes* and *waited/ate treat* independent? Explain how you know.

5 Two experts were asked to decide (separately) whether sculptures and drawings supposedly by Henry Moore were genuine or questionable. All had been offered on eBay over a 21-month period. The results are given in the table below. Suppose that these results are typical.

Evaluator 2

Evaluator 1		Questionable	Genuine	Total
	Questionable	59	6	65
	Genuine	2	6	8
	Total	61	12	73

Source: Joseph Gastwirth and Wesley Johnson, "Dare You Buy a Henry Moore on eBay?" *Significance*, March 2011, pp. 10–14.

a. If you submit a randomly selected Moore sculpture or drawing from eBay to these two evaluators, what is the probability that they both think it is genuine?

b. Which evaluator would you say is more suspicious about the authenticity of the sculptures and drawings? Justify your answer.

c. What is the probability that Evaluator 1 thinks a Moore sculpture or drawing is genuine, given that Evaluator 2 thinks it is genuine?

d. Are the events *Evaluator 1 thinks a Moore sculpture or drawing is genuine* and *Evaluator 2 thinks it is genuine* independent? Show the computations needed to justify your answer.

e. Are the variables *Evaluator 1* and *Evaluator 2* independent? Justify your answer.

f. *Should* the variables *Evaluator 1* and *Evaluator 2* be independent or associated in this situation? Explain your reasoning.

g. Fill in the cells of a copy of the following table so that the variables *Evaluator 1* and *Evaluator 2* are independent

Evaluator 2

Evaluator 1		Questionable	Genuine	Total
	Questionable			65
	Genuine			8
	Total	61	12	73

6 Researchers wanted to establish whether irritable bowel syndrome (IBS) and generalized anxiety disorder (GAD) are associated maladies. They collected the following information from a random sample of 2,005 people contacted by phone.

	GAD	No GAD	Total
IBS	18	91	109
No IBS	63	1,833	1,896
Total	81	1,924	2,005

Source: S. Lee *et al.* Irritable Bowel Syndrome Is Strongly Associated with Generalized Anxiety Disorder: A Community Study, *Alimentary Pharmacology & Therapeutics*, Vol. 30, 2009, pp. 643–651; onlinelibrary.wiley.com/doi/10.1111/j.1365-2036.2009.04074.x/full#t1

  a. Complete a table of expected frequencies, rounding to the nearest tenth, for a chi-square test of independence.

  b. Compute the chi-square statistic χ^2.

  c. To what critical value should your value of χ^2 from Part b be compared?

  d. The article concludes that IBS and GAD are strongly associated. Do you agree? Explain why or why not.

  e. Write a conclusion in context describing the relationship between the variables.

7 North Carolina keeps track of traffic accidents. The table below shows all 150 crashes in 2010 involving both a teen driver and a death. The accidents are sorted on two variables, whether alcohol was involved and whether the accident was in town or near town.

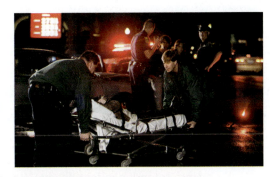

	In Town	Near Town	Total
No Alcohol Involved	30	91	121
Alcohol Involved	10	19	29
Total	40	110	150

Source: buffy.hsrc.unc.edu/crash/datatool.cfm

  a. Complete a table of expected frequencies, rounding to the nearest tenth, for a chi-square test of independence.

  b. Compute the chi-square statistic χ^2.

  c. To what critical value should your value of χ^2 from Part b be compared?

  d. Is the association statistically significant? Explain.

  e. Write a conclusion in context describing the relationship between the variables.

Image Source/Getty Images

8 This chart shows the 36 equally likely outcomes when rolling a pair of six-sided fair dice. The entry (1, 2), for example, represents the outcome *first die landed with a 1 on top* and *second die landed with a 2 on top*.

(1, 1)	(1, 2)	(1, 3)	(1, 4)	(1, 5)	(1, 6)
(2, 1)	(2, 2)	(2, 3)	(2, 4)	(2, 5)	(2, 6)
(3, 1)	(3, 2)	(3, 3)	(3, 4)	(3, 5)	(3, 6)
(4, 1)	(4, 2)	(4, 3)	(4, 4)	(4, 5)	(4, 6)
(5, 1)	(5, 2)	(5, 3)	(5, 4)	(5, 5)	(5, 6)
(6, 1)	(6, 2)	(6, 3)	(6, 4)	(6, 5)	(6, 6)

Suppose that you roll this pair of dice.

a. Find $P(sum\ is\ 6)$.

b. Find $P(sum\ is\ 6\ |\ doubles)$.

c. Find $P(doubles\ |\ sum\ is\ 6)$.

d. Find $P(sum\ is\ 6\ and\ doubles)$.

e. Find $P(sum\ is\ 6\ or\ doubles)$.

9 In Investigation 2, you used this definition of independent events: Events A and B are independent if $P(A\ and\ B) = P(A) \cdot P(B)$.

a. Refer back to your work in Connections Task 8. Are the events *doubles* and *sum is 6* independent according to this definition? Show your work.

b. Are the events *doubles* and *get a 1 on the first die* independent according to this definition? Show your work.

c. You may have learned the following equivalent way of showing that two events are independent: Events A and B are independent if $P(A) = P(A\ |\ B)$ (assuming $P(B) \neq 0$). Describe in words the meaning of this definition.

d. Using the definition in Part c, determine if the events *doubles* and *sum is 6* are independent. Use *doubles* as event A.

e. Using the definition in Part c, determine if the events *doubles* and *get a 1 on the first die* are independent. Use *doubles* as event A.

f. There is a third equivalent way of showing that two events are independent: Events A and B are independent if $P(B) = P(B \mid A)$ (assuming $P(A) \neq 0$). Describe in words the meaning of this definition.

g. Using the definition in Part f, determine if the events *doubles* and *sum is 6* are independent. Use *doubles* as event A.

h. Using the definition in Part f, determine if the events *doubles* and *get a 1 on the first die* are independent. Use *doubles* as event A.

10 Use the definition of independent events in Connections Task 9 Part c to determine whether the following events are independent. Interpret your results.

a. The events *teen driver* and *speed related* in the table below, which shows all crashes with fatalities in North Carolina in 2010

	Not Speed Related	Speed Related	Total
No Teen Driver	793	281	1,074
Teen Driver	90	60	150
Total	883	341	1,224

Source: buffy.hsrc.unc.edu/crash/datatool.cfm

b. The events *junior* and *going to the homecoming dance* in the table below

	Going to the Homecoming Dance	Not Going to the Homecoming Dance	Total
Junior	250	150	400
Senior	125	75	200
Total	375	225	600

11 Use the following table of hypothetical data to think about the difference between a chi-square test of homogeneity and a chi-square test of independence.

	Teen Driver	Driver Not a Teen	Total
Speeding Involved	42	46	88
Speeding Not Involved	20	34	54
Total	62	80	142

a. Suppose that these data were collected by taking a random sample of 62 accidents from accidents involving teen drivers and a random sample of 80 accidents from accidents not involving a teen driver. Do you use a chi-square test of homogeneity or a chi-square test of independence? Why?

b. Now suppose that these data were collected by taking a single random sample of 142 accidents and sorting them according to the two categorical variables. Do you use a chi square test of homogeneity or a chi-square test of independence? Why?

c. A friend says that $\frac{62}{142}$, or about 43.7%, is a reasonable estimate of the percentage of all accidents that involve a teen driver. Is that correct if the data were collected as described in Part a? Is that correct if the data were collected as described in Part b?

REFLECTIONS

12 Often a screening test is judged positive or negative based on a *cut-off value*. For example, a fasting blood glucose test for diabetes is considered positive if it is at or above the cut-off value of 126 mg/dL. Where to place the cut-off value is a matter of experience and good judgment.

Suppose that a higher measurement on a test is more evidence of the condition than a lower measurement. A lab suggests lowering the cut-off value. What is likely to be the effect of lowering the cut-off value on the sensitivity? On the specificity? On the PPV? On the NPV?

13 In a good screening test, should the two variables of *condition present/absent* and *test positive/negative* be independent or associated? Explain your reasoning.

14 Describe the similarities and the differences between a chi-square test of homogeneity and a chi-square test of independence.

EXTENSIONS

15 In your work in Investigation 2 and in Connections Task 9, you saw three equivalent ways to verify that events A and B are independent.

$$P(A \text{ and } B) = P(A) \cdot P(B)$$
$$P(A) = P(A \mid B)$$
$$P(B) = P(B \mid A)$$

Each way is based on the definition of **conditional probability**:

$$P(A \mid B) = \frac{P(A \text{ and } B)}{P(B)}$$

a. Refer to Connections Task 8. Use the definition of conditional probability to find $P(doubles \mid sum \text{ is } 4)$. Then use it to find $P(sum \text{ is } 4 \mid doubles)$.

b. Use the definition of conditional probability to show that if it is true that $P(A) = P(A \mid B)$, then it also is true that $P(A \text{ and } B) = P(A) \cdot P(B)$.

c. Use the definition of conditional probability to show that if it is true that $P(A \text{ and } B) = P(A) \cdot P(B)$, then it is also true that $P(A) = P(A \mid B)$.

16 Sometimes one or both of the categorical variables has more than two categories. In other words, the table is larger than 2 rows and 2 columns. For example, the following table gives information about a random sample of traffic crashes in Virginia in 2011.

	Alcohol Related	Not Alcohol Related	Total
Speed Limit Exceeded	26	103	129
Safe Speed Exceeded	6	109	115
No Speed Violation	48	858	906
Total	80	1,070	1,150

Source: Based on Commonwealth of Virginia, *2011 Virginia Traffic Crash Facts*, page 3.

a. Compute a table of expected frequencies, in the usual way, for a chi-square test of independence.

b. Compute the value of χ^2 in the usual way.

c. The larger the table, the larger the critical value will be. To find the critical value in such cases, you compute the *degrees of freedom* (*df*): Multiply the number R of rows minus 1 times the number C of columns minus 1.

$$df = (R - 1)(C - 1)$$

Then look up the critical value in the following table. Note that the critical values are the same as in the table in Lesson 2. However, this table uses *Degrees of Freedom* rather than *Number of Categories*.

Degrees of Freedom	1	2	3	4	5	6	7
Critical Value	3.84	5.99	7.81	9.49	11.07	12.59	14.07

Is the value of χ^2 from Part b statistically significant?

d. Where does the largest difference between the observed and expected frequency occur? What can you conclude from this test?

17 A random sample of college students were asked to identify their favorite season of the year and whether they preferred to spend most of their time outside, inside, or did not care. The table below is based on their responses.

	Outside	Inside	Did Not Care	Total
Spring	127	162	152	441
Summer	95	143	129	367
Fall	49	75	83	207
Winter	35	62	54	151
Total	306	442	418	1,166

a. Compute a table of expected frequencies for a chi-square test of independence.

b. Compute the value of χ^2.

c. Use the rule in Extensions Task 16 to compute the degrees of freedom? Use the table in that task to find the critical value.

d. Is the value of χ^2 from Part b statistically significant?

e. Where does the largest difference between the observed and expected frequency occur? Write a conclusion for this situation.

18 In this unit, when you used technology to determine the value of χ^2, you may have noticed that you were also given the degrees of freedom *df* and a *P*-value *p*. The **P-value** is the probability of getting a value of χ^2 as large or larger than the one you computed for your sample if it had been taken from a population where the two variables are independent. If the *P*-value is smaller than 0.05, the value of χ^2 is statistically significant.

a. Conduct a chi-square test for the data in Extensions Task 16 using technology. What are the degrees of freedom?

b. What is the *P*-value? Is the *P*-value smaller than 0.05?

c. Is the value of χ^2 statistically significant?

19 Refer to Extensions Task 18. Conduct a chi-square test for the data in Extensions Task 17 using technology.

   a. What are the degrees of freedom?

   b. What is the P-value? Is the P-value larger than 0.05?

   c. Is the value of χ^2 statistically significant?

REVIEW

20 Angles of rotation can be measured in degrees, radians, and revolutions. Fill in each blank with the appropriate value. Recall that
1 revolution $= 360° = 2\pi$ radians.

   a. $270° = $ _____ radians $= $ _____ revolution

   b. _____ $° = \dfrac{5\pi}{4}$ radians $= $ _____ revolution

   c. _____ $° = $ _____ radians $= \dfrac{5}{6}$ revolution

   d. _____ $° = 3\pi$ radians $= $ _____ revolutions

21 Chet has recently accepted a new job as a salesperson. He will earn a base annual salary of \$36,000 plus 3% commission on his sales during the year.

   a. Explain why the function rule $P(s) = 36{,}000 + 0.03s$, where s is Chet's total annual sales, can be used to determine Chet's annual earnings.

   b. Determine the value of $P(130{,}000)$ and explain what it tells you.

   c. For what value of s will $P(s) = 43{,}200$? What does this tell you about Chet's earnings?

   d. Write a function rule that could be used to determine Chet's monthly pay (salary and commission) for any amount of sales during a month.

22 Recall that a coordinate system is divided into four quadrants that are numbered in a counterclockwise direction as shown in the diagram at the right. If the terminal side of an angle in standard position is determined by a counterclockwise rotation of its initial side, the angle has positive measure. If the rotation is clockwise, the angle has negative measure.

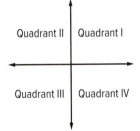

Quadrant II | Quadrant I

Quadrant III | Quadrant IV

a. Consider the following measures of angles in standard position. Based on the location of the terminal side of each angle, place each angle measure in the correct column of a copy of the table below.

$\frac{3}{5}$ revolution $-\frac{1}{12}$ revolution $\frac{1}{5}$ revolution $\frac{5}{12}$ revolution

120°   330°   −110°   72°

$\frac{3\pi}{8}$ radians $\frac{10\pi}{6}$ radians $\frac{2\pi}{3}$ radians $\frac{6\pi}{5}$ radians

Quadrant I	Quadrant II	Quadrant III	Quadrant IV

b. In Quadrant I, which of the angles have terminal sides in the same position?

c. In Quadrant II, which of the angles have terminal sides in the same position?

d. In Quadrant III, which of the angles have terminal sides in the same position?

e. In Quadrant IV, which of the angles have terminal sides in the same position?

23 Without using technology, evaluate each of the following. Then check your answer using technology.

a. log 10,000

b. 3 log 100

c. $\log 10^8$

d. log 0.01

e. $10^{\log 50}$

24 In one class, students were asked to solve the equation $x(x + 7) = x(2x + 3)$. Victoria and Taylor came up with two different solutions to this equation. Their work is shown below. Who is correct? What mistake did the other student make?

Victoria

$x(x + 7) = x(2x + 3)$

$\dfrac{x(x + 7)}{x} = \dfrac{x(2x + 3)}{x}$

$x + 7 = 2x + 3$

$4 = x$

Taylor

$x(x + 7) = x(2x + 3)$

$x^2 + 7x = 2x^2 + 3x$

$0 = x^2 - 4x$

$0 = x(x - 4)$

$x = 0 \text{ or } x = 4$

25 The table and graph below provide information about invoices for new roofs installed by Piedmont Roofing last month.

Roof Area (in sq. ft)	Invoice Total
600	$3,773
750	$5,457
1,000	$5,825
896	$3,906
1,500	$6,923
1,000	$3,380
1,295	$6,083
729	$2,849
1,720	$9,663
1,175	$8,014

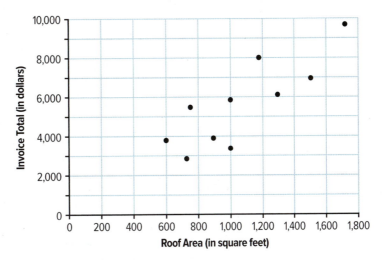

a. Without using technology, estimate what it would cost someone to have Piedmont Roofing install a new roof with an area of 1,225 square feet. Be prepared to explain how you determined your estimate.

b. Using technology such as *TCMS-Tools* data analysis, determine the equation of the least squares regression line. Then explain what the slope and y-intercept of the least squares regression line tell you about the cost of having Piedmont Roofing install a new roof.

26 Miguel is riding on a Ferris wheel that has a 10-m radius and turns in a counterclockwise direction. Miguel is currently in the "3 o'clock" position. Consider Miguel's position in relation to the *vertical* line through the center of the wheel.

a. Determine his directed distance from the vertical line after he has rotated 90° counterclockwise.

b. Determine his directed distance from the vertical line after he has rotated 225° counterclockwise.

c. For what rotations (between 0° and 360°) will Miguel be 5 meters to the left or right of the vertical line through the center of the wheel?

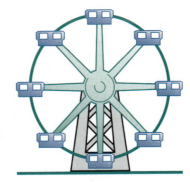

d. Which graph at the right shows Miguel's directed distance from the vertical line through the center of the wheel as he travels through one revolution?

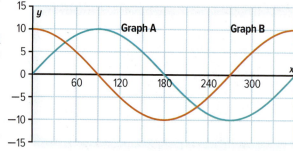

e. Which of the following function rules matches the graph you chose in Part d?

Rule I $f(x) = 10 \cos x$

Rule II $f(x) = \cos 10x$

Rule III $f(x) = \cos x + 10$

27 At the beginning of an experiment to test a bacteria-killing substance, 8,000 bacteria were present. The day after the substance was introduced, only 6,000 bacteria were present.

a. What percent of the bacteria were still present after one day? By what percentage did the number of bacteria decrease?

b. If the same percent change continues each day, how many bacteria will be present at the end of the second day?

c. If *NOW* represents the number of bacteria present on any day and *NEXT* represents the number of bacteria present on the following day, write a rule involving *NOW* and *NEXT* that gives the relationship between the two quantities.

d. Assume that the number of bacteria present is an exponential function of the number of days *d* since the substance was introduced. Which function rule best models this situation? Be prepared to explain your reasoning.

    I. $f(d) = 8,000(0.25^d)$

    II. $g(d) = 8,000(0.75^d)$

    III. $h(d) = 8,000 + 0.75^d$

    IV. $j(d) = 8,000 + 0.25^d$

e. When will the total number of bacteria present be 1,000? Explain how you determined your answer.

Looking Back

In this unit, you learned the meaning of terms often used in the media to compare the proportion of people in different groups who have some characteristic. These terms include absolute risk, absolute risk reduction, relative risk, and statistically significant. You also learned why it is important not to rely on anecdotal evidence, but to look for evidence from well-designed experiments. You used the chi-square statistic to decide whether the difference in the proportions that fall into each category in two random samples is statistically significant. Finally, you used a chi-square test to decide whether it is plausible that two categorical variables are independent in the population from which the random sample was taken.

The tasks in this final lesson will help you review and solidify your understanding of key ideas and methods for making sense of categorical data.

1 About 62% of Americans have a pet, with women more likely than men to have one. A Harris Poll of 601 male adult pet owners and 753 female pet owners asked if they considered their pet to be a member of the family. Eighty-five percent of the males said yes, 12% said no, and 3% were not sure. Of the female pet owners, 95% said yes, 3% said no, and 2% were unsure. You may consider these independent random samples of male and female pet owners. (**Source:** "Pets Really Are Members of the Family," The Harris Poll, June 10, 2011.)

a. Summarize the information in a two-way table of observed frequencies. Let the columns be men and women pet owners.

b. Make a bar graph that best compares the proportion of men and women pet owners who gave the different responses. Does the difference in the responses appear to be significant or do the two groups appear almost homogeneous?

c. To show what homogeneous samples of male and female pet owners would have looked like, make a table of expected frequencies. Round to the nearest tenth.

d. Compute χ^2 to summarize the difference between male and female responses.

e. How many categories are there for each sample? What critical value should be used for comparison?

f. Is the difference in the proportions of male and female pet owners who gave the different responses statistically significant? Explain how you know, and what it means in this context.

2 Tonsils, at the back of the throat, are organs of the lymphatic system. They may help the immune system fight disease. Sometimes, when they are chronically inflamed, doctors remove them, with a surgical procedure called a tonsillectomy. Sweden keeps a register of surgery done on people under the age of 20. During a 15-year period, 27,284 young people had a tonsillectomy. For each of them, five young people the same age, sex, and county of residence were randomly selected to serve as controls. The control group ended up with 136,401 young people. After about 23 years, 47 of the tonsillectomy group and 169 of the control group had had a premature heart attack. (**Source:** Imre Janszky, et al. "Childhood Appendectomy, Tonsillectomy, and Risk for Premature Acute Myocardial Infarction—A Nationwide Population-Based Cohort Study," *European Heart Journal*, online access June 1, 2011.)

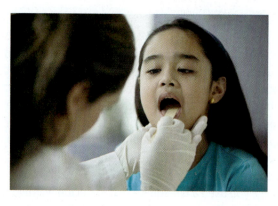

a. What is the explanatory variable in this study? What is the response variable? Was this study an experiment? Explain why or why not.

b. Make a table of observed frequencies.

c. What is the absolute risk for each group? Why is the incidence of heart attack so low in both groups?

d. Compute the absolute risk reduction of a heart attack. Use this in a sentence for parents.

e. Compute the relative risk of a heart attack. Use this in a sentence for parents.

f. Assume that you can consider the two groups equivalent to randomly selected samples from all young people with and without tonsillectomies. Compute χ^2.

g. Is the difference in the proportions of people in the two groups who have heart attacks statistically significant? Explain how you know.

h. Write a short article for your school newspaper about this study. Use and explain the terms absolute risk, relative risk, and statistical significance.

3 Post-traumatic stress disorder (PTSD) is a highly anxious state that can develop after exposure to psychological trauma. Researchers at the U.S. Department of Veterans Affairs (VA) collected data on 852 veterans who screened positive for PTSD.

Blend Images

The following table shows some of their results. You may consider these a random sample of all veterans diagnosed with PTSD.

	Received Minimally Adequate Treatment	Did Not Receive Minimally Adequate Treatment	Total
Served In Iraq or Afghanistan	115	280	395
Served Elsewhere	165	292	457
Total	280	572	852

Source: Mary W. Lu, *et al.* "Correlates of Utilization of PTSD Specialty Treatment Among Recently Diagnosed Veterans at the VA," *Psychiatric Services*, Vol. 62, 2011, pp. 943–949. Frequencies estimated from percentages.

   a. If you select one of these 852 veterans at random, what is the probability that he or she served in Iraq or Afghanistan? What is the probability that he or she served in Iraq or Afghanistan given that he or she received minimally adequate treatment?

   b. According to the mathematical definition of independent events, are the events *served in Iraq or Afghanistan* and *received minimally adequate treatment* independent in this sample? Explain what your conclusion means in practical terms.

   c. Using the marginal totals in the table above, complete a table of expected frequencies for a chi-square test of independence. Round to the nearest whole number.

   d. Compare the expected frequencies and the observed frequencies. Do the differences seem relatively large or small?

   e. Compute the chi-square statistic χ^2.

   f. To what critical value should your value of χ^2 be compared? Is the value of χ^2 statistically significant?

   g. Write a conclusion that can be drawn from your analysis.

SUMMARIZE THE MATHEMATICS

Reports in the media often contrast two groups or discuss the association between two conditions.

a. What is the difference between absolute risk reduction and relative risk?

b. What are the characteristics of a well-designed experiment? Why is it imperative to conduct an experiment rather than rely on anecdotal evidence when deciding how well a treatment, medical or otherwise, works?

c. Describe the importance of each of the following in an experiment: control group, placebo, single blind, double blind.

d. What does it mean for two groups to be homogeneous? What does the stacked bar graph (percent on the vertical axis) look like if the groups are homogeneous?

e. In a test of homogeneity, what does χ^2 measure? How do you use proportional reasoning to compute expected frequencies?

f. Suppose you have two random samples, each classified into the same categories. How can you tell whether the difference in the proportions that fall into each category is statistically significant? What does it mean when the difference is statistically significant?

g. What statistics can be used to evaluate the effectiveness of a screening test? Which is the one you would want to know if you tested positive for some condition? Which is the one you would want to know if you tested negative?

h. What does it mean if two categorical variables are independent? How can you use the definition of independence to compute expected frequencies for a chi-square test of independence?

i. Describe the similarities and differences between a chi-square test of homogeneity and a chi-square test of independence.

Be prepared to share your ideas and reasoning with the class.

✔CHECK YOUR UNDERSTANDING

Write, in outline form, a summary of the important statistical concepts and methods developed in this unit. Organize your summary so that it can be used as a quick reference in your future work.

Functions Modeling Change

Digital Resources at
ConnectED.mcgraw-hill.com

Watch Tools Audio eBook

In your previous mathematical studies, you saw that functions can be used to describe and reason about variables in a wide variety of problem situations. For example, linear and quadratic functions are used to model various income and profit possibilities for a new business. Exponential functions are used to model population change and growth in investments. The sine and cosine functions are used to model periodic phenomena such as the motion of passenger capsules on amusement park rides.

In the three lessons of this unit, you will extend your skill in use of basic function families to model more complex relationships among variables.

LESSONS

1 Function Models Revisited

Review properties of situations in which linear, exponential, power, quadratic, and circular functions are useful models of data patterns and relationships between quantitative variables. Review table, graph, and symbolic rule patterns of those basic function families.

2 Customizing Models by Translation and Reflection

Develop skill in modifying rules for functions to produce models for data patterns whose graphs are related to those of familiar functions by vertical and horizontal translations and reflections.

3 Customizing Models by Stretching and Compressing

Develop skill in modifying rules for functions to produce models for data patterns whose graphs are related to those of familiar functions by vertical and horizontal compressing and stretching.

Photographer's Choice RF/Getty Images

Function Models Revisited

One of the most important and controversial problems in environmental and atmospheric science today is measuring, understanding, and predicting global warming. There is concern that the average annual surface temperature on Earth has been increasing over the past century. This change will have important consequences for industry, agriculture, and personal lifestyles.

The graph that follows shows the pattern of change in average world temperature over the past 162 years. While the average global temperature has increased by less than a degree, this is still a large amount relative to historical data. This recent temperature increase is four to five times faster than any other climate change in the past millennium.

Annual Deviation from the 1961–1990 Average Global Temperature

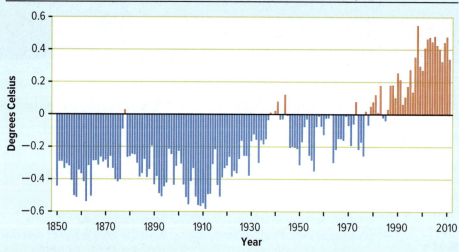

Source: www.cru.uea.ac.uk/cru/info/warming/

Many scientists believe that the most likely variable contributing to the increase in world temperature is greenhouse gases that reduce radiation of energy from Earth's surface into space. The graph below gives data on change in atmospheric greenhouse gases over more than 1,000 years.

Atmospheric Concentrations of Carbon Dioxide, 1000–2011

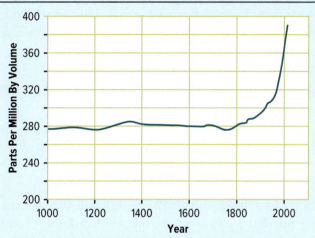

Source: www.earth-policy.org/datacenter/xls/book_fpep_ch8_3.xlsx

THINK ABOUT THIS SITUATION

A challenge for atmospheric scientists is deciding how current trends in greenhouse gas amounts and world temperature change should be projected into the future. Different projections imply different corrective actions.

a. Examine the Annual Deviation from the 1961–1990 Average Global Temperature graph on the previous page. What do you notice? What questions do you have about the information displayed?

b. Based on the data given in that graph, what strategy for projecting change in global temperature would make most sense to you?

c. Examine the Atmospheric Concentrations of Carbon Dioxide graph above. What do you notice? What questions do you have about the displayed information?

d Based on the data given in the graph above, what strategy for projecting change in atmospheric carbon dioxide makes most sense to you?

In your previous mathematics courses, you studied several important **families of functions** that are useful in describing and predicting patterns of change. In the investigations of this lesson, you will review key properties of the most important function families. Then in subsequent lessons, you will learn ways to modify those basic functions to model more complex situations.

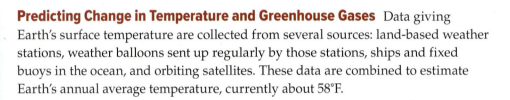

INVESTIGATION 1

Modeling Atmospheric Change

Different teams of scientists have studied the historical records of temperature and carbon dioxide data. They have proposed different scenarios for the future of global warming. Each is based on certain assumptions about the best models for patterns of change. You can get an overview of the issues by visiting: www.ncdc.noaa.gov/cmb-faq/globalwarming.html.

As you work on the problems of this investigation, look for answers to this question:

What problem conditions and data patterns suggest use of linear, exponential, power, and inverse variation functions in modeling different aspects of atmospheric change?

Predicting Change in Temperature and Greenhouse Gases Data giving Earth's surface temperature are collected from several sources: land-based weather stations, weather balloons sent up regularly by those stations, ships and fixed buoys in the ocean, and orbiting satellites. These data are combined to estimate Earth's annual average temperature, currently about 58°F.

1 The rate at which the average Earth temperature is changing is controversial. Many scientists believe it is rising, but estimates vary from an increase of about 0.05°F to 0.15°F *per decade*.

  a. Write function rules that predict the annual average temperature *x* decades from now for three different rate-of-increase estimates: 0.05°F, 0.10°F, and 0.15°F per decade. Compare your function rules with those of your classmates. Resolve any differences.

  b. Draw sketches of the various models on the same coordinate system, paying careful attention to the scales on the axes. Indicate clearly the rules corresponding to each graph.

  c. Use the low (0.05°F per decade) and high (0.15°F per decade) rate-of-change rules to write calculations, equations, or inequalities whose solutions would answer the following questions. Then answer each question using methods that seem appropriate—estimation using tables and graphs or exact solution using algebraic reasoning or a computer algebra system (CAS). Compare your methods and predictions with your classmates.

   i. Predict the average Earth temperature 50 years from now.

   ii. Predict the average Earth temperature 65 years from now.

   iii. When will the average Earth temperature reach 60°F?

   iv. How long will the average Earth temperature remain below 59°F?

Getty Images/Stockbyte

2 Atmospheric carbon dioxide (CO_2) is believed to be a primary factor in global warming. Levels of CO_2 are increasing because human activities, such as burning fossil fuels, send more CO_2 into the atmosphere than natural biological processes remove. Estimates in 2012 suggested that Earth's atmosphere contained 393 parts per million (ppm) of CO_2 with another 2 ppm added each year. (**Source:** co2now.org)

a. Based on findings of the 2012 study, what function could be used to estimate atmospheric CO_2 at any time x years after 2012?

b. Using your model from Part a:

   i. what level of atmospheric CO_2 can be expected in the year 2025?

   ii. in what year can we expect atmospheric CO_2 to reach 400 ppm?

c. Suppose that when the atmospheric CO_2 reaches 400 ppm, a way is found to reduce CO_2 emissions from human activity and increase biological processes that extract CO_2 from the atmosphere.

   i. What linear rate of change would be necessary to bring atmospheric CO_2 back to the 2012 level in 20 years?

   ii. What function could be used to estimate atmospheric CO_2 at any time x years after the corrective action began?

3 Data suggest that the rate of increase in atmospheric CO_2 has *not* been constant. Suppose that the recent annual increase of about 2 ppm is expressed as a percent and that future annual increases (from the 2012 level of 393 ppm) occur at that same percent rate.

a. What is the current percent rate of increase?

b. Assume growth from 393 ppm at the constant percent rate you found in Part a.

   i. What level of atmospheric concentration of CO_2 can be expected in 2020?

   ii. What function could you use to estimate the CO_2 concentration x years after 2012? Compare your function rule with those of your classmates. Resolve any differences.

c. Compare estimates of the increase in atmospheric CO_2 for years 2019 and 2029 under assumptions of the two different models—increase at a constant rate of about 2 ppm per year versus increase at the percent rate calculated in Part a.

d. Based on your model comparisons in Part c, explain why you think you might trust predictions of one model over the other. Explain why neither model should be trusted for very long-term predictions.

Carbon Dating of Past Events It has been estimated that in the 10,000 years since the end of the last ice age, the annual average temperature of Earth has increased by about 9°F. Atmospheric carbon dioxide has increased by at least 50%. Scientists arrived at such estimates by analyzing material that has been trapped deep in very old glaciers and on the floors of lakes and oceans for thousands of years.

One of the interesting problems in such work is estimating the age of deposits that are uncovered by core drilling. A common technique is called *carbon dating*. Carbon occurs in all living matter in several forms. The most common forms (carbon-12 and carbon-13) are stable. A third form, carbon-14, is radioactive and decays at a rate of 1.2% per century.

By measuring the proportion of carbon-14 in a scientific sample and comparing that figure to the proportion in living matter, it is possible to estimate the time when the matter in the sample was last alive. Despite the very small amounts of carbon-14 involved (less than 0.000000001% of total carbon in living matter), modern instruments can make the required measurements.

4 Suppose that drilling into a lake bottom produces a piece of wood which, according to its mass, would have contained 5 nanograms (5 billionths of a gram) of carbon-14 when the wood was alive. Use the fact that this radioactive carbon decays continuously at a rate of about 1.2% per century to analyze the sample.

  a. How much of that carbon-14 would be expected to remain:

   i. 1 century later?

   ii. 2 centuries later?

   iii. x centuries later?

  b. Estimate the **half-life** of carbon-14; that is, the length of time for half of the substance to decay.

  c. Estimate the time when the wood was last alive if the sample contained:

   i. only 3 nanograms of carbon-14.

   ii. only 1 nanogram of carbon-14.

Glaciers, Polar Ice Caps, and Global Warming

Glacier formations hold clues to the past. One of the ominous and spectacular predictions about global warming is that the melting of polar ice caps and expansion of ocean water will cause sea levels to rise and flood cities along all ocean shores. According to the Environmental Protection Agency, the most likely scenario is a 34-cm rise in sea levels by the year 2100. Such a change would flood large parts of low-lying countries like the Netherlands and Bangladesh and areas such as coastal Florida.

Estimates of such a rise in the sea level depend on measurements of glacier volumes and ocean surface areas. Earth is approximately a sphere with the following properties.

- Oceans cover approximately 70% of Earth's surface.

- The Greenland and Antarctic ice sheets cover nearly 6 million square miles.

- Those ice sheets contain about 7 million cubic miles of ice.

- The water in those ice sheets is only 2% of all water on the planet.

5 In making estimates of the size of Earth (and other spherical planets as well), it is useful to have formulas showing the circumference, surface area, and volume of a sphere as functions of the diameter or radius. Sometimes it is useful to modify those relationships to show the radius or diameter required to give specified circumference, surface area, or volume.

a. Which of the following functions gives *circumference* of a sphere in terms of the radius r? Which gives *surface area*? Which gives *volume*? Be prepared to explain the clues that help in matching each function to the corresponding measurement.

    i. $f(r) = 4\pi r^2$ **ii.** $g(r) = \frac{4}{3}\pi r^3$ **iii.** $h(r) = 2\pi r$

b. What patterns would you expect in graphs of the functions above? Describe each pattern by providing specific information such as slope, intercept(s), line symmetry, and half-turn symmetry where appropriate.

c. Earth is not a perfect sphere, but nearly so, with average radius of about 4,000 miles.

    i. What is the approximate surface area of Earth's oceans? (Oceans cover about 70% of Earth's surface.)

    ii. What volume of water would be required to raise the level of those oceans by 3 feet? Assume that raising the level would not change the surface area of the ocean significantly.

d. What rise in ocean levels would be caused by the total melting of the Greenland and Antarctic ice sheets? Again, assume that the surface area of the oceans would not change significantly. (Assume the two ice sheets contain 7 million cubic miles of ice.) The volume of water from melting ice is approximately 92% of the volume of the ice.

e. Earth is the fifth largest planet in the solar system. The largest planet, Jupiter, has a radius roughly 11 times the radius of Earth. The radius of Mars is roughly half that of Earth.

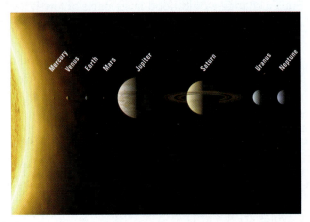

    i. How do the circumference, surface area, and volume of Jupiter compare to the corresponding measures of Earth?

    ii. How do the circumference, surface area, and volume of Mars compare to the corresponding measures of Earth?

Gravitation On and Near Earth's Surface The silent force of gravity influences almost every aspect of life on and near the surface of Earth. The gravitational force that holds all of us anchored to Earth's surface diminishes as one moves up into the atmosphere.

6 In general, the gravitational force of attraction between any two masses is directly proportional to the product of the masses and inversely proportional to the square of the distance between their centers.

a. Describe the pattern of change in gravitational force as:

    i. the distance between two planetary bodies increases.

    ii. one or both of the bodies increase in mass.

b. Which of the following formulas matches the given information about the force between masses m_1 and m_2 with centers located at a distance d apart?

$$F_1 = k(m_1 m_2 - d^2) \qquad F_2 = k\left(\frac{m_1 m_2}{d^2}\right) \qquad F_3 = -k\left(\frac{m_1 m_2}{d^2}\right)$$

Compare your choice with that of others and resolve any differences.

c. How will the gravitational force of attraction between two bodies change:

    i. if the mass of one body increases by a factor of c?

    ii. if the distance between the masses increases by a factor c?

SUMMARIZE THE MATHEMATICS

In this investigation, you modeled aspects and consequences of atmospheric change and gravitational force with various types of functions.

a. Which contexts considered in this investigation involved examples of:

    i. linear functions?

    ii. exponential functions?

    iii. direct variation functions?

    iv. inverse variation functions?

b. What problem conditions or data patterns suggested each type of function in Part a as probably the most appropriate model for the relationship between variables?

Be prepared to share your ideas and reasoning with the class.

✔ CHECK YOUR UNDERSTANDING

Coyotes are mammals similar to wolves and dogs that are most commonly found in wild habitats of the western United States. However, they are very adaptable carnivores and now appear in urban areas as far east as Washington, D.C., and New York.

There is limited data on the actual numbers of coyotes now living in east coast states, but suppose there are now about 1,500 such animals living in Delaware, Maryland, and Washington, D.C.

a. Assume that the coyote population in these areas increases by 15% per year. What function will predict the population *n* years from now?

b. Assume instead that the population increases by 250 coyotes per year. What function will predict the population *n* years from now?

c. Coyotes are predators that feed on other wild animals like fox, geese, raccoons, and deer. It might be reasonable to assume that the population of those prey species would be inversely proportional to the population of coyotes. Under that assumption, which of the following functions could provide a model for the relationship between the population of deer *d* and the population of coyotes *c* around Washington, D.C., over the next few years? Explain your reasoning.

$$d(c) = 10,000 - c \qquad d(c) = \frac{10,000}{c} \qquad d(c) = 10,000c$$

Modeling Change in Business Prospects

Music is a major form of entertainment in our society. Production and sale of music are also big businesses. All across the country, there are thousands of individuals and groups practicing hard and hoping to make it big with a best-selling single or album. The table below shows how a band can sell 200,000 CDs and still make no profit. Discuss the information in the chart with your classmates.

What an Unknown Band Might Expect for its First Album on a Major Label

ADVANCES and RECOUPABLE COSTS	
The label advances money to the band, and it also deducts some of its own expenses from initial royalties:	
• Recording advance	$210,000
• One video	$75,000
• Touring	$40,000
• Independent promotion	$40,000
Total advances and recoupable costs	**$365,000**

EARNINGS (based on 200,000 CDs sold)	
A CD might cost $15, but the label makes deductions before calculating band royalties.	
Original CD price	$15.00
• 20 percent off for packaging and pressing cost	−$3.00
• 15 percent off for discounts to retailers	−$2.25
• Other discounts (11 percent)	−$1.65
Base CD price	**$8.10**
The typical band earns 10 percent per CD.	
$8.10 × 0.1 = $0.81 per CD	
$0.81 per CD × 200,000 CDs sold =	**$162,000**
If the band writes all of its own songs, it can expect another $0.52 per CD for publishing rights.	**$104,000**

THE BOTTOM LINE (for 200,000 CDs sold)	
Total band income from CD sales	$266,000
Minus label's advances and costs	−$365,000
Amount needed before band sees profit	**$99,000**

As you work on the problems of this investigation, make notes of answers to these questions:

What problem conditions suggest use of linear and quadratic functions in modeling change in business income and profit?

Function rules can often be written in different, but equivalent forms. In modeling a situation, how do you decide the form to use?

1 Consider these business conditions when a band recorded its first album of original songs with a major label (record company).

- Expenses of $365,000 for the recording advance, video production, touring, and promotion (to be repaid out of royalties)

- Income of $0.81 per CD from royalties

- Income of $0.52 per CD from publishing rights

The *band's profit* depends on the number x of CDs sold. The group's business manager could express the profit function in several different ways. Here are two possibilities.

$$P_1(x) = (0.81x + 0.52x) - 365,000$$

$$P_2(x) = 1.33x - 365,000$$

a. Explain why both function rules express the correct relation between CDs sold and profit.

b. Which of the two rules do you believe expresses the band's profit function in the most informative way? Why?

c. What is an advantage of the other function rule?

d. What would the band's profit be if only 50,000 copies were sold? If 500,000 copies were sold?

e. How many CD copies must be sold for the band to break even?

2 *Profit for the record company* is also a function of the number of CDs sold. For the band in Problem 1, the label had the following production and distribution conditions to consider.

- Studio, video, touring, and promotion expenses of $365,000

- Pressing and packaging costs of $3.00 per CD

- Discounts to music stores of $2.25 per CD

- Other discounts of $1.65 per CD

Until the band's share of income repays the $365,000 for studio, video, touring, and promotion expenses, the band receives no share of the $15 per CD of sales income.

a. Use the preceding information to write two equivalent expressions for the function that shows how the record company's profit depends on the number of CDs sold while the band is still repaying its advance. (The company would have other expenses, such as staff salaries and office expenses like electricity and rent. For this problem, though, you should ignore those expenses.)

   i. Write one rule that shows how each income and cost item enters the overall calculation of profit.

   ii. Write another rule that will give profit for any number of CDs sold with the simplest possible calculation.

b. How could you use tables and graphs to check that the two rules you wrote in Part a are equivalent?

c. What profit would the company make on sales of 50,000 CDs? On 100,000?

d. How many copies of this CD must be sold for the record company to break even?

e. Once the band has repaid its advance of $365,000, how much must the record company pay the band per CD sold? Explain.

f. The company's profit function changes when the band has repaid its advance. Write two different expressions for profit as a function of CD sales once the advance is paid off:

   i. one that shows each income and expense item.

   ii. another that is the simplest to use for actual calculation.

g. Explain several ways to show the two rules in Part f are equivalent.

As you have seen, many bands often do not make much money on sales of their recordings. However, they do make money from concert tours and sales of T-shirts and other merchandise.

3 Suppose a band with a moderate following is preparing to go on a four-week tour. A market research survey gives the following information about prospects for a single show on the tour.

- Number of tickets sold x and ticket price p are related by $p(x) = -0.04x + 30$. The venue pays the band 10% of the income from ticket sales. Income from ticket sales is $I(x) = (-0.04x + 30)x$.

- About 15% of ticket buyers will also buy a T-shirt, giving the band $5 profit for each sale.

- About 5% of ticket buyers will buy a poster, giving the band $3.50 profit for each sale.

- All the band's other expenses for each show will average about $300.

Because one of the band's main goals is to make a profit, the band manager might combine all the given information in a single relation showing profit as a function of number of tickets sold at a show.

a. One possibility is the rule:

$$P(x) = 0.10(-0.04x + 30)x + 5(0.15x) + 3.50(0.05x) - 300$$

How does each part of this function rule relate to the given information? Compare your answer with that of your classmates and resolve any differences.

b. Rewrite this rule for the profit function in a simpler, yet equivalent, form. Check to see that the simpler function rule you wrote is equivalent to the original.

c. What is the maximum profit that the band could make from the show? How many purchased tickets will produce that maximum profit? What ticket price will produce that maximum profit?

SUMMARIZE THE MATHEMATICS

In this investigation, you modeled income and profit (loss) possibilities in the music business. You found several cases in which quite different symbolic rules expressed the same relation between variables.

a. How can you tell from a business situation if profit will be a linear or quadratic function of the number of items sold?

b. In the case of a quadratic profit function, how can you find:

   i. the break-even point(s)?

   ii. the maximum profit and the corresponding number of items sold?

c. What does it mean if someone says two function rules are *equivalent*?

d. How can you check the equivalence of different function rules?

e. In the case of income and profit functions, what two forms of equivalent rules are most useful when analyzing a business situation? Why?

Be prepared to explain your ideas and reasoning to the class.

The price of $70 for a pair of athletic shoes in a retail store is based on several factors.

- Average manufacturer's income is about $20 per pair of shoes.

- Average wholesaler's income is about $15.50 per pair of shoes.

- Average retailer's operating costs are about $25.50 per pair of shoes sold.

(**Source:** "Why It Costs $70 for a Pair of Athletic Shoes," *Washington Post*)

a. One way to express retailer's profit as a function of the number of pairs x of shoes sold is $P(x) = 70x - (20x + 15.50x + 25.50x)$. Write this function rule in several other equivalent forms, including one that you think is the simplest possible.

b. How much profit does the retail store make on each pair of $70 shoes sold? How is the profit shown in your simplest-form answer to Part a?

c. Suppose the shoe store manager estimates that price p and monthly number of pairs sold x of its most popular shoe model are related by $x = 500 - 4p$.

   i. Write in words how the store manager would calculate *monthly income I* from sales of that shoe model as a function of the price p charged.

   ii. Write a symbolic rule that will give monthly income I from sales of the shoe model as a function of the price p charged: $I(p) = $ _____.

d. Write the income function rule in Part cii in two equivalent forms. One form which is most informative. Another that you think is the simplest form.

e. Suppose the average retailer's operating costs per pair of the most popular athletic shoe sold are the same as the average for the $70 shoes described at the beginning of this task.

   i. Write in words how the store manager could calculate the store's *monthly profit P* on the most popular model in terms of price p.

   ii. What function rule will give the store's monthly profit P in terms of price p of the most popular model alone?

f. Write the monthly profit function from Part eii in several equivalent forms, including one that you think is simplest for calculations.

g. What is the maximum monthly profit of the most popular shoe? What shoe price will produce the maximum monthly profit? How many pairs of shoes must be sold to produce that maximum monthly profit?

Modeling Periodic Change

Increasing carbon dioxide in the atmosphere has been suggested as the most likely variable contributing to the recent observed increase in Earth temperatures. If such global warming continues, polar and glacial ice may melt and ocean waters may expand, resulting in a rise in sea level. However, not all changes in Earth's climate and geography are strictly increasing. Some important variables are **periodic**— they change in regular patterns that repeat over constant intervals of time. The length of the shortest interval for which the repeating pattern occurs is called the **period**.

As you work on the problems of this investigation, make notes of answers to these questions:

How can you determine the period of a periodic function?

How can you determine the amplitude of a periodic function?

Why are the sine and cosine functions often used to model periodic change?

Periodic Change Variations of the *circular functions* sine and cosine are frequently used to model periodic change such as depth of tidal waters, height above ground of a Ferris wheel rider, or predator-prey phenomena. In such models, radian measure is used.

1 Begin by reviewing features of the graphs of the basic functions $y = \cos x$ and $y = \sin x$ for $-2\pi \leq x \leq 2\pi$. Check that your technology's graphing tool is set in radian mode.

   a. Discuss with your classmates how the graphs of the two functions are similar. How are they different?

   b. What is the period of $y = \cos x$? Of $y = \sin x$?

2 Shown below is a graph of $s(x) = 2 \sin x$ and a graph of $c(x) = 2 \cos x$. In each case, $0 \leq x \leq 4\pi$. (4π is approximately 12.56 radians.)

Graph I

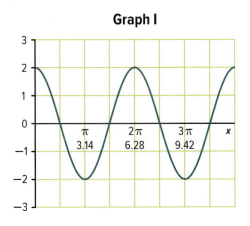

Graph II

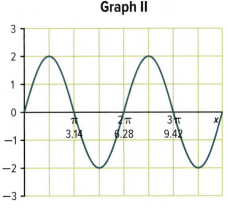

   a. Without using technology, decide which graph is that of $s(x) = 2 \sin x$. Explain how you know.

   b. What is the period of $s(x) = 2 \sin x$? Of $c(x) = 2 \cos x$? Explain.

c. How are the graphs of $s(x) = 2 \sin x$ and $c(x) = 2 \cos x$ similar? How are they different?

d. To quickly sketch the graph of a periodic function such as those above, it is important to know the general patterns of periodic change and critical points on the graphs: x- and y-intercepts and maximum and minimum points. Identify these points for $s(x) = 2 \sin x$. For $c(x) = 2 \cos x$.

e. Use the idea of critical points to quickly sketch graphs of $f(x) = 4 \cos x$ and $g(x) = 4 \sin x$ for $0 \le x \le 4\pi$. Compare your graphs with those of others. Resolve any differences.

3 Now refer back to the graphs in Problems 1 and 2 of $y = \cos x$, $y = \sin x$, $c(x) = 2 \cos x$, $s(x) = 2 \sin x$, $f(x) = 4 \cos x$, and $g(x) = 4 \sin x$.

a. For the graph of each function, what is the distance of the maximum (or minimum) points from the **midline** (the x-axis in these cases)?

b. The distances you determined in Part a are called the *amplitudes* of the respective functions. In general, the **amplitude** of a circular function is: $\frac{1}{2}|maximum\ value - minimum\ value|$. Verify that this formula gives the same values that you found in Part a.

c. How is the amplitude of each of these functions revealed in its function rule?

4 How is the graph of $f(x) = 2 \sin x$ similar to, and different from, the graph of $h(x) = \sin 2x$? How can you explain the difference(s)? Be prepared to explain your reasoning to the class.

5 Sine Function Models If you ride on a Ferris wheel, your height above the ground will vary as the wheel turns. Suppose the wheel starts spinning for a ride when your seat S is at the "3 o'clock" position. This position is 24 feet above ground.

The wheel has a radius of 20 feet and spins so that you move counterclockwise from the indicated position, making a full turn every 6.28 (approximately 2π) seconds.

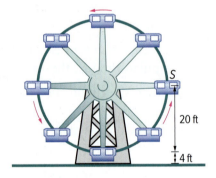

a. Sketch a graph showing the pattern of change in your height above the ground during one complete revolution of the Ferris wheel.

b. Write a rule $h(t) =$ _____ that gives your height (in feet) above ground level at any time t during the ride (in seconds). Compare your rule with that of others.

c. Evaluate and explain the meaning of each of the following.

    i. $h(0)$ **ii.** $h(1.57)$ **iii.** $h(3)$

    iv. $h(4.7)$ **v.** $h(6.28)$ **vi.** $h(7.85)$

d. For what values of the input variable $t \leq 12.56$ seconds is $h(t)$ equal to 24?

e. What do the numbers 20 and 24 in the function rule tell about the situation being modeled?

f. What patterns do you find in tables and graphs of this function rule? How do they relate to the motion of the Ferris wheel?

g. What numbers make sense as input values for the variable t? What numbers would you expect as output values for the function $h(t)$?

6 How do you think you could modify the function rule in Problem 5 Part b if the Ferris wheel made a complete revolution in 12.56 seconds? Compare your conjecture with those of your classmates and resolve any differences.

7 Cosine Function Models

Suppose you lived near the Atlantic Ocean or a tidal bay and tracked the depth of the water on a retaining wall every three hours.

If you began recording data at high tide of five feet, the data might yield a pattern like the one shown in the plot below.

Ocean Depth at Retaining Wall

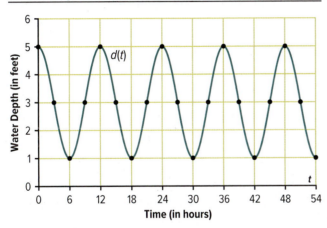

a. Extend a copy of the graph to show the predicted pattern of change for the previous 24 hours.

b. What is the period (length of the smallest time interval between high tides) of this cyclical pattern?

c. What is the midline of the graph? The amplitude?

d. Students in a class at Aurora High School tried to find a rule to model the water depth in Part a as a function of time t beginning at high tide. They assumed radian measure for the input variable t. Different groups of students proposed different rules as shown below.

Use your knowledge of *symbolic rule-graph connections* to explain why:

  i. $d(t) = 2 \cos t$ is not a correct model.

  ii. $d(t) = 2 \cos t + 3$ is not a correct model.

  iii. $d(t) = 2 \cos\left(\frac{\pi}{6}t\right)$ is not a correct model.

  iv. $d(t) = 2 \cos\left(\frac{\pi}{6}t\right) + 3$ is a correct model for the water depth as a function of time.

e. Why do you think each group used the cosine function as the building block for its model?

8 Suppose melting of ice caps caused a four-inch rise in ocean levels.

a. How would this change be reflected in the plot of water depths in Problem 7?

b. Write a rule that models the water depth in Part a as a function of time beginning at high tide. Compare your rule with others. Resolve any differences.

9 Refer to Problem 7. Suppose you began tracking the depth of the water on the retaining wall every three hours, beginning midway between low tide and high tide, as the tide is rising.

a. Make a graph of the expected pattern of change in water depth over the first 48 hours.

b. How is your graph similar to, and different from, the graph of $y = \sin t$?

c. Write a rule using the sine function that models the water depth in this situation. Compare your sine function rule with others. Resolve any differences.

d. When is the depth of the water four feet?

10 Work with your classmates to brainstorm a list of other phenomena in the world around you that change in periodic patterns. Pick one of the phenomena and sketch a graph of the expected pattern of change. Estimate the period and amplitude.

SUMMARIZE THE MATHEMATICS

In this investigation, you revisited the circular functions $s(x) = \sin x$ and $c(x) = \cos x$ and explored variations of those functions to model periodic change.

a. What condition in a problem situation suggests possible use of a periodic function model?

b. What clues in problem situations suggest using the sine function as the basic building block? What clues suggest using the cosine function?

c. Suppose you are modeling a periodic phenomenon using the function $f(x) = a \cos bx$.

   i. How would you determine the value of a? The value of b?

   ii. What is the amplitude of this function?

   iii. What is the period of this function?

Be prepared to explain your ideas and reasoning to the class.

✓CHECK YOUR UNDERSTANDING

Weather conditions and the interdependence of animal species sometimes produce periodic fluctuations in animal populations. Suppose the rabbit population in the Sleeping Bear Dunes National Forest is at a minimum of approximately 4,000 rabbits in January. By July, the population reaches a maximum of about 12,500. It returns to a low of around 4,000 in the following January, completing the annual cycle.

a. Make a plot of the rabbit population for a two-year period, starting with January.

b. In making your plot, what assumption did you make about how the rabbit population varies from minimum to maximum?

c. What are the amplitude and period of your graph?

d. How is your plot of the rabbit population as a function of time similar to, and different from, the graph of $c(x) = \cos x$? Of $s(x) = \sin x$?

Function Families

Solving the problems of Investigations 1–3 required use of a variety of different functions to model patterns of change in variables that are central to Earth's environment, to business decision-making, and to maintenance of ocean harbors. In each case, the function rules involved specific numerical constants to match specific problem conditions.

For example, the linear models proposed to predict change in average global temperature in Investigation 1 included $T(n) = 0.05n + 58$ and $T(n) = 0.15n + 58$, with n representing number of decades from now. The first function assumes the low-end rate of increase of 0.05°F per decade. The second function assumes the high-end rate of increase of 0.15°F per decade. Both functions assumed a value of 58°F for average global temperature.

The two models for predicting global temperature are *linear functions*. All members of that family of functions can be represented with rules in the form $y = ax + b$. The values of a and b that define any specific member in the family are called **parameters**. As you know from prior studies, the specific members of other function families—like exponential, quadratic, circular, and polynomial functions—are also defined by specifying values of parameters.

As you complete this investigation, look for answers to these questions

What are the key features of the basic function families?

How do the parameters in the basic function families provide tools for matching function models to specific problem conditions, data patterns, and graphs?

To explore how the parameters in the basic function families affect graphs and tables of values, it is helpful to use software like the *TCMS-Tools* computer algebra system (CAS) that accepts function definitions like $f(x) = ax^2 + bx + c$, allows you to move "sliders" controlling the values of the parameters, and quickly produces corresponding graphs and tables.

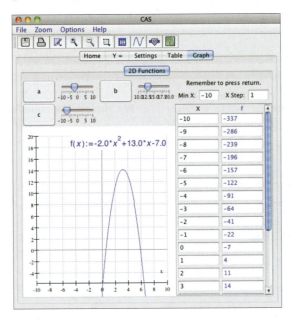

1 On an extended copy of the following functions toolkit table, for each function family indicated:

a. describe the domains and ranges of typical functions in the family.

b. produce graphs of specific examples.

    i. Describe how the patterns in graphs of $(x, f(x))$ are related to parameters in the rules.

    ii. Note key points of the graphs that would help you make quick sketches of the graphs.

    iii. Record sketches of typical functions in each family.

c. describe the patterns of change to be expected in tables of $(x, f(x))$ values and the relationship of those patterns to parameters in the rules.

d. describe any maximum or minimum points of the graphs.

e. describe any symmetries of the graphs.

f. How would you use the words "increasing" and/or "decreasing" to describe the patterns of change exhibited in your sample graphs and tables of values?

Function Family Features	
I. Linear Functions $f(x) = ax + b$ domain: range: maximum/minimum value(s) (if any): symmetries (if any):	**II. Exponential Functions** $f(x) = a(b^x)$, $a \neq 0$, $b > 1$ or $0 < b < 1$ domain: range: maximum/minimum value(s) (if any): symmetries (if any):
III. Power Functions $f(x) = ax^n$, $a \neq 0$ and n a positive integer domain: range: maximum/minimum value(s) (if any): symmetries (if any):	**IV. Inverse Variation Functions** $f(x) = \dfrac{a}{x^n}$, $a \neq 0$ and n a positive integer domain: range: maximum/minimum value(s) (if any): symmetries (if any):
V. Quadratic Functions $f(x) = ax^2 + bx + c$, $a \neq 0$ domain: range: maximum/minimum value(s) (if any): symmetries (if any):	**VI. Circular Functions** $s(x) = a \sin x$, $a \neq 0$ domain: range: maximum/minimum value(s) (if any): symmetries (if any):
VII. Logarithmic Functions $f(x) = \log_a x$, $a > 0$, $a \neq 1$ domain: range: maximum/minimum value(s) (if any): symmetries (if any):	$c(x) = a \cos x$, $a \neq 0$ domain: range: maximum/minimum value(s) (if any): symmetries (if any):

2 The CAS display below shows the graph of the inverse variation function $f(x) = \dfrac{1}{x}$ for $-10 \leq x \leq 10$.

   a. How can the overall pattern in the graph of $f(x)$ be predicted just by examining the function rule?

   b. As values of x get increasingly large, either positive or negative, what do you notice about the graph of $f(x)$? Why does this make sense?

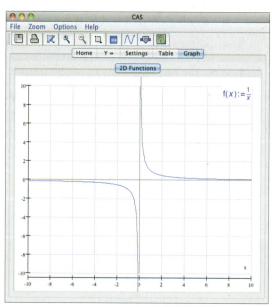

c. The graph of $f(x) = \dfrac{1}{x}$ is said to have the x-axis as a **horizontal asymptote**. As the values of x decrease without a lower bound and increase without an upper bound, the graph of $f(x)$ gets closer and closer to the x-axis but never reaches it. This graph also has a **vertical asymptote**, which is the y-axis. Explain why this is the case.

d. For each function family in the chart on page 124, describe the asymptotes (if any) of its graphs. Record your responses on your copy of the chart.

e. What conditions in problem situations suggest function models with one or more asymptotes?

SUMMARIZE THE MATHEMATICS

In this investigation, you analyzed in a systematic way characteristics of several important families of functions. What are the most striking similarities and differences that appear when comparing rules, tables, and graphs of the following pairs of function families?

a. Linear and exponential functions

b. Linear and quadratic functions

c. Exponential and inverse variation functions

d. Exponential and power functions

e. Sine and cosine functions and all other function families

f. Logarithmic and exponential functions with the same base

Be prepared to explain your comparisons and reasoning with the class.

✓CHECK YOUR UNDERSTANDING

Identify families of functions that should be considered as models for relationships between variables in the following cases. Explain reasons for each model choice.

a. Variables x and y are related as shown by data in this table.

x	−3	−2	−1	0	1	2	3
y	11.1	8.2	5.3	2.4	−0.5	−3.4	−6.3

b. The graph of $y = f(x)$ is a curve that is symmetric about the vertical line $x = 2$, with a maximum point at $(2, 5)$.

c. The graph of $y = f(x)$ is a curve that has the lines $y = 2$ and $x = 0$ as asymptotes.

d. The graph oscillates continuously between 3 and −3 at intervals of length 10.

ON YOUR OWN

APPLICATIONS

Public health officials in many countries are worrying about the possibility that diseases like SARS and avian influenza might spread from mammals and birds to humans. The disease could then spread in deadly fashion around the world. Suppose that data in the following table give numbers of cases in an outbreak of bird flu. Use the data to answer the questions in Tasks 1 and 2. To put the number of bird flu cases in perspective, the world population is about 7 billion.

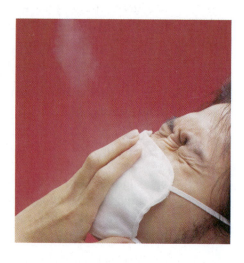

Time t (in weeks)	0	1	2	3	4	5	6
Cases $C(t)$	2,500	3,100	3,900	4,900	6,100	7,600	9,500

1 Explore consequences of using a linear function to model the bird flu data pattern.

   a. What is the best-fit linear model for the change in number of flu cases over time?

   b. How many cases are predicted for times 12, 18, and 24 weeks from the start of the outbreak of this influenza epidemic?

   c. How long will it take for 1,000,000 people to contract the disease?

   d. What do the coefficient of the independent variable and the constant term in your linear model say about the predicted spread of the disease?

   e. What reasons might you have for questioning the validity of the linear model for predicting the rate of spread of the epidemic?

2 Explore consequences of using an exponential function to model the bird flu data pattern.

   a. What is the best-fit exponential model for change in number of flu cases over time?

   b. How many cases are predicted for times 12, 18, and 24 weeks from the start of the outbreak of this influenza epidemic?

   c. How long will it take for 1,000,000 people to contract the disease?

   d. What do the parameters in your exponential model say about the predicted spread of the disease?

   e. What reasons might you have for questioning the validity of the exponential model for predicting the rate of spread of the epidemic?

Cylinders are one of the most common shapes for liquid storage tanks. The volume of a cylinder is a function of its radius r and height h given by the formula $V = \pi r^2 h$. The surface area of a cylinder (including top and bottom) is a function of the same dimensions with the formula $A = 2\pi r h + 2\pi r^2$. Use these relationships to help answer the questions in Tasks 3–5.

3 Suppose that a home oil tank is cylindrical with a radius of 2 feet and a height of 5 feet.

  a. Find the volume of the tank:

   i. in cubic feet.

   ii. in gallons. (There are about 7.5 gallons to a cubic foot.)

  b. Find the surface area of the tank.

  c. Suppose that a large cylindrical oil storage tank has radius of 20 feet and height of 50 feet.

   i. Will its volume be 10 times, 100 times, or 1,000 times as much as the home oil tank with radius of 2 feet and height of 5 feet?

   ii. Will its surface area be 10 times, 100 times, or 1,000 times as much as the home oil tank with radius of 2 feet and height of 5 feet?

   iii. Show how algebraic reasoning can provide answers to parts i and ii without actually calculating the volume or surface area of the larger tank.

4 The functions $V_r = (\pi r^2)(5)$ and $V_h = \pi(2^2)h$ show how the volume of a cylinder depends on radius when height is fixed at 5 feet and on height when radius is fixed at 2 feet.

  a. Suppose the design for a cylinder specifies a radius of 2 feet and a height of 5 feet. Which change in the design will produce the greatest change in volume: (1) increasing the radius by 1 foot; or (2) increasing the height by 1 foot?

  b. How do the algebraic rules for $V_r = (\pi r^2)(5)$ and $V_h = \pi(2^2)h$ help to explain your answer to the comparison question in Part a?

Comstock Images/Alamy

5 Suppose that you need to design a cylindrical tank with fixed volume of 10,000 cubic feet.

   a. Write an equation showing the relationship among radius r, height h, and this fixed volume.

   b. Solve the equation in Part a for r to show how the required radius depends on the choice of height. Then use this equation to find the required radius if the height is 20 feet.

   c. Solve the equation in Part a for h to show how the required height depends on the choice of radius. Then use this equation to find the required height if the radius is 15 feet.

   d. Describe in words the proportionality relationships between r and h that are expressed by the equations in Parts b and c by completing sentences like these.

      i. For a cylindrical tank with volume 10,000 cubic feet, the radius is _____ proportional to _____ with constant of proportionality _____.

      ii. For a cylindrical tank with volume 10,000 cubic feet, the height is _____ proportional to _____ with constant of proportionality _____.

6 The surface area of Earth's oceans is about 361 million square kilometers. The oceans contain about 1.347 billion cubic kilometers of water. Use these facts to answer the following questions about the effects of glaciers on ocean sea levels.

   a. In the greatest ice age, sea levels were about 135 meters lower than they are today. About how much more water could have been contained in the glaciers of that ice age than exists now? (Assume that lowering the depth of the ocean would not appreciably change its surface area. Recall that 1 meter is 0.001 kilometers.)

   b. What is the average depth of the oceans?

Many businesses use vans and small trucks for deliveries, for support of service, and for construction work. Owning such vehicles requires accounting calculations that involve variables like *operating costs*, *depreciation*, and *inflation*.

   Use what you know about functions, equations, and inequalities to answer the questions in Tasks 7–9.

7 Suppose that a company estimated that the annual operating cost of its business delivery van would be $5,000 for insurance plus $0.65 per mile for gas, oil, and maintenance.

   a. What function rule shows how total annual operating cost depends on number of miles driven x?

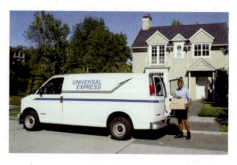

b. Write and solve equations or inequalities that match these questions about the van operating cost.

    i. How many miles can the van be driven if the annual operating cost is to be less than $15,000?

    ii. If the company records show a total operating cost of $20,000 for one year, how many miles must have been driven with the van?

c. Suppose inflation increases the operating cost of the van at a rate of about 2% per year and the current annual operating cost is $12,500. What function could be used to predict the annual operating cost at a time *n* years later? What assumption(s) are you making?

8 When a company buys a new van for $40,000, the resale value of the vehicle can be expected to decrease at a rate of 20% per year.

a. What function rule can be used to predict the resale value of the van *n* years after its purchase?

b. How long can the company keep the van and still expect to resell it for at least $15,000?

9 Accounting guidelines allow companies to recognize expense from wear and tear on equipment like a delivery van by reducing its value in business records each year. This accounting expense, called *depreciation*, reduces the company's taxable income and saves the company money.

a. If a company buys a $40,000 delivery van and plans to depreciate its resale value to $0.00 by a "straight line" method over 8 years, what is the annual depreciation allowance?

b. What function rule gives the depreciated value of the van on the company's books at any time *n* years after its purchase?

10 In 2007, Grammy Award-winning jazz musician Jeff "Tain" Watts completed his album, *Folk's Songs*. Without a record company, Tain recorded the album himself at a recording cost of $15,000 and sold CDs on his Web site at $10 each. Suppose Tain hires a marketing team and a sales/distribution team to help him sell his CDs on his Web site. Each team receives 8% of CD sales. Tain also decides to donate 5% of CD sales to charity. (**Source:** Future of the music business: How to succeed with the new digital technologies. Gordon. 2008.)

a. Write two equivalent function rules to model Tain's profit:

    i. one showing the contribution of various costs and income.

    ii. one in simplified form.

b. How many copies of his CD must Tain sell to break even? To earn a profit of $20,000?

c. Suppose that Tain recorded his album with a record label. He would receive $1.02 per CD. Assume the record label covered the recording, marketing, and sales costs. Write a rule to express Tain's profit in this situation.

d. How many copies of his CD must be sold for Tain to earn a profit of $20,000 when recording with the record label? How does this number compare with the number of CDs Tain would have to sell independently (Part b)?

e. Suppose that the record label's marketing research suggests that the number of CDs sold is related to price p by $s(p) = 20,000 - 1,000p$. Write a symbolic rule that will give income I from sales of CDs as a function of the price p.

f. Write the income function rule in Part e in another equivalent form.

g. What is the maximum income that Tain could make based on the CD demand model?

   i. How many CD sales will produce that maximum income?

   ii. What CD price will produce that maximum income?

11 The following graph shows oscillation of the water depth in a shipping channel of an ocean harbor over two days.

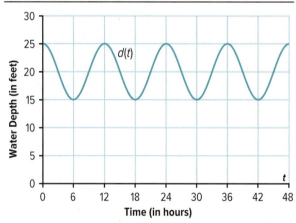

Ocean Harbor Water Depth

a. Based on information in the graph, estimate the maximum and minimum water depths, the period and the amplitude of periodic change in the water depth of the channel.

b. Write a function rule giving the depth of the channel as a function of time.

12 There are many important situations in which variables are related by a function. On the next page are six graphs and descriptions of several such situations. Match the descriptions in Parts a–f to Graphs I–VI that seem to fit them best. Then for each situation:

- explain why the graph makes sense as a model of the relationship between variables.

- describe the function family (if any) that would provide a good model for the relationship. What can you say about the parameters of the model?

Graphs:

I
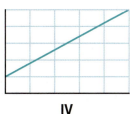

II
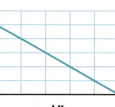

III
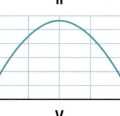

IV
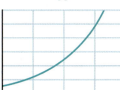

V

VI

Situations:

a. When a football team's punter kicks the ball, the ball's height changes as time passes from kick to catch. What pattern seems likely to relate time and height?

b. The senior class officers at Lincoln High School decided to order and sell souvenir baseball caps with the school insignia and name on them. One supplier said it would charge $150 to create the design and then an additional $6 for each cap made. How would the total cost of the order be related to the number of caps in the order?

c. The number of hours between sunrise and sunset changes as days pass in each year. What pattern seems likely to relate day of the year and hours of sunlight?

d. In planning a bus trip to Florida for spring break, a travel agent worked on the assumption that each bus would hold at most 40 students. How would the number of buses be related to the number of student customers?

e. The Riverside High School sophomore class officers decided to order and sell T-shirts with the names of everyone in their class on the shirts. They checked with a sample of students to see how many would buy a T-shirt at various proposed prices. How would sales be related to price charged?

f. The population of the world has been increasing for as long as records have been available. What pattern of population growth has occurred over that time?

13 Graph V in Applications Task 12 is an example of a **piecewise-defined function**. The rule for the function is different for different intervals of the domain.

$$f(x) = \begin{cases} 5 & \text{for } 0 < x \le 10 \\ \underline{\quad} & \text{for } \underline{\qquad\qquad} \\ \vdots & \end{cases}$$

a. Assume tick marks on the x-axis represent 20 units and tick marks on the y-axis represent 10 units. Complete this rule for the function $f(x)$ whose graph is Graph V.

b. Write the absolute value function $v(x) = |x|$ as a piecewise linear function.

14 A function is called a **one-to-one function** if for any value r in the range there is exactly one x in the domain such that $f(x) = r$.

   a. Explain why every one-to-one function $f(x)$ has an inverse.

   b. Explain why if a function $f(x)$ has an inverse, then $f(x)$ is one-to-one.

CONNECTIONS

15 The diameter of a circle or sphere is double the radius. Use this fact to write formulas for these sphere measurements in terms of diameter d instead of radius r.

   a. Circumference

   b. Surface area

   c. Volume

16 If two rectangular prisms are similar with scale factor k, what can you say about the relationship of:

   a. the perimeters of the bases of the two rectangular prisms?

   b. the heights of the two prisms?

   c. the surface areas of the two prisms?

   d. the volumes of the two prisms?

   e. How would your answers to Parts a–d change if the prism was a right prism with a hexagonal base? Any right prism?

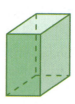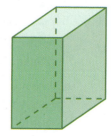

17 Suppose that a game involves tossing two fair coins until both turn up "heads." What is the probability that you will toss the coins:

   a. one time?

   b. two times?

   c. three times?

   d. n times?

   e. If $p(n)$ gives the probability that the first occurrence of "two heads" will be on toss n, to which family of functions does $p(n)$ belong?

18 Radar and sonar devices were first used in warfare to locate enemy airplanes and submarines. The same principles are now used in simple motion detectors that connect to a graphing calculator. If you aim the detector at someone walking toward or away from you, it will produce a sequence of (*time, distance*) data pairs and connect them into graphs like those shown below.

Working with a classmate, describe how you would walk toward or away from a radar device or calculator motion detector to produce graphs with shapes like the following (*time, distance*) graphs. If possible, test your ideas with a motion detector connected to a graphing calculator. Revise your descriptions if necessary.

a.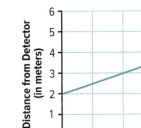

Graph I

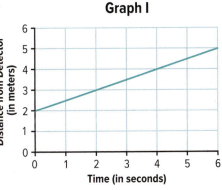

b.

Graph II

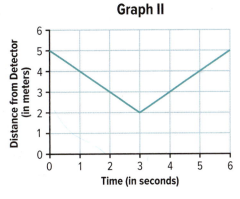

c.

Graph III

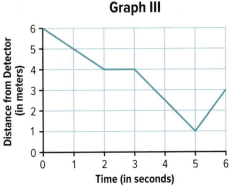

d.

Graph IV

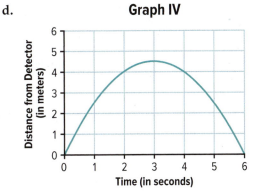

Jim Laser/CPMP

19 Graphs of periodic functions, with domain the set of real numbers, have translation symmetry. Reproduced below is the graph from Investigation 3, Problem 7 (page 119) of the cyclic pattern of depth of water in a tidal bay. Also shown in red is the midline of the graph.

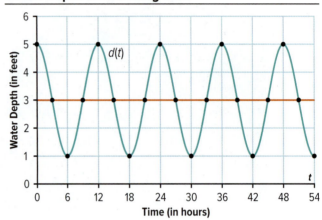

Ocean Depth at Retaining Wall

a. Assuming the pattern of the graph continues to the left and to the right, what is the smallest magnitude of a translation that maps the curve onto itself?

b. What other types of symmetry are present in this graph?

c. What is the amplitude of this function and how is it related to the midline of its graph?

d. What are the midlines of the graphs of $s(x) = 3 \sin x$ and $c(x) = 3 \sin x + 2$? What is the amplitude of each function?

20 Graph $y = \cos x$ and $y = \sin x$, $-6.28 \leq x \leq 6.28$, on the same coordinate axes.

a. The graphs of $y = \cos x$ and $y = \sin x$ are congruent. What geometric transformation will map the graph of $y = \cos x$ onto $y = \sin x$?

b. Write a coordinate rule $(x, y) \rightarrow (?, ?)$ for the geometric transformation in Part a.

c. Explain why the graphs of $y = \sin x$ and $y = 3 \sin x$ are not congruent. Are the graphs similar? Explain.

d. Are the graphs of $y = \cos x$ and $y = \cos x + 3$ congruent? Explain your reasoning.

e. Write a coordinate rule that will map the graph of $y = \cos x$ onto $y = \cos x + 3$.

21 Write a function rule that describes the pattern of change shown in each of the (*time, distance*) graphs in Connections Task 18.

22 You are familiar with *linear scales* such as on a ruler or those on a map. In linear scales, the *difference* between equally spaced points is a constant. The magnitude of earthquakes is reported using the Richter scale, which is a *logarithmic scale*. A Richter scale reading depends on the amount of displacement on a seismogram at a distance of 100 km from the epicenter of the quake. An algebraic rule that gives Richter scale rating R as a function of displacement D (in meters) on the seismogram is $R = 7 + \log D$.

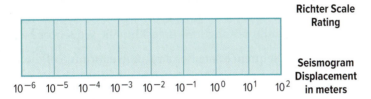

Richter Scale Rating

Seismogram Displacement in meters

10^{-6} 10^{-5} 10^{-4} 10^{-3} 10^{-2} 10^{-1} 10^{0} 10^{1} 10^{2}

a. How are equally spaced displacement values on the logarithmic scale related?

b. On a copy of the logarithmic scale above, compute the Richter scale ratings that correspond to the displacement values shown.

c. The 2011 Sendai earthquake in Japan had a Richter scale rating of 9.0. What would have been the displacement on a seismogram 100 km from the epicenter of the quake?

d. How would the amounts of displacement be related for the 1995 Kobe, Japan, and 1968 Tokachi-oki, Japan, earthquakes with Richter scale ratings of 7.2 and 7.9, respectively?

e. If a seismogram showed a displacement of 0.007 m during an earthquake that is 100 km away, what Richter scale rating would be reported?

Ofunato, Japan, after the magnitude 9.0 earthquake and subsequent tsunami in March, 2011

23 Use technology to make tables and graphs for the exponential function $f(x) = 10^x$ and the base 10 common logarithm function $g(x) = \log x$ on the same set of coordinate axes.

a. Based on an analysis of the graphs of the two functions, how are the graphs of $f(x) = 10^x$ and $g(x) = \log x$ related?

b. Based on an analysis of the two tables, how are the functions $f(x) = 10^x$ and $g(x) = \log x$ related?

c. How could your answer from Part b help you to draw a quick sketch of the base-2 logarithmic function $h(x) = \log_2 x$?

Matthew M. Bradley/US Navy

24 Any function $f(n)$ with domain the set of whole numbers is called a **sequence**.

  a. Any **arithmetic sequence** satisfies a recursive formula of the form:

$$f(n) = f(n-1) + d, \text{ where } f(0) = a.$$

   i. Give an example of the first 8 terms of an arithmetic sequence. What is $f(0)$? What is $f(6)$?

   ii. What general formula shows the value of $f(n)$ for any n directly, when d and $f(0)$ are given?

   iii. To which family of functions do arithmetic sequences belong?

  b. Any **geometric sequence** satisfies a recursive formula of the form:

$$f(n) = r \cdot f(n-1), \text{ where } f(0) = a.$$

   i. Give an example of the first 8 terms of a geometric sequence. What is $f(0)$? What is $f(4)$?

   ii. What general formula shows the value of $f(n)$ for any n directly, when r and $f(0)$ are given?

   iii. To which family of functions do geometric sequences belong?

REFLECTIONS

25 Select and view one of the following videos on global warming accessible on the Internet. Write a short summary report of the video and the mathematics involved.

  • climate.nasa.gov/ClimateReel/TemperaturePuzzle640360/

  • climate.nasa.gov/ClimateReel/KeepingCarbon640360/

  • climate.nasa.gov/ClimateReel/TakingEarthTemp640480/

  • climate.nasa.gov/ClimateReel/SupercomputingClimate640360/

26 What clues do you find most helpful in deciding on the family of functions likely to provide a good model for the relationship between variables in any particular situation?

27 News stories that involve consideration of change in some variable over time or the relation between two or more variables often use phrases like "growing exponentially," "periodic," or "directly or inversely related." What do you think people generally mean when they use each of those descriptive terms? How do the common usages relate to the technical mathematical usage?

28 What is the difference between a variable and a parameter in algebraic expressions like $a + bx$ or $a(b^x)$?

29 Some function families have the property that function values are increasing or decreasing for the entire domain. For other function families, function values increase on some intervals and decrease on others. Sort the function families in the chart on page 124 by these criteria.

30 Look back at the definition of a one-to-one function given in Applications Task 14.

    a. Which of the functions in Applications Task 12 (page 131) are one-to-one functions?

    b. Describe a test you could apply to the graph of a function to determine if it is a one-to-one function.

EXTENSIONS

31 Lighter-than-air transportation such as blimps has been in use for over 100 years. Blimps are now used primarily as air-borne advertising signs and as platforms for televising sporting events.

    Suppose that the shape of a blimp could be approximated as a cylinder with hemispherical caps on each end. A side-view profile consists of a rectangle with congruent semicircles on either end.

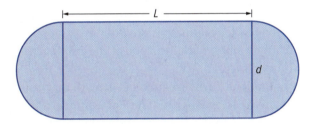

    a. What formula gives the total volume V of such a shape as a function of length L and diameter d as shown above?

    b. What formula gives the total surface area A of such a shape as a function of L and d?

32 The atmosphere of Earth is divided into several regions. For example, the troposphere region extends from Earth's surface to an altitude of about 7 miles. The stratosphere region extends from an altitude of 7 miles to about 30 miles. Recall that the radius of Earth is about 4,000 miles.

    a. What is the approximate volume of Earth plus its troposphere?

    b. What is the approximate volume of the troposphere alone?

    c. What is the approximate volume of the stratosphere alone?

d. The entire atmosphere of Earth extends to an altitude of about 400 miles above Earth's surface. If a satellite in circular Earth orbit were to be at the top of the atmosphere, what would be the length of its orbit?

e. How much larger is the area of the surface formed by the outer boundary of Earth's atmosphere than the surface of Earth at sea level?

33 Piecewise-defined functions (see Applications Task 13) have a number of important applications. For example, income tax owed may be written as a piecewise linear function of income. The 2012 income tax for a single tax filer is determined using the following table.

Taxable Income	Income Tax
$0–$8,700	10% of income
$8,700–$35,350	15% of income over $8,700, plus $870.00
$35,350–$85,650	25% of income over $35,350, plus $4,867.50

a. Last year, James earned $8,700 of taxable income. How much should he pay in income tax?

b. What would his income tax be if he earned $20,000 of taxable income?

c. Write a piecewise linear function $T(x)$ for income tax as a function of taxable income.

d. Sketch a graph of your tax function. What features does your graph have?

34 Exponential models for population growth have graphs that show continually rising population values, with the slope of the graph getting steeper and steeper as time passes.

In many situations, there are limits to growth that cause the rate of population increase to eventually slow. For example, if a colony of ants starts growing in a space with limited food and water, the number of ants at any time w weeks later might be modeled by a function like this:

$$P(w) = \frac{10,000}{1 + 49(0.85^w)}$$

a. What number of ants does the model give for the start of the experiment?

b. What is the shape of the graph of $P(w)$ and what does that shape say about the pattern of increase in the ant population?

c. What seems to be the upper limit of the ant population and how can that limit be found without use of any calculations or graphs, simply analyzing the rule algebraically?

©Digital Vision/Getty Images

35 Velocity measures rate of change—*change in position to change in time.* Students in one class at Plano High School all used the (*time, distance*) data reported below for the walk in Connections Task 18 Part d. Different groups of students came up with different velocity estimates.

Time of Walk (in seconds)	0.0	0.5	1.0	1.5	2.0	2.5	3.0	3.5	4.0	4.5	5.0	5.5	6.0
Distance from Detector (in meters)	0.0	1.4	2.5	3.4	4.0	4.4	4.5	4.4	4.0	3.4	2.5	1.4	0.0

a. Which of these methods of estimating velocity at a time 1.5 seconds into the walk makes most sense to you? Explain your reasoning.

- $\frac{3.4 - 2.5}{1.5 - 1.0} = 1.8$ m/sec (meters per second)

- $\frac{4.0 - 3.4}{2.0 - 1.5} = 1.2$ m/sec

- $\frac{4.0 - 2.5}{2.0 - 1.0} = 1.5$ m/sec

- $\frac{3.4 - 1.4}{1.5 - 0.5} = 2$ m/sec

- $\frac{3.4}{1.5} = 2.2\overline{6}$ m/sec

b. Explain why m/sec is the appropriate unit of measure for this situation.

c. Estimate the walker's velocity at the point 4.5 seconds into the trip.

d. When two groups estimated the walker's velocity at a point 4.5 seconds into the trip, one came up with 1.5 m/sec and the other came up with −1.5 m/sec. How do you suppose each group came up with those estimates? What does a negative velocity mean in this context?

e. What kind of information would help you make more accurate estimates for the walker's velocity at different times?

36 It is increasingly common that consumers are purchasing music recordings through downloads from online music providers. Look back at the production and sale of CDs chart on page 112. Do some Internet research to help you estimate possible modifications to the chart in the case of a first album by an unknown artist or band that is available only as a download from a popular music site. Based on your revised chart, solve Problem 1 (page 113) under the new conditions for production and cost of an album download.

REVIEW

37 Recall that a function is a relationship between two variables where each value of the independent variable corresponds to exactly one value of the dependent variable. Examine the following graphs on an xy-coordinate system.

I

II

III

IV

V

VI

a. Which graphs show relationships between variables in which y is a function of x?

b. For those that are functions, which have inverses?

38 When considering the properties of a function, it is helpful to identify the *domain* and *range* of the function. Recall, the domain of a function is the set of all possible input values for the function. The range of a function is the set of all output values of the function.

a. For each function below, identify the domain and range of the function.

   i. $f(x) = 3x + 4$

   ii. $f(x) = \sqrt{x - 3}$

   iii. $f(x) = x^2 + 4$

b. Use what you know about the graphs of functions to determine a function rule for functions with each of the following characteristics.

i. The domain is all real numbers and the range is all real numbers greater than or equal to 0.

ii. The domain is all real numbers and the range is all real numbers less than or equal to 0.

iii. The domain is all real numbers except 0 and the range is all real numbers except 0.

39 The graph of $f(x)$ below shows Mikayla's height in relation to the ground while she took a ride on a roller coaster.

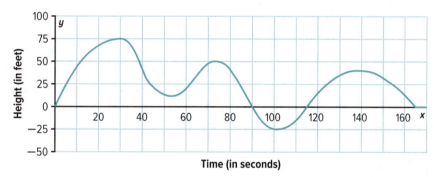

a. What is the value of $f(20)$? What does it tell you about Mikayla's ride?

b. For what x values is $f(x) = 50$? What do these values tell you about Mikayla's ride?

c. What are the zeroes of $f(x)$?

d. What is the minimum value of $f(x)$ and at what point(s) during the ride does it occur?

e. What is the maximum value of $f(x)$ and at what point(s) during the ride does it occur?

40 Use your knowledge of degree and radian angle measure and of coordinates of points on the unit circle to complete a copy of the table below.

θ in degrees		240°	135°	
θ in radians	$\frac{\pi}{6}$			π
cos θ				
sin θ				

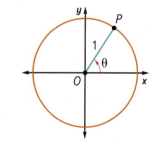

41 Consider the function $y = \log x$.

    a. Without using technology, decide if each statement is true or false. If false, explain why.

        i. $\log 1{,}000 = 3$

        ii. $\log (-1{,}000) = -3$

        iii. $13 < \log 1{,}489 < 14$

        iv. $-2 < \log 0.78 < -1$

    b. Without using a calculator, create a table of values for the function $y = \log x$. Your table should contain at least five pairs of (x, y) values. Be prepared to explain how you decided on the x values to use in your table.

    c. Sketch a graph of the function $y = \log x$.

    d. Does the graph of $y = \log x$ have a vertical asymptote? If so, what is it? If not, explain why not.

    e. Does the graph of $y = \log x$ have a horizontal asymptote? If so, what is it? If not, explain why not.

42 Solve each of the following equations.

    a. $\dfrac{2x}{3} + \dfrac{x + 3}{3} = 11$

    b. $\dfrac{3}{x} + \dfrac{6}{x} = 15$

    c. $\dfrac{x}{2} + \dfrac{x}{8} = 10$

    d. $\dfrac{4}{x} + \dfrac{5}{4x} = 8$

43 Sometimes converting from one unit of measure to another can help you to better understand rates.

    a. Suppose that Teri is driving a car that is traveling at 65 miles per hour. She looks away from the road to read a text message. If reading the text takes her five seconds, how many feet has she traveled while reading the text?

    b. In the United States, an average of approximately 2.5 million plastic bottles are thrown away every hour. How many bottles are thrown away each second? How many bottles are thrown away each year?
    (**Source:** www.smartplanet.com/blog/smart-takes/infographic-americans-throw-away-25-million-plastic-bottles-every-hour/9309/)

    c. A one-gallon can of paint will cover approximately 350 square feet. Suppose that you need to paint a wall that has an area of 100 square yards. How many gallons of paint will you need?

44 Consider the graph of each function and identify any rotational or reflection symmetries of the graph.

a. $y = (x - 3)^2 + 2$

b. $y = x^5$

c. $y = \dfrac{1}{(x - 3)^2}$

45 For each pair of figures below, describe the transformation that will map Figure I onto Figure II and then write a coordinate rule $(x, y) \rightarrow (?, ?)$ for that transformation.

a.

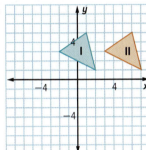

b.

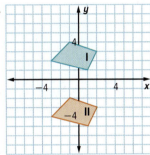

c.

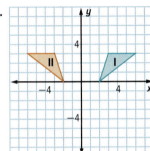

d.

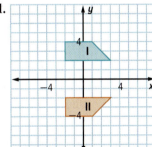

46 Evaluate each absolute value expression.

a. $|-3| + 2$

b. $|-3 + 2|$

c. $-5|-7|$

d. $-|8|$

e. $24 - |35|$

Customizing Models by Translation and Reflection

Sometimes you may want to chill a bottle of water or juice quickly. Often you can put it in the freezer compartment of a refrigerator or in an ice-filled cooler. You might expect the drink bottle to cool steadily from outdoor or room temperature to the temperature of the freezer or the cooler. But surprisingly it turns out that the temperature drops in a pattern like that shown below.

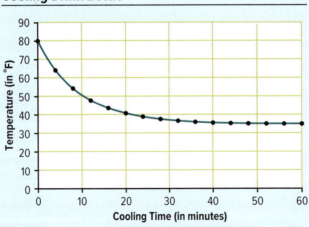

Cooling Drink Bottle

(Graph: Temperature (in °F) on vertical axis from 0 to 90; Cooling Time (in minutes) on horizontal axis from 0 to 60. The curve starts at 80°F and decreases, leveling off near 35°F.)

THINK ABOUT THIS SITUATION

Study the graph on the previous page and use the information it reveals to answer these questions.

a. What does the graph tell about the cooling temperature of the drink bottle?

b. What familiar functions have graphs like that of the cooling drink bottle pattern?

c. How is the cooling graph different from the graphs of functions it most resembles?

d. How could you modify the rule of a familiar function to fit this different, but related, pattern?

In this lesson, you will begin study of strategies for using familiar functions and transformations of their graphs to develop models for patterns that are variations of the basic linear, quadratic, exponential, circular, and absolute value functions. Trial-and-error testing of options and statistical regression methods are often effective strategies for finding function models of data patterns. But it is also helpful to know some general principles for modifying and combining the rules of basic function families to build new models for more complex situations.

INVESTIGATION 1

Vertical Translation

The graph that shows cooling of a drink bottle looks a lot like that of an exponential decay function whose graph has been shifted up about 35 units.

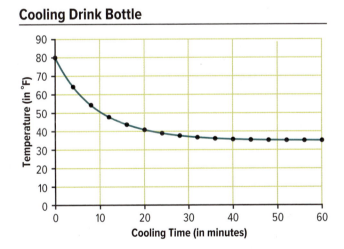

As you work on the problems of this investigation, look for answers to these questions:

What are the connections between rules for functions whose graphs are related by vertical translation?

How can functions with graphs obtained by vertical translation expand the supply of models for important patterns of variation?

To get ideas about how to find models for patterns that are vertical translations of familiar graphs, it helps to look at some simple cases.

1 Shown below is the graph of an important special function called the *absolute value function*.

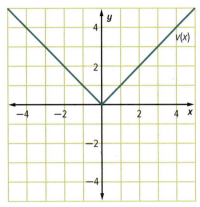

The rule for the **absolute value function** $v(x)$ is expressed symbolically as $v(x) = |x|$. Based on the pattern of values in the graph, describe in words the rule for calculating $|x|$ for any value of x.

2 The next diagram shows the graph of $v(x)$ along with two variations.

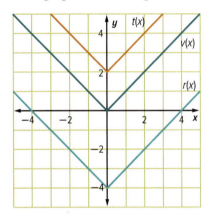

a. Complete a copy of the following table of values for $v(x)$, $r(x)$, and $t(x)$ for the sample values of the independent variable x.

x	−4	−3	−2	−1	0	1	2	3	4
r(x)		−1							
v(x)	4	3	2	1	0	1	2	3	4
t(x)		5							

b. How are the values of $r(x)$ and $t(x)$ related to those of $v(x)$?

c. What rules seem to match the patterns in the graphs of $r(x)$ and $t(x)$?

d. What coordinate rules for geometric transformations $(x, y) \rightarrow (?, ?)$ map the graph of $v(x)$ onto:

    i. the graph of $r(x)$?   **ii.** the graph of $t(x)$?

3 In each part below, modify the given rule for $f(x)$ to produce related functions $g(x)$ and $h(x)$ with the described features.

- The graph of $g(x)$ is congruent to, but translated 5 units upward from, the graph of $f(x)$.

- The graph of $h(x)$ is congruent to, but translated 3 units downward from, the graph of $f(x)$.

Then check your ideas using technology. Be prepared to explain why your rules for $g(x)$ and $h(x)$ do what is requested in each case.

a. $f(x) = x^2 - 4x$

b. $f(x) = 2(1.5^x)$

c. $f(x) = \dfrac{3}{x}$

Properties of Vertical Translations Vertical translation of the graph of a function affects the local maximum and minimum points, x-intercepts, and y-intercept of the graph of the new function in predictable ways.

4 In this problem, you will compare properties of functions whose graphs are related by vertical translation. Consider this roller coaster design graph modeled by a function $f(x)$.

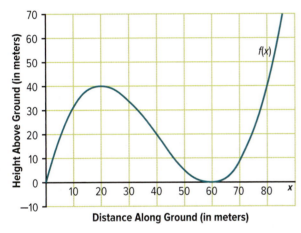

a. On separate grids, sketch graphs of $g(x) = f(x) + 15$ and $h(x) = f(x) - 5$. Interpret your graphs in terms of features of the sections of the proposed roller coaster.

b. Locate the local maximum and minimum points of the graph of each new function and label those points with their approximate coordinates.

c. Locate the x-intercepts of the graph of each new function and label corresponding points with their approximate coordinates.

d. Locate the y-intercept of the graph of each new function and label the corresponding point with its approximate coordinates.

e. Combine results of your work on Parts a–d and your experience in work on Problems 1 and 2 to formulate conjectures that complete these statements. Consider both $k > 0$ and $k < 0$.

    i. When $g(x) = f(x) + k$, the local maximum and minimum points of $g(x)$ … .

    ii. When $g(x) = f(x) + k$, the zeroes of $g(x)$ … .

    iii. When $g(x) = f(x) + k$, the y-intercept of $g(x)$ … .

Be prepared to explain, as precisely as you can, *why* your conjectures are reasonable generalizations of the particular results in Parts a–d. That is, why is it reasonable to believe that each property will hold for *any* functions $f(x)$ and $g(x)$ whose graphs are related by vertical translation?

5 Consider again the cooling drink bottle pattern shown in the graph below.

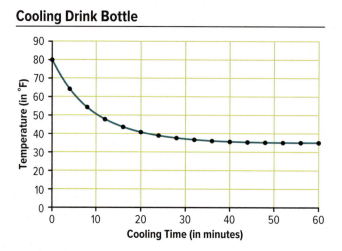

Cooling Drink Bottle

a. The following table gives a sample of (*cooling time, temperature*) values that show how temperature of the drink bottle decreases with the passage of time.

Cooling Time t (in minutes)	0	10	20	30	40	50	60
Temperature $b(t)$ (in °F)	80	51	41	37	35.5	35.2	35.1

Use a curve-fitting tool to find an exponential model for the data pattern. Why is there a relatively poor match between the data and the regression model?

b. Devon and Carissa were still convinced that an exponential function would provide a useful model for the cooling pattern. It looked to them as if such a model would fit if the graph could be translated downward by about 35°F. What evidence in the table and graph supports that reasoning?

Jim Laser

c. Use a curve-fitting tool to find the "best-fitting" exponential model for the adjusted (*cooling time, temperature* − *35°*) data pattern.

d. Use the result of Part b and what you have observed in your work on Problems 1–3 to develop and test a model for the original (*cooling time, temperature*) data pattern. Then explain why your new model is a better match for the shape of the cooling graph than the regression equation found in Part a.

SUMMARIZE THE MATHEMATICS

In this investigation, you explored connections between rules of functions whose graphs are related by vertical translations.

a. If each point on the graph of $h(x)$ is k units above a corresponding point on the graph of $g(x)$, how will the rules for $h(x)$ and $g(x)$ be related? If "k units above" is replaced with "k units below," how does that affect your answer?

b. If $g(x) = f(x) + k$ for all x and $k > 0$, how are the locations of the maximum and minimum points, the x-intercepts, and the y-intercepts of the graphs of the functions $f(x)$ and $g(x)$ related? What if $k < 0$?

c. How can you find models for data patterns that appear to be vertical translations of graphs for familiar functions?

Be prepared to share your ideas and reasoning with the class.

✓CHECK YOUR UNDERSTANDING

Home security systems are a big business in many urban areas. One such service offers a package that costs $99 for installation and $24.50 per month for monitoring.

a. What rule gives the total cost of using that service for x months?

b. How would that rule change if the installation charge was reduced to $49.95? How would the graph of this new cost function be related to the graph of the rule for the original offer?

c. How would the cost function rule change if the monthly fee was reduced to $19.95 (but the installation charge remained $99)? How would the graph of that new cost function be related to the graphs of the rules for the original offer and the modification in Part b?

Reflection Across the x-Axis

In many situations, you will detect patterns of change whose graphs look like graphs of familiar basic functions that have been reflected across an axis. That observation often makes it easy to find symbolic rules for the new relations. Consider the situation of warming a drink bottle. When you take a chilled drink bottle out of a refrigerator or cooler, it immediately begins warming toward room temperature. Once again you might expect that warming to occur at a steady rate, but the temperature will actually increase as shown in the following graph.

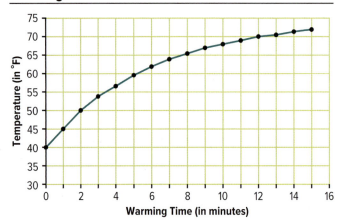

Warming Drink Bottle

Because it is reasonable to assume that warming and cooling should occur in similar patterns, it makes sense to believe that some variation of an exponential decay function would provide a model for the data pattern shown in the graph.

As you work on the problems of this investigation, look for answers to this question:

How can functions with graphs obtained by reflection across the x-axis expand the supply of models for important patterns of variation?

1 Discuss with classmates how you could combine reflection and translation of an exponential decay function graph so that it would fit the pattern of (*warming time, temperature*) data for the warming drink bottle.

2 Once again, it helps to study some relatively simple cases to develop ideas that can be applied in more complex situations. On a copy of the graphs of the functions on the following page:

a. sketch a graph of its reflection image across the *x*-axis.

b. give the rule for the function whose graph is the reflection image of the given function.

Graph I

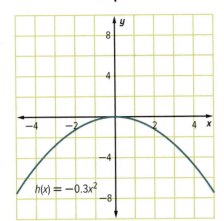

$h(x) = -0.3x^2$

Graph II

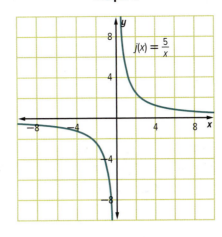

$j(x) = \dfrac{5}{x}$

Graph III

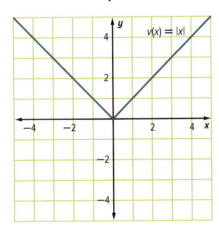

$v(x) = |x|$

Graph IV

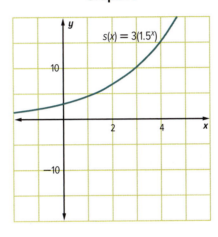

$s(x) = 3(1.5^x)$

Graph V

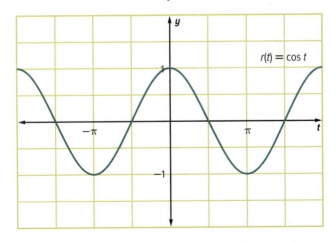

$r(t) = \cos t$

c. What coordinate rule $(x, y) \rightarrow (?, ?)$ maps each graph in Part b onto its reflection image across the x-axis?

3 Use results of your work in Investigation 1 to sketch graphs for the following variations of the functions in Problem 2. Then check your ideas using technology.

    a. $p(x) = -0.3x^2 + 5$ **b.** $w(x) = -\dfrac{5}{x} + 2$

    c. $q(x) = -|x| - 3$ **d.** $u(x) = -3(1.5^x) + 12$

    e. $k(t) = \cos t - 1$

Properties of Reflection Across the *x*-axis Reflection of the graph of a function across the *x*-axis affects the local maximum and minimum points, *x*-intercepts, and *y*-intercept of the graph of the new function in predictable ways.

4 Suppose you are told that functions $f(x)$ and $g(x)$ have graphs that are reflection images of each other across the *x*-axis.

    a. What relationship would you expect between the local maximum and minimum points of the two functions?

    b. What relationship would you expect between the *x*-intercepts of the two functions?

    c. What relationship would you expect between the *y*-intercepts of the two functions?

    d. Illustrate your answers to Parts a, b, and c by using technology to graph $f(x) = x^4 - 3x^3 - x^2 + 3x$. On the same coordinate axes, also graph the function $g(x)$ whose graph is the image of $f(x)$ reflected across the *x*-axis.

    e. What is the rule for $g(x)$? Compare your function rule with those of others. Resolve any differences.

    Be prepared to explain why it is reasonable to believe that each of your conjectures in Parts a–c will hold for *any* functions $f(x)$ and $g(x)$ whose graphs are related by reflection across the *x*-axis.

5 Now consider again the function that models the pattern of change in temperature of a drink bottle after it has been taken from an icy cooler.

Warming Drink Bottle

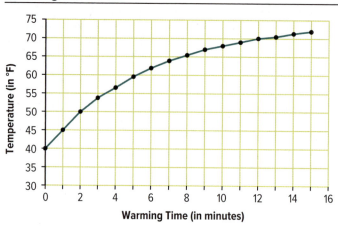

a. The following table shows a representative sample of data points.

Warming Time t (in minutes)	0	3	6	9	12	15
Temperature $d(t)$ (in °F)	40	54	62	67	70	72

The temperature of the drink seems to be headed toward a room temperature of about 75°F. Use technology to plot the adjusted (*warming time, 75° − temperature*) data points.

b. Use an exponential curve-fitting tool to find the rule for a function that models the (*warming time, 75° − temperature*) pattern.

c. Now identify a combination of reflections and vertical translations that will transform the graph of your exponential function in Part b to a shape and location that matches the original drink bottle warming graph. Use those ideas to write a rule for the function $d(t)$ that gives the temperature of the drink t minutes after it has been taken from the cooler.

SUMMARIZE THE MATHEMATICS

In this investigation, you explored the connections between rules of functions whose graphs are reflections of each other across the *x*-axis. You then used those ideas to develop models for variations on familiar data patterns.

a. Suppose $f(x)$ and $g(x)$ are functions whose graphs are reflection images of each other across the *x*-axis. How will the rules of those functions be related?

b. How are the locations of maximum and minimum points, *x*-intercepts, and *y*-intercepts of two functions related if their graphs are reflection images of each other across the *x*-axis?

c. Reflections and vertical translations can be combined to produce models of data patterns that are similar to the basic function families that you know. What clues would you look for to see when such transformations might be useful?

Be prepared to share your ideas and reasoning with the class.

✔CHECK YOUR UNDERSTANDING

For each of the functions below, without use of technology:

- sketch a graph of the function over the specified domain interval.
- sketch the function whose graph is the reflection image across the *x*-axis of the given function. Write its rule.

a. $p(x) = -|x|$ graphed on the interval $[-3, 3]$

b. $h(x) = x^2 - 1$ graphed on the interval $[-3, 3]$

c. $d(x) = -1.5x + 2$ graphed on the interval $[-4, 4]$

d. $j(x) = \frac{4}{x^2}$ graphed on the interval $[-5, 5]$

Horizontal Translation

Slingshots have been used as hunting weapons for thousands of years, but a new version is now available as a toy for propelling water balloons.

Suppose that the following graph shows the trajectories of a series of water balloon shots.

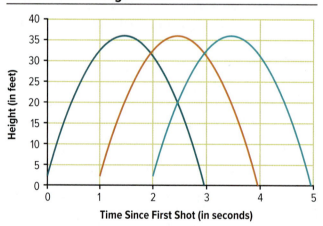

Water Balloon Heights Over Time

The three graphs look identical in shape but translated horizontally from each other.

As you work on the problems of this investigation, look for an answer to this question:

> *How are the rules of functions f(x) and g(x) related if their*
> *graphs are related by horizontal translation?*

Once again, it helps to analyze some simple examples of functions with graphs related by horizontal translation to get some ideas about the patterns relating rules of any such pairs of functions.

1 Consider the absolute value function $f(x) = |x|$ and two variations $g(x) = |x + 3|$ and $h(x) = |x - 3|$.

a. Graphs of $f(x)$, $g(x)$, and $h(x)$ are shown below. Before completing Parts b–d, make a conjecture about the match of function rules and graphs.

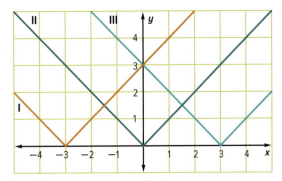

b. Complete a copy of this table that compares values for the three functions.

x	−4	−3	−2	−1	0	1	2	3	4
$f(x) = \lvert x \rvert$									
$g(x) = \lvert x + 3 \rvert$									
$h(x) = \lvert x - 3 \rvert$									

c. Match the functions $f(x)$, $g(x)$, and $h(x)$ to the graphs in the diagram in Part a. Then give a rule in the form $(x, y) \rightarrow (?, ?)$ for a geometric transformation that maps the graph of $f(x)$ onto the graph of $g(x)$. That maps the graph of $f(x)$ onto the graph of $h(x)$.

d. If k is some positive number, how do you think the graphs of the following functions will be related to the graph of $y = |x|$? Test your ideas using technology. Use your answers to Parts b and c to explain why things happen as they do.

   i. $y = |x + k|$

   ii. $y = |x - k|$

2 Now consider variations on the basic quadratic function $q(x) = x^2$. Match these functions with their graphs in the diagram at the right. Then test your ideas to see if you can explain why things happen as they do.

a. $f(x) = x^2$

b. $g(x) = (x - 4)^2$

c. $h(x) = (x + 4)^2$

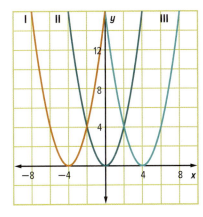

3 Now look back at the graphs of water balloon shots at the beginning of this investigation (page 154). Suppose the height of the first shot is a function of time in flight with rule $h(t) = -16t^2 + 48t$. The second water balloon was shot 1 second after the first and the third balloon was shot 2 seconds after the first.

   a. The function $j(t)$ gives height over time of the second water balloon shot. What is the rule for $j(t)$?

   b. The function $k(t)$ gives height over time of the third water balloon shot. What is the rule for $k(t)$?

4 Next consider further variations on the basic quadratic function $q(x) = x^2$. Match these functions with their graphs in the diagram at the right. Then test your ideas to see if you can explain why things happen as they do.

   a. $f(x) = -x^2 + 4$

   b. $g(x) = -(x - 4)^2 + 4$

   c. $h(x) = -(x + 4)^2 + 4$

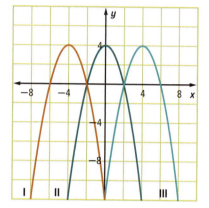

5 If h and k are positive numbers, how do you think the graphs of the following functions will be related to the graph of $y = x^2$? Test your ideas using a computer algebra system that allows you to define functions like $f(x) = -(x + h)^2 + k$ and dynamically adjust values of h and k.

   a. $y = -(x + h)^2 + k$ **b.** $y = -(x - h)^2 + k$

   c. $y = -(x - h)^2 - k$ **d.** $y = -(x + h)^2 - k$

6 The next diagram shows graphs of four other variations on the basic quadratic function $q(x) = x^2$. Use ideas that you have developed from your work on Problems 2–5 to write symbolic rules for each new function graphed below. Be prepared to explain how you could develop each rule using reasoning alone, without a CAS or a curve-fitting program.

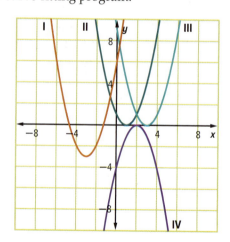

7 From your many encounters with quadratic functions, you know that the graph of $y = ax^2 + bx + c$ with $a > 0$ has a minimum point when $x = -\frac{b}{2a}$.

    a. Why must the graph of $y = x^2$ have its minimum point at $(0, 0)$?

    b. If h is a positive number, what is the minimum point on the graph of $y = (x - h)^2$? How does what you learned in this investigation justify your answer?

    c. Expand the expression $(x - h)^2$ to standard polynomial form and explain how it and the expression $-\frac{b}{2a}$ confirm your answer to Part b.

    d. If h and k are positive numbers, what is the minimum point on the graph of $y = (x - h)^2 + k$? How do results from your work in Investigation 1 and this investigation justify your answer?

    e. Expand the expression $(x - h)^2 + k$ to standard polynomial form. Then explain how it and the $-\frac{b}{2a}$ rule confirm your answer to Part d.

8 Assume h and k are positive numbers. Based on your work in Problems 1–6, how do you think the graphs of the following functions will be related to that of $f(x)$?

    a. $p(x) = f(x + h) + k$ **b.** $q(x) = f(x - h) + k$

    c. $r(x) = f(x - h) - k$ **d.** $s(x) = f(x + h) - k$

9 Explain, as precisely as you can, why the graph-rule translation patterns you observed in examples involving $y = |x|$ and $y = x^2$ seem likely to work with other functions as well.

10 Test your ideas about the graphical relationships of $f(x)$ and $f(x \pm h) \pm k$ on these pairs of functions. In each case:

    • make a sketch of what you think the two graphs will look like.

    • check your prediction using technology.

    a. $f(x) = x^3$ and $g(x) = f(x + 3) + 4$ graphed on $[-6, 6]$

    b. $f(x) = x^3$ and $g(x) = f(x - 3) - 5$ graphed on $[-6, 6]$

Properties of Horizontal Translations Horizontal translation of the graph of a function affects the local maximum and minimum points, x-intercepts, and y-intercept of the graph of the new function in predictable ways.

11 Suppose you are told that functions $f(x)$ and $g(x)$ have graphs that are images of each other under a horizontal translation $(x, y) \rightarrow (x \pm h, y)$ with $h > 0$.

    a. Explain the relationship between the local maximum and minimum points of the two functions.

    b. Explain the relationship between the x-intercepts of the two functions.

c. Explain the relationship between the *y*-intercepts of the two functions.

d. Illustrate your answers to Parts a–c by sketching graphs of the functions:

- $f(x) = x^3 - x^2 - 4x + 4$
- $g(x) = (x - 1)^3 - (x - 1)^2 - 4(x - 1) + 4$

Label key points on each graph with their approximate coordinates.

SUMMARIZE THE MATHEMATICS

In this investigation, you discovered connections between the rules of functions when their graphs are related by horizontal translations. Assume *h* is a positive number.

a. Suppose the graph of a function *g(x)* is the image of the graph of *f(x)* after translation of *h* units to the right. How can the rule for *g(x)* be derived from the rule for *f(x)*?

b. Suppose the graph of a function *j(x)* is the image of the graph of *f(x)* after translation of *h* units to the left. How can the rule for *j(x)* be derived from the rule for *f(x)*?

c. Suppose the graph of a function *g(x)* is the image of the graph of *f(x)* under translation of *h* units to the right. How are locations of maximum and minimum points and *x*-intercepts of the two functions *f(x)* and *g(x)* related? What if the translation is *h* units to the left?

Be prepared to share your ideas and reasoning with the class.

✓ CHECK YOUR UNDERSTANDING

For each of these functions, without using technology:

- sketch a graph of the function and the specified translation of that graph.
- give the rule for the function whose graph is the translation image of the original function.

a. $p(x) = -|x|$ graphed on the interval $[-8, 8]$ and then translated 4 units to the right

b. $h(x) = x^2$ graphed on the interval $[-6, 6]$ and then translated 3 units to the left

c. $d(x) = -1.5x$ graphed on the interval $[-6, 6]$ and then translated 2 units to the right

1 For each function below, find a rule for the function $g(x)$ whose graph is the image of the graph of $f(x)$ under the indicated transformation.

   a. $f(x) = x^2 + 4x - 5$; transformation $(x, y) \rightarrow (x, y + 3)$

   b. $f(x) = 4x - 3$; transformation $(x, y) \rightarrow (x, y - 4)$

   c. $f(x) = 30(2.5^x)$; transformation $(x, y) \rightarrow (x, y - 7)$

2 For each function below, find the transformation $(x, y) \rightarrow (?, ?)$ that maps the graph of $f(x)$ onto the graph of $g(x)$.

   a. $f(x) = -x^2 + 4x - 5$ and $g(x) = -x^2 + 4x$

   b. $f(x) = |x| - 5$ and $g(x) = |x| + 4$

   c. $f(x) = \dfrac{1}{x} + 2$ and $g(x) = \dfrac{1}{x}$

   d. $f(x) = 7(0.5^x)$ and $g(x) = 3 + 7(0.5^x)$

3 The next diagram shows the graph of $j(t) = \sin t$ and two related functions $k(t)$ and $m(t)$.

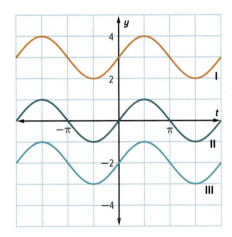

   a. Which is the graph of $j(t)$ and how do you know?

   b. Label the remaining two graphs $k(t)$ and $m(t)$. What rules for $k(t)$ and $m(t)$ would produce the other two graphs?

   c. What coordinate rules $(t, y) \rightarrow (?, ?)$ for geometric transformations would map the graph of $j(t)$ onto:

      i. the graph of $k(t)$?

      ii. the graph of $m(t)$?

4 Consider functions $f(x)$ and $g(x)$ for which $g(x) = f(x) - 3$ for all values of x.

a. How are the graphs of $f(x)$ and $g(x)$ related geometrically?

b. Suppose $f(x)$ has a local minimum at $(-4, -2)$ and a local maximum at $(3, 7)$. What (if anything) can be said about local minimum and maximum points for $g(x)$?

c. Suppose $g(x)$ has y-intercept $(0, -1)$. What is the y-intercept of $f(x)$?

d. Suppose $f(x)$ has zeroes at $x = 7$ and $x = -1$. What (if anything) can be said about the zeroes of $g(x)$?

5 River Crossing Amusement Park predicts that daily net income (in dollars) from tickets, food, and gift shop sales will be related to the price x (in dollars) charged for admission by

$$I(x) = -100x^2 + 3{,}000x + 100{,}000.$$

a. Sketch a graph of $I(x)$ for $0 \leq x \leq 50$. Label the maximum income point, the y-intercept, and any zeroes with their coordinates.

b. Assume that the park has fixed daily expenses of $45,000. How does that information and the income function $I(x)$ combine to give a function $P(x)$ that tells the daily profit of the park as a function of admission price?

c. Add the graph of $P(x)$ to your sketch in Part a. Label the maximum profit point, the y-intercept, and any zeroes with their coordinates. Then explain how (if at all) those key points on the profit function graph are related to corresponding points on the net income graph.

6 Restaurants cook pizza in very hot ovens. As soon as the pizza comes out of the oven, it begins cooling toward room temperature. Data in the following table illustrate the kind of cooling pattern that might be expected.

Cooling Time (in minutes)	0	2	4	6	8
Temperature (in °F)	575	325	200	140	110

a. Suppose that the temperature is 80°F in the room where the pizza is cooling. What transformation of the given temperature data seems likely to yield a pattern that could be modeled by an exponential decay function?

b. What exponential function matches the pattern in the transformed data?

c. What function is a good model for the pattern in change over time of the original pizza temperature data?

d. What temperature does your model from Part c predict for a time 15 minutes after the pizza comes out of the oven?

e. At approximately what time will the pizza reach a temperature of 212°F?

7 In each part, find a rule for the function $g(x)$ whose graph is the image of the graph of $f(x)$ under the indicated transformation.

   a. $f(x) = x^2 + 4x - 5$; transformation $(x, y) \rightarrow (x, -y)$

   b. $f(x) = 4x - 3$; transformation $(x, y) \rightarrow (x, -y)$

   c. $f(x) = 30(2.5^x)$; transformation $(x, y) \rightarrow (x, 7 - y)$

   d. $f(x) = -x^2 + 4x + 5$; transformation $(x, y) \rightarrow (x, -y - 4)$

8 In each part, find the transformation $(x, y) \rightarrow (?, ?)$ that maps the graph of $f(x)$ onto the graph of $g(x)$.

   a. $f(x) = -x^2 + 4x$ and $g(x) = x^2 - 4x$

   b. $f(x) = |x| - 5$ and $g(x) = -|x| + 5$

   c. $f(x) = \dfrac{1}{x} + 2$ and $g(x) = \dfrac{1}{x}$

   d. $f(x) = 7(0.5^x)$ and $g(x) = 10 - 7(0.5^x)$

9 Consider functions $f(x)$ and $g(x)$ for which $g(x) = -f(x)$ for all values of x.

   a. How are the graphs of $f(x)$ and $g(x)$ related geometrically?

   b. Suppose $f(x)$ has a local minimum at $(-4, -2)$ and a local maximum at $(3, 1)$. What (if anything) can be said about local minimum and maximum points for $g(x)$?

   c. Suppose $f(x)$ has y-intercept $(0, -4)$. What is the y-intercept of $g(x)$?

   d. Suppose $f(x)$ has zeroes at $x = 7$ and $x = -1$. What (if anything) can be said about the zeroes of $g(x)$?

10 A Charleston Street snack vendor takes an ice cream bar out of the freezer (which keeps frozen snacks at 0°F) for a customer. When the customer changes her mind, he forgets to put the bar back into the freezer. If the air temperature is about 85°F, the temperature of the melting ice cream bar will increase in a pattern like that shown in this table.

Warming Time (in minutes)	0	5	10	15	20	25	30	35
Temperature (in °F)	0	35	55	67	75	79	81	83

   a. Use a regression routine in your calculator or computer software to find a good-fitting model for the relationship between time and the *difference* between the air temperature and the ice cream bar temperature.

b. Use the model from Part a and what you know about translation and reflection of graphs to devise a rule for the function that gives actual temperature of the ice cream bar at any time after it has been taken from the freezer.

c. Compare the shape of the graph for your function model in Part b and a plot of the given (*warming time, temperature*) data. Explain why any observed differences between the two patterns occur.

11 In each part, find a rule for the function $g(x)$ whose graph is the image of the graph of $f(x)$ under the indicated transformation.

a. $f(x) = 4x - 3$; transformation $(x, y) \rightarrow (x - 5, y)$

b. $f(x) = x^2 + 4x$; transformation $(x, y) \rightarrow (x + 3, y)$

c. $f(x) = |x|$; transformation $(x, y) \rightarrow (x - 2, 4 - y)$

12 Consider functions $f(x)$ and $g(x)$ for which $g(x) = f(x - 5)$ for all values of x.

a. How are the graphs of $f(x)$ and $g(x)$ related geometrically?

b. Suppose $f(x)$ has a local minimum at $(-4, -2)$ and a local maximum at $(3, 1)$. What can be said about local minimum and maximum points for $g(x)$?

c. Suppose $f(x)$ has y-intercept $(0, -4)$. What is the y-intercept of $g(x)$?

d. Suppose $f(x)$ has zeroes at $x = 7$ and $x = -1$. What (if anything) can be said about the zeroes of $g(x)$?

13 To win a prize for the most creative Punkin' Chunker, one group designed a rapid-fire machine that could fire 5 pumpkins at 1-second time intervals. The trajectory of each pumpkin is the same, with height in feet at any time t seconds after firing given by the quadratic function

$$h(t) = -16t^2 + 144t + 20.$$

a. Sketch a graph of $h(t)$ and label with their coordinates the y-ntercept, the maximum height point, and the point where the pumpkin hits the ground.

b. What rule for function $h_2(t)$ will describe the height of the *second* pumpkin at any time t seconds after the first pumpkin is shot? Add a sketch of the graph of that function to the diagram of Part a.

c. What rule for function $h_5(t)$ will describe the height of the *fifth* pumpkin at any time t seconds after the first pumpkin is shot? Add a sketch of the graph of that function to the diagram of Part a.

CONNECTIONS

14 Suppose that the mean of test scores on a mathematics exam is 75.6 and the standard deviation is 5.9. The teacher adds 10 points to every score to correct for a problem all students missed because the wording was unclear. How will that rescoring change:

a. the class mean?

b. the class standard deviation?

15 For each of the following functions:

- write an equivalent function rule in vertex form, $y = (x - h)^2 + k$.

- find the coordinates of the maximum or minimum points on the graphs of the functions.

- explain how the coordinates of the maximum or minimum points could be found by thinking about transforming the graph of $y = x^2$.

a. $y = x^2 + 6x + 8$

b. $y = x^2 - 4x - 3$

16 The diagram at the right shows a geometric figure on a coordinate grid. Explain how each of the following attributes of that figure will or will not change under the transformations described in Parts a–c.

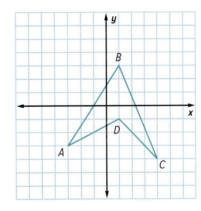

- perimeter

- area

- sum of angle measurements

a. $(x, y) \rightarrow (x, y + 3)$

b. $(x, y) \rightarrow (x + 3, y)$

c. $(x, y) \rightarrow (x, -y)$

17 What are the rules for the inverses of each of these geometric transformations?

Transformation	Inverse
a. $(x, y) \rightarrow (x, y + 3)$	$(x, y) \rightarrow (?, ?)$
b. $(x, y) \rightarrow (x + 3, y)$	$(x, y) \rightarrow (?, ?)$
c. $(x, y) \rightarrow (x, -y)$	$(x, y) \rightarrow (?, ?)$

18 Combine your knowledge about function transformations and solving inequalities. For each inequality below:

- sketch the graphs of the functions involved in the inequality.

- locate the intersection points.

- record the solution using inequality symbols and interval notation.

a. $|x - 6| \leq 4$ **b.** $|x + 6| \geq 4$

c. $|x - 2| < 0.5$ **d.** $|x - 2| \leq \frac{1}{2}x + 2$

19 Shown below is the graph of the **square root function** $f(x) = \sqrt{x}$ for $0 \leq x \leq 20$. On a copy of the diagram, sketch and label graphs of the following variations of the square root function, and then check your ideas. With another color pen or pencil, note any needed corrections.

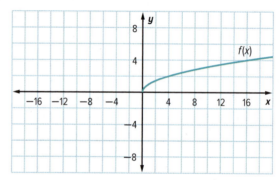

a. $g(x) = -\sqrt{x}$ **b.** $h(x) = \sqrt{x} + 5$

c. $j(x) = -\sqrt{x} + 5$ **d.** $k(x) = \sqrt{x + 4}$

REFLECTIONS

20 A focus of this unit and future units is the process of *mathematical modeling*. The diagram below summarizes the process.

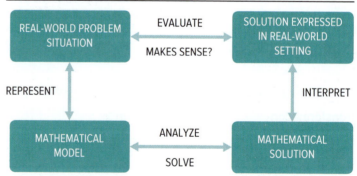

Process of Mathematical Modeling

Choose one example of your work in mathematical modeling from Lesson 1 or Lesson 2. Describe how your example illustrates each part of the diagram.

21 The diagram at the right shows graphs of $y = x^2$ and $y = x^2 - 4$. In this lesson, you learned that the graphs of these two functions are related by a vertical translation. The difference between corresponding y values should be 4 for every x. But somehow the graphs seem to get closer together as $|x|$ increases. How can you explain this apparent contradiction?

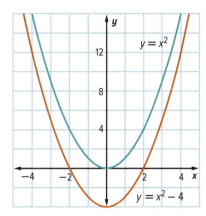

22 Why is it reasonable to say that the graphs of each pair of functions are congruent?

a. $f(x) = |x|$ and $g(x) = -|x| + 7$

b. $h(x) = -x^2$ and $j(x) = (x + 4)^2 - 5$

23 Look back at Connections Task 18. How could you use similar reasoning to solve the following inequalities? Record the solutions using inequality symbols and interval notation.

a. $\sqrt{x} - 4 \leq 0$

b. $\sqrt{x + 5} < 3$

24 Many students find it difficult to believe (or remember) that the graph of $y = (x - 3)^2$ can be found by translating the graph of $y = x^2$ three units to the *right*. Or that the graph of $y = (x + 3)^2$ can be found by translating the graph of $y = x^2$ three units to the *left*. How do you think about the relationship of rules and graphs in the general case where $g(x) = f(x \pm h)$ with $h > 0$? How can you check the correctness of your thinking?

25 Look back at your work in this lesson. Give specific examples where you found the following mathematical practices helpful and explain how they were helpful.

a. Making sense of problems and persevering in solving them

b. Reasoning both quantitatively and algebraically

c. Searching for and making use of patterns or structure in mathematical situations

EXTENSIONS

26 The *Custom Gear* company designs and makes baseball caps and shirts with clever messages and sketches for groups that want personalized items. For each personalized baseball cap, they charge $250 for creating the pattern and then an additional $15 to produce each individual cap.

a. What function gives the cost of an order for x caps?

b. What function gives the cost per cap of an order for x caps?

c. How can the function in Part b be expressed with an equivalent rule in the form $y = m + \dfrac{b}{x}$?

d. Sketch a graph of the cost-per-cap function and describe the horizontal asymptote for that graph.

e. Explain what the horizontal asymptote tells about the cost per cap as the number of caps ordered increases.

f. Explain how you can identify the horizontal asymptote by analyzing the expression $m + \dfrac{b}{x}$.

27 A circle with center O and radius r is the set of all points at distance r from O.

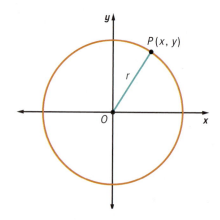

a. Using the sketch at the right, explain why the equation for a circle of radius r centered at the origin of a coordinate system is $x^2 + y^2 = r^2$.

b. How does the reasoning developed in Investigation 3 for horizontal translation of graphs explain why the equation for a circle of radius r and center $(h, 0)$ is $(x - h)^2 + y^2 = r^2$?

c. How does similar reasoning explain why the equation for a circle of radius r and center (h, k) is $(x - h)^2 + (y - k)^2 = r^2$?

28 The graph shown is of the piecewise-defined function with rule:

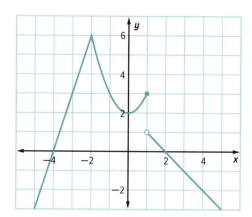

$$f(x) = \begin{cases} 3x + 12 & \text{for } x \le -2 \\ x^2 + 2 & \text{for } -2 < x \le 1 \\ -x + 2 & \text{for } x > 1 \end{cases}$$

a. On a coordinate grid, copy the graph of $f(x)$.

b. On the same coordinate grid, sketch the graph of $g(x) = f(x + 1)$.

c. Write the symbolic rule for $g(x)$.

29 Shown at the right is the graph of $y = \sin t$ for $-2\pi \le t \le 2\pi$. On a copy of this graph, sketch the graphs of the variations of the $\sin t$ function in Parts a–c.

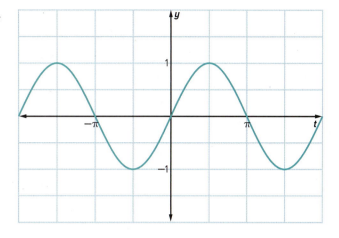

a. $y = -\sin t$

b. $y = 1 - \sin t$

c. $y = \sin (t - 2)$

REVIEW

30 Rewrite each algebraic expression in an equivalent form with the fewest number of terms possible.

a. $6(x + 7)(x + 2)$ **b.** $(2y - 5)^2$

c. $(5x + 4y)(2x - y)$ **d.** $(3x + 7)(3x - 7)$

e. $(t + 2) - (4t + 9) - t$

31 Represent the solution to each inequality or equation on a number line.

a. $3(x + 7) > -2x - 13$ **b.** $x(x - 4) = x^2 - 12$

c. $(2x - 3)^2 \le 25$ **d.** $(x + 4)(-2x + 5) < 0$

32 David and Victoria are exploring what happens when you combine reflections and translations. David first reflects a triangle across the x-axis and then translates the image down five units. Victoria begins with the same triangle but first translates the figure down five units and then reflects the image across the x-axis. Will David's final triangle be in the same position as Victoria's final triangle? Explain your reasoning.

33 Suppose that in September, a random sample of 422 high school seniors in Texas was surveyed about whether they had their own car and if they had a summer job the previous summer. Some of the results are shown in the table below.

	Had Summer Job	No Summer Job	Total
Has Own Car			
Does Not Have Own Car	31		92
Total		75	442

a. Fill in the rest of the table.

b. Which type of chi-square test would you use with these data: a chi-square test of independence or a chi-square test of homogeneity? Explain your reasoning.

c. Compute the chi-square statistic and determine if the value of χ^2 is statistically significant.

d. Write a conclusion in context describing the relationship between the variables.

34 Consider the function: $h(x) = \begin{cases} 3x - 3 & \text{for } x \geq 2 \\ -(x - 1)^2 + 4 & \text{for } x < 2 \end{cases}$

a. Evaluate $h(4)$ and $h(-5)$.

b. Find the zeroes of $h(x)$.

c. For what values of x is $h(x) = -12$?

d. For what values of x is $h(x) = 10$?

35 Find the coordinates of the vertex, the x-intercepts, and the y-intercept of the graph of each quadratic function.

a. $y = x^2 - 6x$

b. $y = x^2 - 8x + 15$

c. $y = (x + 5)^2 - 3$

36 The box plots below display the heights (in inches) of the boys and girls in a senior English class of 32 students. All measurements were made to the nearest half-inch.

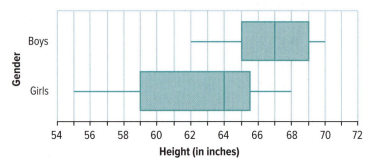

a. What is the difference in height between the tallest and the shortest person?

b. Is the median height of the girls or the boys greater? Explain.

c. Is the mean height of the girls greater than or less than the median height of the girls? Explain.

d. What is the interquartile range of the boys' heights?

e. Is the standard deviation of the girls' heights greater than or less than the standard deviation of the boys' heights? Explain your reasoning.

37 Write each expression in a simpler equivalent exponential form using only positive exponents.

a. $(3x^{-4})x^2$

b. $\dfrac{x^4(2x^3)^2}{x^{-2}}$

c. $-8x^4y(4x^{-2}y)$

d. $3(5x)^{-1}$

38 It is about 6 miles from Uma's house to the South Haven Pier on Lake Michigan.

   a. If Uma walks an average of 4 miles per hour, about how long will it take her to walk to the Pier?

   b. Erina took 40 minutes to run to the Pier from Uma's house. What was Erina's average running speed in miles per hour?

   c. Claudio lives further from the South Haven Pier than does Uma. He rides his bike at an average speed of 12 miles per hour and arrives at the Pier in 50 minutes. How far does Claudio live from the Pier?

39 Consider the following matrix representation of $\triangle ABC$.

$$\triangle ABC = \begin{bmatrix} -4 & 2 & 2 \\ 2 & 2 & -3 \end{bmatrix}$$

   a. On separate grids for each part, sketch and label $\triangle ABC$ and its image $\triangle A'B'C'$ under each of the following transformations.

   i. $(x, y) \rightarrow \left(\frac{1}{2}x, y\right)$

   ii. $(x, y) \rightarrow \left(x, \frac{1}{3}y\right)$

   iii. $(x, y) \rightarrow (-2x, -2y)$

   b. Determine whether each of the image triangles in Part a is similar, congruent, or neither similar nor congruent to $\triangle ABC$. Explain your reasoning.

40 Write an equation for the line satisfying each set of conditions.

   a. Contains the points $(-1, -3)$ and $(-7, 12)$

   b. Contains the points $(5, 0)$ and $(0, 7)$

   c. Is parallel to the graph of $y = \frac{3}{2}x + 2$ and contains the point $(6, 0)$

Customizing Models by Stretching and Compressing

In Lesson 1, you analyzed world climate over many decades. One important aspect of climate change is temperature change. Meteorologists measure temperature change continuously, but often report it in terms of daily or monthly high and low temperatures. The following table and plot show the pattern of change in average monthly Fahrenheit temperature for Des Moines, Iowa. A function model has been fitted to the data.

Average Monthly Temperatures for Des Moines, IA, 2010

Month	Temperature (in °F)
Jan	16.7
Feb	20.0
Mar	41.8
Apr	58.7
May	62.9
June	74.1
July	78.2
Aug	78.0
Sept	66.6
Oct	57.6
Nov	40.9
Dec	23.5

Source: www.crh.noaa.gov/images/dmx/2010 DSM MonthlyTables.pdf

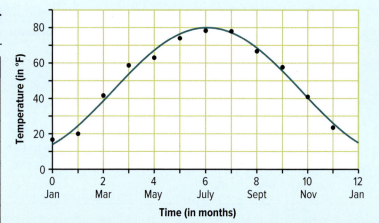

THINK ABOUT THIS SITUATION

Study the graph and use the information it reveals to answer these questions.

a. What does the graph tell about the pattern of change over time in average monthly temperature for Des Moines?

b. What familiar functions have graphs similar to that of the average monthly temperature graph?

c. How is the average monthly temperature graph different from graphs of the functions it most resembles?

d. How do you think you could modify the rule for a familiar function to fit this somewhat different pattern?

In this lesson, you will continue to study strategies for using familiar functions and transformations of their graphs to develop mathematical models. The focus will be on techniques for modeling patterns that are produced by vertical or horizontal stretching or compressing of familiar graphs.

INVESTIGATION 1

Vertical Stretching and Compressing

The graph of average monthly temperature looks a lot like that of a portion of the graph of the function $y = \sin x$.

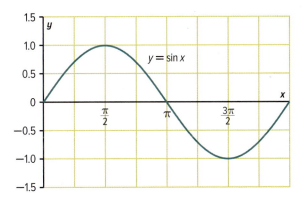

However, the temperature graph shows variation between 16.7°F and 78.2°F (not between −1 and 1) and the pattern of variation repeats in cycles of 12 months (not $2\pi \approx 6.28$ months).

As you work on the problems of this investigation, look for answers to these questions:

What are the connections between rules for functions whose graphs are related by vertical stretching or compressing?

How can those connections be used to build related models for important data patterns?

1 The diagram below shows graphs of $y = \sin x$ and two functions $f(x)$ and $g(x)$ with graphs related to that of $y = \sin x$ by *vertical stretching*—away from the x-axis.

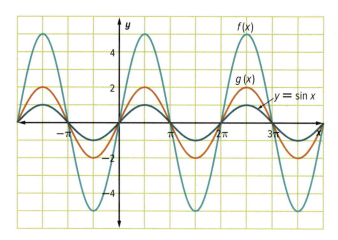

a. Complete a copy of the following table showing a sample of approximate (x, y) values for the three functions. (Remember: $\frac{\pi}{2} \approx 1.57$, $\pi \approx 3.14$, $\frac{3\pi}{2} \approx 4.71$, and $2\pi \approx 6.28$.)

x	$-\frac{3\pi}{2}$	$-\pi$	$-\frac{\pi}{2}$	0	$\frac{\pi}{2}$	3π	$\frac{7\pi}{2}$
$f(x)$							
$g(x)$							
$\sin x$							

b. What geometric transformations $(x, y) \rightarrow (?, ?)$ do you think would map the graph of:

i. $y = \sin x$ onto $f(x)$?

ii. $y = \sin x$ onto $g(x)$?

iii. $g(x)$ onto $y = \sin x$?

Compare your transformation rules with those of your classmates and resolve any differences.

c. What rules would you expect to model the patterns shown in the graphs of $f(x)$ and $g(x)$? Check your ideas using technology.

2 The next diagram shows the graph of $y = \cos x$ and another function $h(x)$ that is related to it by *vertical compressing*—toward the *x*-axis.

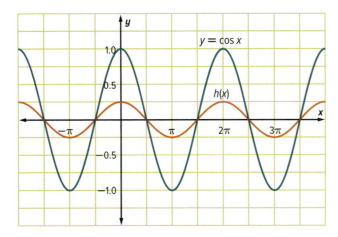

a. What geometric transformation $(x, y) \rightarrow (?, ?)$ do you think would map the graph of $y = \cos x$ onto the graph of $h(x)$?

b. Write a rule for $h(x)$.

3 Now reconsider the average monthly temperature plot shown at the start of this lesson. What variation of the rule $s(x) = \sin x$ will define a function that varies between 16.7°F and 78.2°F, giving a range of values comparable to those in the plot?

Applications of Vertical Stretching and Compressing The ideas you have developed about vertical stretching and compressing of $y = \sin x$ and $y = \cos x$ graphs can also be used in work with other families of functions.

4 In your previous mathematics courses, you may have used quadratic functions to model the relationship between time and distance of moving objects (soccer balls, skate boarders, luge riders, and others) as they respond to the pull of gravity. The rules for those functions were all derived from the simplest quadratic $y = x^2$. For instance, if you drop a ball from the top of a 60-foot tall building, its height in feet above the ground at any time t seconds later will be given by $h(t) = 60 - 16t^2$.

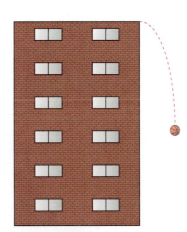

The diagram below shows a graph of $y = x^2$.

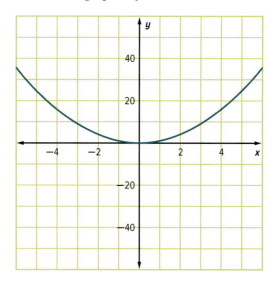

a. On a copy of this diagram, sketch and label the graphs for these functions.

$$f(x) = -x^2 \qquad g(x) = -16x^2 \qquad h(x) = 60 - 16x^2$$

b. Study the following sequences of transformations to see which would lead to a correct graph for $h(x) = 60 - 16x^2$. In each case, provide algebraic reasoning that justifies your answer.

   i. Stretch the graph of $y = x^2$ vertically by a factor of 16. Then reflect across the x-axis. And then translate up 60 units.

   ii. Stretch the graph of $y = x^2$ vertically by a factor of 16. Then translate up 60. And then reflect across the x-axis.

   iii. Reflect the graph of $y = x^2$ across the x-axis. Then translate up 60 units. And then stretch vertically by a factor of 16.

5 In many science museums, one of the most popular exhibits is a Foucault pendulum. With a large mass attached to a long cable, it swings in a way that demonstrates the rotation of Earth.

You may view this phenomenon by searching for videos of a Foucault pendulum on the Internet.

The photo on the previous page shows a Foucault pendulum at the Indiana State Museum in Indianapolis. The functions in Parts a and b describe possible patterns of displacement from vertical as such a pendulum swings.

Draw graphs of each function on $[0, 4\pi]$. Then explain what each rule and graph tell about motion of the pendulum with passage of time t (in seconds).

a. $f(t) = 10 \cos t$ **b.** $g(t) = 25 \cos t$

6 One of the simplest strategies for individuals and organizations to earn income on savings is to invest in mutual funds. Individual contributions are spread over many different common stocks, bonds, or other assets. At the start of one year, a club invested $5,000 in such a fund and the profit on that investment varied over the next 12 months as shown in the following graph.

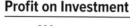

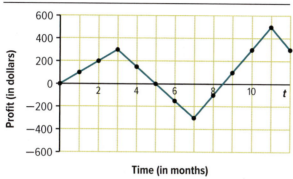

Profit on Investment — Time (in months)

Suppose that the club had instead invested $10,000 in the same mutual fund.

a. What would the maximum and minimum profit figures have been during the year and when would they have occurred?

b. At what times during the year would the profit have been $0?

c. What would the profit have been at the end of the year?

d. How would the profit graph differ from the graph tracking profit for the $5,000 investment?

7 Suppose two regular six-sided dice are rolled and the sum of the dots on the upper faces is computed. The plot below shows the probabilities associated with the possible sums.

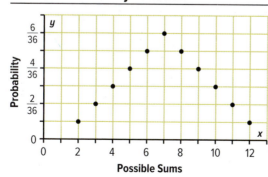

Discrete Probability Function — Possible Sums

a. Discuss with your classmates why the scale on the x-axis is labeled as it is. The scale on the y-axis.

b. Choose three possible sums and verify that the corresponding plotted points are accurate.

c. What basic function type is suggested by the pattern in the plot?

d. Use your understanding of graph transformations to build a function rule that models that pattern in the data.

e. Compare your function rule with those built by others. Resolve any differences.

f. What are the domain and range of your modeling function?

Properties of Vertical Stretching and Compressing Vertical stretching and compressing of the graph of a function affect the local maximum and minimum points, *x*-intercepts, and *y*-intercept of the graph of the new function in predictable ways.

8 Consider the following graph of a function *f(x)*.

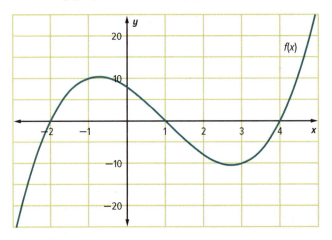

On a copy of the diagram:

a. sketch and label the graphs of $g(x) = 1.5f(x)$ and $h(x) = -2f(x)$.

b. locate all local maximum and minimum points of the three functions and label those points on the graphs with their approximate coordinates.

c. locate all *x*-intercepts of the three functions and label those points on the graphs with their approximate coordinates.

d. locate the *y*-intercepts of the three functions and label those points on the graphs with their approximate coordinates.

9 Look over the results of your work on Problem 8. Formulate conjectures that answer these questions about vertical stretching and compressing of graphs for any function $f(x)$. Be prepared to explain your reasoning.

   a. How are local maximum and minimum points of functions $f(x)$ and $g(x)$ related if $g(x) = kf(x)$?

   b. How are x-intercepts of $f(x)$ and $g(x)$ related if $g(x) = kf(x)$?

   c. How are y-intercepts of $f(x)$ and $g(x)$ related if $g(x) = kf(x)$?

   d. Test your ideas in Parts a–c by comparing $f(x) = x^2 + x - 6$ and $g(x) = 3f(x)$.

      i. What is the standard polynomial form of the rule for $g(x)$?

      ii. How are the local maximum and minimum points, x-intercepts, and y-intercepts of the two functions related?

SUMMARIZE THE MATHEMATICS

In this investigation, you explored connections between rules of functions whose graphs are related by vertical stretching or compressing.

   a. Suppose the y-coordinate of each point on the graph of $g(x)$ is exactly k times the y-coordinate of the point just below or above it on the graph of $f(x)$. What is the connection between the rules of the two functions?

   b. What is the coordinate rule $(x, y) \rightarrow (?, ?)$ that will map the graph of $f(x)$ onto the graph of $kf(x)$ by vertical stretching or compressing with a scale factor of k?

   c. If $g(x) = kf(x)$ for all x, how are the graphs of $g(x)$ and $f(x)$ related when $|k| > 1$? When $|k| < 1$?

   d. How are the local maximum and minimum points, x-intercepts, and y-intercept of a function $g(x)$ related to those of $f(x)$ if $g(x) = kf(x)$ for all x when $k > 0$? When $k < 0$?

Be prepared to explain your ideas to the class.

✓ CHECK YOUR UNDERSTANDING

For each of these functions, without using technology:

- sketch a graph of the function and the specified stretching or compressing of that graph.

- give a rule for the function whose graph is the image after stretching or compressing of the original graph.

a. $f(x) = -|x|$ graphed on $[-5, 5]$ and then stretched by a factor of 1.5 away from the x-axis

b. $g(x) = x^2$ graphed on $[-5, 5]$ and then compressed by a factor of 0.5 toward the x-axis

c. $h(x) = \frac{1}{x}$ graphed on $[-5, 5]$, stretched by a factor of 2 away from the x-axis, and then translated vertically 4 units up

Horizontal Stretching and Compressing

Recall from Investigation 1 that the average monthly temperature data from Des Moines, Iowa, can be modeled by a variation of the graph of $y = \sin x$. In Problem 3 (page 173), you began to build a function rule for the model: $f(t) = 30.75 \sin t + 47.45$. This function has a graph with amplitude 30.75, the same as that of the average monthly temperature function.

The function $f(t) = 30.75 \sin t + 47.45$ graphed below has *period* 2π (or approximately 6.28 months). If we assume the 12 months of a year are of equal length, the period of the average monthly temperature function is 12.

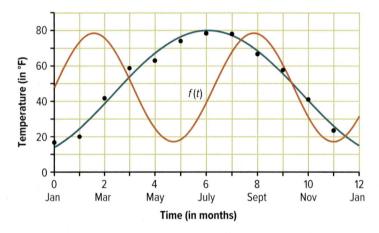

Finding the rule for a function whose graph better fits the data requires a modification of $f(t) = 30.75 \sin t + 47.45$ that corresponds to a horizontal stretching of its graph *away* from the y-axis.

As you work on the problems of this investigation, look for answers to these questions:

What are the connections between rules for functions whose graphs are related by horizontal stretching or compressing?

How can those connections be used to build related models for important data patterns?

As is often the case, it helps to develop a strategy for answering these questions by working on somewhat simpler problems first. The next diagram shows the graph of $y = \sin x$ and a function $g(x)$ whose graph was obtained by horizontal compressing of the graph of $y = \sin x$ by a scale factor 0.5. Think about what might be a suitable rule for $g(x)$.

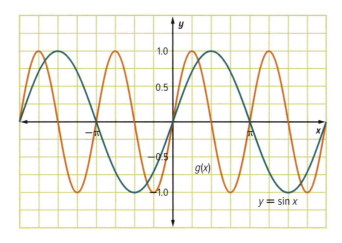

1 Answer the following questions to help develop a strategy for finding a rule for $g(x)$ in this and similar cases.

a. When one group of students in a Sitka, Alaska class worked on this problem, they noticed that $g(x)$ seems to track through the same pattern of values as $y = \sin x$, but twice as fast. They wondered how to make a rule for $g(x)$ so that when x increases from 0 to π, the function $g(x)$ actually operates on inputs from 0 to 2π. Can you see a way to do that?

b. Write the geometric transformations $(x, y) \rightarrow (?, ?)$ that map:

   i. the graph of $y = \sin x$ onto the graph of $g(x)$.

   ii. the graph of $g(x)$ onto the graph of $y = \sin x$.

   How do these answers suggest a way to modify the rule for $y = \sin x$ to get a rule for $g(x)$?

c. Now examine how the graphs of the functions below are related to the graph of $y = \sin x$ for various values of k.

   i. $y = \sin x + k$

   ii. $y = \sin (x + k)$

   iii. $y = k \sin x$

   What do the patterns in these cases suggest about developing rules for functions whose graphs are related by horizontal stretching or shrinking?

d. What do your answers to Parts a–c suggest as a way to modify the function $y = \cos x$ so that its graph is the cosine graph stretched or compressed horizontally by a scale factor of 1.5? By a scale factor of k?

2 Find the rule, amplitude, and period of functions whose graphs are related to $y = \sin x$ by:

a. horizontal stretching with scale factor 2.

b. horizontal compressing with scale factor $\frac{1}{3}$.

c. horizontal stretching with scale factor $\frac{1}{\pi}$.

d. horizontal compressing with scale factor $\frac{2}{3}$ and vertical stretching with scale factor 5.

3 Draw on your understanding of translations and stretches of graphs and the cyclic nature of the seasons to build a function model for the pattern of change in average monthly temperature for Des Moines, Iowa shown at the beginning of this lesson.

a. In Investigation 1, Problem 3 (page 173), you built a variation of the sine function, $f(t) = 30.75 \sin t + 47.45$, that varies between the minimum and maximum values of the average monthly temperature for Des Moines, IA. How could you modify the model to have a period of 12 months? Compare your rule with those of others and resolve any differences.

b. What transformation of the graph of the modified $f(t)$ function will produce an even better model that fits the data? Write a function rule $T(t) = $ _____ for the improved model. Compare your final rule with those of your classmates. Resolve any differences.

c. Use technology and its sinusoidal regression feature to find a model that best fits the average monthly temperature data for Des Moines. Compare the technology-produced model with the one you developed.

d. Use each of the two mathematical models to estimate the average temperature for the month of April. Compare your answers with the data on page 170.

4 The graph representing average monthly temperature for a city is similar to graphs of many other physical phenomena. For example, if you live in or visit an ocean seaport city or beach resort town, you will have a chance to observe the tidal patterns of change in water depth that are caused by the gravitational pull of the Moon. A 24-hour graph of water depth at a point near the shore of a harbor in Boston, New York, Miami, San Diego, San Francisco, or Seattle might produce a graph like that in the next diagram.

Harbor Water Depth Over Time

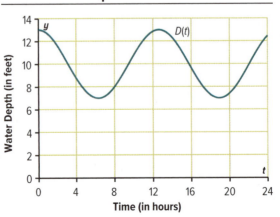

Suppose that this tidal pattern has period $4\pi \approx 12.57$ hours. What function $D(t)$ will model the pattern of change in water depth over time? You might find it helpful to think about the following questions as you build a function rule.

- What variation of $y = \cos t$ has amplitude 3 feet?

- What variation of the result in Part a will have period $4\pi \approx 12.57$ hours?

- What variation of the result in Part b will have range [7, 13] feet?

Properties of Horizontal Stretching and Compressing In previous investigations, you have seen graphs of two functions that are related by a horizontal or vertical translation or by vertical stretching or compressing. It is often possible to infer locations of maximum and minimum points, x-intercepts, and the y-intercept of one function from those of the other. In Problems 5 and 6, you will investigate if similar inferences can be made when two functions have graphs related by horizontal stretching or compressing.

5 Consider the cubic polynomial function $f(x) = x^3 + x^2 - 9x - 9$ with graph shown here.

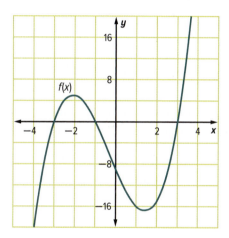

a. Use the graph to estimate the following critical points.

   i. Coordinates for the local maximum and minimum points on the graph of $f(x)$

   ii. The x-intercepts of $f(x)$

   iii. Coordinates for the y-intercept on the graph of $f(x)$

b. Now consider the function $g(x) = f(2x)$. Use the relationship between $g(x)$ and $f(x)$ to estimate each of the following features of $g(x)$. Check your ideas using technology.

   i. Coordinates for the local maximum and minimum points on the graph of $g(x)$

   ii. Coordinates for the x-intercepts of the graph of $g(x)$

   iii. Coordinates for the y-intercept of the graph of $g(x)$

c. Next consider the function $h(x) = f(0.5x)$. Use the relationship between $h(x)$ and $f(x)$ to estimate each of the following features of $h(x)$. Check your ideas using technology.

   i. Coordinates for the local maximum and minimum points on the graph of $h(x)$

   ii. Coordinates for the x-intercepts of the graph of $h(x)$

   iii. Coordinates for the y-intercept of the graph of $h(x)$

6 Suppose that two functions $c(x)$ and $d(x)$ are related by the equation $d(x) = c(kx)$ for some positive number k. Use results of your work in Problem 5 and your earlier exploration of horizontal stretching and compressing of sine and cosine functions to answer these questions.

a. How will coordinates of the local maximum and minimum points on graphs of $c(x)$ and $d(x)$ be related?

b. How will coordinates of the x-intercepts of the two functions be related?

c. How will coordinates of the y-intercepts of the two functions be related?

Be prepared to explain why your answers make sense based on the effects of horizontal stretching and compressing of the function graphs.

SUMMARIZE THE MATHEMATICS

In this investigation, you explored connections between rules of functions whose graphs are related by horizontal stretching or compressing.

a. How will the graph of $g(x) = \cos kx$ be related to the graph of $y = \cos x$ when $k > 1$? When $0 < k < 1$?

b. What do the values of a, b, and c tell about the relationship of $y = \sin x$ to $f(x) = a \sin bx + c$?

c. Suppose $f(x)$ and $g(x)$ are functions such that $g(x) = f(kx)$ for all x. How are the local maximum and minimum points, x-intercepts, and y-intercept of a function $g(x)$ related to those of $f(x)$?

Be prepared to explain your ideas to the class.

✔CHECK YOUR UNDERSTANDING

For each of these functions, without using technology:

- sketch a graph of the function and the specified stretching or compressing of that graph.
- give a rule for the function whose graph is the image after stretching or compressing of the original graph.

a. $p(x) = -\cos x$ graphed on $[-4\pi, 4\pi]$ and then stretched by a factor of 1.5 away from the y-axis

b. $h(x) = -\sin x$ graphed on $[-2\pi, 2\pi]$ and then compressed by a factor of 0.5 toward the y-axis

APPLICATIONS

1 In each part, find a rule for the function $g(x)$ whose graph is the image of the graph of $f(x)$ under the indicated transformation.

   a. $f(x) = 3x^2 + 4$; transformation $(x, y) \rightarrow (x, 0.5y)$

   b. $f(x) = 3 \cos x$; transformation $(x, y) \rightarrow (x, 2.5y)$

   c. $f(x) = 5 \sin x$; transformation $(x, y) \rightarrow (x, 4y + 1)$

   d. $f(x) = |x|$; transformation $(x, y) \rightarrow (x + 2, 0.5y)$

2 In each part, find the transformation $(x, y) \rightarrow (?, ?)$ that maps the graph of $f(x)$ onto the graph of $g(x)$.

   a. $f(x) = 3x^2 + 4$ and $g(x) = 9x^2 + 12$

   b. $f(x) = 3x^2 + 4$ and $g(x) = 9x^2 + 15$

   c. $f(x) = \sin x$ and $g(x) = 4 \sin (x - 3)$

   d. $f(x) = |x|$ and $g(x) = 7 - 2|x + 4|$

3 Suppose that two functions are related by the equation $h(x) = 7j(x)$ for all values of x.

   a. Assume $j(x)$ has a local maximum point at $(-3, 2)$ and a local minimum point at $(5, -3)$. What can be said about local maximum and minimum points of $h(x)$?

   b. If $h(x)$ has y-intercept $(0, -2)$, what can be said about the y-intercept of $j(x)$?

   c. If $j(x)$ has zeroes -5, -1, and 8, what can be said about the zeroes of $h(x)$?

4 When distance is measured in meters and time in seconds, the function $d(t) = 4.9t^2$ tells the approximate distance fallen in t seconds by an object dropped from a high place anywhere on Earth.

   a. Gravitational force near the surface of the Moon is one-sixth that near the surface of Earth. What distance function $d_M(t)$ is implied for falling objects on the Moon?

   b. Consider an object dropped from a point 50 meters above the surface of the Moon. What function tells its altitude at any time t after it is dropped?

   c. When will the dropped object described in Part b reach the Moon's surface?

5 If the temperature in a restaurant's pizza oven is set at 500°F, it will actually vary above and below that setting as time passes. Suppose that the actual temperature is a function of time in minutes with rule in the form $D(t) = a \sin t + b$.

a. What are the values of a and b if the oven temperature varies from 490°F to 510°F?

b. What are the values of a and b if the oven temperature varies from 500°F to 510°F?

6 On a hilltop overlooking the city of Prague in the Czech Republic, there is a giant metronome that swings back and forth during daylight hours. The metronome was built in 1991 to celebrate freedom from communist rule and Prague's proud musical heritage. To see and learn more about the giant metronome, visit atlasobscura.com/place/prague-metronome.

Suppose that the metronome is set to swing 30° either side of vertical, making one complete swing every $6\pi \approx 18.85$ seconds.

a. What function tells the angle from vertical of the metronome at any time t seconds after it passes the vertical position?

b. What function tells the angle from vertical of the metronome at any time t seconds after it leaves its extreme left position?

7 The Santa Monica Pier in California reaches out into the Pacific Ocean. There are restaurants, arcade games, and amusement park rides on the pier, making it a popular spot for weekend entertainment.

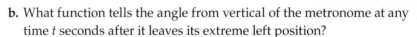

Suppose that the water halfway out along the pier is 20 feet deep at high tide and that water depth varies 4 feet from high to low tide.

a. The depth of water (in feet) at that point off the Santa Monica Pier varies according to a function in the form $d(t) = a \cos 0.5t + b$, where t represents time (in hours) after high tide. What values of a and b will give a function rule that matches the given information about water depth and tide range?

b. Suppose that at the end of the pier, typical water depth is 25 feet at high tide, but the range from high to low is the same 4 feet. What function $d_1(t)$ gives the typical depth of water at the end of the Santa Monica Pier at any time t hours after high tide?

c. Tidal variation tends to be greater as one moves away from the equator, but the time from high to low is the same around the world. Suppose that at the end of a long pier extending into Cook Inlet in Alaska, the water is typically 45 feet deep at high tide, and that the range from high to low tide is 15 feet. What function $d_2(t)$ gives the depth of water at the end of the Cook Inlet pier at any time t hours after high tide?

d. Suppose the mean high tide depth at a point closer to shore along the Cook Inlet pier is only 25 feet. What function $d_3(t)$ would give the depth of water as a function of time since high tide at this point?

8 The following diagrams show three periodic relationships that are different from the sine and cosine functions, but useful in modern electronics. For each graph:

a. identify which of the relationships are *not* functions of t. Explain your reasoning.

b. identify the period and amplitude of $r(t)$.

c. sketch and label graphs of these two variations: $v(t) = 2r(t)$ and $w(t) = r(2t)$

d. identify the period and amplitude of each variation.

I. Square Wave

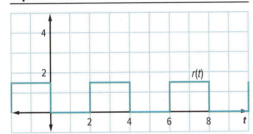

II. Triangle Wave

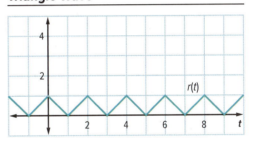

III. Saw Tooth Wave

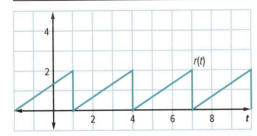

9 Find values of k so that each of the following functions has the indicated period.

a. $f(t) = \cos kt$; period π

b. $g(t) = \cos kt$; period $\dfrac{\pi}{2}$

c. $h(t) = \cos kt$; period 6π

d. $j(t) = \sin kt$; period 10

e. $p(t) = \sin kt$; period 0.1

f. $s(t) = 4 \cos kt + 15$; period $\dfrac{1}{60}$

10 The London Eye, also called the Millennium Wheel, was (until 2006) the largest observation wheel in the world. It stands 135 meters (443 feet) tall in the Jubilee Gardens of London, England. Passengers ride in 32 air-conditioned sealed capsules that are mounted on the wheel's exterior.

It takes about 30 minutes for each complete rotation of the wheel. Assume that the wheel rotates at a constant speed. As you ride in one of the capsules, your position changes in two directions—horizontally and vertically.

a. If you step into a capsule at the bottom of the Millennium Wheel, your distance from the *vertical axis* of symmetry will be given at any time t (in minutes) by a function $d(t) = a \sin kt$.

  i. What value of k will guarantee that the distance function has period 30 minutes?

  ii. What value of a is implied by the fact that the wheel has diameter 135 meters?

b. If you step into a capsule at the bottom of the wheel, your height above the bottom of the wheel at any time t will be given by a function $h(t) = a(1 - \cos kt)$.

  i. What value of k will guarantee that the height function has period 30 minutes?

  ii. What value of a is implied by the fact that the wheel has diameter 135 meters?

c. Use results of your work on Parts a and b to find the location of your capsule (distance from vertical and height above the bottom of the wheel) at the following times after boarding at the bottom. Check each answer by reasoning with principles from geometry and trigonometry and the fact that the capsule turns through an angle of about 12° each minute.

  i. 30 minutes   **ii.** 10 minutes

  iii. 15 minutes   **iv.** 22.5 minutes

11 The periodic patterns of change in many physical and physiological situations are often monitored by instruments that produce electronic displays like those seen on oscilloscopes. For example, an instrument monitoring pressure created by the beating of a newborn baby's heart might produce a readout that looks like that shown here.

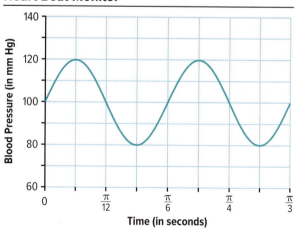

Heart Beat Monitor

Cynthia L. Black

a. The graph has period of $\frac{\pi}{6}$ seconds.

    i. What values of a, b, and c will produce a model in the form $f(t) = a \sin bt + c$ that matches the graph on the readout shown?

    ii. What pulse rate in beats per minute is indicated by the graph and rule?

b. Suppose that the pulse rate doubled.

    i. How would the displayed graph change?

    ii. How would you modify the rule for $f(t)$?

c. Suppose that the pulse rate slowed to half the rate shown in the graph.

    i. How would the displayed graph change?

    ii. How would you modify the rule for $f(t)$?

12 For a simple pendulum, the time it takes to swing from side to side depends only on the length of the pendulum—not the weight attached or the point from which it is released.

Since we usually set a pendulum in motion from its maximum displacement point, it makes sense to model pendulum motion with variations of $y = \cos t$. Displacement is naturally measured by degrees left or right of vertical and time is naturally measured in seconds.

    What variation of $y = \cos t$ will describe motion of a pendulum with maximum displacement 20° on both sides of vertical and period of 10 seconds (that is, one complete swing from right to left and back again every 10 seconds)?

13 The mean of the average monthly temperatures in degrees Fahrenheit for Baltimore, MD, over 40 years are given in the table below.

Mean of Average Monthly Temperatures for Baltimore, MD, 1971–2010

Month	Jan	Feb	Mar	Apr	May	June	July	Aug	Sept	Oct	Nov	Dec
Temperature (in °F)	33.4	35.9	44.4	54.2	63.5	72.7	77.4	75.9	68.7	56.7	47.1	37.5

Source: www.erh.noaa.gov/lwx/climate/bwi/bwitemps.txt

a. Explain what the temperature entry for January means.

b. Plot the data beginning with January as month 0.

c. Find a variation of the sine function whose graph fits the data well.

d. How would you modify your modeling method in Part c to use a variation of the cosine function?

e. Use technology and its sinusoidal regression feature to find a model that best fits the mean of average monthly temperature data for Baltimore.

f. Compare the technology-produced model with the one you constructed in Part c.

CONNECTIONS

14 Consider again the absolute value function $v(x) = |x|$. The diagram below shows a graph of $v(x)$ and two variations of that basic function.

a. What are the rules for $r(x)$ and $s(x)$?

b. What coordinate rule $(x, y) \rightarrow (?, ?)$ will map the graph of $v(x)$ onto the graph of $r(x)$?

c. What coordinate rule $(x, y) \rightarrow (?, ?)$ will map the graph of $v(x)$ onto the graph of $s(x)$?

d. When asked to describe the relationship of the graph of $r(x)$ to those of $v(x)$ and $s(x)$, most people say that the graph of $r(x)$ is "narrower." Why does the visual descriptor "narrower" imply "increasing at a faster rate" as values of x move away from 0 in both directions?

15 Suppose that scores on a 40 question, multiple-choice American history test are recorded as number correct. The mean number correct for a class is 28, with standard deviation 4.3. Suppose that each student's number correct score is converted to a percent score.

  a. Why can that transformation of scores be accomplished by multiplying each score by 2.5?

  b. What is the mean percent correct?

  c. What is the standard deviation of percent correct scores?

16 Imagine a Japanese language class in which the majority of students failed a unit test. The grading scale was from 0 to 100. Rather than have the students repeat the test, the teacher used a correction factor $f(x) = 10\sqrt{x}$, where x is the original grade and $f(x)$ is the improved grade.

  a. What is the improved grade for a student who scored 49 on the original test?

  b. Will this correction factor improve the grades of all students? Explain your answer:

  i. using graphs.

  ii. using algebraic reasoning.

  c. Will some students gain more points than others? Explain your reasoning.

  d. Why do you think the parameter 10 was chosen in defining the correction factor?

17 The diagram at the right shows a geometric shape on a coordinate grid. Explain how each of the following attributes of that figure will or will not change under the transformations described in Parts a–c.

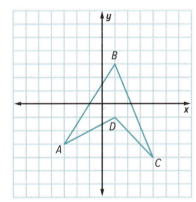

  • perimeter

  • area

  • sum of angle measurements

  a. $(x, y) \rightarrow (x, 3y)$

  b. $(x, y) \rightarrow (3x, y)$

  c. $(x, y) \rightarrow (3x, 3y)$

18 For what values of k will the following functions $g(x)$ have graphs that are geometrically congruent to the graph of $f(x)$?

  a. $g(x) = kf(x)$ **b.** $g(x) = f(kx)$

  c. $g(x) = f(x) + k$ **d.** $g(x) = f(x + k)$

19 Suppose two regular octahedral dice are rolled and the sum of the numbers on the top faces is calculated.

a. Make a table and plot of:
(*possible sum, probability of sum*) values.

b. Use your understanding of the basic function types and graph transformations to construct a function rule that models the pattern in the plot.

c. Look back at your function model for the six-sided dice in Problem 7 (page 175) of Investigation 1. How are the function rules similar and how are they different?

REFLECTIONS

20 Marie graphed the function $f(x) = x(x + 2)(x - 3)$ as shown on the left below. Dario graphed $3f(x)$ as shown on the right. Then he looked at Marie's graph and decided since they were the same that his graph was incorrect. Assuming that the graph of $3f(x)$ is correct, what should Marie say to Dario?

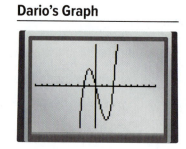

Marie's Graph   **Dario's Graph**

21 As was the case with horizontal translation of graphs, the connection between geometric action and function rule for graphs that are horizontal stretches or compressions of each other does not seem natural. Stretching the graph horizontally means multiplying the independent variable by a number less than 1; compressing the graph horizontally means multiplying by a number greater than 1.

a. How do you think about the problem of matching transformed rules to transformed graphs so that you can get the right connections?

b. Why do you think this task proves to be challenging?

22 Look back at the tidal pattern modeling problem in Investigation 2 (page 180). Would it be possible to model the harbor water depth over time t using transformations of the basic sine function? Explain your reasoning.

Jim Laser

23 In previous work, you have used linear (exponential) functions to model patterns of change in which the dependent variable changes by a constant amount (by a constant factor), for each unit change in the independent variable. What connections do you see between that work and the use of translations versus stretches in Lessons 2 and 3?

24 When customizing a basic function to model a more complex situation:

a. why is any needed vertical translation of the graph applied *after* any vertical stretch?

b. why is any needed horizontal translation applied *after* any horizontal compression?

25 Look back at your function models for the Des Moines, IA, average monthly temperature data for 2010 (page 170) and for the Baltimore, MD, temperature data (Applications Task 13). Why would the approach of using the mean of average monthly temperatures over an extended period of time likely lead to a better-fitting model?

EXTENSIONS

26 The number of hours between sunrise and sunset at any location on Earth varies in a predictable pattern with a period of 365 days. In the Northern Hemisphere, the date of maximum sunlight hours is on or around June 21 (the Summer solstice) and the date of minimum sunlight hours is on or around December 21 (the Winter solstice). The number of hours of sunlight is midway in the range at the Spring and Fall equinoxes—around March 21 and September 21, respectively.

In Chicago, Illinois, the "longest day" has about 15 hours of sunlight and the "shortest day" has about 9 hours of sunlight.

a. Using the Spring equinox as a starting point, one can model the number of hours of sunlight in Chicago with a function in the form $S(t) = a \sin bt + c$, where t is time in days after March 21. What values of a, b, and c will produce a model that meets the conditions described above?

b. Use the questions below to help build a variation $S_1(d)$ of the function from Part a that will model the change in sunlight hours in a way that d represents *day of the year* beginning with January 1 as day 1. Assume your year of analysis is not a leap year.

   i. On what day t will your function in Part a first reach its maximum value?

Design Pics/Don Hammond

   ii. On what day d do you want your new function $S_1(d)$ to first reach its maximum value (i.e., what day of the calendar year is June 21)?

   iii. What transformation of the graph of $S(t)$ will place its maximum point at the desired day for the new function $S_1(d)$?

   iv. What rule for $S_1(d)$ is implied by the answer to part iii?

27 Research the average *daily* Fahrenheit temperature for the town/city in which you live or reside closest to.

   a. Use technology to plot the data.

   b. Transformations of what basic function would most likely model the data? Explain your reasoning.

   c. Use technology to find a model that fits the pattern in the data well.

   d. Use your model to predict the temperature on Thanksgiving Day.

REVIEW

28 In many different board games, each player moves a token according to the results of the sum of the numbers rolled on two die.

   a. Complete a copy of the table below that shows all possible outcomes for the sum of numbers on two die.

Die 1

Die 2	1	2	3	4	5	6
1						
2						
3						
4						
5						
6						

   b. Use the information in the table to determine the following probabilities.

   i. *P(sum is 4)*

   ii. *P(sum is less than 4)*

   iii. *P(sum is odd and greater than 3)*

   iv. *P(sum is not 3)*

29 Rewrite each expression as a product of linear factors.

    a. $x^2 + 12x + 20$

    b. $25x^2 - 16$

    c. $3x^2 - 18x + 27$

    d. $x^2 + 7x - 30$

30 Dorothy is building a dollhouse that is a scale model of her house. The scale factor between the dollhouse and the real house is 10.

    a. If a door in the real house has dimensions 30 in. by 84 in., what should be the dimensions of the corresponding door in the dollhouse?

    b. A rectangular window opening in the dollhouse is 1.5 in. by 3 in. What are the dimensions of the corresponding window in Dorothy's house?

    c. The rug in Dorothy's room has area 15 sq. feet. How much material will she need if she wants to make a similar rug for the dollhouse?

31 Write a function rule for the inverse of each function.

    a. $f(x) = 9x + 3$

    b. $g(x) = \dfrac{5x - 4}{3}$

    c. $h(x) = \dfrac{3}{x} - 7$

32 Find the indicated angle measure and side lengths.

    a. Find $m\angle B$.

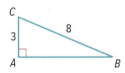

    b. Find AC.

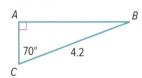

    c. Find BC and AC.

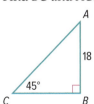

    d. Find AB.

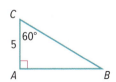

33 The table below indicates the number of minutes Regina spent practicing the piano each day for the last several weeks. Find the mean and median number of minutes she practiced during this time period.

Minutes of Practice	0	10	20	30	45	60
Number of Days	5	2	4	6	2	1

34 Write each sum of rational expressions in equivalent form as a single algebraic fraction. Then simplify the result as much as possible.

a. $\dfrac{x}{3} + \dfrac{2x + 3}{3}$

b. $\dfrac{6x}{5} + \dfrac{x + 1}{2}$

c. $\dfrac{3x + 5}{x} + 2x$

35 Points A, B, and C are on the circle with center O and radius 8 cm.

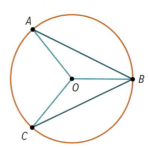

In addition, m$\angle AOC = 100°$ and $AB = BC$. Find each of the following.

a. m$\overset{\frown}{AC}$

b. m$\angle ABC$

c. m$\overset{\frown}{AB}$

d. BC

Looking Back

In this unit, you revisited a variety of basic types of functions and developed strategies for modifying those functions to build models for more complex relationships. As a result of that work, you developed greater skill in using linear, exponential, quadratic, power, and circular functions to model data patterns and problem conditions. You also learned how to modify the rules of basic function types to model and analyze data patterns that are related to familiar functions by vertical and horizontal translation, by reflection across the *x*-axis, and by vertical and horizontal stretching or compressing.

The tasks in this final lesson give you a chance to review your skill and understanding of function families and transformations.

Thinking in Millennia New Year's Day is celebrated in cultures and countries around the world. But January 1, 2000 was a very special date, because it marked the beginning of a new millennium, or thousand-year time period. The occasion prompted many comparisons with life at the start of the previous millennium in the year 1000. At that time:

- Earth's human population was about 250 million and growing at a rate of 0.1% per year.

- one fourth of the population lived in China, and the world's largest city was Cordoba, Spain, with a population of 450,000.

- half of all children died before the age of five.

By the year 2000, the world's population had increased to about 6 billion (6,000 million) and it was growing at an annual rate of 1.7%.

1 About how many people were added to the world population in the year 1000? In the year 2000?

JLImages/Alamy

2 The relatively low world population growth rate in the year 1000 continued until the 1700s, when more modern medicine and improved water and sewage systems emerged.

    a. Suppose that world population growth had continued at an annual rate of 0.1% from 1000 to 1700. What function would this condition imply as a model for estimating world population $P(t)$ in year $1000 + t$?

    b. What world population does your model in Part a predict for 1700? The actual world population in 1700 is estimated to have been about 640 million.

    c. Suppose that world population had continued to increase by the same number of people in each year after 1000. What function would this condition imply as a model for estimating world population in year $1000 + t$?

    d. What world population does your model in Part c predict for 1700?

    e. If world population growth had continued at the rates in the year 1000 until the year 2000, what would the population models have predicted for the year 2000:

        i. using the 0.1% growth rate condition in Part a?

        ii. using the constant number-of-people-per-year growth rate condition in Part c?

    f. If world population grows beyond the year-2000 figure of 6 billion at the rate of 1.7% per year:

        i. what function would this condition imply as a model for predicting world population in year $2000 + t$?

        ii. what world population does the model in part i predict for the year 2050?

        iii. what reasons can you imagine for doubting that the prediction in part ii will actually occur?

Planetary Motion Whenever scientists report an unusual astronomical event, we are reminded that our Earth is a very small planet in a very large universe. For example, when the Hale-Bopp comet flew within sight of Earth during 1996 and 1997, there was considerable discussion about the chances that other comets and asteroids might actually enter Earth's atmosphere. Some scientists even made estimates of the damage that would result from such an event.

One theory predicts that if an asteroid with a diameter of only 3 miles were to land in the middle of the North Atlantic Ocean, it would send a 300-foot tsunami crashing on the shores of North America and Europe. Fortunately, such an event is estimated to occur only once every 10,000,000 years!

3 Comets and asteroids have irregular shapes, but most can be approximated as spheres.

  a. If an asteroid has average diameter d miles, what function rules give the:

    i. disk area of the cross section at a diameter of the asteroid?

    ii. total surface area of the approximately spherical body?

    iii. volume of the approximately spherical body?

  b. The diagram below shows graphs of the three measurement functions in Part a on the interval $0 \leq d \leq 10$.

    i. Match the functions and graphs and explain how you know you are correct.

    ii. Explain what the relative shape of the three graphs says about the rates at which disk area, surface area, and volume change as asteroid diameter increases.

Asteroid Measurements

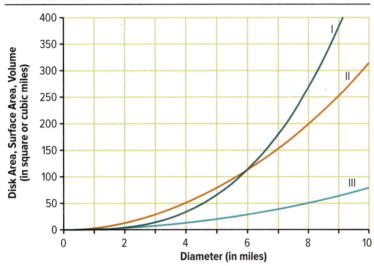

  c. Comet Hale-Bopp appeared recently near Earth. Prior to that appearance, it last came near Earth over 4,200 years ago. The gravitational pull of the planet Jupiter will cause it to return near Earth again in about 2,400 years. The gravitational attraction of two large bodies is directly proportional to the product of their masses and inversely proportional to the square of the distance between their centers.

    i. If m_J and m_H represent the masses of Jupiter and comet Hale-Bopp respectively and d represents the distance between the centers of those masses, what is the form of the rule for the function $g(d)$ telling the gravitational attraction of those bodies at any distance d?

    ii. What will a graph of $g(d)$ look like?

4 The visible Moon varies in size from a full moon to a new moon (not visible at all) and back to a full moon in a cycle that takes roughly 30 days. Dates of many important religious and cultural events are set by reference to lunar calendars.

  a. What function family seems likely to be the best starting point in building a model that tells visible area at any time during its 30-day cycle of phases:

  i. if you assume that the cycle starts with a full moon?

  ii. if you assume that the cycle starts with a half-moon on its way toward a full moon?

  b. What particular members of the function family described in your response to Part a are likely to be good models of change in the visible moon if we assume that a full moon is 100%, a new moon is 0%, the cycle is 30 days long, and

  i. the cycle starts with a full moon?

  ii. the cycle starts with a half-moon on its way toward a full moon?

Matching Function Rules and Graphs In the lessons of this unit, you discovered that when building models of data patterns, it helps if you can identify a likely function rule by inspecting the graph.

5 Match each of the functions given in Parts a–j with their graphs in the following diagrams without using technology.

  a. $y = x - 2$ **b.** $y = (x + 2)^2 - 5$

  c. $y = 1.5^x - 2$ **d.** $y = 1.5x - 2$

  e. $y = 3 \cos x + 1$ **f.** $y = 3 \cos x - 1$

  g. $y = (x - 2)^2 - 5$ **h.** $y = 3 \cos 2x - 1$

  i. $y = -2(x + 1)$ **j.** $y = -2|x| + 2$

Graph I

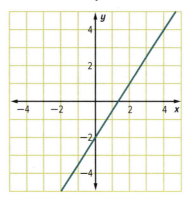

Graph II

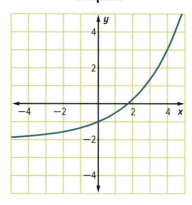

Graph III

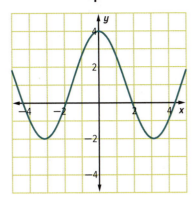

Graph IV

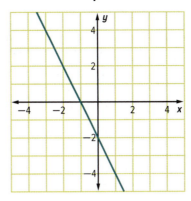

Graph V

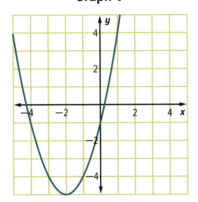

Graph VI

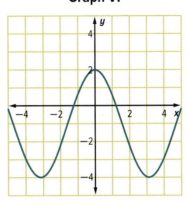

Graph VII

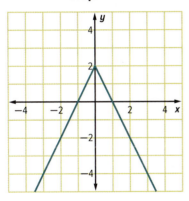

Graph VIII

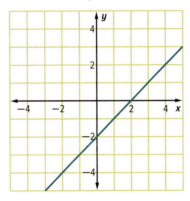

Graph IX

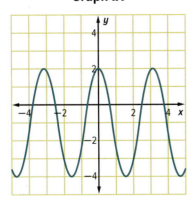

Graph X

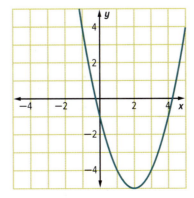

SUMMARIZE THE MATHEMATICS

In this unit, you investigated a variety of situations in which rules for familiar functions had to be modified to model patterns in data plots and conditions in particular problems.

a. What table and graph patterns and problem conditions are clues to use the following families of functions as models?

   i. Linear   **ii.** Exponential

   iii. Quadratic   **iv.** Direct power

   v. Inverse power   **vi.** Absolute value

   vii. Sine   **viii.** Cosine

   ix. Square root   **x.** Base-10 logarithm

b. What are the general forms of rules for each of the types of functions listed in Part a? What do the values of the parameters in those rules tell you about the expected patterns in tables and graphs of those particular functions?

c. How can you adjust the rule for a function $f(x)$ so that its graph matches the graph of a function $g(x)$ related by these transformations?

   i. Vertical translation   **ii.** Horizontal translation

   iii. Vertical stretching/compressing   **iv.** Horizontal stretching/compressing

   v. Reflection across the x-axis

d. How can the basic sine and cosine functions be modified to give functions with amplitude A and period p?

Be prepared to share your responses and reasoning with the class.

✔ CHECK YOUR UNDERSTANDING

Write, in outline form, a summary of the important mathematical concepts and methods developed in this unit. Organize your summary so that it can be used as a quick reference in future units.

Counting Methods

Digital Resources at
ConnectED.mcgraw-hill.com

Watch | Tools | Audio | eBook

Counting methods are part of an area of mathematics called *combinatorics*. They are used to answer the question, "How many?" This question arises in everyday life, in many applied contexts, and in more abstract settings. For example, how many boys with GPA greater than 3.0 are in your graduating class? How many Internet addresses (IP numbers) are possible? How many cell phone numbers can have the prefix 919? Or even, how many terms are in the expansion of $(a + b)^6$ and what is the coefficient of a^2b^4 in that expansion?

In this unit, you will develop skill in systematic counting. You will learn concepts and methods that will help you solve counting problems in many different contexts. You will also develop skill in combinatorial reasoning.

These ideas are developed in the following three lessons.

LESSONS

1 Systematic Counting

Develop skill in systematic counting through careful combinatorial reasoning with systematic lists, tree diagrams, the Multiplication Principle of Counting, and the Addition Principle of Counting.

2 Order and Repetition

Develop the understanding and skill to count the number of possible selections from a collection of objects, including those that involve combinations or permutations.

3 Counting Throughout Mathematics

Use counting strategies to help solve probability problems in which all outcomes are equally likely and enumeration problems in geometry and algebra. Develop and apply the Binomial Theorem and its connection to Pascal's triangle.

Jim Laser

Systematic Counting

It seems like everywhere you look—in school, at the mall, at movie theaters and concerts, on college campuses, and at work—denim jeans have become a wardrobe staple. According to a recent representative survey of 6,000 consumers aged 13–70, 60% female and 40% male, conducted by Cotton Incorporated's Lifestyle Monitor™, 96% of consumers own denim jeans—seven pairs on average! Consumers in the survey reported wearing denim jeans an average of four days a week. (**Source:** www.cottoninc.com)

Just Jeans, a store in local malls, specializes in denim jeans and carries several of the more popular brands for both males and females. Maintaining proper inventories requires understanding the market and mathematics. Consider, for example, that one popular brand of jeans for males is available in six styles: Slim, Bootcut, Straight Leg, Relaxed, Original, and Loose; nine finishes, including stone washed, raw indigo, Texas crude, and vintage; 14 waist sizes, 28"–46" (37", 39", 41", 43", and 45" excepted); and nine inseam lengths, 28"–36".

The same brand also manufactures women's denim jeans. Women's jeans are available in seven styles: Legging Jeans, Skinny, Real Straight, Curvy, Bootcut, Natural, and Long and Lean; two finishes, basic indigo and black-dyed; 11 sizes, 0 (XS), 2 (XS), 4 (S), 6 (S), 8 (M), 10 (M), 12 (L), 14 (L), 16 (XL), 18 (XL), and 20 (XXL); and 12 inseam lengths, 29"–37", including some "half" lengths such as 32½".

Jim Laser/CPMP

You can see that there are many different types of jeans possible! Knowing just how many different types is important for the store manager, who must order and stock the jeans. In this unit, you will learn how to count in situations like this and many others.

THINK ABOUT THIS SITUATION

Think about the popularity of denim jeans and inventory questions the manager of Just Jeans and her employees face.

a. What struck you about the consumer survey on wearing denim jeans?

b. Assuming your school does not require school uniforms, how many students in your mathematics class are wearing jeans today?

c. Are jeans more popular among boys or girls in your math class today? Why should you report your answer as a percentage rather than a count?

d. Considering the styles, finishes, waist sizes, and inseam lengths for the male jeans listed on the previous page, how many different male jean options do you think are available from the manufacturer? Explain your reasoning.

e. Considering the styles, finishes, sizes, and inseam lengths for the brand's female jeans, how many different female jean options do you think are available from the manufacturer? Explain your reasoning.

f. Just Jeans would likely not stock all the options available from the manufacturer. However, the store would stock more than one sample of male and female jeans that tend to be more popular. Given this additional information, how do you think that would affect your answers and counting strategies in Parts d and e?

As you have seen, answering the question, "How many?" can, at times, be more difficult than you might think. The branch of mathematics that deals with systematic methods of counting is called **combinatorics**. In this unit, you will learn some of the basic concepts, strategies, and reasoning methods important in combinatorics. You may be surprised at the variety of contexts in which strategic counting is useful.

Methods of Counting

In the Think About This Situation, you brainstormed about denim jeans selections and store inventories involving careful counting, its importance, and some possible counting strategies. In this investigation, you will begin a careful analysis of counting concepts, methods, and their applications.

As you work on the problems in this investigation, look for answers to this question:

What are some useful methods for systematic counting?

1 Some Web sites, as you may know, require users to have passwords so that access can be controlled and security can be maintained. The number and type of characters allowed in a password can vary.

   a. Suppose a password consists of only two characters: one letter from A to D, followed by one digit from 0 to 2.

   i. How many different passwords are possible? Explain the method you used to get your answer. Describe at least one other method that could be used.

   ii. Do you think this password format is practical? Why or why not?

   b. For one Web site, passwords consist of five characters: three digits from 0 to 9, followed by two letters from A to Z.

   i. How many different passwords are possible? Compare your answer with that of your classmates. Resolve any differences.

   ii. Suppose an unauthorized user tries to gain access to the system simply by trying different passwords. If the person can try one password every second, and the system does not cut her off, what is the longest it could take to get into the system? How many minutes? Hours? Days? (For this question, ignore the need to also know the correct username.)

2 Personal Identification Numbers (PINs) used in Automated Teller Machines (ATMs) are similar to computer passwords. You insert your card and then enter the correct PIN in order to get access to your bank account and withdraw or deposit money. A PIN often consists of four digits, 0 to 9. Some ATMs will capture your card if you enter the wrong PIN too many times. Use a counting argument to help explain why this is done.

There are many different methods you might use to solve counting problems. Make note of different methods as you solve Problems 3 and 4.

3 Outdoor Adventure Clothing and Gear stocks windproof ski jackets in a different style for men and women. For each of these styles, there are 4 colors (purple, teal, yellow, blue), in each of 3 sizes (small, medium, large).

a. How many different jacket options do you think are possible?

b. Students in one class at Pioneer High School proposed the following methods for answering this question.

Carefully study each method. Which methods work? Compare your answer with those of your classmates. Resolve any differences.

Julia's Method
(tree diagram)

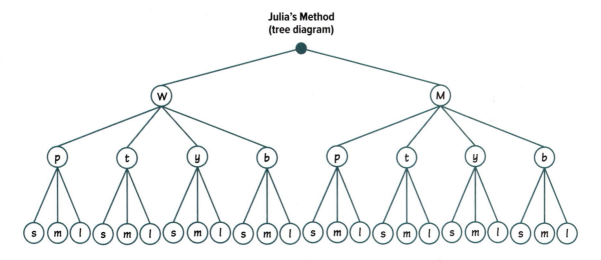

Alicia's Method

M-b-s	W-b-s
M-b-m	W-b-m
M-b-l	W-b-l
M-y-s	W-y-s
M-y-m	W-y-m
M-y-l	W-y-l
M-t-s	W-t-s
M-t-m	W-t-m
M-t-l	W-t-l
M-p-s	W-p-s
M-p-m	W-p-m
M-p-l	W-p-l

Kaya's Method

W-b-l   M-p-s   W-y-s   W-t-s
W-t-m   M-y-s   M-y-m
M-t-l   W-p-s   M-p-l   W-t-l
W-b-s   W-p-l   M-y-l
M-b-s   W-p-m   M-t-m
W-b-m   W-y-m   M-t-s
M-b-l

Alonzo's Method

The jackets are for men or women.
There are 12 options for Men and
12 options for Women.
So, there are 12 + 12 = 24 different
jacket options.

Ryan McVay/Getty Images

Monica's Method

2 × 4 × 3 = 24 different
jacket options

2 choices: 4 choices   3 choices
men or   for color   for size
women

Samuel's Method

There are 2 genders
plus 4 colors
plus 3 sizes.
So, there are
2 + 4 + 3 = 9
different jacket
options possible.

Sandra's Method

There are 4 colors for
both men and women;
that makes 4 × 4 = 16.
Then there are 3 sizes;
that makes 16 × 3 = 48.
So, there are 48
different jacket
options possible.

c. For those methods that work, describe connections among them. For example, how does Monica's multiplication method relate to Julia's tree diagram? Find connections among the other correct methods.

4 Use methods similar to those in Problem 3 that were correct or invent your own method to solve the following counting problems.

a. Suppose the men's and women's jackets from Problem 3 are available in four sizes—small, medium, large, and extra large and the same four colors as before. How could you modify the correct methods in Problem 3 to determine the number of different jacket options that are now possible? What is the total count now?

b. Suppose that the men's jackets are available in five sizes including an extra, extra large (XXL), while the women's jackets are available in four sizes (and the same four colors as before). If each different jacket will be displayed on a separate hanging rack in the store, how many racks are needed? Explain your method.

c. Suppose a password consists of only two characters—one letter from A to D, followed by the digit 1 or 2. Count the number of passwords in this situation using each correct method from Problem 3.

d. Suppose a password consists of six characters—four digits from 0 to 9, followed by two letters from A to Z. Which of the methods in Problem 3 would be most effective for determining how many different passwords are possible? Why? Use that method to determine the answer.

SUMMARIZE THE MATHEMATICS

In this investigation, you used systematic counting to answer the question, "How many?" in several different contexts.

a. List four correct counting methods you have seen or used in this investigation.

b. For each method:

- describe how it works.

- give an example of how to use it.

- discuss some advantages and disadvantages of the method.

c. Look back at the counting situations you have considered. Find an example where you might use more than one counting method to solve the problem.

Be prepared to explain your methods, examples, and thinking to the class.

✓ CHECK YOUR UNDERSTANDING

The telephone number for a local landline has seven digits, like 472-5555. The first three of these digits are called the *prefix*. In some small towns, there may be only one prefix for all the telephone numbers of residents. When a new prefix is needed, it can be big news. For example, on October 1, 2011, a fourth prefix, 974, was added for the city of Iqaluit, Nunavut (Canada). Read the notice below from the local telephone company announcing the addition of a new prefix.

Northwestel Introduces New Telephone Number Prefix in Iqaluit

Northwestel customers in Iqaluit will soon be dialing 974 when making local calls. The telephone company will introduce a new phone number prefix in Iqaluit on October 1st.

There are already three prefixes, also known as office codes or NXX numbers, being used in Iqaluit. They are 979, 975, and 222.

Northwestel provides a new prefix in a community when the available numbers begin to be depleted, to ensure that the community does not run out of phone numbers.

Source: nwtel.ca

a. Assuming no restrictions on the digits, how many different phone numbers can be created with the 974 prefix?

b. How many different phone numbers are available in Iqaluit using all four prefixes?

Principles of Counting

Some of the counting methods in Investigation 1 are based on fundamental principles of counting. A common element of these principles is that they enable you to count without counting!

As you work on the problems in this investigation, look for answers to the following questions:

What are some fundamental principles of counting?

How and when are these principles useful in solving counting problems?

1 Multiplication Principle of Counting Monica's reasoning in Problem 3 of Investigation 1 illustrates the *Multiplication Principle of Counting.*

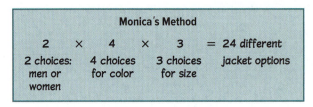

Monica's Method

2	×	**4**	×	**3**	=	24 different
2 choices: men or women		4 choices for color		3 choices for size		jacket options

In general, if you want to count all the outcomes from a sequence of tasks, count how many outcomes there are from each task and multiply those numbers together.

If you can picture the counting situation as a *counting tree* as shown below, then the Multiplication Principle probably applies. More formally, you can use the Multiplication Principle when you want to count all the *combined outcomes* from a *sequence* of tasks, where the numbers of outcomes from each task are *independent* of each other as elaborated below.

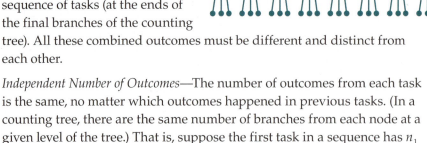

- *Sequence of Tasks*—You want to count all the possible outcomes that result from a sequence of tasks. (In a counting tree, the sequence is seen as the branches of the tree.)

- *Distinct Combined Outcomes*—You want to count all the combined outcomes at the end of the sequence of tasks (at the ends of the final branches of the counting tree). All these combined outcomes must be different and distinct from each other.

- *Independent Number of Outcomes*—The number of outcomes from each task is the same, no matter which outcomes happened in previous tasks. (In a counting tree, there are the same number of branches from each node at a given level of the tree.) That is, suppose the first task in a sequence has n_1 outcomes; then for each of these, the second task has n_2 outcomes; and for each of these, the third task has n_3 outcomes; and so on.

If these three conditions are met, then the number of possible combined outcomes from the entire sequence of tasks is $n_1 \times n_2 \times n_3 \times \ldots$.

a. Identify two examples from your solutions of previous counting problems where you used the Multiplication Principle of Counting.

b. For each of your examples in Part a, describe how the counting problem satisfies the three conditions at the bottom of page 208.

2 Examine the structure of each of the following problems in terms of requirements for use of the Multiplication Principle of Counting. Then solve each problem.

a. Consider a password that consists of just two characters. One character is a letter from A to C and the other character is a digit from 0 to 4. The two characters can be in either order: letter-digit or digit-letter. How many such passwords are there?

b. A Web site password consists of any 7 letters followed by any 2 digits.

    i. How many passwords are possible?

    ii. Suppose that no letters or digits can be repeated. In this case, how many passwords are possible?

c. Suppose a password consists of 4 letters and 1 digit in any order. Letters can be repeated in a given password. How many passwords are possible?

3 With the popularity of cell phones, people are using more and more phones. This creates a demand for new telephone numbers. In most of North America, a phone number looks like 641-555-0136. There is a three-digit area code, then a three-digit local prefix, then the final four digits. One way to create more phone numbers is to create more area codes. There are rules for how to create an area code. The rules changed in 1995.

a. Examine the area code rules below from before 1995. How many area codes were possible at that time?

Area Code Rules Before 1995:

- The first digit cannot be 0 or 1, since these digits indicate special numbers for the phone company, like calling the operator.

- The second digit must be 0 or 1.

- The third digit can be any digit 0–9.

b. The old area codes were running out fast. So, now the restriction on the second digit has been lifted. The rules today are the same except that the second digit can be any digit 0 to 9. How many different area codes are possible today?

4 Addition Principle of Counting Some counting situations require more than the Multiplication Principle of Counting. Addition also can play an important role. In fact, another counting principle is the **Addition Principle of Counting**—The total number of outcomes from two tasks is the sum of the number of outcomes from each task (minus the number of outcomes that are common to both tasks, if there are any).

a. Identify two examples from your solutions of previous counting problems where you used the Addition Principle of Counting. Explain how the principle was used.

b. Look back at the student counting methods from Problem 3 in Investigation 1. Which student(s) used the Addition Principle of Counting? Explain.

5 Counting problems are sometimes based on information presented in a table. For example, the number of men and women in the U.S. Senate in 2014 who were Democrats or Republicans is shown in the table below.

	Men	Woman	Total
Democrat	37	16	53
Republican	41	4	45
Total	78	20	98

a. Suppose a brochure was sent to every U.S. Senator who is a Republican or a woman. How many brochures were sent so that every Republican and every woman received exactly one brochure?

b. Study George's answer to Part a. Do you agree with George's solution? Why or why not?

> *I need to count everybody who was a Republican or a woman. There were 45 Republicans and 20 women. So, by the Addition Principle of Counting, the answer is 45 + 20 = 65.*

SUMMARIZE THE MATHEMATICS

In this investigation, you studied two fundamental principles of counting—the *Multiplication Principle of Counting* and the *Addition Principle of Counting*.

a. State each principle in your own words.

b. These two principles are meant to express fundamental common sense ideas about counting. For each principle, explain why it makes sense.

Be prepared to explain your statements and reasoning to the entire class.

Think about how many different automobile license plates can be made under certain constraints. For example, in 2012 in the state of New York, an Empire license plate had seven characters. The first three characters were letters and the last four were numbers.

a. How many different Empire license plates were possible in New York in 2012?

b. Suppose that no letters or numbers can be repeated. How many Empire license plates are possible?

c. Suppose that the three letters are used to represent the county, with MON used for all Empire license plates issued to residents of Monroe County. Do you think this is a good plan? Explain.

d. License plate configurations do not always include all possible letters and numbers. Suppose New York started their lettering with ACA-1000 and counted up, leaving out letters "I," "O," and "Q" to avoid confusion with 1 and 0. How many plates are possible?

e. In some states, for a special fee, you can request a personalized plate. Suppose that in Michigan, you can order a personalized plate with five characters drawn from all possible letters or the digits 0 to 9 with repetitions allowed. How many personalized plates could be issued in Michigan?

f. How many personalized license plates could be issued in your state?

1 Every device on the Internet is assigned a unique number, known as an *IP address*, like 188.165.140.31. If you check the network settings on your computer, you will probably find a number like this. The most commonly used version of IP addresses as of 2014 was IP version 4 (IPv4). At the right is a diagram showing an IPv4 address.

An IPv4 address (dotted-decimal notation)

172 . 16 . 254 . 1

10101100.00010000.11111110.00000001

One byte = Eight bits

Thirty-two bits (4 • 8), or 4 bytes

a. An IPv4 number consists of four decimal numbers separated by dots (which is called "dotted-decimal notation"), like 172.16.254.1 in the figure above. The actual number as stored in a computer is a 32-bit binary number. A "bit" (binary digit) is a 0 or a 1. Thus, an IPv4 number is a string of 32 bits. How many IPv4 numbers are possible, assuming no restrictions on the bits?

b. As with telephone numbers, there are some restrictions on IPv4 numbers. The restrictions get rather technical, but the most common restrictions put limitations on the use of the decimal numbers 0 and 255. The decimal number 0 is represented in binary as 00000000; the decimal number 255 is represented as 11111111. Suppose that the only restrictions for an IPv4 number are that the decimal numbers 0 and 255 are not allowed for the 1st or 4th bytes. In this situation, how many IPv4 numbers are possible?

c. Due to the rapidly growing numbers of Internet users and Internet devices, there are not enough IPv4 numbers. Another type of IP address is IPv6. An IPv6 number has 128 bits. How many times bigger is the number of possible IPv6 numbers than the number of possible IPv4 numbers (assuming no restrictions on bits)?

2 Web site passwords sometimes have very specific requirements. For example, a password for some non-classified U.S. military Web sites in 2010 had requirements similar to the following. The password must contain exactly 10 characters in the following order: 2 upper-case letters, 2 lower-case letters, 2 digits, 2 special characters, and 2 final characters that can be of any of the four previous types. A "special character" must be one of the "shift" characters from the row that includes the numbers (and characters such as the hyphen)on a standard computer keyboard. How many different passwords are possible with these requirements?

3 Suppose you toss a fair coin three times, and make a note of *heads* or *tails* on each toss.

a. Construct a tree diagram showing all possible results.

b. List the ways to get at least two heads.

Robert Dant/E+/Getty Images

4 AT&T was the only telephone company in the United States until 1984. This company devised the original telephone numbering policies in the 1940s. The original policies are listed below. Under these restrictions, how many phone numbers were possible?

Original Telephone Numbering Policies

- A phone number consists of ten digits: 3 digits for the area code, 3 digits for the local prefix, and four digits for the local number.

- 0 cannot be used as the first digit of an area code or a local prefix, since dialing 0 is reserved for reaching the operator.

- Phone numbers beginning with 1 are reserved for internal use within the telephone system, so 1 cannot be used as the first digit of an area code or a local prefix.

- Early phone numbers included letters, for example PYramid4-1225 instead of 794-1225. In order to be able to dial letters, most numbers on a telephone have letters associated with them. However, to avoid confusing the numbers 0 and 1 with the letters O and I, the numbers 0 and 1 on a telephone do not have associated letters. So, 0 and 1 cannot be the second digit of a local prefix.

- The telephone system used the second digit to distinguish an area code from a local prefix. Since the second digit of a local prefix can be anything except 0 and 1, the second digit of an area code must be 0 or 1.

- The third digit of an area code can be any number. The third digit of a local prefix can be any number except 0 or 1.

- The four digits of the local number can be anything.

5 The "call sign" of a radio station is the set of letters by which it is identified, like KOA or WGBH. The call signs of most regular broadcast radio stations in the United States have 3 or 4 letters.

a. Generally, the call signs of radio stations west of the Mississippi river have letters starting with K. How many different 3- and 4-letter call signs like this are possible? In your answer, are you assuming that letters can be repeated or not? Whatever you decide, give a brief argument supporting your decision.

b. If all radio call signs begin with K or W, how many different call signs are possible?

Jim Laser/CPMP

6 Dominos are rectangular tiles used to play a game. Each tile is divided into two squares with a number of dots in each square, as in the figure below.

a. The standard set of dominos has from 0 to 6 dots in each square. How many different standard dominos are possible?

b. A deluxe set of dominos has from 0 to 9 dots in each square. How many different deluxe dominos are possible?

c. How many dominos are possible in a set that has 0 to *n* dots in each square?

7 The original plan for assigning telephone numbers that you investigated in Applications Task 4 was implemented in 1947. At that time, the supply of numbers was expected to last for 300 years. However, by the 1970s the numbers were already starting to run out. So, the numbering plan had to be modified. In this task, you will count the number of different phone numbers that were available in 2012.

a. For three-digit area codes, the first digit cannot be a 0 or a 1. Assuming no additional restrictions, how many three-digit area codes are possible under this plan?

b. Certain area codes are classified as "Easily Recognizable Codes" (ERCs). ERCs designate special services, like 888 for toll-free calls. The requirement for an ERC is that the second and third digit of the area code must be the same. The first digit again cannot be a 0 or a 1. How many ERCs are there?

c. Consider the seven digits after the area code. As with the area code, the first digit of the three-digit local prefix cannot be a 0 or a 1. The remaining six digits for the local number have no restrictions. How many of these seven-digit phone numbers are possible?

d. Assuming only the 0 and 1 restrictions in Parts a and c, how many ten-digit phone numbers are possible?

8 The password format for a particular Web site has five letters and three digits. How many different passwords are possible in each of the following situations?

a. The letters follow the digits, and the letters and digits can be repeated.

b. The letters follow the digits, and the letters and digits cannot be repeated.

c. The three digits can be anywhere in the password, and the letters and digits can be repeated.

9 The table below shows information about age and income for employees in a manufacturing business.

	$30,000 or Less	More Than $30,000	Total
25 or Younger	19	8	27
Older Than 25	34	31	65
Total	53	39	92

a. How many employees make $30,000 or less?

b. How many employees are older than 25 and make more than $30,000?

c. How many employees are older than 25 or make more than $30,000?

d. Which of Parts a, b, or c uses the Addition Principle of Counting?

CONNECTIONS

10 Using your class data from Part c of the Think About This Situation (page 203), determine if wearing denim jeans is independent of gender.

11 Recall that a matrix is a rectangular array of numbers. The 3×2 matrix below has 3 rows and 2 columns.

$$\begin{bmatrix} 0 & 0 \\ 1 & 1 \\ 0 & 1 \end{bmatrix}$$

a. How many 3×2 matrices that have only 0 or 1 as entries are possible?

b. How many $n \times m$ matrices that have entries of only 0 or 1 are possible?

12 In Investigation 2 Problem 5, George incorrectly counted the number of brochures needed so that every U.S. Senator who is a Republican or a woman was sent exactly one brochure. A Venn diagram representing this counting task is shown at the right.

a. Analyze the diagram by comparing it to the table on page 210. How does the Venn diagram help avoid George's mistake?

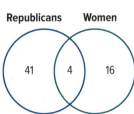

Republicans   Women

41   4   16

b. Draw a Venn diagram to help count the number of brochures needed to send exactly one to each U.S. Senator who is a Democrat or a man. How many brochures are needed?

13 At many private and some public schools, students are required to wear uniforms. Suppose the boys must wear navy or khaki pants; a short-sleeve or long-sleeve white shirt; and a sweater, vest, or blazer.

a. Explain how the Multiplication Principle of Counting can be used to reason that in any group of at least 13 boys, two will be wearing the same type of uniform.

b. How big would the group need to be to guarantee that three boys would be wearing the same type of uniform?

REFLECTIONS

14 How would your answer to Part d of the Think About This Situation (page 203) change if the particular brand manufacturer added a Flair style for males with the same options for finish, waist size, and inseam length as their other jeans?

15 Examine the license plates from Michigan and Minnesota shown below.

Typical Michigan and Minnesota license plates issued in 2012 and 2008, respectively

a. What appear to be the license identification schemes for the two states?

b. Why do you think the two states have different schemes?

c. How is the license plate identification in your state similar to, and different from, those for Michigan and Minnesota?

16 Recall or listen to the song "The Twelve Days of Christmas." How many gifts have accumulated after the 12th day?

17 Two useful mathematical practices are:

• Making sense of problems and persevering in solving them

• Reasoning abstractly and quantitatively

For each of these practices, give one example from this lesson where you used that practice. Explain how you used the practice.

18 Find a counting problem in your daily life or in a newspaper whose solution involves the Multiplication Principle of Counting or the Addition Principle of Counting. Describe the problem and explain how it can be solved.

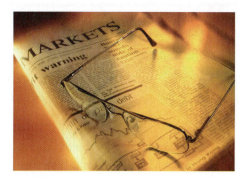

19 In Applications Task 1, you were introduced to binary numbers. The decimal number 2 is written as 10 in binary notation. The decimal number 3 is written as 11 in binary notation. If necessary, do some research to explain the conversion of the decimal number 255 to the binary number 11111111.

20 In the book *The Man Who Counted: A Collection of Mathematical Adventures* by Malba D. Tahan, translated by Leslie Clark and Alastair Reid, a story is told of Beremiz Samir, a man with amazing mathematical skills. Read at least the first chapter of this book. Then describe and explain at least one counting feat performed by Beremiz.

21 Fifty-five seniors at Hackett High School were surveyed about their food preferences, with regard to fish, chicken, and beef. Here are the results.

> 23 like fish
> 17 like chicken
> 17 like beef
> 6 like beef and chicken
> 8 like beef and fish
> 10 like chicken and fish
> 2 like all three

a. Represent this situation with a Venn diagram. Include the appropriate numbers in all regions of the diagram.

b. How many students like chicken and fish, but not beef?

c. How many students like beef and fish, but not chicken?

d. How many students like only beef, and not chicken or fish?

e. How many students do not like any of the three?

22 Think about all possible non-negative integers with their standard ordering; that is, 0, 1, 2, 3, 4, 5, … . What proportion of all non-negative integers contains the digit 3?

(t)©Danilo Calilung/Corbis, (b)Jess Alford/Getty Images

23 When using a calculator to multiply large or small numbers, the result may be displayed in scientific notation. For each number below, rewrite the number using standard decimal notation and using scientific notation.

a. 250^5

b. $(5 \times 10^4)(12 \times 10^6)$

c. 0.02^5

d. $\dfrac{1}{(100)(25^2)}$

24 The Venn diagram below represents some characteristics of the 33 students in one of Mrs. Chen's mathematics classes.

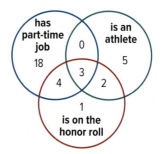

Suppose that you were to randomly pick a student from this class for an interview. Determine each of the following probabilities.

a. The student is an athlete.

b. The student is on the honor roll and an athlete.

c. The student has a part-time job and is not an athlete.

d. The student does not have a part-time job and is not on the honor roll.

25 Each of the following graphs is a transformation of the graph of $f(x) = x^2 - 4$. For each graph, describe the transformation and then write a function rule that matches the graph.

a.
Graph I

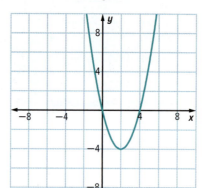

b.
Graph II

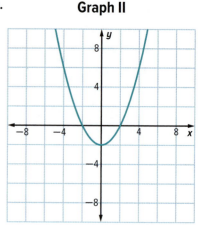

c.
Graph III

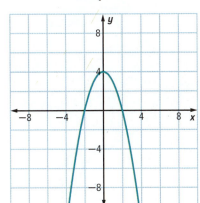

d.
Graph IV

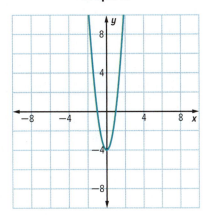

26 In March of 2011, the journal *Pediatrics* reported on a study of obesity in preschool-age children. Susanna Huh and her colleagues considered the relationship between obesity and the age at which formula-fed babies were introduced to solid food. There were 279 formula-fed babies in their study. Of the 91 babies who were introduced to solid food before they were four months old, 23 of them were obese at age 3. Out of the 188 babies who were introduced to solid foods after they were four months old, 12 were obese at age 3.

(**Source:** Susanna Y. Huh, Sheryl L. Rifas-Shiman, Elsie M. Taveras, Emily Oken and Matthew W. Gillman, "Timing of Solid Food Introduction and Risk of Obesity in Preschool-Aged Children," *Pediatrics*, Vol 127 (2011): e544–e551; pediatrics.aappublications.org/content/127/3/e544.full.pdf)

a. What are the two groups under study here? What is the response variable?

b. Summarize the information in a two-way table, with the numbers for each group in separate rows.

c. Compute and interpret the difference in absolute risk of obesity at age 3.

d. Compute and interpret the relative risk of obesity at age 3.

e. Write a conclusion based on this evidence.

27 Consider a right cylinder that has diameter 6 cm and height 12 cm.

a. Will this cylinder hold more or less than a 355-ml can of soda? (Recall that 1 ml = 1 cm³.) Explain your reasoning or show your work.

b. Could you place a pencil that is 12.5 cm long completely inside this cylinder? Explain your reasoning.

28 Without using technology, write a function rule that matches each description.

a. A linear function whose graph contains the points (1, 7) and (3, −5)

b. A linear function whose graph is parallel to the graph of $g(x) = \frac{1}{2}x - \frac{2}{5}$ and containing the point (7, 2)

c. A quadratic function with vertex (2, −3) and x-intercept (5, 0)

Order and Repetition

Two important issues that often arise in counting problems are *order* and *repetition*. You have already seen these issues in your counting work in Lesson 1, such as whether letters or digits could be repeated in a password or not. Now consider how these issues are treated in the *Education Week* article below.

Take Note: Counting Their Chickens

A national restaurant chain that boasts about its tasty chicken is eating crow after a high school mathematics class cried foul over a television ad.

The ad shows Joe Montana, the National Football League quarterback, standing at the counter at a Boston Chicken restaurant puzzling over side-dish choices when an announcer says that more than 3,000 combinations can be created by choosing three of the restaurant's 16 side dishes.

But Bob Swaim, a math teacher at Souderton Area High School near Philadelphia, and his class did the math and told the Colorado-based chicken chain that there were only 816 combinations.

"We goofed," said Gary Gerdemann, a spokesman for Boston Chicken, explaining that the restaurant had confused "combinations" with "permutations."

"Apparently we didn't listen to our high school math teachers," Mr. Gerdemann said.

The company has, however, listened to Mr. Swaim and corrected its ads. For their eagle eyes, the students were awarded free meals and $500 to expand the math menu at Souderton.

Source: *Education Week.* Vol. 14, No. 20, p. 3

McGraw-Hill Education

THINK ABOUT THIS SITUATION

Think about the mathematics in the situation involving side-dish choices described at the bottom of the previous page.

a. Consider order and repetition in choosing three of the restaurant's 16 side dishes.

    i. Are repetitions allowed in this counting situation? Explain your thinking.

    ii. Do two different orderings count as two different possibilities? Explain.

b. The announcer claimed that, "more than 3,000 combinations can be created by choosing three of the restaurant's 16 side dishes." How do you think this number was determined?

c. What error in reasoning likely explains the restaurant's inflated claim of more than 3,000 combinations?

d. How do you think Mr. Swaim's mathematics class came up with the number of 816 combinations of three side-dish meals?

In this lesson, you will apply the Multiplication Principle of Counting while taking into account order and repetition to obtain formulas for counting *permutations* and *combinations*. More generally, you will learn concepts and methods for counting the number of possible selections from a collection of objects. You will see how these problems are similar to, and different from, counting problems involving a sequence of tasks, which were the focus of Lesson 1.

INVESTIGATION 1

Permutations and Combinations

In the last lesson, you solved counting problems using fundamental counting principles. In this investigation, you will apply the Multiplication Principle of Counting, while taking into account the issues of order and repetition, to count *permutations* and *combinations*. These arise when you make selections from a collection of objects.

As you analyze the counting situations in this investigation, look for answers to the following questions:

Does the situation involve counting the number of selections from a collection of objects?

Are order and repetition important in the situation?

What strategies and formulas are useful in solving counting problems in which order and repetition are important?

Considering Order and Repetition in Counting To help you understand order and repetition in counting, consider the number of different types of groups chosen from a club. Suppose one club is selecting *officers*—President, Vice-President, and Treasurer. Another club is choosing a *committee*.

- Officers—The French Club will select three officers: President, Vice-President, and Treasurer. There are 15 members in the club from which to select these officers.

- Committee Members—The Ski Club will choose three people for a committee. There are 15 members in the club from which to choose these committee members.

1 Think about order and repetition in each of the above situations.

a. French Club Officers

  i. Is repetition allowed when selecting President, Vice-President, and Treasurer? Explain.

  ii. Do different orderings count as different possibilities when selecting President, Vice-President, and Treasurer? Explain.

b. Ski Club Committee Members

  i. Is repetition allowed when choosing three members of a committee? Explain.

  ii. Do different orderings count as different possibilities when choosing three members of a committee? Explain.

c. Which is larger—the number of possible groups of three officers or the number of possible groups of three committee members? Why does that make sense?

2 Now think about the number of different possibilities in each situation.

a. French Club Officers

How many different selections for President, Vice-President, and Treasurer are possible? Show the calculation you used to determine how many. Compare your reasoning and answer with that of your classmates. Resolve any differences.

b. Ski Club Committee Members

How many different three-person committees are possible? Show the calculation you used to determine how many. Compare your reasoning and answer with that of your classmates and resolve any differences.

3 Below are responses from two students related to the club election problems. Carefully read each student's response.

a. Analyze Amy's reasoning about the number of French Club officer selections.

> *There are 15 choices for President. Once the President is chosen, then there are 14 members left who could be chosen for Vice-President. Once the President and Vice-President have been chosen, then there are 13 members left who could be Treasurer. So, there are 15 × 14 × 13 = 2,730 different possibilities for the three officers. This counts different orders as different possibilities. For example, ABC is different than BAC.*

This reasoning is essentially correct. Compare Amy's reasoning and answer with yours in Problem 2 Part a. If necessary, revise your work.

b. Analyze Latricia's reasoning about the number of Ski Club committees.

> *There are 15 choices for the first committee member. This leaves 14 club members who could be chosen as the second committee member, and 13 choices for the third committee member. So far, this is 15 × 14 × 13 possibilities. However, in this situation, order doesn't matter. For example, a committee of ABC is the same as a committee of BAC. In fact, for a committee consisting of A, B, and C, there are 6 different orderings that make the same committee. So, you must divide the first calculation by 6. This gives a final answer of $\frac{15 \times 14 \times 13}{16} = 455$ possible committees.*

This reasoning is essentially correct. Compare Latricia's reasoning and answer to yours in Problem 2 Part b. If necessary, revise your work.

4 Whenever an idea is particularly common and important in mathematics, a definition is made to capture that idea. There are a couple of definitions related to order and repetition. Specific mathematical terms are used for what you have been counting in the club situations.

a. A **permutation** is an arrangement in which order matters and repetitions are not allowed.

   i. Explain why counting the number of possible three-officer selections is a permutation problem.

   ii. Look again at the French Club officer selection problem in Problem 3 Part a. You now know this is a permutation problem. It also uses the Multiplication Principle of Counting. Where did Amy use the Multiplication Principle of Counting in her reasoning?

Image Source

b. A **combination** is an arrangement in which order does *not* matter, that is, different orderings are *not* counted as different possibilities. Also, repetitions are not allowed.

    i. Explain why counting the number of three-person committees is a combination problem.

    ii. The word "combination" in mathematics has a very specific meaning. It means something different than, for example, the combination of a dial locker or bicycle lock. Explain why the combination of a lock is *not* a mathematical combination.

    iii. Look again at the Ski Club committee problem in Problem 3 Part b. You now know that this is a combination problem. The Multiplication Principle of Counting is *not* used in this problem. Latricia correctly says that:

> *I did not use the Multiplication Principle in my reasoning. But it's like I started to use it, and then adjusted.*

    Where did Latricia start to use the Multiplication Principle, and at what point didit not work, so she adjusted?

c. In general, for a collection of people or identified objects, are there more permutations or combinations?

    i. Explain why your answer makes sense in terms of order and repetition.

    ii. Explain why your answer makes sense in terms of the computations used for counting permutations and combinations in the previous problems.

Formulas for Counting Permutations and Combinations

Formulas can be very helpful in computations and in understanding concepts and connections among them. There are several different yet equivalent formulas for both combinations and permutations.

Some of the formulas use *factorials*. **Factorial notation** is a compact way of writing certain products of consecutive non-negative integers. For example, $5 \times 4 \times 3 \times 2 \times 1 = 5!$, which is read as "5 factorial." In general, when n is a positive integer, $n! = n \times (n - 1) \times \cdots \times 2 \times 1$. By convention, $0!$ is defined to be 1.

5 Consider possible formulas for counting the number of permutations of k objects selected from n objects.

a. When Javon was working on Problem 2 Part a, his calculations suggested a general approach.

> *To count the number of permutations of k objects selected from n objects, start with n and carry out a factorial-type computation using exactly k factors.*

Describe how Javon's formula-in-words works for counting the number of French Club officer selections in Problem 2 Part a.

b. Elise proposed the following algebraic formula for counting permutations:

$$P(n, k) = n(n - 1)(n - 2)\cdots(n - k + 1)$$

    i. What do you think Elise meant by the notation $P(n, k)$?

    ii. Explain how her formula fits the formula-in-words from Part a.

c. Amy proposed a somewhat different formula for counting permutations, where $P(n, k)$ is the number of permutations of k objects selected from n objects.

$$P(n, k) = \frac{n!}{(n - k)!}$$

She described her formula as follows.

> I'm thinking of filling k boxes. In each box, I will put one of n objects. No repetitions are allowed and order matters. To count the number of ways to do this, I start with n possibilities for the first box, then $n - 1$ possibilities for the second box, then $n - 2$ for the next box, and I keep going like that until I've used up all k boxes. So, it's like a factorial but you cut it off after k factors. I can cut it off after k factors by dividing by $(n - k)!$.

    i. Use Amy's formula to find the number of French Club three-officer selections from 15 people. Make sure your answer agrees with that in Problem 2 Part a.

    ii. Discuss Amy's reasoning with your classmates. Does it make sense to you?

    iii. Explain how her formula fits with the formulas in Parts a andb.

d. Use any of the methods and formulas for counting permutations to help answer these questions.

    i. How many different officer slates fora club are possible if there are four officers (President, Vice-President, Treasurer, and Secretary) selected from 20 club members?

    ii. How would your reasoning and answer in part i change if there were 25 club members?

6 Javon's, Amy's, and Latricia's teacher next challenged the class to try to find a general formula for $C(n, k)$, the number of different combinations when choosing k objects from n objects, like choosing three Ski Club committee members from 15 people. Sometimes $C(n, k)$ is read as "n choose k." Working together, with much effort, they proposed the formula below to their class.

$$C(n, k) = \frac{n(n - 1)(n - 2)\cdots(n - k + 1)}{k!}$$

a. Latricia described their thinking as follows.

> Combinations are different than permutations because order doesn't matter. My formula counts permutations and then adjusts. So, I start by counting the permutations—that's the numerator. That gives me too many because it counts different orderings as different possibilities. So, I'll adjust for the fact that order doesn't matter by dividing by all the ways that k things can be ordered—that's the denominator.

i. Use Latricia's formula to find the number of three-person Ski Club committees chosen from 15 people. Make sure your answer agrees with that in Problem 2 Part b.

   ii. Discuss Latricia's reasoning with your classmates. Does it make sense to you?

   iii. In her description, she claims that k objects can be ordered in $k!$ ways. Explain why this is so.

b. As you know from your previous studies, formulas can be written in different, but equivalent, forms. Below is another common formula for combinations.

$$C(n, k) = \frac{n!}{k!(n - k)!}$$

Explain why this formula is equivalent to the formula given above Part a.

c. Alex gave this brief description of how to count combinations.

> **Start with _n_ and carry out a factorial-type computation using exactly _k_ factors. But then you have counted too many, so divide by _k_!.**

Explain how Alex's description fits with the formulas in Parts a and b.

d. Use any of the methods and formulas for counting combinations to help answer these questions.

   i. How many different possible four-member committees can be chosen from a club with 20 members?

   ii. How would your answer to part i change if there were 30 club members?

7 Use the methods and formulas you have learned to help solve these counting problems.

a. A sample of 10 cell phones will be selected from a shipment of 200 phones to test for flaws. How many different samples of 10 can be chosen?

b. Five students in a music class will be chosen to perform individually in a recital. There are 18 students in the class. How many different groups of five students can be chosen to perform?

c. Think again about the recital in Part b. How many different programs for the recital can be created using five students from the class of 18? Each student will perform individually, as in Part b. How is this problem different from Part b?

SUMMARIZE THE MATHEMATICS

Many counting situations can be analyzed in terms of order and repetition. In this investigation, you focused on permutations and combinations.

a. The cells in the following table show four possibilities for counting situations in terms of order and repetition. On a copy of this table, fill in two of the cells by writing "permutation" or "combination" in the appropriate cells.

Counting with Order and Repetition

	No Repetitions	Repetitions Okay
Different Orderings Count as Different Possibilities		
Different Orderings Do Not Count as Different Possibilities		

b. In the "permutation" and "combination" cells, enter this additional information:

- a relevant example

- a relevant formula, using factorials

- a brief description in words of how to carry out the calculations indicated by the formula

c. Explain how the Multiplication Principle of Counting is used in reasoning strategies to derive formulas for counting permutations and combinations.

Be prepared to explain your ideas and examples to the class.

✓CHECK YOUR UNDERSTANDING

The Union High School marching band has developed a repertoire of 10 music pieces for this semester. Due to time constraints, four pieces can be performed at an upcoming event.

a. How many different collections of four music pieces could the director choose?

b. Four music pieces will be put together to create a program for the event. How many different concert programs are possible?

Collections, Sequences, This or That

In the previous investigation, you focused on permutations and combinations. In each case, repetition is *not* allowed. In this investigation, you will study counting situations in which repetition *is* allowed.

As you work on the problems of this investigation, look for answers to the following questions:

What are similarities and differences among the four types of problems involving order and repetition?

What are similarities and differences among methods that count the number of selections from a collection of objects (Lesson 2) and methods that count the outcomes from a sequence of tasks (Lesson 1)?

1 Begin by considering the following two counting situations that you experience in your daily life.

a. In which cell of the table below does the "Pick 5" situation belong? Enter this into your copy of the table.

Counting with Order and Repetition

	No Repetitions	Repetitions Okay
Different Orderings Count as Different Possibilities		
Different Orderings Do Not Count as Different Possibilities		

b. Think about a counting problem related to the Nile.com password situation in the figure above. For example, how many different Web site passwords are possible if a password consists of 4 digits chosen from 0–9? In which cell of the table does this problem belong? Enter it into the table.

2 Look back at the article on page 220, "Take Note: Counting Their Chickens." The counting problem in this article is:

How many different three-side-dish orders can be made from the restaurant's 16 side dishes?

a. The spokesman explained that the restaurant had confused combinations with permutations. However, neither combinations nor permutations are correct in this situation. Explain why.

b. Place this restaurant-side-dish problem in the appropriate cell of your table.

c. The solution given by the Souderton math class is 816. Verify that this is correct. You may need to work hard to do this, but the math students in the article got it and so can you.

Sides	
(served after 10:30 A.M. until 8:00 P.M.)	
Tomato Bisque	Macaroni and Cheese
Caesar Salad	Black Beans and Rice
House Salad	Sweet Potato Fries
Potato Salad	Cornbread
Green Beans	Mashed Potatoes
Vegetable Medley	Cranberry Relish
Cole Slaw	Cinnamon Apples
Creamed Spinach	Fruit Salad

Now apply your understanding of counting methods to help solve the following problems. Permutations and combinations may be involved directly, indirectly, or not at all. You may use any counting principles or methods from Lessons 1 and 2. Be prepared to justify your solutions.

3 How many different 13-card hands can be made from a deck of 52 cards?

4 Information is stored in computers as strings of 0s and 1s (because 0 and 1 can be interpreted as "off" and "on" settings for switches inside the computer). Recall that 0 and 1 are called binary digits or *bits*. A sequence of bits is called a *binary string*. For example, 10110 is a binary string with five bits. A binary string with eight bits is called a *byte*. For example, 11010110 is a byte but 010 is not.

   a. How many different bytes are possible?

   b. How many bytes can be created that contain exactly two zeroes?

5 How many different ways are there for win, place, and show (first, second, and third, respectively) positions in a horse race with seven horses?

6 Seven people are running for three unranked positions on the school board. In how many different ways can these three positions be filled?

7 A **set** is any well-defined collection of objects. A *subset* of a set is, roughly, a smaller set inside the set. Precisely, set A is a **subset** of set B if every element of A is an element of B. For example, if $B = \{1, 2, 3\}$, then *some* of the subsets of B are

$$\varnothing, \{1\}, \{1, 3\}, \{2, 3\}, \{1, 2, 3\}.$$

The symbol $\varnothing$ denotes the **empty set**—the set with no elements. The empty set is a subset of every set. The whole set (in this case, $B = \{1, 2, 3\}$) is also considered a subset of itself.

   a. There are more subsets of B than those listed above. List all the subsets of $B = \{1, 2, 3\}$. How many are there?

   b. How many subsets of set B above have two elements? You can of course just look at your list in Part a to answer this question. In addition, explain how you could use combinations to answer this question.

   c. Suppose a set has five elements. How many subsets of three elements does this set have?

   d. For a set with five elements, how many subsets are there?

   e. Now try to generalize your thinking about sets and subsets. If a set has $n > 0$ elements, how many subsets are there? Compare your answer with that of others and resolve any differences.

SUMMARIZE THE MATHEMATICS

In this investigation, you refined and extended your understanding of counting methods. You considered four types of counting problems related to issues of order and repetition. You solved problems that require several different counting strategies.

a. When you are presented with a complex counting problem, what questions should you ask yourself as a start to solving the problem? What might be the next step in your solution strategy?

b. In solving counting problems, it is often helpful to detect the underlying structure of the problem as outlined below. Give an example of each of these three types of problem structures. Then describe how you might solve problems of each type.

    i. Count the number of selections from a collection of objects.

    ii. Count the number of outcomes from a sequence of tasks.

    iii. Count the number of outcomes from one situation or another situation.

c. What are similarities and differences among the four types of counting problems involving order and repetition?

d. Permutations and combinations can also be thought of in terms of sets and subsets.

    i. Explain why a subset of a set is a combination.

    ii. Explain why an ordered sequence of distinct elements of a set is a permutation.

Be prepared to explain your examples and reasoning to the class.

✔ CHECK YOUR UNDERSTANDING

Complete each of the following counting tasks, making notes on how you decided what method or methods to use.

a. Four of the 11 members of a championship gymnastics team will be chosen at random to stand in a row on stage during an awards ceremony. Assuming it is more prestigious to stand closer to the podium, how many different arrangements of four gymnasts on stage are possible?

b. A particular bicycle lock has four number-dials. Each dial consists of the digits from 0 to 9.

    i. How many different lock combinations are possible? Explain, including any assumptions you are making about how the lock works.

    ii. Are these lock combinations actually "combinations" in the mathematical sense of the word? Are they permutations? Explain.

1 Carlos's hockey team has ten players, not counting the goalie. Five non-goalie players need to be selected for the starting line-up.

   a. How many different starting line-ups (not including the goalie) are possible if positions are not assigned?

   b. How many starting line-ups are possible if positions are assigned?

2 A group of four students is working on a collection of eight simple tasks. They decide they will share the work and each do two of the tasks. How many ways are there for them to divide the work?

3 Major League Baseball teams maintain a 25-man roster and also a 40-man roster. Players on the 25-man roster may play in official games throughout the season. The additional 15 players on the 40-man roster may play in games starting September 1.

   a. How many 9-player batting orders are possible from a 25-man roster? A 40-man roster?

   b. In 2012, world population was estimated at about 7 billion. If all the distinct batting orders for a 25-man roster were written down and distributed equally to the people of the world, how many batting orders would each person receive?

4 A researcher distributes a questionnaire to 80 participants. To gain additional insight into participant responses, 14 of the 80 participants are randomly selected to be interviewed.

   a. How many different ways could she pick a group of 14 participants to interview?

   b. A stack of 500 sheets of paper is about 5 cm high. Assume she writes the names of a single selection of a group of 14 interview participants on a sheet of paper.

      i. How high would the stack of papers containing all possible selections stand?

      ii. How does your answer compare to the average distance from Earth to the Sun, roughly 150 billion meters?

5 Suhayla is making a music playlist for Jahanna. She has narrowed the possible songs down to 30 and wants to make a mix with exactly 20 songs.

   a. If she carefully chooses the order of the songs, how many possible mixes could she make for him?

Lawrence M. Sawyer/Getty Images

b. The total surface area of Earth is nearly 150 million square kilometers. Though sand grains vary considerably in size, a reasonable approximation for the average volume of a grain of sand is 10^{-12} m^3. If Suhayla had one grain of sand for each possible playlist, spread out over the land surface area of Earth, how deep would the sand be?

6 Examine the following portion of a recent television commercial.

Customer:	So what's this deal?
Pizza Chef:	Two pizzas.
Customer:	[Looking towards a four-year-old boy.] Two pizzas. Write that down.
Pizza Chef:	And on the two pizzas choose any toppings—up to five [from the list of 11 toppings].
Customer:	Do you ...
Pizza Chef:	... have to pick the same toppings on each pizza? No!
4-Year-Old Boy:	Then the possibilities are endless.
Customer:	What do you mean? Five plus five are ten.
4-Year-Old Boy:	Actually, there are 1,048,576 possibilities.
Customer:	Ten was just a ballpark figure.
Pizza Chef:	You got that right.

a. Do you think the customer's "ballpark figure" is too low? Explain your reasoning.

b. Suppose you order just one pizza and you must choose exactly 5 different toppings from 11 choices. How many different pizzas are possible?

c. Suppose you order just one pizza and you must choose exactly 3 different toppings from 11 choices. How many different pizzas are possible?

d. Suppose you order just one pizza and you can choose from 0 to 5 different toppings. How many different pizzas are possible?

e. In the TV commercial, does the 4-year-old boy have the correct answer? If so, explain how to compute his answer. If not, explain why it is incorrect and determine the correct answer.

f. Belinda reasoned as follows.

> There are 1,024 possibilities for one pizza. Since 2 pizzas are ordered, that makes $(1{,}024)^2$ possibilities for a two-pizza order. But order does not matter for the two pizzas, so divide by 2. Thus, the correct answer is 524,288.

Explain the error in Belinda's reasoning.

ON YOUR OWN

CONNECTIONS

7 Use counting methods to help answer each of the following geometric questions.

  a. Given a set of n points, how many distinct line segments can be formed with two of the n points as endpoints? Does it make any difference if all of the points are not in the same plane? Explain your reasoning.

  b. How many points of intersection are formed by n coplanar lines if no two are parallel and no three intersect in a common point?

  c. Using combinations, explain why the questions in Parts a and b are essentially the same.

8 The notation used for the number of combinations is not completely standardized. One of the most common notations is the one used in this lesson, namely, $C(n, k)$. As you use technology and read other books or Websites you might see notations like $_nC_k$ or $\binom{n}{k}$. Practice interpreting the different notations by computing each of the following. Compute each by hand and then check your answer using technology.

  a. $C(12, 5)$

  b. $_8C_3$

  c. $\binom{12}{7}$

9 The following formula shows one way that $C(n, k)$ and $P(n, k)$ are related.

$$C(n, k) = \frac{P(n, k)}{k!}$$

  a. Explain this relationship in words (by reasoning about combinations, permutations, order, and repetition).

  b. Justify this relationship using factorials and other formulas.

10 As you learned in this lesson, a general formula for the number of permutations is

$$P(n, k) = \frac{n!}{(n - k)!}.$$

A general formula for the number of combinations is

$$C(n, k) = \frac{n!}{(n - k)!k!}.$$

What restrictions must be placed on n and k for these formulas to make sense?

11 Develop a general formula for those problems in which you are counting the number of possible selections of k objects from a collection of n objects when order matters and repetitions are allowed. (An example of a problem of this type is counting the number of different Web site passwords possible if a password consists of six letters and repetitions are allowed.) Enter this formula into the appropriate cell of your Counting with Order and Repetition table.

REFLECTIONS

12 As you have learned, two important issues to consider in counting situations are order and repetition. It can be tricky sometimes to decide how order and repetition are involved.

  a. How do you decide whether order should be considered in counting? Give an example.

  b. How do you decide whether repetitions are involved in a counting situation? Give an example.

13 Combinatorics is sometimes described as "methods for counting without counting." In what sense is this an apt description of the mathematics you have been doing in this lesson?

14 The restaurant-side-dish problem (Investigation 2, page 229) is not a direct combination or permutation problem. As you discovered, you cannot simply refer to the Counting with Order and Repetition table, choose a formula, and compute. Often problems involve several counting ideas. Using the restaurant-side-dish problem as an example:

  a. explain how the Multiplication Principle of Counting is used, or could be used, in solving this problem.

  b. explain how the Addition Principle of Counting is used, or could be used, in solving this problem.

  c. explain how combinations are used, or could be used, in solving this problem.

15 You can often think about a counting situation in several ways. Consider this question.

> How many different passwords are possible that consist of five lowercase letters, with no letters repeated?

  a. Explain how you can answer this question using the Multiplication Principle of Counting.

  b. Explain how you can answer this question using permutations.

16 Suppose you have n objects with which you can sequentially fill k slots. Are there more or fewer possible k-slot sequences if repetitions are allowed when filling the slots? Justify your answer.

17 An important mathematical practice is to: *Construct viable arguments and critique the reasoning of others.* Look back at the work you did in Investigation 1. Find at least two instances where you used this mathematical practice. Explain how you used the practice in each instance.

18 RNA (ribonucleic acid) is a messenger molecule associated with DNA (deoxyribonucleic acid). RNA molecules consist of a chain of bases. Each base is one of four chemicals: U (uracil), C (cytosine), A (adenine), and G (guanine). It is difficult to observe exactly what an entire RNA chain looks like, but it is sometimes possible to observe fragments of a chain by breaking up the chain with certain enzymes. Armed with knowledge about the fragments, you can sometimes determine the makeup of the entire chain. One type of enzyme that breaks up an RNA chain is a "G-enzyme." The G-enzyme will break an RNA chain after each G link. For example, consider the following chain:

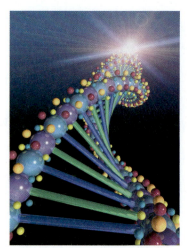

AUUGCGAUC

A G-enzyme will break up this chain into the following fragments:

AUUG   CG   AUC

a. What fragments result when a G-enzyme is applied to the following chain?

CGUUGGAUCGAU

b. Unfortunately, the fragments of a broken-up chain may be mixed up and in the wrong order. However, you can use reasoning to figure out the right order. For example, suppose a chain is broken up by a G enzyme and the fragments are out of order. Explain why a fragment that does not end in G must be the last fragment in the chain.

In Parts c–e, you will use information about fragments to reconstruct the complete chain. Suppose you have the fragments of an unknown RNA chain of 10 bases.

c. Suppose the complete RNA chain of 10 bases is broken by a G-enzyme into the following fragments (although not necessarily in this order):

AUG   AAC   CG   AG

How many different ways can these fragments be combined into a complete RNA chain of 10 bases?

d. Another enzyme, the U-C enzyme, breaks up an RNA chain after every U and every C. This enzyme breaks the unknown RNA chain in Part c into the following fragments.

GC   GAAC   AGAU

How many different ways can these fragments be combined into a single RNA chain of 10 bases?

e. So far, just by counting, there are many possible ways to recombine the fragments into a complete 10-base chain. However, if you reason about the fragments resulting from the two enzymes, by examining the fragments in Parts c and d, there is only one possible complete chain. What is the complete chain?

19 In Problem 2 of Investigation 2 (page 229), you verified the solution to the restaurant-side-dish problem from the newspaper article. That is, you determined that the number of different 3-side-dish selections that can be made from 16 side dishes is 816. There are several possible solution methods. Read and fill in the missing pieces of the following correct solution.

For simplicity, label the 16 side dishes A through P. Think about an order form that could be used to record someone's order, like the following.

```
A  B  C  D  E  F  G  H  I  J  K  L  M  N  O  P
|XX|  |  |  |  |  |  |  |  |  |  |X|  |  |
```

a. To help in the counting process, think about this order form as a sequence of marks, without the letter labels. Each mark is either a dividing line to separate the different side dishes or an "X" to indicate a selection.

   i. What 3-side-dish order is represented by the completed order form above?

   ii. Why are there three Xs?

   iii. Why are only 15 vertical dividing marks needed?

b. You can count the number of orders by thinking about sequences of marks, as follows.

   So, there are 18 total marks. Think about 18 slots. In each of those slots, you will put a mark, either a dividing line or an X.

   You want 3 Xs in this sequence of marks. The sequence of marks thus created will determine an order.

   So to count the number of different orders, you just count the number of ways you can choose 3 of the 18 slots for the Xs. Thus, you are choosing 3 slots from 18 slots; order does not matter in the selection of slots; and none of the 18 slots can be repeated.

   So, the solution is $C(18, 3)$, which is 816.

   Go back through this reasoning process and restate it in your own words, so that you understand it and could explain it to someone else.

c. Suppose the restaurant had 10 side dishes and you could choose 4. How many 4-side-dish orders are possible?

d. Suppose you want to select k objects from a collection of n objects, where order does not matter and repetitions are allowed. Explain why the number of possible selections in this case is $C(n + k - 1, k)$. Enter this formula into the appropriate cell of your Counting with Order and Repetition table.

20 Look back at the "Pick 5" part of Problem 1 (page 228) of Investigation 2. How many different "Pick 5" orders are possible?

21 Determine the number of all possible three-number combinations for a dial combination lock with 20 numbers on the dial. The combination is entered by rotating the dial to the right, then left, then right again, but the left number cannot be the same as either of the right numbers.

22 Counting problems may look simple but they often require very careful thinking. Consider this counting problem.

> How many different flags of 8 horizontal stripes contain at least 6 blue stripes if each stripe is colored red, green, or blue?

Here is one student's proposed solution.

> *Each stripe on the flag must be labeled with a color—red, green, or blue. Because at least 6 blue stripes are required, one can begin by choosing 6 stripes to color blue. There are $C(8, 6)$ ways to do this. Once the minimum of 6 blue stripes is fulfilled, the remaining 2 stripes (whichever ones they are) can each be colored in any one of 3 ways (red, green, or blue). Therefore, there are $(3)(3) = 9$ ways to finish coloring the stripes, giving a total of $(28)(9) = 252$ different patterns of color on the flag's stripes.*

This solution may seem plausible, but it is incorrect. Explain why it is wrong. Then solve the problem correctly. (**Source:** Annin, Scott A. and Kevin S. Lai. "Common Errors in Counting Problems." *Mathematics Teacher*, 103, no. 6 (February 2010): 402–409. Reprinted with permission from *Mathematics Teacher*, copyright 2010 by the National Council of Teachers of Mathematics. All rights reserved.)

REVIEW

23 Daphne needs to earn $1,225 in the next eight weeks in order to be able to buy the electric bicycle she wants. She has two ways in which to earn the money. When she babysits, she earns $10 per hour. When she cleans houses with her mother, she makes $13 per hour.

a. Will Daphne meet her goal if she averages 4 hours of cleaning and 11 hours of babysitting per week?

b. Suppose that Daphne's mother anticipates she will need 6 hours of help each week. Along with this house cleaning, how many hours of babysitting will Daphne need to average per week in order to meet her goal?

c. Let c represent Daphne's average weekly hours of cleaning and b represent the average weekly hours of babysitting. Write an expression representing the amount of money Daphne earns during this 8-week period.

d. Draw a graph that shows all combinations of average weekly hours cleaning and babysitting that will allow Daphne to meet her goal.

24 For each quadratic expression, find an integer value of k so that the expression can be factored into a product of two binomials. Then write the factored form of your trinomial.

a. $x^2 + 7x + k$

b. $x^2 + kx + 9$

c. $2x^2 - 5x + k$

25 Consider the following two functions:

$$f(x) = x + a, a \neq 0$$

$$g(x) = a^x, a > 0$$

Determine whether each statement is true or false. Provide reasoning to support your answer.

a. $2f(x) = f(2x)$ **b.** $f(x) + 2 = f(x + 2)$

c. $2g(x) = g(2x)$ **d.** $g(x + 2) = g(x) + 2$

26 In the diagram at the right, $\ell \parallel m \parallel n$. Determine the following measures.

a. HD

b. EB

c. $m\angle HAF$

d. $m\angle H$

e. $m\angle FEB$

f. $m\angle BED$

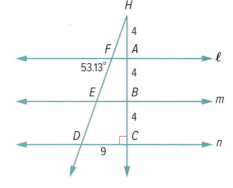

27 Write each product or sum as a fraction in simplest form.

a. $\frac{3}{4} \cdot \frac{2}{3} \cdot \frac{1}{2}$

b. $\frac{130}{405} \cdot \frac{129}{404}$

c. $\frac{3}{5} + \frac{1}{8}$

d. $-\frac{3}{4} + \frac{3}{8} + \frac{1}{2}$

28 Write each product in standard polynomial form.

a. $(2x + 3)(5x - 1)$

b. $(x^3 - 4)(x^3 + 4)$

c. $(x - 7)^2$

d. $(3x + 5)^2$

e. $(x + 3)(x + 8)(x + 2)$

29 Suppose that you roll a regular tetrahedral die that has the letters F, N, T, and M on its faces and then spin a spinner that is divided into two equal parts with the letter O on one side and the letter Y on the other side. You make note of the letter on the bottom face of the tetrahedral die and the letter of the region in which the pointer lands.

 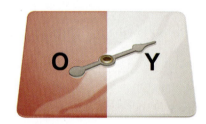

a. Construct a sample space of all possible two-letter outcomes where the first letter is from a roll of the die and the second letter is from a spin of the spinner. What is the probability of each outcome?

b. What is the probability that your outcome spells a word?

c. What is the probability that both of the letters come before S in the alphabet?

d. What is the probability that your outcome spells a word if you spin an O?

e. Are the two events of rolling the die and spinning the spinner independent events? Explain your reasoning.

Counting Throughout Mathematics

In the last two lessons, you saw that counting problems arise in many different contexts, for example, menu choices, telephone numbers, automobile license plates, Web site passwords, club committees, card games, IP addresses, and even DNA strings. The counting methods you learned are also applied throughout mathematics.

For example, one of the most famous mathematicians of the last century, Paul Erdös, often used clever methods of counting in his research. Erdös published well over 1,000 papers, almost always working with collaborators. In fact, there is a counting number based on working with Erdös, called an Erdös number. If you collaborated with Erdös to write a paper, your Erdös number is 1. If you collaborated with a collaborator, then your number is 2, and so on. For instance, some of the authors of this book have Erdös number 4 and thus others have Erdös number at most 5.

Photo of Paul Erdös taken at Cambridge University in 1991 by George Csicsery for his documentary film "N is a Number: A Portrait of Paul Erdös" (1993). ©1993. All Rights Reserved. www.zalafilms.com

THINK ABOUT THIS SITUATION

Think about some of the ways counting occurs throughout mathematics.

a. Suppose you write $(a + b)^2$ in expanded form. How many terms are in the expansion? How many terms would be in the expansion of $(a + b)^3$? In the expansion of $(a + b)^4$?

b. Do you notice anything special about the coefficients of the terms in each of the above expansions? Explain.

c. Given five points, no three of which are collinear, how many different triangles can be formed?

d. Does your answer to Part c depend on whether or not the five points are contained in the same plane? Explain your reasoning.

e. How many diagonals can be drawn in a regular polygon with 12 sides?

f. Suppose you flip a fair coin three times. How many different sequences of heads and tails are possible? How many of the possible sequences have exactly two heads? What is the probability of getting exactly two heads when you flip a coin three times?

In this lesson, you will apply counting methods to some of the branches of mathematics in which Erdös worked, including probability and discrete mathematics, and also algebra and geometry.

INVESTIGATION 1

Counting and Probability

In your previous studies, you used counting to help calculate probabilities. In this investigation, you will review and extend your understanding of probability and counting.

As you work on the problems in this investigation, look for answers to these questions:

> *How can counting methods be used in determining probabilities?*

> *How are the Multiplication Rules for probability similar to, and different from, the Multiplication Principle of Counting?*

Counting is especially useful in determining probability when there are a finite number of outcomes, all of which are equally likely. In this case, the **probability of an event** A, denoted $P(A)$, can be defined as

$$P(A) = \frac{\textit{number of outcomes corresponding to event } A}{\textit{total number of possible outcomes}}.$$

Thus, when all the outcomes are equally likely, you can determine the probability of an event by counting the number of outcomes corresponding to the event and dividing by the total number of possible outcomes.

1 Suppose you roll two fair dice, one red and one blue. An outcome is the number of spots showing on the top face of each die. For example, (3, 5) denotes the outcome of getting 3 on the red die and 5 on the blue die. Note that (3, 5) is a different outcome than (5, 3).

a. Are all the outcomes equally likely? What is the total number of possible outcomes?

b. Consider the *event* of getting a sum of 5 on the two dice. How many outcomes correspond to this event?

c. What is the probability of getting a sum of 5 when rolling two dice? The probability of getting a sum of 8?

2 In a certain state lottery, a player fills out a ticket by choosing five "regular" numbers from 1 to 45 and one PowerBall number from 1 to 45. The goal is to match the numbers with those drawn at random at the end of the week. The regular numbers are not repeated and they do not have to be in the same order as those drawn. The PowerBall number can be the same as one of the regular numbers.

a. How many different ways are there to fill out a ticket?

b. A player wins the jackpot by matching all five regular numbers plus the PowerBall number. This is called "Match 5 + 1." Since all the numbers must match, there is only one way to fill out a ticket that is a "Match 5 + 1" winner. What is the probability of the event "Match 5 + 1"?

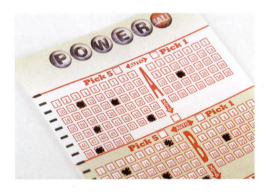

c. A player wins $100,000 by matching the five regular numbers but not the PowerBall number. This is called "Match 5." What is the probability of getting a "Match 5" winner?

d. A player wins $5,000 for "Match 4 + 1" (match exactly four of the regular numbers plus the PowerBall number). What is the probability of getting a "Match 4 + 1" winner?

e. Look back at your solutions to Parts a–d. Describe at least two different counting methods from Lessons 1 and 2 that you used in your solutions.

Counting Methods used to Determine Probabilities You used counting to help find the probabilities in Problems 1 and 2. For example, you may have used combinations and the Multiplication Principle of Counting. Think about which counting methods you use as you solve the rest of the problems in this investigation.

3 Suppose you toss a fair coin four times and record the sequence of heads and tails.

a. How many possible outcomes are there? Describe two ways of determining this number. Are the outcomes equally likely?

b. Find the following probabilities.

    i. *P(four heads)*

    ii. *P(exactly one head)*

    iii. *P(at least three heads)*

c. You may recall that the **Multiplication Rule for independent events** states that if *A* and *B* are independent events, then *P(A and B)* = *P(A)* × *P(B)*. Show how to calculate the three probabilities in Part b using the Multiplication Rule.

4 Suppose the names of six boys and four girls written on individual slips of paper are placed in a hat. You draw two names, in succession and with replacement. That is, you draw a slip of paper, record the name, return the slip of paper to the hat. Shake the hat to re-mix the slips of paper. Then draw again and record the name.

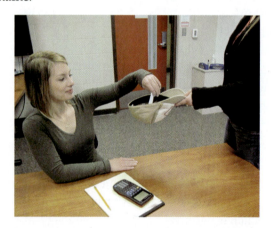

a. Find the probability that the first name drawn is a girl's name and the second name is a boy's name.

b. Show how you used, or could use, the Multiplication Rule for independent events to find this probability.

c. You now have a multiplication rule for probability and a multiplication principle for counting. They are closely related. Find the probability in Part a using the Multiplication Principle of Counting and the definition of probability given at the beginning of this investigation. (Or just explain how you did it if you have already solved the problem that way.)

5 Suppose again the names of six boys and four girls are written on individual slips of paper and placed in a hat. This time you draw two names *without replacement*. That is, you draw one name, you do *not* return the slip of paper to the hat, then you draw a second name.

  a. Find the probability that the first name drawn is a girl's name and the second name is a boy's name.

  b. Explain why the answer to Part a is *not* $\frac{4}{10} \times \frac{6}{10}$.

  c. Show how you can find the probability in Part a using the Multiplication Principle of Counting and the definition of probability given at the beginning of this investigation.

  d. To find the probability in Part a, you can also use the **General Multiplication Rule** for any two events:

  If A and B are events, then $P(A \text{ and } B) = P(A) \times P(B \mid A)$.

  The notation $P(B \mid A)$ is read "probability of B given A." This means you find the probability of B assuming that you know A happened. Show how to use the General Multiplication Rule to find:

  $P(girl's\ name\ on\ first\ draw$ and $boy's\ name\ on\ second\ draw)$.

6 Suppose you draw four names without replacement from the hat containing six boys' names and four girls' names on slips of paper.

  a. Find the probability that all names drawn are those of girls. That is, find $P(all\ four\ are\ girls'\ names)$.

  b. There are several methods for finding the probability in Part a. Compare your solution to those of other students until you find at least one other solution method. Explain your solutions to each other.

7 There are 50 people in a jury pool and 15 of them are Native Americans. You select two jurors at random from this pool.

  a. What is the probability that the two you select are a Native American and someone who is not a Native American?

  b. Find the probability of this event using a different method.

SUMMARIZE THE MATHEMATICS

In this investigation, you applied counting methods and probability rules to determine probabilities.

a. Under what conditions can you calculate the probability of an event by using the following ratio?

$$\frac{number\ of\ outcomes\ corresponding\ to\ the\ event}{total\ number\ of\ possible\ outcomes}$$

b. Give one example from this investigation where you used each of the following to determine a probability.

- The Multiplication Principle of Counting
- Combinations
- Permutations
- The General Multiplication Rule for probability

c. How are the Multiplication Rule and the General Multiplication Rule for probability similar to the Multiplication Principle of Counting? How are they different?

Be prepared to explain your ideas and reasoning to the class.

✓CHECK YOUR UNDERSTANDING

Consider the following probability and counting situations.

a. Revisit the names-in-a-hat situation one more time, to help you pull together what you have learned. You have the names of six boys and four girls on slips of paper in a hat. This time, you reach into the hat and pull out two names *at the same time*. What is the probability you get a boy's name and a girl's name?

b. In Iowa in 2011, a license plate had six characters. The first three characters were numbers and the last three were letters. Suppose that no numbers or letters can be repeated. What is the probability that a randomly chosen Iowa license plate has three odd numbers?

Combinations, the Binomial Theorem, and Pascal's Triangle

In your previous work in algebra, you rewrote powers of binomials like $(x + y)^2$ and $(x + y)^3$ in equivalent expanded form. In this investigation, you will explore some of the properties of combinations and their applications in calculating binomial expansions.

As you complete the problems in this investigation, look for answers to these questions:

> *What are some connections among combinations, Pascal's triangle, and expansions of binomial expressions of the form $(a + b)^n$?*
>
> *How can you explain and prove some of those connections?*

1 Think about expanding $(a + b)^n$. In particular, think about the coefficients of the terms in the expansion. For example, $(a + b)^2 = (a + b)(a + b) = a^2 + 2ab + b^2$. The coefficients are 1, 2, and 1. There is an important connection between combinations and the coefficients of the terms in the expansion of $(a + b)^n$. Investigate this connection by expanding $(a + b)^n$ for several values of n, as follows.

  a. Without using technology, expand $(a + b)^n$ for $n = 0, 1, 2, 3,$ and 4. See the computer algebra system (CAS) output below for the cases of $n = 5$ and $n = 6$.

  b. Examine and organize the coefficients of the terms of the expansions. Describe any patterns in the coefficients. Describe any connections you see to combinations.

  c. What do you notice about the exponents on a and b for successive terms in the expansions?

Connections Between Pascal's Triangle and Expanding $(a + b)^n$

The coefficients of the terms in the expansion of $(a + b)^n$ have a close connection to an array of numbers called Pascal's triangle. In the next few problems, you will explore this connection.

2 You might organize your work from Problem 1 as follows.

coefficients of $(a + b)^0$					1			
coefficients of $(a + b)^1$				1		1		
coefficients of $(a + b)^2$			1		2		1	
coefficients of $(a + b)^3$		1		3		3		1
coefficients of $(a + b)^4$	1		4		6		4		1

   a. Continue this array of numbers using the coefficients of the terms in the CAS expansions of $(a + b)^5$ and $(a + b)^6$.

   b. Describe how you could compute the numbers in a specific row of the array by using the numbers in the previous row.

   c. Based on the pattern in the array, what do you think the coefficients are in the expansion of $(a + b)^7$? Check your conjecture.

The triangular array of numbers in Problem 2 is called *Pascal's triangle*. It is named for the French philosopher and mathematician Blaise Pascal (1623–1662). He explored many of its properties, particularly those related to the study of probability. Although the triangle is named for Pascal, other mathematicians knew about it much earlier. For example, the triangular pattern was known to Chu Shih-Chieh in China in 1303.

Blaise Pascal

Pascal's Triangle

row 0				1		
row 1			1		1	
row 2		1		2		1
row 3	1		3		3		1
row 4	1		4		6		4		1

The rules for constructing **Pascal's triangle** are as follows: The top row, which is the top vertex of the triangle, consists of the single number 1. Each succeeding row starts and ends with 1. The remaining entries are constructed by looking at the row above. Specifically, each number in a given row is found by computing this sum:

   (*the number just above and to the left*) + (*the number just above and to the right*).

This is illustrated by the connector lines between 2, 1, and 3 in Pascal's triangle above.

3 You can add rows to Pascal's triangle indefinitely. Use the rules above to add rows 5 and 6 to the triangle. Compare to the rows of coefficients of the terms in the expansions of $(a + b)^5$ and $(a + b)^6$ that you determined in Problem 2. Resolve any differences.

Connections Between Pascal's Triangle and Combinations In Problems 2 and 3, you saw a remarkable connection. On the one hand, you have the coefficients of the terms in the expansion of $(a + b)^n$, which can be computed using algebraic multiplication. On the other hand, you have the numbers in Pascal's triangle, which are computed using the specific arithmetic rules given on the previous page. You have seen that these two very different procedures generate the same rows of numbers! Later in this lesson, you will see a reason for this connection. But first, consider a related connection between Pascal's triangle and combinations.

4 Notice that the rows of Pascal's triangle (shown on the previous page) are numbered starting with row 0. The entries in a given row can also be numbered beginning with 0. So, the initial entry in each row is labeled "entry 0," the next entry is labeled "entry 1," and so on.

a. Compute $C(4, 2)$. Where is this number found in Pascal's triangle (which row and which entry)? What is the coefficient of a^2b^2 in the expansion of $(a + b)^4$?

b. Compute $C(6, 4)$. Where is $C(6, 4)$ found in Pascal's triangle? What is the coefficient of a^2b^4 in the expansion of $(a + b)^6$?

c. Now try to generalize your work in Parts *a* and *b*. Describe how to find $C(n, k)$ in Pascal's triangle. Describe where in Pascal's triangle you can find the coefficient of $a^{n-k}b^k$ in the expansion of $(a + b)^n$.

Connections Between Combinations and Expanding $(a + b)^n$ So far in this investigation, you have studied connections among three seemingly different mathematical topics: coefficients in the expansion of $(a + b)^n$, numbers in Pascal's triangle, and values of $C(n, k)$. One of the most important of these connections involves using combinations to expand $(a + b)^n$.

5 Study the following reasoning used by a group of students in Bertie STEM High School who were challenged to find the coefficient of $a^{54}b^{46}$ in the expansion of $(a + b)^{100}$. Discuss the students' reasoning with your classmates. Expand or clarify the reasoning as needed so that everyone understands.

> $(a + b)^{100} = (a + b)(a + b)(a + b)\cdots(a + b)$ (100 factors). To carry out this multiplication, you multiply each term in the first factor, that is, *a* and *b*, by each term in the second factor, then by each term in the third factor, and so on. You must multiply through all 100 factors. To get $a^{54}b^{46}$, you need to multiply by *b* in 46 of the factors. That is, you must choose 46 of the 100 factors to be those where you use *b* as the multiplier (and in the other factors, *a* will be the multiplier). So, the total number of ways to get $a^{54}b^{46}$ is the number of ways of choosing 46 factors from the 100 factors, which is $C(100, 46)$. So, the coefficient of $a^{54}b^{46}$ in the expansion of $(a + b)^{100}$ is $C(100, 46)$.

a. Use similar reasoning to find the coefficient of $a^{29}b^{71}$ in the expansion of $(a + b)^{100}$.

b. Based on this reasoning, use combinations to find the coefficients of the terms in the expansion of $(a + b)^5$. Confirm that your coefficients match those in the CAS display on page 247.

c. Use similar reasoning to find the coefficient of a^3b^5 in the expansion of $(a + b)^8$.

d. Now think about the general term in a binomial expansion. What is the coefficient of $a^{n-k}b^k$ in the expansion of $(a + b)^n$?

6 Your work in Problem 5 suggests the following general result, called the **Binomial Theorem.**

For any positive integer n,

$$(a + b)^n = C(n, 0)a^n + C(n, 1)a^{n-1}b + C(n, 2)a^n - 2b^2 + \cdots + C(n, k)a^{n-k}b^k + \cdots + C(n, n-2)a^2b^{n-2} + C(n, n-1)ab^{n-1} + C(n, n)b^n.$$

a. Use the Binomial Theorem to expand $(a + b)^4$. Verify that you get the same answer as in Problem 1.

b. Use the Binomial Theorem to find the coefficient of a^3b^5 in the expansion of $(a + b)^8$. Compare to the answer you found using combinatorial reasoning in Part c of Problem 5.

c. Explain why the sum of the exponents of a and b in each term of the expansion of $(a + b)^n$ is n.

d. Explain why the coefficient of $a^{n-k}b^k$ is the same as the coefficient of a^kb^{n-k}.

e. Use the Binomial Theorem to expand $(2x - 3y)^5$.

Pascal's Triangle and Properties of Combinations Now that you have observed that the entries in Pascal's triangle are values of $C(n, k)$, you can make conjectures about properties of combinations by looking for patterns in Pascal's triangle.

7 Carefully write the first 10 rows of Pascal's triangle. Based on the symmetry and other patterns in Pascal's triangle, make at least two conjectures about properties of combinations. State your conjectures using $C(n, k)$ notation. Compare your conjectures to those of other classmates.

8 Consider the line symmetry in Pascal's triangle.

a. If you have not already done so in Problem 7, use the line symmetry to make a conjecture about the precise relationship between $C(n, k)$ and $C(n, n - k)$. You might find it helpful to examine a few examples using specific values of n and k.

b. State the relationship from Part a in the specific instance when $n = 8$ and $k = 3$. Prove this specific relationship in the following two ways:

    i. Using a factorial formula for combinations and algebraic reasoning.

    ii. Using combinatorial reasoning. That is, carefully explain how to choose and count combinations. In this case, you might find it helpful to think about how choosing 3 objects from 8 objects is the same as *not* choosing a particular number of objects.

  c. Now prove the general property $C(n, k) = C(n, n - k)$, in two ways.

    i. Using factorial formulas and algebraic reasoning.

    ii. Using combinatorial reasoning by thinking about ways of choosing objects.

  d. Which of the arguments in Part c was most convincing for you? Why?

SUMMARIZE THE MATHEMATICS

In this investigation, you explored several connections among combinations, the expansion of algebraic expressions of the form $(a + b)^n$, and Pascal's triangle.

a. Describe these connections.

b. Explain how you can reason with combinations to find the coefficient of a^2b^5 in $(a + b)^7$.

c. Describe how to use Pascal's triangle to find the coefficient of a^2b^5 in $(a + b)^7$.

d. Describe how to use Pascal's triangle to find $C(7, 5)$.

e. Use combinatorial reasoning to justify that $C(16, 4) = C(16, 12)$.

Be prepared to share your descriptions and reasoning with the class.

✓ CHECK YOUR UNDERSTANDING

Think about the relative advantages of algebraic, visual (Pascal's triangle), and combinatorial approaches to binomial expansions as you complete these tasks.

a. Expand $(x + 2)^3$ in the following three ways.

    i. Multiply by hand.

    ii. Use Pascal's triangle.

    iii. Use the Binomial Theorem.

b. Find the coefficient of a^4b^2 in the expansion of $(a + b)^6$ in the following three ways.

    i. Reason with combinations.

    ii. Use Pascal's triangle.

    iii. Use the Binomial Theorem.

1 Monograms on jewelry, clothing, and other items consist of the initials of your name. Examine the three-initial monogram offer below.

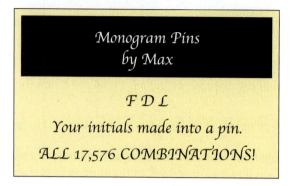

*Monogram Pins
by Max*

*F D L
Your initials made into a pin.
ALL 17,576 COMBINATIONS!*

a. Is the number of different three-initial monograms given in the ad correct? Explain.

b. Are these really "combinations" in the mathematical sense of the word?

c. What is the probability that a randomly selected three-initial monogram has all three initials the same?

d. Now consider the case where two of the initials are the same.

   i. How many three-initial monograms are possible if the first two initials are the same and the third is different?

   ii. How many three-initial monograms are possible if any two initials are the same and the other is different?

   iii. What is the probability that a randomly selected three-initial monogram will have two initials the same and the other initial different?

e. What is the probability that a randomly selected three-initial monogram has all three initials different?

f. What should be true about the probabilities in Parts c, d, and e?

2 Suppose you have 10 blue socks and 8 white socks in a drawer.

a. You select two socks at random, in succession, and with replacement. Find the probability that both socks are blue.

b. You reach in and pull out two socks at the same time, at random. Consider the following two possible solutions to finding the probability that both socks are blue.

   i. Mariam gives this correct solution: $\dfrac{C(10, 2)}{C(18, 2)}$. Describe reasoning that supports this solution.

   ii. John gives this correct solution: $\dfrac{10 \times 9}{18 \times 17}$. Describe reasoning that supports this solution.

c. Explain why the two solution methods in Part b are equivalent.

3 The Chess Club at Asiniboyne High School consists of 6 seniors and 11 juniors. Presently, the club president and the club secretary are both seniors. (They must be different people.) If the students were selected randomly for these offices, what is the probability both would be seniors?

   a. Show how to find the answer to this question using the General Multiplication Rule for probability.

   b. Show how to find the answer using the Multiplication Principle of Counting and the definition of probability.

4 Consider the experiment of flipping a fair coin four times and counting the number of heads.

   a. Out of the possible sequences of heads and tails, how many sequences contain no heads? How many contain exactly one head? Exactly two heads? Exactly three heads? Four heads?

   b. Show or describe where you can find the answers to these questions in Pascal's triangle.

   c. Explain how you can reason about combinations to find the answers to the questions in Part a.

   d. Find the probability of getting more than three heads.

5 There are many interesting patterns in Pascal's triangle. For example, consider the sum of each row of Pascal's triangle.

row 0						1					
row 1					1		1				
row 2				1		2		1			
row 3			1		3		3		1		
row 4		1		4		6		4		1	
row 5	1		5		10		10		5		1

a. Compute the sum of each of the first five rows of Pascal's triangle. Describe any patterns you see. What kind of sequence is formed by the sums of the rows?

b. Make a conjecture about the sum of row n in Pascal's triangle.

c. Consider row 3 of Pascal's triangle. Express each entry as a combination; then express the sum of the entries as a sum of combinations. This should suggest a property of combinations. State that property.

6 Anaba is designing an agricultural experiment. The factors of interest to her are Fertilizer (F), Herbicide (H), and Pesticide (P), each of which can be either *present* or *absent* for the duration of the growing season. She is concerned that pairs of factors may lead to additional effects. For example, using both fertilizer and herbicide may lead to an effect that would not have been present with only fertilizer or with only herbicide.

a. If one of her plots must be a control (no factors present), how many plots of land must she obtain access to in order to account for all single factors (such as $\{P\}$) and all paired factors (such as $\{F, H\}$)?

b. Explain how Anaba could solve this problem by thinking in terms of four factors, one of which is an "empty factor," E.

c. How many plots will she need if she has n factors, instead of three?

CONNECTIONS

7 Re-examine the General Multiplication Rule: $P(A \text{ and } B) = P(A) \times P(B \mid A)$, where A and B are events.

a. You may recall that $P(B \mid A)$ is called a *conditional probability*. $P(B \mid A)$ is often found directly from information in the problem situation. Explain why $P(B \mid A)$ can also be found by calculating $\dfrac{P(A \text{ and } B)}{P(A)}$, provided $P(A) \neq 0$.

b. Use the result in Part a to show that if A and B are independent events, then $P(B \mid A) = P(B)$.

8 Justify that $C(2n, 2) = 2C(n, 2) + n^2$ for $n \geq 2$, using the two methods below.

    a. The factorial formula for $C(n, k)$ and algebraic reasoning

    b. Combinatorial reasoning

9 In Investigation 2, you saw that there are close connections among combinations, the coefficients of the expansion of $(a + b)^n$, and the entries in row n of Pascal's triangle. These connections allow you to use technology to quickly generate any row of Pascal's triangle.

    a. Using combinations, express the entries in row 4 of Pascal's triangle.

    b. How could you use the sequence command on your calculator to generate row 4 of Pascal's triangle?

    c. Use the sequence command to generate row 10 of Pascal's triangle.

    d. Use a CAS to expand $(a + b)^n$, for the appropriate value of n, to generate row 10 of Pascal's triangle.

10 In this lesson, you have primarily studied counting situations in probability and algebra. Combinatorial questions also arise naturally in geometry.

    a. Given a set of n points, no three of which are collinear, how many distinct triangles can be formed with three of the n points as vertices? Does it make any difference if all the points are not in the same plane? Explain your reasoning.

    b. Given a set of n points, no three of which are collinear and no four of which are coplanar (that is, no four of which lie in the same plane), how many distinct tetrahedra can be formed using four of the n points as vertices?

    c. Suppose a map is formed by drawing n lines in a plane, no two of which are parallel and no three of which intersect in a common point. What is the fewest number of colors needed to color this map so that no two regions with a common boundary are the same color?

11 Describe connections between the General Multiplication Rule for probability and the ideas of permutations and combinations.

12 In their article, "The Evolution with Age of Probabilistic, Intuitively Based Misconceptions," in the *Journal for Research in Mathematics Education* (January 1997, pp. 96–105), Efraim Fischbein and Ditza Schnarch reported on a survey in which 100 high school students were asked the following question: "When choosing a committee composed of 2 members from among 10 candidates, is the number of possibilities smaller than, equal to, or greater than the number of possibilities when choosing a committee of 8 members from among 10 candidates?"

a. What is the correct answer? State this answer using combinations.

b. Eighty-five percent of the high school students surveyed in the study stated that the correct answer is "greater than." Why do you think that so many students believed that there are more two-member committees than eight-member committees?

13 At the beginning of this lesson, you were introduced to Paul Erdös, one of the most famous recent mathematicians in the area of combinatorics. Read a few chapters about Erdös in one of the following books. Write a short report about what you have read.

- Paul Hoffman, *The Man Who Loved Only Numbers: Mathematical Truth*. New York: Hyperion, 1998.

- Bruce Schecter, *My Brain is Open: The Mathematical Journeys of Paul Erdös*. New York: Simon and Schuster, 1998.

14 A useful mathematical practice is to: *Look for and express regularity in repeated reasoning*. Look back at the work you did in Investigation 2 "Combinations, the Binomial Theorem, and Pascal's Triangle." Identify at least one instance where you used this mathematical practice. Explain how you used the practice.

EXTENSIONS

15 Poll your class to find the number of students who have been to the movies in the last week. Suppose you select two students at random from your class using the following method: Write the name of each student on a slip of paper. Mix up the slips of paper. Select a slip at random. Do not replace that slip. Select another slip at random.

a. What is the probability that both students you select have been to the movies in the past week? If you had replaced the first slip of paper, what would be the probability that each name you select is a student who has been to the movies in the past week?

b. In the remainder of this task, you will investigate whether the probability of selecting a student who has been to the movies in the last week is the same on the first draw as on the second draw if you do *not* replace the slip of paper.

Make a conjecture about whether the probability of selecting a student who has been to the movies in the last week is the same on the first draw as on the second draw, when you do not replace the slip of paper. Then check your conjecture by completing the parts below.

i. What is the probability of selecting a student on the first draw who has been to the movies in the last week?

ii. What is the probability of selecting a student on the first draw who has not been to the movies in the last week and a student on the second draw who has been to the movies in the last week?

iii. What is the probability of selecting a student on the first draw who has been to the movies in the last week and a student on the second draw who has been to the movies in the last week?

iv. Use results from parts ii and iii to find the probability of selecting a student on the second draw who has been to the movies in the last week. Is this the same as, or different from, the first-draw probability you found in part i?

16 A small inland lake is stocked with 100 fish, 20 of which are tagged. Some time later, a fisherman catches five fish.

a. Assuming that all 100 fish are still in the lake when he starts fishing and that this is the total population of the lake, how many ways are there for him to catch five fish of which two are tagged fish?

b. Assuming that the fish are all equally likely to be caught, what is the probability of this event?

c. Problems of this type can be challenging. How could you check your work?

17 Justify that $rC(n, r) = nC(n - 1, r - 1)$ for $n \geq r \geq 1$, using the following two methods.

a. A factorial formula for $C(n, r)$ and algebraic reasoning

b. Combinatorial reasoning (*Hint:* Think about choosing a committee of r people and a chairperson from a group of n people.)

18 Consider the following example of a property of combinations:

$$C(5, 3) = C(4, 2) + C(3, 2) + C(2, 2)$$

a. Show that this statement is true by using a factorial formula for $C(n, k)$.

b. Another example of this property is the following:

$$C(6, 2) = C(5, 1) + C(4, 1) + C(3, 1) + C(2, 1) + C(1, 1)$$

Based on this example and the example in Part a, make a conjecture for a general statement of this property.

c. Explain how this property appears as a pattern in Pascal's triangle.

d. Justify this property using combinatorial reasoning.

e. Justify the property by giving an argument in terms of coefficients in the expansion of $(a + b)^n$.

19 In this lesson, you have seen that there is a close correspondence between the entries of Pascal's triangle and values of $C(n, k)$, but you have only seen this correspondence as a pattern; you have not yet proven it. The reason for the close correspondence is because the construction rule for Pascal's triangle is the same as an important recursive property of combinations.

Rule for Pascal's triangle:	*(number in any row)* =	*(number above and to the left)* +	*(number above and to the right)*
Combination property:	$C(n, k)$ =	$C(n - 1, k - 1)$ +	$C(n - 1, k)$

a. Think about this correspondence. Using the fact that entry k in row n of Pascal's triangle is $C(n, k)$, explain why $C(n - 1, k - 1)$ corresponds to the "number above and to the left" and $C(n - 1, k)$ corresponds to the "number above and to the right."

b. Verify the combination property for some specific values of n and k.

c. Justify the combination property. That is, prove
$C(n, k) = C(n - 1, k - 1) + C(n - 1, k)$, where $k > 0$ and $k < n$.

REVIEW

20 Being able to solve problems that require using ratios, proportions, and percents is important in many different careers. Solve each of the following problems.

a. In January, Erin sold $15,247 worth of furniture. In February, she sold $16,193 worth of furniture. She had a goal of increasing her sales by at least 5%. Did she meet her goal?

b. In one experiment, approximately four out of every seven cauliflower seeds planted in 50°F soil germinated. If a nursery plants 400 cauliflower seeds in 50°F soil, based on the results of this experiment, approximately how many can they expect to germinate?

c. A doctor has given an order for 100,000 units of penicillin. The penicillin on hand is labeled 250,000 units/ml. How many ml of this penicillin should the patient be given?

21 Between what two consecutive integers does each number lie? Determine your initial answer without using technology and then check your answer using technology.

  a. $\sqrt{95}$ **b.** $-\sqrt{20}$

  c. $\sqrt[3]{31}$ **d.** $\sqrt[3]{-5}$

  e. $\log 137$ **f.** $\log 0.03$

22 Make a sketch and determine the period and amplitude of each periodic function.

  a. $f(x) = 3 \sin x$ **b.** $f(x) = \cos 4x$

  c. $f(x) = -\sin x + 5$

23 U.S. students completing a 12th-grade National Assessment of Educational Progress mathematics test were asked if their math work in school is too easy. The responses of these students are summarized in the table below.

Percent of U.S. 12th-graders who say math work is too easy	
Never or Nearly Never	32
Sometimes	48
Often	15
Always or Almost Always	5

  a. There were approximately 51,000 students who participated in the end-of-test survey. How many students gave each answer?

  b. There were 3,500 students in Florida who completed the end-of-test survey. If the distribution of answers for Florida students and those for all 51,000 students were homogeneous, how many students in Florida said that math was often too easy?

  c. The table below shows the survey response data for students from Arkansas and Idaho.

Number of 12th-graders who say their math work is too easy		
	Arkansas	Idaho
Never or Nearly Never	1,008	1,054
Sometimes	1,316	1,457
Often	364	434
Always or Almost Always	112	155

Is there statistically significant evidence that if, in these two states, all 12th-graders had been surveyed, the proportions giving these responses would be different? Explain.

24 Solve each equation.

a. $\dfrac{x}{x+5} = \dfrac{3}{8}$

b. $-5(7-2x) = 4 + (6x-5)$

c. $\dfrac{2x+3}{4} = \dfrac{x-9}{8}$

d. $\dfrac{3}{4} + \dfrac{1}{2}(x-6) = 9$

25 Determine the value of x in each figure.

a.

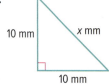

b.

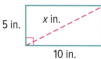

c.

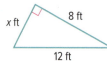

d.

26 For each function listed below, determine:

I $f(x) = 2^x + 3$

II $g(x) = \dfrac{1}{x} + 7$

III $h(x) = -\dfrac{5x}{4} + 18$

IV $k(x) = (x-3)^2$

a. the function family to which it belongs.

b. its domain and range.

Looking Back

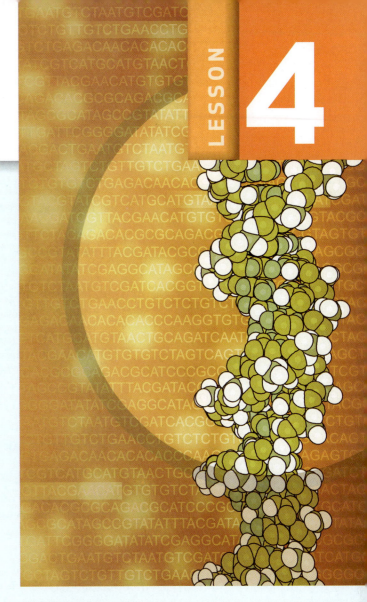

In this unit, you have learned useful counting strategies, including systematic lists, counting trees, the Multiplication Principle of Counting, the Addition Principle of Counting, and careful analysis of order and repetition when counting the number of choices from a collection. You have developed and applied formulas for counting permutations and combinations, and you have started to develop a new type of mathematical reasoning—combinatorial reasoning.

These strategies and skills enable you to solve counting problems both within and outside of mathematics. Outside of mathematics, you encountered applications of counting in many contexts. Within mathematics, your work led to several important results that have wide-ranging applications, including the Binomial Theorem, Pascal's triangle, and the General Multiplication Rule for probability. In this final lesson, you will review and pull together these important ideas and apply them in new contexts.

1 DNA is part of every cell in every living organism. DNA in humans is arranged into 24 distinct molecules called chromosomes. Each chromosome is a pair of intertwined chains twisted into a spiral. Each spiral consists of a long sequence of pairs of bases. Each base is one of four chemicals: T (thymine), C (cytosine), A (adenine), and G (guanine). The order of the base chemicals determines the instructions, the genetic information, embedded in the DNA. (**Source:** www.ornl.gov/sci/techresources/Human_Genome/project/info.shtml)

a. You have probably seen news reports or TV shows in which DNA is used to uniquely identify people. There are only four base chemicals that make up DNA, and yet there are so many sequences of these bases that no two people have exactly the same DNA. According to the Human Genome Project, there are about 3 billion DNA base pairs in the human genome, that is, in all of the 24 distinct chromosomes. Each chromosome has from about 50 million to 250 million base pair locations.

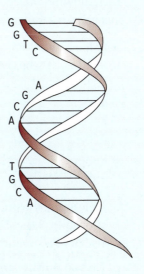

i. A base pair is a pair of the base chemicals: T, C, A, and G. How many base pairs are possible? (The pair CG is different than the pair GC since C is on one chain in the twisted spiral in one case and on the other chain in the spiral in the other case.)

ii. Suppose one chromosome has 120 million base-pair locations. So, there are 120 million locations for pairs of base chemicals to occur. How many different sequences of base pairs are possible for this chromosome?

b. Not all regions of a chromosome directly encode genetic information. Those regions that do are the functional units of heredity, called genes.

i. Genes comprise only about 2% of the human genome, that is, about 2% of the roughly 3 billion base-pair locations in the human genome. How many base-pair locations is this?

ii. It is estimated that there are about 22,000 different genes. Suppose a given gene has about the average number of base-pair locations. How many base-pair locations are in this gene?

iii. How many different sequences of base pairs are possible for this gene? Compare this number to an estimate of the number of atoms in the universe.

2 Examine the following information from a state lottery ticket where each play costs $1 including sales tax.

How to Play

To play, choose two numbers from 1 to 21 from each section (Red, White, and Blue).

How to Win

Every day, two numbers will be drawn from each of three colored ball sets (Red, White, and Blue). Each set has 21 balls numbered from 1 to 21. To win a prize for the following matches, you *must* match both number *and* color.

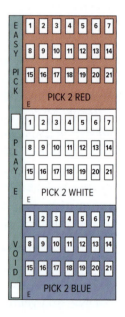

Match	Prize	Number of Winning Choices
6	$1,000,000	1 (out of 9,261,000)
5	$5,000	114 (out of 9,261,000)
4	$100	4,845 (out of 9,261,000)
3	$5	93,860 (out of 9,261,000)
2	$2	828,495 (out of 9,261,000)

a. Verify at least two of the entries in the column entitled "Number of Winning Choices." Explain your reasoning and show your calculations.

b. Determine the probability of each of the different types of matches. State the probability definition that you are using and explain why it is appropriate to use in this situation.

c. Do you think a state lottery like this is an effective way for the state to make money? Do you think it is an effective way for players to make money? Explain.

3 A Congressional international relations committee consists of five Democrats, six Republicans, and four Independents. Some of the committee members will be selected to attend an economic conference in Africa.

a. Suppose three members will be chosen to attend the conference. How many different groups of three people can be chosen?

b. How many groups of three people can be chosen if the group must contain exactly one Democrat?

c. Suppose three people will be chosen for the conference and each will take on a role that could be filled by anyone—one will attend all meetings at the conference related to banking, one will attend all meetings related to energy, and one will attend all meetings related to water. How many different role-specific groups of three people can be selected?

d. Suppose that due to budget cuts, only two members will go to the conference. The two members will be chosen at random. What is the probability that both members chosen are Republicans?

e. Think of another counting problem that might arise in this situation. Pose the problem and give the solution. Compare your problem and solution with those of some other students. Be prepared to answer questions they may have.

4 Consider a club of 10 students. Without doing any computations, explain why the number of possible committees of size 6 is equal to the number of possible committees of size 4.

5 The following identity exhibits several connections among combinatorial ideas.

$$C(n, 0) + C(n, 1) + \cdots + C(n, n) = 2^n \ (n \geq 0)$$

To discover some of these connections, complete at least two of the following parts.

a. Explain how this statement appears as a pattern in Pascal's triangle.

b. Use the Binomial Theorem to justify this statement.

c. Use combinatorial reasoning about the number of subsets of a set with n elements to justify this statement.

SUMMARIZE THE MATHEMATICS

In this unit, you have investigated many situations where systematic counting methods and combinatorial reasoning are useful.

a. Below is a list of some of the combinatorial ideas you have studied. Add other important ideas and topics to the list. Then give a brief explanation and describe an application for each topic on your list.

- Multiplication Principle of Counting

- Combinations

- Binomial Theorem

- Permutations

- Combinatorial Reasoning

- General Multiplication Rule for Probability

b. Why are order and repetition important when deciding how to count the number of possible choices from a collection? How are these ideas related to permutations and combinations?

c. Describe at least one counting application in each of the following areas of mathematics: algebra, probability, and geometry.

d. How are Pascal's triangle and the Binomial Theorem related to each other and to the ideas of counting?

Be prepared to explain your applications and thinking to the class.

✓ CHECK YOUR UNDERSTANDING

Write, in outline form, a summary of the important mathematical concepts and methods developed in this unit. Organize your summary so that it can be used as a quick reference in future units.

Mathematics of Financial Decision-Making

Digital Resources at
ConnectED.mcgraw-hill.com

Watch · Tools · Audio · eBook

Whether your future career is in business, industry, the social sciences, or health and life sciences, understanding saving and borrowing options, and how interest is computed on investments and loans is important to you as a consumer.

The proliferation of personal finance and money management software tools on the Internet attest to this importance. In this unit, you will learn the mathematics that is the basis for such software. You will also learn how to use software tools in analyzing financial trends and in making financial decisions.

In the two lessons of this unit, you will extend your skill in modeling quantitative problems and data patterns with a focus on the mathematics of finance.

LESSONS

1 Financial Decision-Making: Saving

Use linear and exponential functions to develop and apply methods for determining principal, interest, and future value related to various savings plans and for making comparative investment decisions. Use technological tools strategically in analyzing saving options and tracking investments.

2 Financial Decision-Making: Borrowing

Use exponential functions to develop and apply methods for repayment of loans and mortgages and for making comparative borrowing decisions. Compare auto loan, lease, and purchase options. Use technological tools strategically in analyzing borrowing options and tracking amortization of loans and mortgages.

Financial Decision-Making: Saving

Every day you may see stories on television, the Internet, or in newspapers about wealthy people. Many of these people acquired their wealth through inheritance, starting successful businesses, and investments such as stocks and real estate, as well as from earnings in highly paid careers. A news story in the *Washington Post* suggests that it is not impossible for any one of us to become surprisingly wealthy and to do a great deal of good with that wealth.

Oseola McCarty was born in 1908 in rural Mississippi, and she had to quit school after the sixth grade to care for her aunt. Rather than returning to school, she started working. She started out doing laundry at $1.50 a bundle, but she also began setting aside part of her earnings in a savings account. By the time she retired 75 years later, she had accumulated a very substantial amount of money. She gave the majority of her savings for college scholarships to help students have access to the educational opportunities she never had for herself.

When the University of Southern Mississippi announced her gift of $150,000 to endow the Oseola McCarty Scholarship, Miss McCarty said, "I want young people to know that it's okay to save money, to invest it and have lots of money. More important, though, I want them to understand that what they don't need they should give to those who are less fortunate." Oseola McCarty received widespread recognition for her extraordinary gift, including an honorary doctoral degree from Harvard University, carrying the Olympic Torch, a Presidential Citizens Medal, and many newspaper and television interviews.

The University of Southern Mississippi

THINK ABOUT THIS SITUATION

The savings story of Oseola McCarty is very impressive. Even more admirable was her plan to give her savings to those who have been less fortunate.

a. Describe a pattern of savings that Miss McCarty might have followed in order to accumulate $150,000 for the scholarship fund after 75 years. What assumptions did you make?

b. If you wanted to have $1,000,000 in your savings account 50 years from now, how much do you think you would have to invest now and over the years? Explain your reasoning.

c. What technological tools might be helpful in reasoning about the questions in Parts a and b? Explain your thinking.

Work on the problems of this lesson will introduce you to important topics in financial mathematics related to savings and investment plans. In the process, you will extend your quantitative reasoning skills. You will also review and extend your algebraic skills and understandings useful in modeling linear and exponential growth situations.

INVESTIGATION 1

Simple-Interest Investments

Trying to figure out a savings plan that might lead to results like those of Oseola McCarty, or even more ambitious plans of your own, requires exploration of several key variables—the initial amount invested, the amount set aside each month or year, the interest rate that can be earned, and the length of time that the savings strategy is followed.

As you work on the problems in this investigation, make note of answers to the following questions:

What is simple interest and how is it calculated?

How can you determine the future value of an investment paying simple interest and the rate of return on short-term investments such as real estate or stocks?

When you deposit money in an investment like a bank savings account or certificate of deposit (CD), that deposit earns money for you in interest paid by the bank. The interest may be *simple interest* or *compound interest*. Simple interest forms the basis for all calculations involving interest paid on an investment. Begin by reviewing and extending your understanding of this basic idea of the mathematics of finance.

1 Simple Interest If you invest a sum of money (called the **principal**) for a time period t at an interest rate r per period, then the simple interest accrued is given by the formula $I = Prt$, where I is the amount of interest earned (in dollars), P is the principal (in dollars), r is the annual interest rate (expressed as a decimal), and t is time (in this case, years). Note that the units of measurement for r and t must agree. That is, if t is in *years*, then r is *per year*.

The interest for a given period is paid to you, the investor; the principal invested remains the same.

a. Suppose the Ludington High School cycling club invested $2,500 in a two-year certificate of deposit at a simple annual interest rate of 3%. How much interest would be earned at the end of the two years? What would be the total amount accumulated? This amount is called the **future value** at the end of the period earning interest.

b. Suppose P represents the principal (in dollars), r represents the annual interest rate, and t is the length (in years) of a simple-interest investment. If F denotes the future value of the investment, write a formula relating P, r, t, and F.

c. Suppose the bank in Part a paid the interest quarterly. How would this affect your answer to Part a? Explain.

d. Describe an investment strategy that would yield a higher accumulated amount in two years than the scenario in Part c. Estimate the amount and describe the assumptions you made. Compare your strategy and future value estimate with that of your classmates.

2 Suppose you invest $800 for five years in a municipal bond paying 2% simple annual interest. So, the *START* balance of the account is $800. Assume no withdrawals or additional deposits are made.

a. Use the word *NOW* to represent the amount accumulated at the end of any given year. Write a **recursive formula** *NEXT* = _____ that shows how to calculate the accumulated amount at the end of the next year.

b. Use your rule in Part a and the automatic calculation feature of your graphing calculator or CAS to compute the value of your investment at the end of five years. Compare your answer and various keystrokes and commands with your classmates. Resolve any differences.

c. Compare your answer in Part b to that obtained using your future value formula from Problem 1 Part b.

d. Based on her work in Problems 1 and 2, Alicia claimed that, "growth of simple-interest investments is linear." Do you agree? If so, how would you justify the claim? If not, why not?

e. With initial investment fixed at $800, no additional deposits, and a fixed simple annual interest rate of 2%, the amount of interest earned and the future value of the account at any time t (in years) are functions of t.

  i. Write a rule expressing interest $I(t)$ as a function of t.

  ii. Write a rule expressing the future bond value $F(t)$ as a function of t.

  iii. Make a graph of $y = F(t)$. Interpret the meaning of the y-intercept and slope.

3 Tracking Simple Interest Investments Chelsea High School received an anonymous gift of $15,000 to be applied toward the purchase of a set of 30 laptop computers and a cart to be used for mathematics instruction. If that gift is matched by $15,000 in funding from the school board and the funds are invested at 2.5% simple annual interest, how long would it take to accumulate the purchase price of $32,000?

4 Suppose due to an improving U.S. economy, a multi-year $10,000 investment pays 2.5% simple annual interest during the first year and 3% simple annual interest during the second year. By what percent will the investment have grown over this two-year period?

5 Reconsider Problem 4, but instead assume there is a downturn in the economy so that the $10,000 investment earns 3% simple annual interest during the first year and 2.5% simple annual interest during the second year. By what percent will the investment have grown over this two-year period? Compare your answer with that of Problem 4. Why does this finding make sense?

6 Pose and answer a question that extends your discovery in Problems 4 and 5. Share your problem and solution with other classmates. Be prepared to respond to questions they or your teacher might have.

7 Mutual funds and individual stocks sometimes have graphs like the one shown below.

AMZN (NASDAQ) February 12, 2012–February 12, 2013

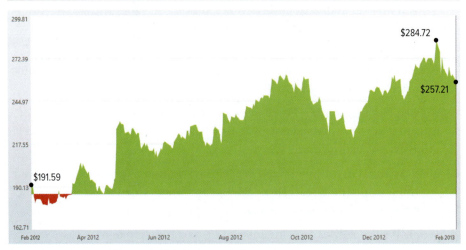

Source: *Microsoft Finance* February 12, 2013. Market data by Morningstar.

a. How would you describe the pattern of change in the price of one share of Amazon.com stock over the one-year period shown?

b. Estimate the equivalent simple annual interest rate earned on a share of Amazon stock purchased on February 12, 2012.

c. Suppose you had purchased 100 shares of Amazon stock on February 12, 2012 and sold it at the highest value of the stock during the year that followed. What equivalent simple annual interest rate would you have earned? What would be the profit on your investment?

8 The Finance Club at a local university makes short-term investments in stock to study trends in various sectors of the stock market. On October 1, 2014, they purchased 100 shares of LeTableaux stock for $6,150. After eight months, the stock shares had increased in value by $135.00 and had paid *dividends* (part of a company's profit paid to stockholders) totaling $142.68. The Club decides to sell the stock at this point and evaluate their investment. Compare their gain to a bank savings plan paying simple annual interest (for 8 months) to determine the interest rate that this gain represents.

9 A farmer in a rural community in Iowa wanted to make an investment that would accumulate to $20,000 in nine months for needed farm equipment. The best available short-term interest rate was 3.5% per year. How much money would need to be invested to yield the desired amount?

SUMMARIZE THE MATHEMATICS

In this investigation, you examined patterns of investment growth involving simple interest. Financial institutions typically pay simple interest on short-term investments, on bonds, and on select certificates of deposit (CDs).

a. The formula for simple interest, $I = Prt$, involves four variables.

   i. What is the meaning of each of the variables?

   ii. Which variables must agree in terms of units of measurement? Explain your reasoning.

   iii. If I, P, and r are given, how can you determine the time required to earn the given interest?

   iv. If I, P, and t are given, how can you determine the interest rate?

b. Why is simple interest an example of linear growth?

c. Suppose d is invested in a 10-year CD paying r% simple annual interest.

   i. Write a recursive formula using *NOW* and *NEXT* showing how the value of the investment changes from one period to the next. What is the initial (or starting) value?

   ii. Write a function rule that gives the future value of the investment $F(t)$ at any time $t \leq 10$ (in years).

Be prepared to explain your ideas and reasoning to the class.

✓ CHECK YOUR UNDERSTANDING

Suppose Rosanna invested $1,000 in a local bank for nine months at a simple annual interest rate of 1.8%.

a. How much interest will she have earned at the end of nine months?

b. What is the future value of the investment at the end of six months?

c. How, if at all, would your answers to Parts a and b change if Rosanna's investment was placed in an account paying the same simple annual interest rate, but interest was paid quarterly?

Compound-Interest Investments

Unlike in the case of a simple-interest investment, in a **compound-interest** investment, the interest for each period is added to the principal before interest is calculated for the next period. The **interest rate per period** is the *annual percentage rate* (APR) divided by the number of interest periods per year. Banks usually use 360 days as the equivalent of a year. Throughout this unit, you should also use 360 days as the equivalent of a year and 30 days as the equivalent of a month.

As you work on the problems of this investigation, look for answers to this question:

How can you determine effective annual interest rates, balances, and yields of investments involving compound interest?

1 Suppose you are considering making an initial deposit of $1,500 in a five-year CD and interest is paid at an APR of 4% under each plan below.

> **Plan I**   Interest is compounded semi-annually.
>
> **Plan II**   Interest is compounded quarterly.
>
> **Plan III**   Interest is compounded monthly.

  a. Calculate and compare the value of the three investment plans after one year. After two years. After five years.

  b. What conclusion(s) can you draw from your comparison of the three investment plans?

  c. Use *NOW* to represent the account balance for each plan at the end of a compounding period. Write a recursive formula *NEXT* = _____ that shows how to compute the balance at the end of the next period.

2 For each condition in Problem 1, write a formula *F* = _____ that would give the value of the above investment after *t* years.

3 Write a formula that gives the **future value** *F* of an investment after *t* years, where the initial deposit is *A* dollars and interest is paid at an annual percentage rate *r* (expressed as a decimal) with compounding *n* times in each year. Compare your formula with that of others. Resolve any differences.

4 It was recently reported in a local newspaper that an Oakland High School teacher had just won the daily lottery from a Michigan lottery ticket that she got as a birthday gift. In a new lottery payoff scheme, the teacher has two payoff choices. One option is to receive a single $20,000 payment now. In the other plan, the lottery promises a single payment of $40,000 ten years from now.

MBI/Alamy

Imagine that you had just won that Michigan lottery prize.

a. Discuss with others your thinking on which of the two payoff methods to choose.

b. Suppose a local bank called with congratulations and said you could invest your $20,000 payment in a special 10-year certificate of deposit (CD), earning 8% interest compounded annually. How would this affect your choice of payoff method?

5 In Problem 4, you investigated the length of time for an investment to double in value when interest is compounded annually. The **Rule of 72** is a rule of thumb, often used by bankers and investors, that gives an easy way to estimate the doubling time of an exponentially growing quantity like an investment with a fixed annual rate that is compounded annually. If $r\%$ is the fixed annual growth rate of a quantity, the Rule of 72 tells you that:

$$doubling\ time \approx \frac{72}{r}.$$

a. Test the Rule of 72 by completing a copy of the table below for a $1,000 investment that grows at the indicated fixed rates compounded annually. Then check your estimates using the appropriate exponential model and base-10 logarithms for the first three estimates. Use *NOW-NEXT* reasoning to check the last three estimates.

Rule of 72 Predicted Doubling Times

Annual Interest Rate (percent)	1	2	6	8	10	12
Predicted Doubling Time (in years)						
Actual Doubling Time (in years)		35		9		

b. Current annual interest rates on money market investments are about 0.10%. How well does the Rule of 72 predict doubling time (in years) for this rate?

c. Investments in the fast-growing Internet-based companies sector may have an annual growth rate of close to 35%. Assume a 35% growth rate for such an investment. Compare the predicted doubling time with the actual doubling time.

d. For what range of annual interest rates does the Rule of 72 provide a reasonably good estimate of doubling time for an investment paying interest compounded annually?

6 Shown at the top of the following page is the performance of Apple stock over the period February 2008– February 2013.

a. How would you describe the pattern of change in the value of Apple stock during this five-year period?

b. What equivalent interest rate, compounded annually, would yield earnings that match the stock gain?

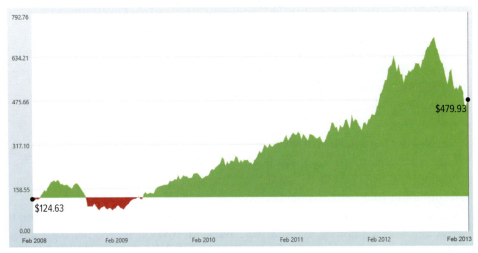

Source: *Microsoft Finance* February 12, 2013. Market data by Morningstar.

Effective Rates of Interest The annual (or *nominal*) percentage rate of interest is the rate most often quoted by banks and local credit unions. Its true worth depends on the number of compounding periods. The APR cannot help you decide, for example, whether a savings account that pays 3.75% interest compounded quarterly is better than a savings account paying 3.7% compounded monthly. For comparison purposes, a better measure of worth is the **effective annual interest rate** or **annual percentage yield** (APY). The APY is the simple interest rate that yields the same amount after one year as the compounded annual percentage rate of interest.

7 Look back at the investment, APR, and compounding periods in Problem 1. Determine the APY of each investment plan.

8 Suppose an investment of A dollars has an annual percentage rate r (expressed as a decimal) compounded n times per year.

    a. Explain why the value of the investment after one year can be represented by $A(1 + \text{APY})$ or by $A\left(1 + \dfrac{r}{n}\right)^{n}$.

    b. Use these two expressions to find a formula for calculating the APY.

        i. Check if your formula produces results consistent with what you calculated in Problem 7. Make any needed adjustments.

        ii. Compare your formula for calculating the APY with that of your other classmates. Resolve any differences.

9 Six months after graduation from college, Jennifer decided to start a savings account for graduate school. She researched her investment options and found the three options below.

Option 1	2.99% APR	compounded monthly
Option 2	3.00% APR	compounded quarterly
Option 3	3.01% APR	compounded semi-annually

Make a recommendation on the option she should choose. Be prepared to explain your reasoning.

10 When Sabrina started college, her grandfather deposited $10,000 in a savings account for her that earned 3.25% annual interest compounded quarterly. Beginning with the second quarter, and continuing each quarter thereafter, Sabrina withdrew $1,500 for tuition. Prepare a table showing the balance in the account for each of the first four quarters. Compare your account balances with those of your classmates and resolve any differences.

SUMMARIZE THE MATHEMATICS

In this investigation, you examined patterns of investment growth involving compound interest.

a. Explain why the future value of an investment paying compound interest can be modeled by an exponential function. Give an example illustrating this fact.

b. Suppose A dollars is invested in a five-year certificate of deposit (CD) paying an APR of r% compounded quarterly. Write a *NOW-NEXT* formula showing how the value of the CD changes from one compounding period to the next. What is the initial (or starting) value?

c. What is the effective annual interest rate (APY) on an investment plan? How does it differ from the commonly advertised annual percentage rate (APR)? Why is knowing the APY more useful than knowing the APR in investment decision-making?

d. If $F = A\left(1 + \frac{r}{n}\right)^{nt}$ gives the value of an investment paying compound interest, what would the values of n, r, and t tell about the situation?

Be prepared to explain your ideas and reasoning to the class.

Chantal opened a credit union account to save for a downpayment for her first house. The savings account has an APR of 1.2% compounded quarterly. She plans to make initial and quarterly deposits of $450 for five years.

a. How much will Chantal have in her account at the end of:

  i. year 1?

  ii. year 2?

  iii. year 5?

b. After five years, what is the total of her investment contributions? What is the total interest earned?

c. Chantal had considered opening an account at a local bank that was offering an APR of 1.15% compounded *daily*, but again investing $450 per quarter. Did she make the correct decision on where to invest her money? Explain your reasoning.

d. Suppose Chantal wants to make a downpayment of 10% on a condominium with a value of $150,000. For how many quarters would Chantal need to contribute $450 to her credit union account to have enough money for the downpayment on the house?

e. When Chantal opened her credit union account, her parents purchased a five-year CD for $5,000 earning interest at 7% compounded monthly. If she combines the future value of this CD with her credit union account, will she have enough for the 10% downpayment on the condominium in five years?

INVESTIGATION 3

Investment Strategies and Tools

Planning financial investments for the present and future requires making especially important decisions in today's ever-changing global economy. Many people save money each month in an effort to build some financial security. Saving by making fixed deposits at regular intervals is a good financial practice.

As you work on the following problems, look for answers to these questions:

How do you determine the future value of an investment earning compound interest and involving fixed deposits at regular intervals?

What technological tools are useful in analyzing such investments?

How do you decide on what technological tool might be most useful in exploring particular investment options? In tracking changes in the future value of an investment?

1 When Lucy was born (August 1), her parents decided to save money each year for her post-secondary education. That day, they opened a savings account with $400 and then deposited $600 every August 1 for the next 18 years. Assume that each year, the deposit is made at the end of the compounding year.

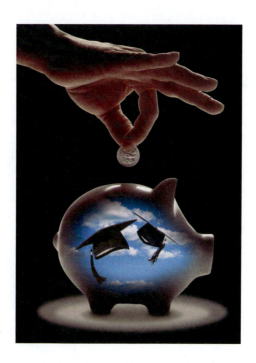

   a. If the bank provides an annual percentage yield of 3.2%, what will be the balance when Lucy turns 1? Turns 2? Turns 3?

   b. How much interest does the college savings account earn at the end of the first year? At the end of the second year? At the end of the third year?

   c. Use the word *NOW* to represent the savings balance in any given year. Write a recursive formula using the words *NOW* and *NEXT* that shows how to compute the balance at the end of the next year. What is the starting value of the account?

   d. How would your answers to Parts a, b, and c change if the savings account was only paying simple annual interest at the equivalent rate?

2 **Using a Spreadsheet in Reasoning About Investments** Lucy's parents may have used a spreadsheet to calculate the growth of their daughter's education investment. Examine this start of a portion of a possible spreadsheet.

	A	B	C	D
1	Age	Deposit	Account Balance	Initial Deposit
2	0	=$D$2	=$D$2	400
3	1	=$D$4	=C2*(1+$D$6)+B3	Recurring Deposit
4	=A3+1	=$D$4	=C3*(1+$D$6)+B4	600
5	=A4+1	=$D$4	=C4*(1+$D$6)+B5	Interest Rate (APY)
6	=A5+1	=$D$4	=C5*(1+$D$6)+B6	0.032

   a. Discuss with a partner how the spreadsheet formula **=C2*(1+$D$6)+B3** in cell **C3** calculates the account balance when Lucy turns 1.

   b. Construct your own spreadsheet using the above template.

   c. What is the balance in the account at the end of the 10th year? The 18th year?

3 A group of students at Waterford High School were curious as to whether there was a formula for predicting the future value of an investment like the one in Problem 1 that involved fixed deposits at regular intervals.

The table below indicates approaches by three different students.

End of Year	Account Balance Jason's Method	Account Balance Bekah's Method	Account Balance Max's Method
1	$(1.032)(400) + 600$	$(1.032)(400) + 600$	$(1.032)(400) + 600$
2	$(1.032)[(1.032)(400) + 600] + 600$	$(1.032)^2(400) + (1.032)(600) + 600$	$(1.032)^2(400) + 600\left(\dfrac{1.032^2 - 1}{1.032 - 1}\right)$
3	$(1.032)[(1.032)^2(400) + (1.032)(600) + 600] + 600$		$(1.032)^3(400) + 600\left(\dfrac{1.032^3 - 1}{1.032 - 1}\right)$
4			
⋮	⋮	⋮	⋮
n			

a. Analyze the thinking of each student. Then complete a copy of the table for calculating the account balance.

b. How do the methods compare to your work in Problem 2?

c. Which of the methods and resulting expressions reveals more about how the money accumulates? Explain your thinking.

d. Which of the methods would be useful if the investment plan was set up for the parents' retirement? Explain your reasoning.

4 The account balance after 18 years that you found in Problem 2 Part c represents the future value of this investment.

a. Explain how the following **future value formula** generalizes the pattern in Max's reasoning.

$$F = I(1 + r)^t + A\left(\frac{(1 + r)^t - 1}{r}\right)$$

b. What do the variables I, A, r, and t represent in this situation?

Jim Laser

c. Using this formula, what is the future value of Lucy's college savings account after 10 years? After 18 years? Compare your results with those calculated using spreadsheet and recursion methods. Explain any differences.

d. Suppose Lucy is considering joining the Peace Corps and delaying college enrollment for two years. Lucy's parents decide to extend their investment for those two years. After the additional two years, what will be the future value of their investment?

5 Suppose Lucy needs $20,000 for the cost of attending her local community college.

a. How many annual deposits are needed before her parents have saved enough money to cover her college costs? Check your answer using the formula in Problem 4.

b. How much money should Lucy's parents invest for each payment after the initial $400 deposit if they plan to save for only 18 years?

c. Suppose Lucy's parents invest the same amount at her birth and annually on her birthday for 18 years in a savings account with an annual percentage yield of 1.4%. Make a prediction for the amount of each deposit to achieve the needed amount of $20,000. Check your conjecture using the formula in Problem 4.

6 The following portion of a spreadsheet can be used to explore how Oseola McCarty (page 266) could have saved $150,000 in 75 years. To limit the size of the spreadsheet, it is assumed that savings and interest on her bank account were deposited only once each year and that the deposit was made at the end of the interest period. This is only the start of the complete spreadsheet.

	A	B	C
1	Year	Bank Balance	Yearly Savings
2	0	25	100
3	=A2+1	=(1+C$4)*B2+C$2	Interest Rate (APR)
4	=A3+1	=(1+C$4)*B3+C$2	0.05
5	=A4+1	=(1+C$4)*B4+C$2	

a. Study the beginning of the spreadsheet and discuss with a partner what each cell and formula represents. How do the formulas indicate that each annual deposit is made at the end of the interest period?

b. Then extend the sheet to cover 75 years. Does this choice of beginning balance, yearly savings, and interest rate result in $150,000 saved after 75 years?

c. Experiment with changes in the entries for initial deposit, yearly savings amount, and interest rate. Give four combinations of beginning balance, yearly savings amount, and interest rate that could produce $150,000 in 75 years.

d. What other factors should be considered if your analysis is to reflect the probable pattern of Miss McCarty's savings and interest earnings? Show how the spreadsheet formulas could be adjusted to account for those factors.

7 For someone starting a savings plan today, it is quite reasonable to set a much higher goal than the $150,000 achieved by Oseola McCarty. Modify the spreadsheet of Problem 6 to explore combinations of initial deposits, annual savings, annual interest rates, and lengths of saving time that could make you a millionaire. Consider at least the following options.

- Suppose first that the same amount of money is saved and invested each year.

- Next consider the possibility that your increasing annual earnings would allow you to *increase* the annual savings amount in a steady pattern— either a constant dollar amount or a constant percent each year.

After you have explored various combinations of annual savings, interest rates, and lengths of saving time, summarize your findings. Be prepared to share your findings and methodology with the class.

8 **Using an Online Financial Calculator in Reasoning about Investments**

There are many financial calculators available on the Internet. They can be very helpful in financial decision-making and exploring "what-if" questions. Suppose you decide to open a savings account with an initial deposit of $100. The account pays an annual interest rate (APR) of 2.4% compounded monthly. You plan to make regular monthly deposits of $50 into the account. Study the investment information provided by the following computer display.

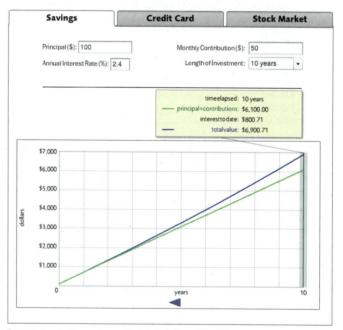

Source: illuminations.nctm.org/ActivityDetail.aspx?ID=172

a. According to the computer display above, what is the balance in your account after 10 years?

b. How much interest did your account earn after 10 years?

c. Examine the graph displayed on the previous page.

    i. What part of the graph represents the total value with interest of your account as a function of time? What clue(s) did you use?

    ii. What part of the graph represents only the money you contributed and no interest as a function of time? Why does the pattern of change shown in the graph make sense?

    iii. How is the interest earned represented on the graph?

d. Access the above online financial calculator to help answer the following questions.

    i. How many months before your account balance will be at least $6,000? At least $10,000?

    ii. What is your best estimate for when you will have earned $1,000 in interest?

9 Reproduced below is the formula for the future value of an investment which involves an initial investment and then fixed annual deposits with interest compounded annually.

$$F = I(1 + r)^t + A\left(\frac{(1 + r)^t - 1}{r}\right)$$

a. Recall what the variables I, A, r, and t represent in the future value formula. How would you modify this formula to calculate the future value of an investment paying a fixed APR compounded monthly? Compare your formula with that of your classmates and resolve any differences.

b. Now use your agreed-upon formula to determine the future value of the account described in Problem 8. Compare the future value calculated by your formula with the total value shown in the computer display? Explain any differences.

c. Discuss what each term in the future value formula represents. Compare the amount calculated by each term of the future value formula for the situation in Problem 8 to the amounts displayed in the yellow box in the computer display on page 280. Why does this make sense?

10 Pose an investment question that is of interest to you. Use the *Illuminations* financial calculator or other online calculator to help answer your question. Be prepared to describe your question and how you answered it to the class.

In the On Your Own set at the end of this lesson, you will have the opportunity to explore the features and use of other Internet-based financial calculators.

SUMMARIZE THE MATHEMATICS

In this investigation, you explored ways to model investments that involved fixed deposits at regular intervals with compounding and how to predict future value.

a. If the formula $F = I\left(1 + \frac{r}{n}\right)^{nt} + A\left(\dfrac{\left(1 + \frac{r}{n}\right)^{nt} - 1}{\frac{r}{n}}\right)$ gives the future value of a savings

investment started with an initial deposit and then increased by fixed deposits at regular intervals with compounding, what would the values I, A, r, n, and t tell about the situation?

b. Why can the formula in Part a be thought of as a function of the length of investment in years?

    i. How would the function rule and graph for the future value change if the initial balance and fixed regular contributions increased? Decreased?

    ii. How about if the interest rate changed?

    iii. What if the number of compounding periods increased?

c. What technological tools are helpful in analyzing and tracking investment options? How do you decide on the tool to use?

Be prepared to explain your ideas to the class.

✔ CHECK YOUR UNDERSTANDING

When Danielle was a freshman at Dexter High School, she decided to use money from her part-time job at a public library to open a savings account with the intent of buying a car when she went to college.

a. What fixed amount must Danielle save each month in order to purchase an $8,000 car in four years if the savings account interest is 4% compounded monthly?

b. How much of the $8,000 is money Danielle deposited and how much is interest?

c. How did you decide on a technological tool to help complete this task?

Compound Interest and the Number *e*

In Problem 1 of Investigation 2, you calculated the value of three accounts with initial deposit of $1,500 paying a 4% annual interest rate after one, after two, and after five years. The interest for one account was compounded semi-annually, the interest for the second account was compounded quarterly, and the interest for the third account was compounded monthly. You found that the account for which the interest was compounded more frequently earned a better effective yield. This process could be continued so that interest is compounded every day or every hour or every minute or even every second.

As you work on the problems of this investigation, look for answers to these questions:

What is the number e and how is it connected to frequent compounding of interest?

How can you predict the future value of an investment with a fixed APR that is compounded continuously?

What Is the Number *e*? It turns out that there is a way to explain the effects of frequent compounding by using a special number as the base. The key to that strategy lies in understanding behavior of the expression $\left(1 + \frac{1}{n}\right)^n$ for large values of *n*.

1 Explain why it makes sense to think of the expression $\left(1 + \frac{1}{n}\right)^n$ as representing the value of a $1 initial investment earning a 100% annual percentage rate that is compounded at *n* equal periods each year.

2 Now evaluate $\left(1 + \frac{1}{n}\right)^n$ for values of *n* in the following table.

n	1	3	5	10	100	500	1,000	10,000	100,000
$\left(1 + \frac{1}{n}\right)^n$									

The number to which values in the above table are converging is a special mathematical constant that is labeled with the letter *e*, in honor of the mathematician Leonhard Euler who discovered some of the most important properties of the number. **The value of *e* is an irrational number, approximately 2.71828.**

3 For increasingly large values of n, the expression $\left(1 + \frac{1}{n}\right)^n \approx e$.

 a. Use the relationship between $\left(1 + \frac{1}{n}\right)^n$ and e and algebraic properties to explain steps in the following connection of e to frequent compounding of interest.

 Step 1. $\left(1 + \frac{1}{\frac{n}{r}}\right)^{\frac{n}{r}} \approx e$

 Step 2. $\left(1 + \frac{r}{n}\right)^{\frac{n}{r}} \approx e$

 Step 3. $\left(1 + \frac{r}{n}\right)^{n} \approx e^r$

 Step 4. $\left(1 + \frac{r}{n}\right)^{nt} \approx e^{rt}$

 b. Complete the following sentence in a way that explains the relationship of the number e and frequent compounding of interest.

 The value of an investment (or debt) that starts at A and is paid (or charged) at an annual percentage rate of r (expressed as a decimal) and is compounded at shorter and shorter periods can be approximated by _____.

 This interest is said to be **compounded continuously**.

4 Continuous Compounding Suppose the future value after t years of $500 compounded continuously is given by the function $F(t) = 500e^{0.02t}$.

 a. What is the continuous percent growth rate?

 b. What is the APY?

 c. What is the balance in the account after two years? After 42 months?

 d. How long would it take for the value of the investment to double?

5 An initial investment in a start-up business of $1,500 is to pay an APR of 4% compounded weekly. The future value of the investment is

 $B(t) = 1,500\left(1 + \frac{0.04}{52}\right)^{52t}$ after t years.

 a. Write a rule for $B(t)$ if the interest is instead compounded continuously.

 b. How does the approximate future value produced using continuous compounding compare to the future value with weekly compounding? Why does this make sense?

 c. Suppose that the interest was compounded daily. Estimate the investment accrual after five years. Check your estimate by calculating this balance using 360 days for a year.

6 Write an algebraic formula for the future value of an account with an initial investment of $15,000 for t years at an APR of 1.75% under each of the following conditions.

 a. Interest compounded quarterly

 b. Interest compounded monthly

 c. Interest compounded continuously

7 How much interest is earned in 48 months under each of the three conditions in Problem 6?

8 Suppose $V(t)$ is an exponential function growing at a continuous percent growth rate. Then V can be expressed as a function of time t (in years) in two ways.

- $V(t) = Ae^{kt}$, where A is the initial investment and k is the continuous percent growth rate.

- $V(t) = A(B^t)$, where A is the initial investment and B is the growth factor.

a. If r is the effective annual interest rate of the investment (expressed as a decimal), explain why $B = 1 + r$.

b. Explain why $B = e^k$.

9 Suppose you invested \$500 in a local credit union and:

- $f(t)$ gives the future value of the investment in t years, if the APR is 2% and interest is compounded quarterly.

- $g(t)$ gives the future value of the investment in t years, if the APR is 2% and interest is compounded monthly.

- $h(t)$ gives the future value of the investment in t years, if the APR is 2% and interest is compounded continuously.

a. Write a function rule for $f(t)$. For $g(t)$. For $h(t)$. Then describe how the rules are similar and how they are different.

b. Based on your understanding of exponential growth, describe how the graphs of $f(t)$, $g(t)$, and $h(t)$ are similar, and how they are different.

c. On the same coordinate grid, use algebraic reasoning to sketch graphs of the three functions.

10 Now examine more closely the function $f(x) = e^x$ from an algebraic perspective.

a. To which function family does $f(x)$ belong?

b. Sketch a graph of $f(x) = e^x$ and state the domain and range of the function.

c. How does the graph of $f(x) = e^x$ compare to the graph of $g(x) = 2^x$? To the graph of $h(x) = 3^x$? To the graph of $j(x) = 10^x$? Explain.

d. Does $f(x) = e^x$ have an inverse that is a function? Explain your reasoning.

e. How would the graph of $j(t) = e^{0.18t}$ compare to the graph of $k(t) = e^t$? Check your conjecture and explain your findings.

SUMMARIZE THE MATHEMATICS

In this investigation, you examined ways that continuous growth at constant percent rates can be modeled by exponential functions with the special base $e \approx 2.71828$.

a. How could you produce an estimate for the value of e using a calculator that does only the four basic operations $+$, $-$, $\times$, and $\div$?

b. How and under what conditions can the function $g(t) = A\left(1 + \dfrac{r}{n}\right)^{nt}$ be approximated by an exponential function using base e?

c. Suppose you invest \$1,000 at 3.4% APR compounded continuously. Write a function $F(t)$ that gives the value of the account as a function of time t in years. What would be the ending balance, to the nearest cent, after five years?

d. How is the graph of $f(x) = e^x$ similar to, and different from, the graph of $g(x) = 10^x$?

Be prepared to explain your ideas and reasoning to the class.

✔CHECK YOUR UNDERSTANDING

When Sophie was born, her parents wanted to open a savings account that would provide money to buy her a car when she turned 16 years of age. They considered two options for their initial deposit of \$2,000.

Option 1: A credit union account paying 2.4% interest compounded monthly

Option 2: A bank account paying 2.04% compounded continuously

a. Which option would you recommend choosing? Explain your reasoning.

b. What is the APY of the credit union account? Of the bank account?

c. Based on your answer to Part a, what is the approximate price of a car that Sophie can purchase at age 16?

d. How much interest would the credit union account earn in 16 years? The bank account?

APPLICATIONS

1 Suppose you sold your car for $4,500, purchased a motor scooter for $1,800, and invested the difference in a two-year CD paying simple annual interest of 3.2%.

   a. How much interest will you earn?

   b. What will be the value of your CD investment when it matures at the end of two years?

2 Governments and corporations use *bonds* to borrow money. A **government** or **corporate bond** represents a loan to the government or corporation that you, the bondholder, will be paid back with interest, much like in the case of a bank account. For some bonds, the interest is paid semiannually or annually. Bonds, like CDs, that take longer to mature generally pay a higher interest rate.

   a. Clarissa bought a $5,000 corporate bond in Striker Corporation paying 4.1% annual interest. How much does Clarissa receive each year from the bond investment?

   b. Government bonds, and some municipal and corporate bonds, have a fixed amount that is paid at maturity. Suppose on June 1, Sasha purchased a 30-year government bond that paid an annual interest rate of 5.5% with a *face value* at maturity of $10,000. What was the initial purchase price of the government bond?

3 As you saw with the Amazon.com stock in Investigation 1, the value of a share of stock varies over time. Investors hope that the stock they have invested in ultimately increases in value over time even if there are periods of time where the value decreases instead.

GOOG (NASDAQ) February 12, 2012–February 12, 2013

Source: *Microsoft Finance* February 12, 2013. Market data by Morningstar.

a. How would you describe the pattern of change in the price of Google stock over the one-year period shown in the graph on the previous page?

b. Estimate the equivalent annual interest rate earned on a share of Google stock purchased on February 12, 2012.

c. Suppose you had purchased 100 shares of Google stock on February 12, 2012 and sold it at the highest value of the stock, $787.90 per share, during the year that followed. Estimate the equivalent annual interest rate you would have earned. What would be the profit on your investment?

4 As you may have read during your years in high school, interest rates fluctuate with the economy. In the 1980s, the highest CD rate was over 16% compounded monthly. In 2012, the highest CD interest rate was approximately 1.8%.

a. What is the APY for a CD with an APR of 16% compounded monthly?

b. What is the APY for a CD with an APR of 1.8% compounded monthly?

c. If $500 is invested in a 3-year CD at 16% interest compounded monthly, what is the ending balance? How much interest was earned?

d. If $500 is invested in a 3-year CD at 1.8% interest compounded monthly, what is the ending balance? How much interest was earned?

5 When banks compete for your savings account, they usually advertise their annual percentage rate quite prominently. Suppose that Alex receives an inheritance of $10,000 from his grandfather.

a. A local bank offers to invest the money at an APR of 5% compounded quarterly.

   i. What formula gives the value of that investment $V(x)$ after x quarters?

   ii. What is the effective annual interest rate (APY)?

b. A credit union is offering the same 5% rate compounded daily.

   i. What function $V(t)$ gives the value of the investment after t years? (Use 360 days as equivalent to a year.)

   ii. What is the APY?

c. Which of the two investment offers should Alex choose? Explain.

6 Suppose Ahmed and his twin brother pool a portion of their summer earnings and invest $2,000 in a four-year CD with a 2.8% APR compounded quarterly. Determine:

a. the number n of compounding periods per year.

b. the number m of compounding periods for the CD.

c. the interest rate for each compounding period.

d. the value of the CD at maturity.

e. the amount of the CD allocated to earned interest.

f. the APY.

7 Derrick opens an account at a local bank with an initial deposit of $200. The account pays 1.1% compounded monthly. Suppose Derrick is able to deposit $100 monthly to the account for each of four years (48 months) prior to going to college.

a. What is the APY for Derrick's investment?

b. How much money is in the account after 48 months?

c. How much interest did the account earn?

8 **Zero-coupon bonds** are investments that do not pay interest during the investment period. They are sold at a price considerably less than their redemption value. Income to the bondholder is the difference between the redemption value and the purchase price at the time of redemption. If the APR for a zero-coupon bond is 2.75% compounded daily, what should be the purchase price of the bond if it will pay $10,000, 10 years from now?

9 A movie theater is saving money monthly for renovations to their video projection and sound equipment. The owner is considering the following two investment plans, each paying interest compounded monthly.

	Plan A	Plan B
Annual Percentage Rate	1.3%	1.2%
Monthly Deposit Amount	$350	$400

a. Write formulas representing the future value for each investment plan.

b. What is the future value for Plans A and B after 1, 2, and 3 years?

c. Which plan would you recommend based upon your results from Part b?

10 The Edson's decided to invest in an annuity to save for retirement. They deposit $580 every month into an account that has an annual percentage rate of 0.13% compounded monthly.

a. What is the value of the annuity after the first five months?

b. What is the value of the annuity at the end of five years? 10 years? 15 years? 30 years? Prepare a table and a graph to show your results.

Digital Vision

11 Maddie wants to save money for her first car. She opens a savings account with an annual percentage rate of 1% compounded monthly. She plans to make *n* equal monthly deposits.

   a. What function models the future value of her investment after *t* years?

   b. Suppose she wants to save $8,000 over three years. What should be the amount of each monthly deposit?

12 Bette is saving money for an upgrade to her laptop computer. She opens a savings account at the local credit union with an annual percentage rate of 1.7%, compounded monthly, and plans to contribute $100 each month.

   a. Write a function that gives the future value of Bette's investment after *t* years.

   b. Suppose she needs a total savings of $4,000 for a new laptop computer in three years. How much money should each monthly deposit be to save enough for Bette's laptop computer by the end of three years?

13 New product sales often follow predictable patterns. Sales of a product can start off slowly and then pick up as more people learn about it, and finally level off once the market is saturated.

$$\text{The } \textbf{logistic function } S = \frac{1}{k + ae^{-bt}}$$

is often used to model growth situations like that above. In this model, S is number of sales (in millions), t is elapsed time (in weeks), and parameters a, b, and k are positive numbers.

   a. Graph the logistic function for $k = 0.2$, $a = 3$, and $b = 0.25$. Then draw a rough sketch of the graph.

      i. Does the pattern of change match the described pattern of sales? Explain.

      ii. Estimate from the graph when 1 million sales can be expected. Estimate the total sales expected for the product over time.

   b. If release of a new product has been highly anticipated, sales of the product can start off rapidly and then level off. Draw a sketch illustrating this pattern of change in sales.

   c. The mathematical models in Parts a and b have many other applications. Match the following phenomena with one or both of the above models.

      i. Learning of a new mathematical topic

      ii. Population growth

      iii. Spread of epidemics

14 In Investigation 1, Problem 1 Part b, you developed a formula $F = P(1 + rt)$ giving the future value of a simple-interest investment in t years.

a. What do t, r, and P represent?

b. Solve this formula for P. What is calculated using this formula?

15 Suppose a simple-interest investment earns $r\%$ one year and $s\%$ the following year.

a. Find an algebraic expression for the account balance after two years, where the interest rate for the first year is $r\%$ and the interest rate for the second year is $s\%$.

b. What can you conclude?

16 Compound interest problems involve five variables:

- P principal of the investment
- r annual percentage rate
- t length of investment in years
- n number of compounding periods per year
- F future value of the investment

a. Write a formula for F in terms of four of the variables.

b. Which of the five variables is necessary to calculate the APY?

c. Write a formula for the APY of the investment.

17 Suppose you deposit a graduation gift of $100 in a savings account with an APR of 6% compounded semi-annually.

a. What would be the future value of your account in five years?

b. Under which of the following conditions would the future value F of your account in five years be double that in Part a?

I initial investment of $100 is doubled

II interest rate doubles

III number of interest paying periods per year doubles

IV number of years doubles

18 In this lesson, you used *NOW-NEXT* formulas to describe the patterns of change in simple-interest investments and in compound-interest investments from one period to the next. These patterns of change can also be described by *recursive formulas* using a_0 as the initial investment, a_n as the value of the investment after n periods, and i as the interest rate per period. Complete each recursive formula.

Simple interest: $a_n =$ _____

Compound interest: $a_n =$ _____

19 In Connections Task 14 Part b, you derived a formula for calculating the **principal of a simple-interest investment** that would yield a specified future value knowing the length of the investment in years and annual rate of return. Look back at the given information in Connections Task 16. Derive a formula for the **principal of a compound-interest investment** that yields a specified future value knowing the length (number of periods) of the investment, compounding period, and APR.

20 Although you may have a financially sound savings plan in place, there is one factor that influences the value of your savings that no one can control. That is the increase in the price of most goods and services over time. The rate at which the general level of prices for goods and services rises (and subsequently, *purchasing power* of your savings falls) is called the **inflation rate**.

a. One way to understand *inflation* is to compute the future price of an item subjected to inflation. For example, a chain saw that costs $100 today would cost $134.39 in 10 years given a 3% annual inflation rate. In 15 years, the same item would cost $155.80, or over 50% more than today. How were the values $134.39 and $155.80 calculated?

b. Another way to understand the impact of inflation is to determine the value of today's dollar in the future. For instance, given a 3% annual inflation rate, $100 today would be worth only $64.19 in 15 years. How do you think the value $64.19 was calculated?

c. Suppose 10 years ago you invested $1,000 in an account paying 5% interest compounded annually and that the average inflation rate for the past 10 years has been about 3% each year.

   i. Is the purchasing power of your investment after 10 years more or less than your $1,000 initial investment? Explain your reasoning.

   ii. What will be the "real" rate of return on your investment when inflation is considered? Explain.

21 After graduation from college, Malcolm and his sister Whitney started their corporate careers in different states. Both understand the importance of saving money in planning for the future. Each started an **ordinary annuity**—a financial plan involving making an initial deposit and then equal deposits of the same initial amount at regular intervals.

Malcolm invests $2,000 of his year-end bonus in an ordinary annuity at the end of each year for 10 years. The annuity pays 3.75% interest compounded annually. After the initial 10 years, Malcolm invested more aggressively in stocks, making no further contributions to his annuity, which continues to earn 3.75% compounded annually.

The best annual annuity rate available to Whitney was 3.15%. She contributed $2,200 of her bonus money at the end of each year for 45 years (to age 67).

a. How much money do Malcolm and Whitney have in their annuity accounts at the end of 45 years?

b. In each case, how much of the account value is money contributed? How much is earned interest?

c. How would you use a recursive *NOW-NEXT* formula to describe the change in the value of Whitney's account from one year to the next?

d. How would you modify the future value formula in Summarize the Mathematics Part a for the case of an ordinary annuity that involves the same yearly investment compounded annually?

e. Make and compare the growth graphs of Malcolm's and Whitney's annuities.

22 The Dow Jones Industrial Average provides one measure of the "health" of the U.S. economy. It is a weighted average of the stock prices for 30 major American corporations. The following table shows the low point of the Dow Jones Industrial Average in selected years from 1965 to 2005.

Year	DJIA Low
1965	840
1970	631
1975	632
1980	759
1985	1,185
1990	2,365
1995	3,832
2000	9,796
2005	10,012

Source: www.analyzeindices.com/dow-jones-history.shtml

a. Find what you believe are the best possible linear and exponential models for the pattern of change in the low value of the Dow Jones Industrial Average over the time period shown in the table (use $t = 0$ to represent 1965). Then decide which you think is the better of the two models and explain your choice. What do each of the parameters in your chosen function rule represent?

b. Use your chosen predictive model from Part a to estimate the low value of this stock market average in 2010. Use the Internet to check your predicted low value.

c. Why should you be cautious about DJIA low values your chosen model predicts for 2015 or 2020?

d. Some stockbrokers who encourage people to invest in common stocks claim that one can expect an average return of 10% per year on that investment. Does the rule you chose to model increase in the Dow Jones average support that claim? Why or why not?

23 The grid at the right shows the graph of $f(x) = e^x$ and its reflection image $g(x)$ across the line $y = x$.

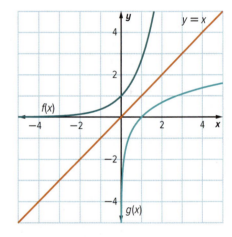

a. How are the functions $f(x)$ and $g(x)$ related?

b. The function $g(x)$ is called the **natural logarithm function** and is denoted $g(x) = \ln (x)$, or simply $g(x) = \ln x$. By definition,

   $\ln a = b$ if and only if $e^b = a$.

   What are the domain and range of $g(x) = \ln x$?

c. Use the definition of the natural logarithm (base-e) function and the key on your calculator to solve the following equations.

   i. $e^x = 20$ **ii.** $e^x = 1$

   iii. $15e^x = 30$ **iv.** $20.7e^{0.03t} = 27$

24 You may recall from your previous course work that for the base-10, or common, logarithm,

   $\log a = b$ if and only if $10^b = a$.

a. For a given nonzero number P, will $\ln P$ be greater than, less than, or equal to $\log P$? Explain your reasoning.

b. What does your answer to Part a say about the relative positions of the graphs of $y = \log x$ and $y = \ln x$?

25 In your previous work with base-10 or common logarithms, you learned that it is often helpful to write given logarithmic expressions in different equivalent forms. It should be reasonable to expect that some of the rules for rewriting base-10 logarithms have analogs in work with base-e logarithms. Write the following expressions in different forms that you suspect will be equivalent to the originals. Then complete the given numerical examples to check your conjectures.

a. $\ln ab = \ldots$, for a and b both positive numbers. For example, $\ln (5 \cdot 4) = \ldots$.

b. $\ln \frac{a}{b} = \ldots$, for a and b both positive numbers. For example, $\ln (20 \div 2) = \ldots$.

c. $\ln a^b = \ldots$, for positive number a and any number b. For example, $\ln (3^4) = \ldots$.

REFLECTIONS

26 Consider the results of these two investments.

I An investment of A dollars earns 3% one year and then loses 2% the next year.

II An investment of A dollars loses 2% one year and then gains 3% the next year.

Make a conjecture as to which investment has the larger balance at the end of the second year. Check your conjecture using algebraic reasoning.

27 Albert Einstein wrote that compound interest was, "… the most powerful thing I have ever witnessed." What do you think would lead him to this conclusion?

28 How do the results of your work on problems in Investigation 2 illustrate the importance of checking the *effective annual interest rate* when considering savings plans that advertise only *annual percentage rates* and compounding periods?

29 Based on your work in this lesson, you observed that more frequent compounding yields greater interest. What can you say about the corresponding pattern of change in future values as the frequency of compounding increases? Illustrate your answer with a specific example.

30 Look back at Problem 1 (page 277) in Investigation 3. Suppose Lucy's parents decide to consider a plan where they deposit more payments throughout the year. Suppose they instead deposit semi-annual payments of $300 in a savings account with 3.2% interest compounded semi-annually. Would this decision be more or less attractive than their original plan? Explain. Be sure to check your understanding by discussing how the future value of the savings balance compares with Lucy's parents' original plan by providing a function model.

31 Look back at Problem 1 (page 277) in Investigation 3. Suppose Lucy's parents decided to organize their investment in a different way as shown in the spreadsheet below.

	A	B	C	D
	Deposit Number	Deposit Amount	Number of Interest Years	Future Value of Deposit
1				
2	1	$400.00	17	=B2*(1+0.032)^C2
3	2	$600.00	16	=B3*(1+0.032)^C3
4	3			
5	4			
6	5			
7	6			
8	7			
9	8			
10	9			
11	10			
12	11			
13	12			
14	13			
15	14			
16	15			
17	16			
18	17			
19	18			

a. Lucy's parents are using the spreadsheet formula **=B2*(1+0.032)^C2** in cell **D2**. Describe fully how this formula is calculating the future value for the first deposit.

b. Complete a copy of the spreadsheet. What is the future value of all deposits and what does it represent? Write a spreadsheet formula to describe this total.

c. Compare this strategy for calculating the future value with the strategy in Problem 2 in Investigation 3.

32 As you have seen in Connections Task 20, inflation affects the future purchasing power of your investments. Suppose you invest an initial amount of $1,000 in a savings account that earns 4.3% annual interest compounded monthly. You plan to make monthly contributions of $100 to this account for five years. Economic forecasters expect an average yearly inflation rate of about 2% for the next five years.

a. Without accounting for inflation, what would be the future value of your investment?

b. To help manage the effects of inflation, what decisions could you make related to your investment plan that would help to retain its future value as found in Part a?

c. Access an online savings and investment calculator such as the one at www.mycalculators.com/ca/savecalcm.html to explore concrete options for adjusting your savings to account for inflation of about 2% per year. What did you learn related to a 2% inflation rate?

d. Historically, inflation rates in the U.S. have varied considerably, from highs of close to 20% in 1947 to lows of close to −16% in 1921. Negative inflation is called *deflation*. (**Source:** www.usinflationcalculator.com/inflation/historical-inflation-rates/)

   Choose some values for investments, interest rates, number of compounding periods, and annual inflation and deflation rates to consider ways to combat the effects of inflation and the influence of deflation on investments. Summarize your findings.

33 Matthew is considering investing his minor league baseball bonus of $20,000 at an APR of 1.72% compounded continuously for four years. To estimate the ending balance, he uses the formula $B = Pe^{rt}$ and enters numbers as shown at the right.

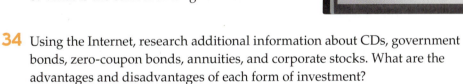

a. Examine the calculator display. How would you respond to Matthew's work?

b. What is the correct ending balance?

34 Using the Internet, research additional information about CDs, government bonds, zero-coupon bonds, annuities, and corporate stocks. What are the advantages and disadvantages of each form of investment?

EXTENSIONS

35 Choose a company or corporation that sells a product you are interested in. Using the Internet, find and record the daily low, high, and close of one share of your chosen company's stock beginning the day you start Unit 4 and ending the day you complete it.

a. Prepare a graph showing the pattern of change in the daily closing price of your chosen stock.

b. Discuss the trends in the daily closing price during the course of this unit.

36 Using the Internet, research the history of the Rule of 72. Prepare a summary report for the class on your findings.

37 Feng-Chiu is considering two investment plans created by her fellow students that use different annual percentage rates, methods for compounding interest, and deposit amounts.

	Cody	Kristin
Annual Percentage Rate	8.2%	0.88%
Method for Compounding Interest	Annually	Monthly
Deposit Amount	$400	$100

a. Write function rules giving future values after t years. Graph the future value functions.

b. Compare Cody's and Kristin's investment plans. Write a recommendation for Feng-Chiu describing conditions under which she should invest with each plan.

38 Look back at Applications Task 13. You saw that the logistic function $S = \dfrac{1}{k + ae^{-bt}}$ is a good model of new product sales that start off slowly, then pick up as more people learn about the product, and finally level off as the market becomes saturated.

The **Gompertz function** $S = k \cdot a^{(b^t)}$

is often used to model sales of a new product that has been highly anticipated. In this case, sales start off rapidly and then level off. Here S is the number of sales (in millions), t is time elapsed (in weeks), and a, b, and k are positive parameters.

a. Graph the Gompertz function for $k = 4$, $a = 0.01$, and $b = 0.8$. Estimate from the graph when 1 million sales will be reached and the total sales expected.

b. Compare the graph from Part a to your sketch in Applications Task 13 Part a and explain any differences.

c. How are the logistic and Gompertz functions related to your work in this lesson?

d. Describe a recent product whose sales might be modeled by a logistic function. By a Gompertz function.

39 Akia was curious about the Rule of 72 in the case of continuous compounding. Study her derivation below giving a reason for, or correcting, each step shown in her presentation.

a. Assume A is the initial deposit of a continuously compounding investment. Then the future value of A could be expressed:

$$F = Ae^{rt}, \text{ where } r > 0 \qquad (1)$$
$$2A = Ae^{rt} \qquad (2)$$
$$2 = e^{rt} \qquad (3)$$
$$\ln 2 = \ln e^{rt} \qquad (4)$$
$$\ln 2 = rt \ln e \qquad (5)$$
$$\ln 2 = rt \qquad (6)$$
$$\frac{\ln 2}{r} = t \qquad (7)$$
$$\frac{0.693}{r} \approx t \qquad (8)$$

b. Akia concluded that for a continuously compounding investment with an APR of $r\%$, the doubling time is $\frac{70}{r}$. She called her result the Rule of 70. Why do you think she preferred a Rule of 70 rather than a rule of 69.3?

c. Calculate the doubling time for continuous compounding at 2%, 6%, and 8% and compare your results with those predicted by the Rule of 72. Do your findings make sense mathematically? Explain.

40 Use the fact that $a = e^{\ln a}$ and $b = e^{\ln b}$ and properties of exponents to complete and give reasons for each step in proofs of your conjectures in Connections Task 25 about properties of the natural logarithm function.

a. For any positive numbers a and b:

$$ab = e^{\ln a}e^{\ln b} \qquad (1)$$

$$= e^{\underline{\qquad}} \qquad (2)$$

$$\text{So,} \quad \ln ab = \underline{\qquad} \qquad (3)$$

b. For any positive numbers a and b:

$$\frac{a}{b} = \frac{e^{\ln a}}{e^{\ln b}} \qquad (1)$$

$$= e^{\underline{\qquad}} \qquad (2)$$

$$\text{So,} \quad \ln \frac{a}{b} = \underline{\qquad} \qquad (3)$$

c. For any positive number a and any number b:

$$a^b = (e^{\ln a})^b \qquad (1)$$

$$= e^{\underline{\qquad}} \qquad (2)$$

$$\text{So,} \quad \ln a^b = \underline{\qquad} \qquad (3)$$

REVIEW

41 Consider the following furniture-purchasing scenarios.

a. You want to purchase a couch that normally sells for $439.99. There is an 8% storewide sale the day you buy the couch. What is the final cost of the couch including 5% sales tax?

b. Last year José bought a chair for $175. When he went back this year to buy a matching chair, the price was $180.25. By what percent did the price of the chair increase?

c. A table that normally sells for $625 is on sale for $531.25. By what percentage is the selling price reduced?

42 Consider the function $f(x) = 3(2^x)$.

    a. Evaluate $f(5)$ and $f(-5)$.

    b. For what value of x does $f(x) = 96$?

    c. For what values of x is $f(x) > 96$?

    d. Does $f(a + 1) = 6^a + 6$? Explain your reasoning.

    e. Does $f(3a) = 3(8^a)$? Explain your reasoning.

43 Recall there are three properties of base-10 logarithms that are useful when solving problems involving common logarithms.

$$\log xy = \log x + \log y$$

$$\log x^n = n \log x$$

$$\log \frac{x}{y} = \log x - \log y$$

If $\log a = 3$ and $\log b = 5$, without using technology evaluate each of the following.

    a. $\log a^4$ **b.** $\log ab^2$

    c. $\log \dfrac{a}{b}$ **d.** $\log \left(\dfrac{a}{b}\right)^4$

    e. $\log \dfrac{a^2}{b} - 5 \log b$ **f.** $(\log a)(\log b)$

44 Recall that a geometric sequence is a sequence of numbers for which the ratio of successive terms is a constant. For example, the geometric sequence 3, 6, 12, 24, 48 … has a common ratio of 2.

    a. Determine the common ratio for each sequence:

        i. 4, 20, 100, 500, …   **ii.** 30, 42, 58.8, 82.32, …

    b. Explain why $4(5^9)$ will give the value of the 10th term of the sequence in Part ai. What expression represents the 15th term of sequence in Part aii?

    c. In general, what expression will represent the nth term of a geometric sequence with a first term of a and a common ratio r?

    d. You may recall that the sum of the first n terms of a geometric sequence can be calculated using the formula $S = a\left(\dfrac{r^n - 1}{r - 1}\right)$. In this formula, a is the first term of the sequence, r is the common ratio and n is the number of terms. Use the formula to find the sum of the first four terms of each sequence in Part a. Then add the first four terms to check that you used the formula correctly.

45 Jonathan conjectured that "a high correlation between two variables means that the variables are linearly related."

a. Find the correlation for the following pairs of values. From your calculation, would you expect the scatterplot to have a linear pattern?

x	2	10	3	9	5	8	0	1
y	0	65	2	50	8	35	3	1

b. Produce a scatterplot for these values. Is the pattern of these values linear?

c. How well does $y = x^2 - 4x + 3.8$ model the scatterplot pattern?

d. Create a set of paired values that has a parabolic shape, but the correlation is 0.

e. How would you respond to Jonathan's conjecture? What else can you say about the relationship between the correlation of paired data and the possible pattern in the plot of the data? Why is this valuable to know?

46 Tamera is planning a singles badminton tournament. She wants to begin by dividing the players into groups where each person in the group will play a game against every other person in the group. Each group will play their games on a different day.

a. The facility for the tournament has four badminton courts and each day Tamera has time slots for a maximum of eight games on each court. What is the maximum number of people she can place in a group so that the group can complete their games in a single day?

b. Are combinations or permutations used in solving the problem in Part a? Explain your reasoning.

c. Could Tamera put the same number of people in a group if she had eight courts and could use each court for a maximum of four game slots? Explain your reasoning.

47 Rewrite each expression in an equivalent form that does not contain parentheses.

a. $3(2x + 6) - 10$

b. $0.03(2x^2 + 25x)$

c. $5(30(1.05^x) + 300)$

d. $(2x + 1)^2 - 5(2x + 1) + 15$

e. $\left(\sqrt{4x + 8}\right)^2$ for $x \geq -2$

48 Find the exact value of each expression if $x = -3$ and $y = 4$.

a. x^2

b. $-y^2$

c. x^{y+1}

d. y^x

e. $(x - y)^{-2}$

f. $\left(\dfrac{x}{y}\right)^y$

Financial Decision-Making: Borrowing

Each year, more and more colleges are requiring entering students to have their own laptops with appropriate software. After careful comparative research, Natalie purchased a laptop, software, case, and extended warranty that she expects will last through her four years of undergraduate studies. She paid for the purchase with her recently acquired credit card.

Study Natalie's credit card billing statement below. The only purchase during the previous billing cycle was the laptop, software, case, and extended warranty.

New Balance	**$1,935.72**
Minimum Payment Due	**$48.39**
Payment Due Date	**02/22/15**

Late Payment Warning: If we do not receive your Minimum Payment Due by the Payment Due Date listed above, you may have to pay a late fee of up to $35.00 and your Purchase APR may be increased to the Penalty APR of 27.24%.

Minimum Payment Warning: If you make only the minimum payment each period, you will pay more in interest and it will take you longer to pay off your balance. For example:

If you make no additional charges each month you pay...	You will pay off the balance shown on this statement in...	And you will pay an estimated total of...
Only the Minimum Payment Due	7 years, 4 months	$3,263
$69	3 years	$2,457 (Savings = $815)

Previous Balance	$450.00
Payments/Credits	−$450.00
New Charges	+$1,935.72
Fees	+$0.00
Interest Charged	+$0.00

New Balance	**$1,935.72**
Minimum Payment Due	**$48.39**
Credit Limit	$5,000.00
Available Credit	$3,064.28
Cash Advance Limit	$1,500.00
Available Cash	$1,500.00
Days in Billing Period: 31	

McGraw-Hill Education

THINK ABOUT THIS SITUATION

Think about all the information provided in the credit card statement on the previous page and the general basics of credit card use.

a. What did you find striking or surprising in Natalie's credit card billing?

b. The minimum payment each month is usually a percentage of the new balance. That percentage varies from one credit card to another. What is the percentage for Natalie's credit card?

c. Natalie's credit card account has a finance charge with an APR of 17.2% compounded monthly on the unpaid balance. What is the monthly interest rate Natalie pays?

d. With what credit cards are you familiar? What do you think are some advantages of using a credit card? Disadvantages?

e. The focus of this lesson is on consumer borrowing. How is credit card use a form of borrowing?

Work on problems in this lesson will introduce you to important topics in financial mathematics related to credit card use, loans, and automobile leases. In the process, you will hone your quantitative reasoning skills. You will also review and extend your algebraic skills and understanding needed in modeling exponential decay situations.

Credit Card Options

Credit cards are almost more common than cash in today's consumer world. People around the world use billions of credit cards every day. Unfortunately, some people also find that their unpaid credit balance is a serious personal finance problem. Credit card companies typically charge interest at annual percentage rates from 7% to 36%, and they usually compound those interest charges when balances are not paid in full at the end of a billing cycle.

As you work on the problems of this investigation, look for answers to these questions:

What is the general strategy for using equations to answer questions about credit card balances modeled by exponential functions?

What technological tools are most useful in analyzing credit card payment plans? How do you decide which tool to use?

Ryan McVay/Getty Images

1 Credit Card Cautions Suppose that a person runs up a credit card bill of $1,500, but then loses his/her job and cannot make the monthly payments.

a. If the credit card company charges interest at an APR of 18% compounded monthly, what will be the outstanding balance, without any late payment fees, in the person's account after:

   i. one month?

   ii. two months?

   iii. three months?

b. What recursive formula shows how to calculate the credit card balance from one month to the next?

c. What formula shows how to calculate the credit card balance after x monthly billing cycles?

d. What will be the balance on the credit card account if no payments are made for an entire year?

e. What is the effective annual interest rate (APY) under this scenario?

2 How would your answers to Problem 1 Parts a–c change if, in addition, the credit card company charged a late fee of $35.00 each month?

3 How would your answers to Problem 1 Parts a–c change if the credit card company charged, in addition, a late fee of $35.00 each month *and* after the second missed minimum payment increased the APR to 27% compounded monthly?

4 Suppose that a credit card account balance starts at B_0, the company charges interest at an annual percentage rate r (expressed as a decimal), and interest is added with monthly compounding on the total unpaid balance. Assume that no payments are made on the account and there are no late payment fees.

a. What formula shows how to calculate the outstanding balance $B(x)$ after x compounding periods?

b. How many compounding periods are there in t years?

c. What formula shows how to calculate the outstanding balance $C(t)$ after t years with initial balance B_0?

Credit Card Basics As you have seen in the previous problems, a credit card balance can grow very fast if monthly payments are not made—even faster if a late payment fee is added and the APR is significantly increased.

In the remainder of this unit, assume there are no late payment fees charged by the particular credit card companies.

As you work on the following problems, also assume:

amount subject to finance charge = previous balance − payments/credits + purchases

5 Suppose Anita has a credit card with an APR of 16% compounded monthly. Her current monthly credit card statement shows a balance of $400, in response to which she pays $80.00. In the following month (month 1), she makes charges amounting to $185. When she receives her next statement, Anita makes a payment of $100.00 to reduce the new balance, and the next month (month 2) she makes charges totaling $75.00. Complete a table like the following.

	Previous Balance	Payments	Purchases	Finance Charge	New Balance
Month 1					
Month 2					

6 Using the column-heading variables above:

a. write a formula for the amount subject to monthly finance charge.

b. write a formula to calculate the monthly finance charge.

c. write a formula to calculate the new balance.

Compare your formulas with your classmates and resolve any differences.

7 A credit card statement typically includes the following information in addition to information on minimum payment required by the federal Credit CARD Act of 2009.

	A	B	C	D	E	F	G
1	Previous Balance	Payments and Credits	New Purchases	Late Payment Fee	Finance Charge	Approved Credit	Available Credit
2							
3							

a. The previous balance after the last billing is represented by **A**, payments by **B**, recent purchases by **C**, late charge by **D**, and finance charge by **E**. Express the relationship among those variables if the new balance is to be zero.

b. Suppose the credit card statement above is modeled by a spreadsheet. Write the formula to calculate available credit in cell **G2**.

Making Only Minimum Payments In general, credit cards require a minimum monthly payment that is a fixed percentage of your current balance. As you can infer from the credit card billing statement on page 302, if you make only the minimum payment and no further charges, your balance will decrease very slowly.

8 Refer back to your responses to the Think About This Situation and Natalie's most recent credit card billing statement (pages 302–303). Assume Natalie makes no further purchases on the credit card and each month makes the minimum payment.

    a. Calculate the new balance after Natalie makes her first minimum payment. Then calculate the minimum payment due for March.

    b. Calculate the new balance after Natalie has made her second minimum payment. Then calculate the minimum payment due for April.

    c. Use reasoning similar to that in Parts a and b to calculate the balance for May and June.

    d. Now calculate the first four monthly finance charges. Why does the pattern of change make sense?

9 Using the same purchase and payment assumptions as in Problem 8, start a spreadsheet that tracks Natalie's monthly minimum payment and credit card balance.

    a. What is Natalie's credit card balance after 48 months? After 60 months?

    b. Describe patterns you see in the spreadsheet column for the minimum monthly payment. In the spreadsheet column for the corresponding monthly balance.

    c. How can the patterns you detected in Part b be explained by the regularity in the spreadsheet calculations?

10 Explain as precisely as you can why it makes sense that a credit card balance (with no further purchases) is a decreasing exponential function of the number of minimum monthly payments.

Minimum Payment Balance Formula The regularity in your calculations in Problems 8–10 suggest the following formula. If r is the APR for a credit card and m is the minimum monthly payment as a fixed percent of the balance (both r and m in decimal form), then

$$\text{balance after n minimum payments} = \text{initial balance} \times \left(\left(1 + \tfrac{r}{12}\right)(1 - m)\right)^{n}.$$

The product $\left(1 + \tfrac{r}{12}\right)(1 - m)$ should *not* be rounded in the calculation. In Connections Task 14, you will use algebraic reasoning to justify this formula.

11 Suppose after a year of college, you have a balance of $4,500 on your credit card and that with fulltime summer employment you make no more charges. Assume your credit card has a finance charge of 18% APR compounded monthly and that each month you make only the minimum payment of 3% of the unpaid balance.

   a. Write a formula that gives the account balance for this situation after n monthly payments.

   b. What will the balance be after 30 months?

   c. What will the balance be after five years?

   d. Determine how long it will take for you to reduce your balance to less than $50.

   e. After working on this problem, Samantha claimed that, "paying the credit card balance in this fashion will never get your balance to exactly $0." How would you respond to her claim?

Credit Card Purchases and Payments As you have seen in the previous problems, a credit card balance can grow very fast if monthly payments are not made. If only minimum monthly payments are made each month, the credit card balance is not paid off very quickly.

   Credit card companies address the issue raised by Samantha in Problem 11 with a requirement that the *minimum payment due is 3% of the balance or $50, whichever is greater*. The percentage and fixed dollar amount vary among credit card companies.

12 Consider again Natalie's credit card debt. Recall that her account has a finance charge of an APR of 17.2% compounded monthly on the unpaid balance. Suppose the terms of her credit card agreement specify a minimum monthly payment of either 2.5% of the unpaid balance or $35, whichever is higher.

   a. If Natalie chooses to make a fixed monthly payment of $69, what will be the total cost for her laptop, software, case, and extended warranty purchase?

   b. If Natalie chooses to finance her purchase with the minimum monthly payment option as outlined in her credit card agreement above, what will be her total cost?

   c. Suppose Natalie finds a well-paying, part-time job while at college. She decides instead to make a $100 payment per month on her credit card bill. How much would she save on the total cost of her purchase when compared with the $69-per-month payment plan?

13 Reston paid off one of his credit card balances in full last month in planning to purchase a new car. The card had an APR of 19.8% compounded monthly. His plans are to use that credit card to make a downpayment of $5,000 on the new car and not use that card for any future purchases. He plans to pay $200 each month toward the outstanding balance on the card. Use technology tool(s) of your choice to aid in answering the following questions.

a. How many months will it take Reston to reduce the balance to zero?

b. What is the total amount of the money he will have paid the credit card company?

c. Of the amount in Part b, how much is interest paid?

d. What technology tool(s) did you use in answering the above questions? Explain your choice(s).

14 The *Illuminations* financial calculator (page 280) also permits analysis of credit card payments. This calculator assumes that interest is compounded monthly.

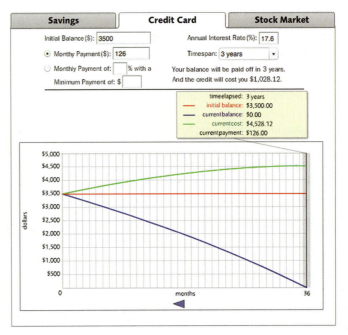

Source: illuminations.nctm.org/ActivityDetail.aspx?ID–172

a. Study the pattern of change in the credit card balance reported above for a recent purchase by Emily of new kitchen appliances.

i. What consumer information is provided by this financial calculator?

ii. Explain why the patterns in the graphs make sense.

iii. Pose a question related to Emily's credit card balance. Use the online calculator to help answer your question.

b. Use this online financial calculator to check your answers to Problem 13.

SUMMARIZE THE MATHEMATICS

In this investigation, you examined ways to model growth and reduction of credit card debt.

a. Some people pay off their entire credit card balance each month. However, more people only pay a portion of their balance each month. In this case, to avoid any late fee, a payment equal to or greater than the minimum monthly payment must be made.

    i. How is the minimum monthly payment calculated?

    ii. How is the monthly finance charge calculated?

    iii. How is the new balance calculated?

b. Suppose you have a balance on your credit card and decide to stop any further charging and pay off the balance by making the minimum payment each month. How can you determine the balance after *n* minimum payments using algebraic reasoning?

c. How could you determine the balance after *n* minimum payments using technology?

d. How could you determine the number of minimum monthly payments required to pay off a credit card balance, assuming no additional purchases are made?

e. What technology tools are helpful in tracking credit card debt? How do you decide which tool(s) to use?

Be prepared to explain your ideas and reasoning to the class.

✓CHECK YOUR UNDERSTANDING

Alicia has a $2,600 debt on her credit card that has an APR of 19% with interest compounded monthly. Her budget allows her to make payments of $125 at each monthly billing cycle (and make no new charges).

a. What will be her credit card balance after 12 months?

b. How many months will it take Alicia to pay off the credit card debt?

c. What was the total amount she paid the credit card company to pay off the $2,600 balance?

d. How much interest did she pay the credit card company?

e. How would your answers to Parts a–c change if Alicia was able to make monthly payments of $225?

Borrowing and Lending

It is common for businesses, families, and individuals to borrow money from a bank or credit union to make large purchases such as new manufacturing equipment, a new house, or a new car. Just as you invest money to earn interest, lending institutions lend money and collect interest on its use. Most loans are *installment loans* or *mortgages* that involve the borrower making equal (usually monthly) payments of a fixed amount (equal to a portion of the unpaid principal plus the interest on the unpaid balance for the period).

The process of repaying the loan is called **amortization**.

As you work on problems of this investigation, look for answers to these questions:

How can you determine the monthly payment, unpaid balance, and total cost of a home mortgage or loan?

What functions and technological tools are most useful in analyzing installment purchases with fixed rates involving compounding?

Using a Monthly Payment Table Monthly loan or mortgage payments are computed using a formula. Payment information is often arranged as in the table below to make it easier for lending institutions and realtors to help customers consider options in decision-making on purchases. Note that the table represents principal and interest only. The years columns refer to the length of the mortgage loan.

Monthly Payment Table for Each $1,000 Borrowed

Interest Rate	15 Years	20 Years	30 Years
4.0%	$7.40	$6.06	$4.77
4.5%	$7.65	$6.33	$5.07
5.0%	$7.91	$6.60	$5.37
5.5%	$8.17	$6.88	$5.68
6.0%	$8.44	$7.16	$6.00
6.5%	$8.71	$7.46	$6.32
7.0%	$8.99	$7.75	$6.65
7.5%	$9.27	$8.06	$6.99
8.0%	$9.56	$8.36	$7.34
8.5%	$9.85	$8.68	$7.69
9.0%	$10.14	$9.00	$8.05
9.5%	$10.44	$9.32	$8.41
10.0%	$10.75	$9.65	$8.78

Source: publications.usa.gov/epublications/low_down/low_down.htm

1 You can use this table to calculate your monthly payment on a home mortgage, not including property taxes and insurance.

a. What is the monthly payment on a 30-year home loan (*mortgage*) of $110,000 with an APR of 4.5%?

b. Assuming you made a downpayment of $30,000 as a condition for the $110,000 loan, what will be the total amount paid for the home at the end of 30 years? How much interest was paid at the end of the mortgage term?

c. Suppose final payment of college loans enables you to reduce the length of the term of a home mortgage. What is the monthly payment on a 15-year mortgage of $110,000 with no downpayment and an APR of 4.5%?

d. Again, suppose you made a downpayment of $30,000 as a condition for the loan, what will be the total amount paid for the home at the end of 15 years? How much interest was paid at the end of the mortgage term?

e. Compare your answers to Parts b and d.

2 The amount of the loan that is paid off each month increases throughout the life of the loan. Each monthly mortgage payment first pays off any accumulated interest and then the remainder is used to decrease the balance of the mortgage. Consider a 30-year loan of $110,000 with an APR of 4.5% compounded monthly.

a. How is the first monthly payment divided between paying off accumulated interest and paying off part of the mortgage balance? What about for the second month?

b. Explain why the amount of money credited toward the mortgage balance increases each month over the life of the mortgage?

c. Which of the following graphs do you think matches the overall shape of the graph of the mortgage balance as a function of the number of monthly payments made? Explain your reasoning.

Mortgage Balance

Mortgage Balance

3 Home interest rates dropped dramatically between 2008 and 2012. On January 15, 2000, the Lopez family took out a 20-year mortgage for $150,000 with monthly payments at an APR of 6.5%.

a. Assuming the Lopezes make each monthly payment, calculate the unpaid balance of the mortgage on March 15, 2000.

b. Calculate the unpaid balance of their mortgage on January 15, 2001.

c. Given this comparatively high interest rate, they have decided to continue making their monthly payments and additionally save and make a lump sum payment on January 15, 2015, to complete payment on the mortgage. What is the amount of the lump sum payment that must be made? How much money in interest did the Lopez family save?

4 When working with his group on Problem 3 Part c, Josh sensed that the underlying structure of finding the mortgage balance owed after 15 years by the Lopez family was similar to finding the future value of an investment with an APR compounded monthly that included a fixed monthly deposit. Since this approach was quite different than the approach taken by his group, Josh decided to test his thinking at home. His insight into the common mathematical structure of two quite different problems led to a striking discovery. The next day, he presented his ideas to his class for their critique.

a. Examine and then discuss the soundness of Josh's thinking as described below.

> It seems like finding the balance owed on a mortgage (in the future) is like finding the future value of an investment or savings account with regular deposits of a fixed amount. So, I looked back at the future value formula
>
> $$F = I\left(1 + \frac{r}{12}\right)^{12t} + A\left(\frac{\left(1 + \frac{r}{12}\right)^{12t} - 1}{\frac{r}{12}}\right)$$ that we discovered in Lesson 1, Investigation 3.
>
> I asked myself, "Could I use this formula to find the future balance on a mortgage by subtracting the expression representing the fixed monthly savings contributions?" I was thinking of the expression $A\left(\dfrac{\left(1 + \frac{r}{12}\right)^{12t} - 1}{\frac{r}{12}}\right)$ as
>
> the sum of the monthly payments for t years. That should work if the first expression $I\left(1 + \frac{r}{12}\right)^{12t}$ represents the cost of the mortgage for t years.

b. Using Josh's idea, how much interest could be saved by paying off the 20-year mortgage as a lump sum on January 15, 2015?

c. Now compare your answer for the interest saved to the amount you calculated in Problem 3. What might account for any difference?

d. Discuss the various methods you have now used to analyze the Lopez mortgage situation. What are the advantages of each method? Be prepared to share your ideas with the class.

5 Look back over your work in Problems 1 and 3.

a. Complete the recursive formula below for the loan amortization in Problem 1 Part a. Repeat for Problem 1 Part b. Here B_n represents the mortgage balance *after* the nth monthly payment.

$$B_0 = \$110,000$$
$$B_n =$$

b. Write a recursive formula for the loan amortization in Problem 3.

c. Now try to generalize your work. If P dollars are borrowed at an APR of r (expressed as a decimal) and R dollars are paid back monthly, write a recursive formula for calculating the new balance B_n from the previous balance B_{n-1}.

Using Computer-based Tools in Reasoning about Loans The spreadsheet and online financial calculator methods you used in analyzing investments can be used in a similar way to analyze loans and loan payments.

6 An **amortization schedule** is a chart showing the amount of each loan payment that is interest, the amount that is applied toward the principal of the mortgage, and the remaining balance.

Use an online mortgage calculator or a spreadsheet to create an amortization schedule for the mortgage described in Problem 1 Part a.

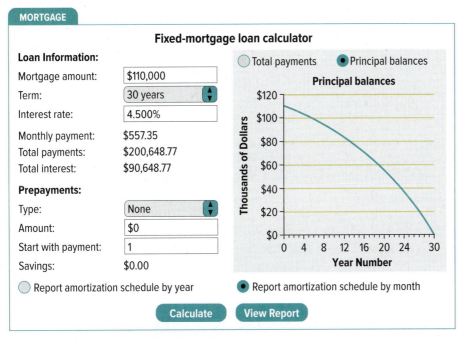

Source: www.bankrate.com/calculators/mortgages/mortgage-loan.aspx

a. Determine the balance of the mortgage after 60 monthly payments. After 120 monthly payments.

b. At what point in the loan payoff is the amount paid to principal greater than the amount paid to interest?

c. What is the total amount that was paid for the home? Of that amount, how much was paid in interest?

d. Repeat Parts a, b, and c above for the mortgage in Problem 1 Part b.

7 Brian is considering taking out a car loan for $12,000 for a term of five years at an APR of 4.75% compounded monthly.

    a. What will be the amount of his monthly payment?

    b. How much would Brian owe on the car after one year? After two years?

    c. What is the total amount Brian will pay for the car?

    d. What is the total amount of interest he will pay?

8 **Comparing Loans and Mortgages** Suppose Ford Motor Co. is offering a choice of a 2.88% APR loan compounded monthly for 36 months or $1,500 cash back on a $24,000 Mustang.

    a. If you take the 2.88% loan offer, how much will be your monthly payment?

    b. If you take the $1,500 cash back offer and borrow the remainder of the price of the car from a local credit union at an APR of 2.99% compounded monthly for three years, how much will your monthly payment be?

    c. Which offer is the better offer for you? Explain.

9 The recession in the United States that began in 2008 was due to many factors. One of those was the housing crisis. Housing prices were declining precipitously and soon many people owed more money on their home mortgage than the property was worth. This was a result, at least in part, to interest-only mortgages.

    At the peak of the recession, there were significant numbers of home foreclosures in many states across the country. This was due, in part, to the inability of home purchasers to afford the larger mortgage payments at the end of the interest-only periods.

With an **interest-only mortgage**, the homebuyer's monthly payment for a certain number of years (usually five or ten) consists only of interest payments. At the end of the interest-only period, the monthly payment is determined by the number of months remaining in the mortgage.

Consider a 30-year mortgage of $450,000 at an APR of 3.2% compounded monthly where the mortgage payment is interest-only for the first five years.

a. What is the monthly payment during the first five years?

b. What is the monthly payment during the last 25 years?

c. What is the total amount paid for the home?

d. What is the total amount of interest paid for the home?

e. Suppose the couple purchasing the home had sufficient savings for a downpayment to take out a conventional 30-year loan, reducing the mortgage to $360,000, with an APR of 3.2% compounded monthly. What is the monthly payment for this mortgage?

f. How do the total amount paid for the home and the total amount of interest paid under the terms in Part e compare with that of the interest-only mortgage?

SUMMARIZE THE MATHEMATICS

In this investigation, you examined amortization of home mortgages and auto loans.

a. What are the four factors that influence the amortization process?

b. How is calculating a mortgage or loan balance similar to, and different from, calculating an investment balance?

c. Suppose P dollars are borrowed at an APR of r (expressed as a decimal) compounded monthly and payments of R dollars are made each month. How would you set up the cell formulas of a spreadsheet to track the balance each month, time to payoff the loan, and determine the total interest paid on the loan?

d. Suppose the monthly payment on a 30-year home mortgage is $820. If $R(t)$ represents the amount of a monthly payment that is apportioned to the mortgage repayment as a function of time t in years and $I(t)$ represents the amount of the payment apportioned to interest as a function of time t in years, how would you illustrate the pattern of change in these two functions over 30 years? Summarize your illustrations in words.

e. What technological tools are helpful in analyzing loan options and in tracking loan balances? How would you decide which tool(s) to use in a particular situation?

Be prepared to explain your ideas to the class.

In March 2014, a car dealership was offering incentives to potential buyers. Ginny was interested in possibly purchasing a Scion TC for $22,561 and was offered the choice of 0% financing for 48 months or $2,500 cash back on the purchase of the vehicle.

a. If Ginny takes the 0% loan offer, what will her monthly payment be?

b. If Ginny takes the $2,500 cash back offer and can borrow money from a local credit union at 3.3% interest compounded monthly for four years, how much will her monthly payment be?

c. Which of the two offers is the better choice for Ginny? Explain.

INVESTIGATION 3

Purchasing vs. Leasing

In the previous investigation, you examined credit card usage and bank or credit union loans where the borrowed amount plus accrued interest were paid back over a certain period of time with fixed regular payment amounts. For a loan, the fixed regular payment amount was determined based on the amortization of the loan so that the ending loan balance, including all interest accrued over the loan period, was $0.

An alternative to securing a loan to purchase a new car is to instead *lease* the car, usually with significantly lower monthly payments. Under a **lease**, ownership of the car does not transfer from the auto dealership (for example) to the person leasing the car. Instead, the person is in essence borrowing the vehicle for the duration of the lease; all the while, the value of the car depreciates. At the end of the lease period, there is often an option of purchasing the car in full by paying the *residual value*. The **residual value** is the estimate of the car's value at the end of the lease period.

There are many factors that go into deciding whether a car should be purchased using a loan or using a lease and auto dealerships and lending institutions often provide incentives to entice perspective buyers. This can make the choice between purchasing and leasing less clear.

As you work on problems of this investigation, look for answers to these questions:

What is the depreciated value of a purchased automobile and how is it calculated?

What is the residual value of a leased automobile and how is it calculated?

How can you determine whether an automobile loan or a lease will be the best choice in a given situation?

1 In a typical car loan situation, you need to know the total amount that needs to be borrowed to purchase the car, the APR that the financing institution offers, and the duration of the loan. The amount financed is determined based on things like the negotiated cost of the car (typically less than the manufacturer's suggested retail price, or MSRP), any additional costs (e.g., tax, title, registration, license, acquisition fee), and downpayment amount.

Shown below is a sample 36-month lease for a new 2013 FIAT 500 with amounts rounded to the nearest dollar. Analyze this sample lease and the descriptions of the calculations involved.

MSRP	$16,000	(manufacturer's suggested retail price)
Base Capitalized Cost	$15,750	(negotiated cost of the car)
Additional Lease Costs	$1,000	(extra costs like title and registration)
Capitalized Cost Reductions	$500	(cost reductions like downpayment)
Adjusted Capitalized Cost	**$16,250**	
Residual Value	$7,520	(projected value of the car at lease end)
Money Factor	0.001246	(*money factor* × 2,400 = *APR*)
Term (in months)	36	(duration of the lease)
Sales Tax (in %)	6	(state sales tax)
Depreciation	$8,730	(*adjusted capitalized cost − residual value*)
Finance/Rent Charge	$1,066	(*(adjusted cap cost + residual)(money factor)(term)*)
Total Sales Tax	$588	(*(depreciation + finance charge)(sales tax)*)
Total of All Payments	**$10,384**	
Monthly Payment	$272	(*(depreciation + finance charge) ÷ term*)
Sales Tax on Monthly Payment	$16	(*total sales tax ÷ term*)
Total Monthly Payment	**$288**	

a. Create a spreadsheet for this sample lease that automatically calculates results based on user-entered information using the formulas above. Save your Car Lease Calculator spreadsheet for use in other problems in this unit.

b. The projected value of the car at lease end is often calculated as a percentage of the MSRP. For the above sample lease, what percent would be reported for the residual?

c. What is the APR for the sample lease?

d. If there were no downpayment of $500, what would be the new total monthly payment? Is there a benefit to including a downpayment of $500? If so, what is it?

2 Suppose an alternative to the 36-month lease for the 2013 FIAT 500 in Problem 1 is a 36-month loan with an APR of 4.74%. The amount that needs to be borrowed to purchase the car is the same as the lease's adjusted capitalized cost plus 6% sales tax on that cost.

a. What is the total monthly payment on the loan? How much does it cost to own the vehicle if it is purchased with this loan?

b. Suppose that after the 36-month lease has ended, an option to purchase the car for the residual value is offered. How much does it cost to own the vehicle if it is leased? Which financing option is the best?

c. Lease deals often restrict the mileage put on the car during the lease period. The sample lease in this problem has a restriction of no more than 10,000 miles per year. If the car is driven more than the allowed number of miles in a given year, a per-mile charge of $0.25 is assessed on the overage. How many miles over 10,000 miles per year must be driven before the cost of the lease is the same as the cost of the loan?

3 Tina is interested in buying a similar 2013 FIAT 500 (MSRP of $16,000) from a local auto dealership and narrows down her available options to two choices. She lives in a state that does not have a sales tax on new automobiles.

- Three-year loan of $20,000 with an APR of 3.08%

- Three-year lease with adjusted capitalized cost of $20,000, 55% residual, and money factor of 0.00129

a. If the auto dealership allows Tina to purchase the car for the residual value at the end of the lease, which financing option should Tina choose? Explain your reasoning.

b. Suppose that before Tina can sign the paperwork based on her selected choice from Part a, the auto dealership discontinues the three-year loan option. A new five-year loan option is offered for $20,000 with an APR of 3.48%. What is the total monthly payment for this five-year loan? Should Tina make a different choice from that determined in Part a?

c. Dealerships often provide incentives to entice potential buyers. Suppose Tina's dealership offers a $1,000 cash-back incentive for the three-year lease and a 2%-off the APR incentive for the five-year loan. If Tina applies the $1,000 cash back as a downpayment under the three-year lease option (and intends to pay the residual when the lease is over), should she choose the three-year lease or the five-year loan with 2% off the APR?

SUMMARIZE THE MATHEMATICS

In this investigation, you examined the financial advantages and disadvantages of purchasing a vehicle with a loan vs. leasing the vehicle.

a. What information is necessary to know in order to analyze the cost of a potential car loan?

b. What information is necessary to know in order to analyze the cost of a potential car lease?

c. When determining whether a loan option or a comparable lease option is the best choice, what reason(s) might there be for choosing the lease option over the loan option?

Be prepared to explain your ideas and reasoning to the class.

✓CHECK YOUR UNDERSTANDING

After graduation from high school, Toby plans to get a new 2013 Scion iQ for college and start building his credit. He does not have money for a downpayment and his parents agree to help him secure either a loan or a lease from a local car dealership.

The dealership offers Toby a 48-month lease with 12,000 miles allowed per year, an adjusted capitalized cost of $17,470, residual of $6,973, and money factor of 0.0006208. He lives in a state that has an 8% sales tax. Since he recently graduated from high school, Toby qualifies for the car dealership's offer of a $1,000 graduation rebate towards the adjusted capitalized cost.

a. What is Toby's total monthly payment if he chooses this deal? What is the total cost to purchase the car if Toby is given the option to pay the residual at lease end and he stays within the allowed 12,000 miles per year?

b. The car dealership offers a 48-month loan of $17,470 with an additional $1,000 graduation rebate plus sales tax at an APR of 1.49%. If Toby intends to own the car, which of the two offers is his best choice? Explain your reasoning.

c. Now, assume Toby was unable to stay within 12,000 miles per year. The car dealership charges an additional $0.20 per mile over each year's limit. Suppose Toby drove an additional 4,000 miles before the lease was up. Should Toby have chosen the loan or the lease? Explain.

McGraw-Hill Education

ON YOUR OWN

1 Suppose you start with a balance of $3,500 on your credit card that has an APR of 15% compounded monthly. Each month you make only the minimum payment of 3% of the balance. During the first month, you charge purchases totaling $350. During the second month, you charge $600 in purchases. Complete a copy of the following table.

	Previous Balance	Purchases	Finance Charge	New Balance	Minimum Payment
Month 1					
Month 2					

2 Suppose you have a balance of $4,000 on your credit card that has an APR of 18% compounded monthly and you make no further purchases. Each month, you make only the minimum payment of 5% of the balance.

   a. Write a formula that gives the credit card balance after t monthly payments.

   b. What will be the balance after 30 months?

   c. At what balance do you begin making payments of $20 or less?

   d. How many months will it take to reduce the remaining balance to the value in Part c?

3 Riel used his credit card to purchase some furniture for his apartment. The amount charged to the card was $2,583. The card has an APR of 21.6% compounded monthly.

   a. Riel's budget allows him to make monthly payments of $100. Assume he makes no further charges on this credit card until the furniture is paid off. Determine:

     i. the number of months to pay off the furniture purchase.

     ii. the total amount of money paid to the credit card company for the furniture purchase.

     iii. the total interest paid to the credit card company.

   b. If Riel chooses to use his option of paying the higher of a 3% minimum payment on the unpaid balance or $45 per month, how will the total interest paid to the credit card company compare to his plan in Part a?

4 Many American students pay for at least part of their college education by taking out loans. Typical student loans require no repayment until after graduation. Most loans, except those with special government sponsorship, allow interest to accrue while the borrower is in school.

Suppose Mae borrows $5,000 for tuition from a lender who charges interest at an annual percentage rate of 4% compounded monthly.

a. What formula gives the balance of her loan at any time x months later assuming no payments are made and late fees are not added?

b. What is the outstanding loan balance after:

    i. one month?

    ii. two months?

    iii. one year?

c. What is the effective annual interest rate (APY) of the loan?

5 If a student borrows money at the start of college, repayment might not begin until as much as 5 years later. Consider the case of Renee who borrows $8,000 at an annual percentage rate of 4% compounded monthly. Write function rules or recursive formulas that can be used to calculate the balance on Renee's loan after:

a. 4 years (assuming no monthly payments while in college).

b. $t \geq 5$ years (assuming fixed monthly payments of $60).

6 Logan and his wife are considering purchasing a home for $370,000. They intend to put $40,000 down and then finance the rest at an APR of 6.3% interest compounded monthly for 25 years.

a. What will be the amount of their monthly payment?

b. How much would they owe after five years? After 10 years?

c. What is the total amount Logan and his wife will pay for the home?

d. What will be the total amount of interest they will pay?

e. Suppose 20 years after the mortgage was taken out, Logan inherits some money and they are able to pay off the remaining mortgage at that time. What will be the total amount paid for the home? How much interest will they have saved?

7 There are functions denoted PPMT and IPMT that are helpful when planning financial decisions using spreadsheets. For a given period, the PPMT(*rate, per, nper, pv*) and IPMT(*rate, per, nper, pv*) functions calculate the amount paid on the principal and interest, respectively. They report *negative values* and are based on the following arguments.

- *rate* interest rate per period
- *per* period for which you want to find the payment or interest
- *nper* total number of payment periods
- *pv* present value (principal)

	A	B	C	D
1	Principal	$12,000		
2	Total Number of Payments	60		
3	APR	4.75%		
4				
5	Payment Period	Amount Paid to Principal	Interest	Remaining Balance
6	1			=B$1+B6
7	2			
8	3			

a. Using the PPMT and IPMT functions, create a spreadsheet like the one above that calculates the payment on the principal and the interest payment for each month using the annual percentages rates and years given in Investigation 2 Problem 6. Describe fully how cell **D6** calculates the remaining balance after the first payment period. What should be written in cell **D7**?

b. After how many payments will the remaining balance be below $10,000?

c. The function PMT(*rate, nper, pv*) calculates the payment for a loan based on fixed payments at a constant interest rate and regular intervals. Write a formula for PMT in terms of PPMT and IPMT for a given payment period. Check your work using your spreadsheet.

d. Modify your spreadsheet from Part a for Logan's home mortgage in Applications Task 6. At what point does the amount paid to the principal exceed the interest on the remaining balance?

8 Since many people do not have enough money to purchase a new car, most people seek one of two options—obtaining a loan for a new or previously used car or leasing a car. Suppose Carrie is interested in buying a 2013 MINI Cooper S with a list price of $23,300. Following are some optional packages she is considering for her new customized car.

Options	Price
Cold Weather Package	$750
Leather Interior Package	$1,000
Sport Package	$1,250
Technology Package	$1,750

a. Carrie has been approved for a 36-month loan with 2.99% APR compounded monthly in the amount of the total price of the car. How much more is her monthly payment if she includes all four packages in the loan than if she does not add any packages?

b. Suppose she wants to keep her monthly payment below $725. What package options are within her budget?

c. Suppose Carrie finds a used MINI Cooper S with a price of $15,000. What is her monthly payment if she is approved for a 36-month loan with 3.4% APR compounded monthly?

9 Tyrese has recently enlisted in the U.S. Military and was given a $6,000 enlistment bonus. He had always wanted to own an American muscle car and decided to look at financing options for a new 2013 Chevrolet Camaro. The city in which he plans to buy the car has an 8.75% sales tax. His local credit union offers the following two plans.

Plan 1: Lease First, Purchase Later
36-month lease with adjusted capitalized cost of $25,245 (MSRP is $23,345), residual of 62%, and money factor of 0.001825

Plan 2: Purchase Plan Financed by a Loan
36-month loan of $25,245 plus sales tax with APR of 4.74% compounded monthly

a. If Tyrese does not use any of his enlistment bonus for a downpayment, what is the total monthly payment for Plan 1? For Plan 2?

b. Suppose instead that Tyrese uses all $6,000 of his enlistment bonus for a downpayment. If he plans to pay the residual at lease end, how much money does Tyrese save under Plan 1 by using his bonus for a downpayment? Under Plan 2?

c. Which financing plan would you recommend Tyrese take advantage of? Does your recommendation change if Tyrese puts money on a downpayment or chooses not to? Explain your reasoning.

d. If the credit union offers Tyrese a savings account for his $6,000 enlistment bonus at a high APR of 5.1% compounded monthly, should he put his bonus in the savings account or use it for a downpayment? Does it make a difference which plan Tyrese chooses? Explain.

CONNECTIONS

10 Write a recursive formula that models the pattern of change in each situation.

a. The balance on a credit card account with a current balance of $2,500 for which no monthly payments are made with an APR of 18% compounded monthly charged on the unpaid balance

b. The current balance of $4,500 on a credit card that has a finance charge of 22.8% compounded monthly with minimum monthly payments of 5% of the unpaid balance

c. The monthly balance on a four-year car loan of $29,000 that has an APR of 3.25% compounded monthly

11 Suppose you have a $500 balance on a credit card with an APR of 17.2% compounded monthly and you make no further charges. You wish to pay off the balance in six months by making equal payments each month. What should be your monthly payment?

12 Suppose a bank gives you a special introductory offer on their credit card. It has an APR of 9.8% compounded monthly and requires a minimum monthly payment of 2% of the balance. Your monthly budget allows you to pay at most $180 each month. How much can you afford to charge? Assume the credit card has a limit of $10,000 on purchases.

Ingram Publishing/SuperStock

13 In Investigation 1, you calculated the declining balances in various credit card accounts where only the minimum monthly payment was made and no further charges were added to the account. The regularity in those calculations suggested the Minimum Payment Balance formula:

$$\text{balance after } n \text{ minimum payments} = \text{initial balance} \times \left(\left(1 + \frac{r}{12}\right)(1 - m)\right)^n,$$

where r is the APR compounded monthly and m is the minimum monthly payment as a fixed percentage of the new balance. Both r and m are expressed as decimals.

  Provide reasons for each step in the following derivation of the formula. Assume you have made a series of payments and B_c is the current balance on your monthly statement. Let B_n be the new (next) account balance after you make the minimum payment.

Balance on next statement:

$$B_n = B_c - mB_c + \frac{r}{12} \times (B_c - mB_c) \qquad (1)$$

$$= B_c(1 - m) + \frac{r}{12} \times B_c(1 - m) \qquad (2)$$

$$= B_c \times \left(\left(1 + \frac{r}{12}\right)(1 - m)\right) \qquad (3)$$

So, the new balance each month is the current balance multiplied by $\left(1 + \frac{r}{12}\right)(1 - m)$. $\qquad (4)$

That is,

$$\text{balance after } n \text{ minimum payments} = \text{initial balance} \times \left(\left(1 + \frac{r}{12}\right)(1 - m)\right)^n. \qquad (5)$$

14 Look back at the Future Value Formula in the Lesson 1 Summarize the Mathematics on page 282. How could you modify this formula to calculate the balance on an automobile loan? Pose a problem and solution illustrating an application of your formula.

15 Nathan's great-grandmother has agreed to lend him $5,000 at a very low APR of 2% compounded monthly to purchase a used car. They agree that Nathan will pay the loan back in monthly installments for six years. Nathan needs to determine how much the monthly payment would be to decide if he will be able to afford the monthly payments. He decided to use the Future Value Formula to find the monthly payment A, as shown below.

$$6{,}000\left(1 + \frac{0.02}{12}\right)^{60} = A\left(\frac{\left(1 + \frac{0.02}{12}\right)^{60} - 1}{\frac{0.02}{12}}\right)$$

  a. What is the value of Nathan's loan at the end of six years?

  b. How does the answer to Part a help justify Nathan's equation above?

  c. Nathan determined that the monthly payment for his loan would be about $105. Explain why $105 seems reasonable.

  d. Find the monthly payment to the nearest cent that Nathan should make.

16 With continuing Federal debates about the future solvency of the Social Security Fund, it is prudent to start an individual retirement account (IRA) as early as possible after graduation. This might be done by contributing to an annuity program (see Lesson 1, Connections Task 21 on pages 28–29). At retirement, the annuity can supplement your income from other sources. This annuity pays you a fixed amount of money at regular intervals (usually each month). Payments end when the funds in the account are depleted.

In Connections Task 21, you calculated the *future value* for ordinary annuities and developed formulas for their calculation. The balance of funds in the annuity at the time of retirement (when contribution stops) is called the **present value** of the annuity. From then on, the annuity will be drawn down in equal payments until the funds are depleted.

a. Among the important considerations to consider are:

  i. How many years should you plan for the annuity to last after you retire? Make an estimate in your case.

  ii. What annuity payment amount at the end of each month would be sufficient for your retirement needs and plans? Make an estimate in your case.

  iii. How large a retirement "nest egg" would be needed to achieve your desired annuity yield? Make a conjecture.

b. Anna, a *Transition to College Mathematics and Statistics* student at Black River Charter High School in Michigan, gave these estimates for Part ai and Part aii, respectively: 25 years with $4,000 per month. Her annuity pays 3% monthly. To answer Part aiii, she reasoned as follows. Carefully check Anna's procedure and answer. Correct any mistakes.

$$my\ needed\ nest\ egg = 4{,}000\left(\frac{\left(1+\frac{0.03}{12}\right)^{300}-1}{\frac{0.05}{12}\times\left(1+\frac{0.05}{12}\right)^{300}}\right)$$

$$= 4{,}000\left(\frac{(1.0025)^{300}-1}{0.0041\overline{6}\times(1.0041\overline{6})^{300}}\right)$$

$$\approx 4{,}000(76.869395)$$

$$= \$307{,}477.58$$

c. Explain as precisely as you can why Anna's procedure makes sense.

d. Write a **present value formula** for an annuity retirement account that is paying an APR of 4% compounded monthly to enable you to retire with a 20-year annuity that yields $5,000 per month. Compare your formula with that of your classmates. Resolve any differences.

Beth Ritsema

17 In all the problems of Lesson 2 involving credit card usage, the monthly finance charge was always less than the minimum payment (as a percent). In fact, this is always the case. What would happen if the minimum payment were less than the monthly finance charge?

18 It is common practice for some credit card companies to waive a minimum monthly payment (usually for the month of November) as a "holiday gift." Who benefits most from such a waiver—the credit card holder or the credit card company? Explain your reasoning.

19 Credit cards are relatively new. Visit www.creditcards.com/credit-card-news/credit-card-industry-facts-personal-debt-statistics-1276.php and write a brief report on the most interesting facts about credit card use, both in terms of the individual consumer and the U.S. economy. Be prepared to discuss your findings with the class.

20 Many people find the following pitch enticing.

Need some cash to tide you over until the next payday?
We'll loan up to $500 for two weeks.

For example, in Virginia during 2006, over 400,000 people took out over 3.5 million payday loans with total value of over $1.3 billion. (Source: Anita Kumar, "Pressure Mounts on Va. Payday Lenders," *The Washington Post*, Monday, December 3, 2007, page B5.)

What unsuspecting customers of payday loan stores often fail to realize is that the fee of "only" $15 per $100 loaned amounts to a very large annual percentage rate.

a. If you pay $15 upfront to borrow $100 for two weeks, what is the annual percentage rate?

b. In response to complaints by consumer advocates, the U.S. Congress passed a bill limiting to 36% the annual percentage rates charged on payday loans to military personnel. What does this limit imply about the cost to military personnel of borrowing $100 for two weeks?

21 Look back at Investigation 1 Problem 11 Part e (page 307). Research how credit card companies resolve the issue of a cardholder reducing a card account balance to $0. Write a brief report summarizing your findings.

22 Look back at your work for Applications Task 4.

a. Suppose in Applications Task 4, Mae borrows $8,000. What is the effective annual interest rate in this case?

b. Based on your work in Applications Task 4 and your answer to Part a, what appears to be true about the initial balance of a loan and the effective annual interest rate? Why does this make sense?

23 A young couple buying a downtown loft borrows $165,000 for 25 years at 3.4% compounded monthly.

a. What is the amount of the monthly payment?

b. How much interest could the couple save if they contributed an extra $50 to the principal each month?

c. How much could the couple save if they contributed $100 more to the principal each month?

d. Under the payoff plan in Part c, how long would it take to pay off the mortgage?

EXTENSIONS

24 As you saw in Investigation 1 with credit card debt, it is important to pay down a card balance with regular payments of a sufficient amount, otherwise the balance will increase due to factors such as the credit card's annual interest rate and late payment fees. Many people have credit card debt, but they may not be aware that all U.S. citizens share an equal portion of our U.S. national debt as well.

Source: www.usdebtclock.org (February 22, 2013)

a. The Web site www.usdebtclock.org provides real-time changes to all of the numbers shown above. Study the screen and verify that "Debt Per Citizen" is being calculated correctly. It may seem shocking that each U.S. citizen would need to pay $52,504 to cover the U.S. national debt at the time this screen was captured.

b. In February 2013, the U.S. national debt was estimated to be increasing at a rate of 8.627% per year and the U.S. population was estimated to be increasing at a rate of only 0.969% per year (or, in actuality, decreasing). (**Source:** www.wolframalpha.com) Under these conditions, what is the projected debt per citizen in February 2023?

c. Imagine that from February 2013 to February 2023, the U.S. population continues to change by 0.969% per year as in Part b. What annual percent change must the U.S. national debt be reduced to so that the projected debt per citizen in February 2023 is the same as it is in February 2013, $52,504? Why does this result make sense?

d. Now assume the same U.S. national debt growth rate is 8.627%, but that there is no change to the U.S. population from February 2013 to February 2023. How much must each U.S. citizen pay per year to reduce the debt per citizen to $0 by February 2023? How much more would each citizen end up paying after those 10 yearly payments than if they had each paid $52,504 in February 2013 instead?

25 In Investigation 1, you used a simplified mathematical model for tracking credit card balances. Today, it is standard practice to calculate finance charges on the *average daily balance*.

AJ's credit card account showed the following daily balances for his last billing cycle:

> four days at $99.78
> ten days at $345.56
> nine days at $515.32
> eight days at $580.64

AJ's credit card has a 19.2% APR compounded monthly and a minimum payment of 2% of the monthly balance.

a. What will be the finance charge shown on his billing statement assuming AJ made the minimum payment the previous month?

b. What will be his minimum payment to avoid any penalties?

c. If AJ makes no further purchases and makes the minimum payment due each month, how long will it be before his payment drops below $10.00?

Thomas Northcut/Getty Images

26 In January 2013, an upstart digital media company took out a business loan of $100,000 with an APR of 6.5% for a period of 15 years. The table at the right shows the amortization of the loan by year. Here 2013 is year 1.

Year Number	Loan Balance
1	$95,926.78
2	$91,580.78
3	$86,943.71
4	$81,996.09
5	$76,717.12
6	$71,084.61
7	$65,074.87
8	$58,662.65
9	$51,821.00
10	$44,521.15
11	$36,732.41
12	$28,422.05
13	$19,555.12
14	$10,094.36
15	$0.00

  a. Make a scatterplot of the (*year, loan balance*) data.

  b. What type of function model appears to be a good fit for these data? Explain your reasoning.

  c. Find a quadratic regression model and graph it on the scatterplot.

  d. Find a cubic regression model and graph it on the scatterplot.

  e. Write a brief summary of your findings modeling loan balance with polynomial functions.

27 As you have seen in this lesson, home mortgages have a stated annual interest rate compounded monthly. Some home loans also carry **points** or **discount points**. Discount points are actually prepaid interest on the mortgage loan. The more points you buy, the lower the interest rate. The price to buy discount points is 1% of the stated loan amount.

This has the effect of reducing the interest rate by about 0.25% per point.

  Compare a 30-year $250,000 mortgage at an APR of 4.8% compounded monthly with one in which the borrower decides to pay two points to lower the interest rate. Under what conditions is the standard mortgage a good option? The reduced-interest option?

28 Suppose B_n is the balance of a car loan after n payments of R dollars, i is the interest rate per period expressed as a decimal, I_n is the interest portion of the nth payment, and Q_n is the portion of the nth payment applied to the balance.

  a. Write a recursive formula for B_n.

  b. Show that $B_{n-1} = \dfrac{R - Q_n}{i}$ and $B_n = \dfrac{R - Q_{n+1}}{i}$.

c. Using the results in Part b, derive the recursive formula $Q_{n+1} = Q_n(1 + i)$. Then write in words what this formula tells you.

d. Suppose that for the tenth payment on a one-year installment loan at 1% interest per month for furniture for a college apartment, $200 was applied to the principal. How much of the eleventh and twelfth monthly payments will be applied to the principal?

29 Identify a car model that you might like to purchase or lease after completing college or an apprentice program.

a. Using automobile manufacturer Web sites and www.edmunds.com: (1) "build" and price out your car of choice, (2) find information on best dealer pricing on the car in your geographical region, and then (3) determine the downpayment, APR, and amount of the monthly payments for 60 months to purchase the car.

b. What is the Kelley Blue Book (www.kbb.com) value of your chosen car at the end of the loan?

c. Contact a dealer for your automobile choice and possibly a credit union for the necessary information to repeat Parts a and b for the case of a lease.

d. Write a report summarizing your research and justifying your decision to purchase or lease.

REVIEW

30 Determine if each table of values shows a linear or an exponential pattern of change. Then find a function rule that matches each table of values.

a.
x	−5	0	5	10	15
f(x)	10	5	0	−5	−10

b.
x	2	3	4	5	6
f(x)	12	36	108	324	972

c.
x	−4	−1	2	5	8
f(x)	1.5	3	4.5	6	7.5

d.
x	−2	0	2	4	6
f(x)	640	160	40	10	2.5

31 Find the perimeter of each polygon described below.

   a. An isosceles triangle with base of length 8 cm, and with each leg 3 cm longer than twice the length of the base

   b. A right triangle with hypotenuse that is 12 cm long and one leg that is 5 cm long

   c. A square with an area of 36 in²

   d. A rectangle with an area of 112 cm², and the lengths of the longer sides are seven times the lengths of the shorter sides

32 A production line fills 5-gallon bottles of spring water at a rate of 1,500 bottles per hour. The production line runs continuously from 8 A.M. to 4 P.M. every day (including weekends).

   a. How many gallons of spring water are needed to fill the bottles each day?

   b. How many bottles can be filled each week?

   c. Suppose an order for 10,000 bottles of spring water is started at 8:00 Monday morning. If the production line experiences no delays and operates on its normal schedule, when is the earliest the bottles will be filled?

33 Solve and check each equation.

   a. $\dfrac{x+5}{x} = 10$ **b.** $2x - 7 = \dfrac{8}{x-3}$ **c.** $\dfrac{3x+1}{x-2} = \dfrac{x+4}{2x+6}$

34 Kentwood is electing a new mayor of the city *and* is voting on a referendum about building a new high school. The results of a random poll of 850 people who were asked about the two ballot items are shown in the table below.

		Mayor		
		Kennedy	Menendez	Total
New High School	Yes	341	178	519
	No	209	122	331
	Total	550	300	850

   a. Suppose you were to randomly pick one of these people to interview.

      i. What is the probability that the person favors Kennedy for mayor?

      ii. What is the probability that the person favors Kennedy for mayor given that the person supports building a new high school?

   b. Is there statistically significant evidence that, in this city, people's choices for mayor and whether or not they support a new high school are independent? Explain your reasoning.

   c. Is it correct to conclude that people who intend to vote for Kennedy are more likely to support building a new high school? Explain your reasoning.

35 Match each equation to the most appropriate graph.

Graph I

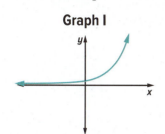

Graph II

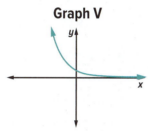

Graph III

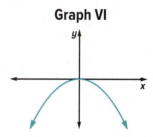

Graph IV

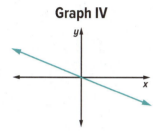

Graph V

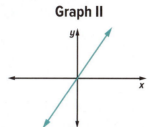

Graph VI

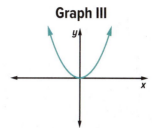

a. $y = kx, k > 0$

b. $y = kx, k < 0$

c. $y = \left(\frac{1}{k}\right)^x, k > 1$

d. $y = k^x, k > 1$

e. $y = kx^2, k > 0$

f. $y = kx^2, k < 0$

36 The summer reading list for senior English includes 10 different books. Over the summer, each student must read and write reports on four of the books. If Beatriz and James randomly choose the books they will read, determine the probabilities that:

a. Beatriz and James choose the same four books.

b. none of the books that they choose to read are the same.

c. at least one of the books that Beatriz chooses, James also chooses.

d. each of them chooses from the list the book *Wuthering Heights* by Emily Bronte.

Looking Back

In this unit, you investigated some of the important ideas and methods in the mathematics of finance. In particular, you developed an understanding of the methods and technological tools used to determine simple and compound interest, future value of investments, future and present value of annuities, and amortization of debts such as credit card debt, car loans, and home mortgages. You also explored systematic methods for making decisions with respect to purchasing or leasing an automobile.

In those contexts, you strengthened your quantitative and algebraic reasoning skills. You revisited and deepened your understanding of linear and exponential models, formulating and interpreting algebraic expressions and formulas, and writing and solving equations to help answer questions related to a variety of financial situations.

You will be faced with major financial decisions upon graduation from high school, and then again after post-secondary education upon starting a career. In this final lesson, you will review and pull together key ideas and methods in the mathematics of finance that will be useful today and into the future.

1 Saving Money: The Power of Compounding Suppose you wanted to put money in a bank savings account and saw the following advertisement from an aggressive bank seeking deposits.

> # We Make Your $ Work
> ## 4% APR Compounded 24/7/365

It appears that this bank will pay interest in a way that is compounded every hour of the day, every day of the week, and every week of the year.

a. What expression shows how to calculate the value of a $1,500 investment in such an account at any time *h* hours later?

b. How many compounding periods will occur in a year, and to what value will the original $1,500 deposit grow in that time?

c. What is the effective annual interest rate of the savings account?

2 Matia's parents purchased a four-year CD in the amount of $5,000 the summer before he started high school to help him save for college. The CD earns an APR of 4.5% compounded monthly.

a. What will be the future value of the account after 48 months?

b. Estimate the doubling time using the Rule of 72.

c. Use an exponential model and algebraic reasoning to calculate the *exact* doubling time. Round your answer to one decimal place.

3 Suppose two years after graduation from college, you invest $4,000 in a savings account with an APR of 3% compounded quarterly. A coworker invests $3,500 in an account with an APR of 2.95% compounded continuously.

a. Will your coworker's balance ever exceed yours? If so, when?

b. What function models the future value of your coworker's investment after *t* years?

c. Suppose your coworker wants to save $8,000 over 3 years. What should be the amount of his initial deposit?

4 Paying Off Consumer Debt One major U.S. bank determines a consumer's minimum monthly credit card payment as the higher of either:

4% of the current balance or $40.00

Suppose Katrina just received her monthly credit card statement. Last month she made the minimum payment of 4% on her outstanding balance. She decides to make no additional purchases or payments beyond the minimum monthly payment until the card is paid off. This month's statement indicates a current balance of $1,545. The finance charge for her credit card is 15.6% APR compounded monthly.

a. What was the balance on Katrina's card after the sixth minimum monthly payment? How much interest did she pay during those six months?

b. After the sixth payment, Katrina decides she needs to pay the bill off more quickly. If she decides to pay $75 per month, how long will it take her to pay off the credit card debt? How much would the credit from this point on cost her?

c. Using the payment strategies of Parts a and b, what is the total cost of the credit for Katrina's purchase of $1,545?

5 Suppose you have negotiated to buy a new 2014 Sportster 1200 Custom motorcycle for $12,000 after taxes.

You make a 10% downpayment and pay the balance remaining with a loan from the manufacturer's finance division at an APR of 3% compounded monthly. You have a choice of three payment plans.

Plan I	36 months
Plan II	48 months
Plan III	60 months

a. For each plan, determine your monthly payment on the motorcycle.

b. For each plan, calculate the total interest you will pay the manufacturer's finance division.

c. For each plan, calculate the total cost of the motorcycle.

6 Jasen is researching how best to purchase a computer bundle from Better Buys Electronics. He is considering three different purchasing plans based on his personal finances. Examine the plans below.

	Cash	Store Finance	Bank Loan
Need Credit Check	No	No	Yes
Price Including Tax	$3,000	$45 per week	$3,000
Payment Terms	None	78 weeks	$220 per month
Annual Interest Rate	0%	Not Applicable	12%, compounded monthly

   a. What technology tool(s) would you use to help answer the following questions? Explain your thinking.

   b. How much is the cost of the computer bundle under each plan?

   c. Compare the three plans based upon a credit check, price including tax, payment terms, and interest rate conditions. What are the advantages and disadvantages of each plan for Jasen?

7 Suppose a year after getting married, Andy and Megan purchase their first home. In addition to making the monthly mortgage payments, they remodeled the home and installed new landscaping, leaving them with a balance of $10,000 on their credit card, which has an APR of 18% compounded monthly. The card requires a minimum payment of 5% of the balance. At this point, they stop charging and continue to make minimum payments until the balance is below $100.

   a. Write a formula that gives their balance after t monthly payments.

   b. Determine their balance after five years of minimum payments.

   c. How long, including the final payment of less than $100, will it take for Andy and Megan to pay off their credit card balance?

   d. What was the total amount of money they paid for their home remodeling and landscaping?

8 Planning for the Future Elana deposited a $12,000 bonus into a new retirement account that earns an APR of 5.2% compounded quarterly. She makes additional deposits of $500 at the end of each quarter until she retires in 20 years. At her retirement, Elana plans to supplement her retirement benefit by making withdrawals at the end of each quarter for the next 15 years (at which time, the account balance will be $0).

   a. How much money did Elana deposit in the account?

   b. How much money is in her retirement account after the last deposit?

   c. What is the amount of money of each withdrawal at retirement?

   d. What is the total amount of money Elana withdraws?

9 Recursion and Financial Mathematics You used recursive thinking often in this unit to describe patterns of change resulting from different saving and borrowing instruments. If B_i represents the balance in an account, r the APR, A an initial amount of money, and R a monthly payment, identify the financial instrument that is described by each recursive formula below.

a. $B_n = B_{n-1} + Ar$

b. $B_n = \left(1 + \frac{r}{n}\right)B_{n-1}$

c. $B_n = \left(1 + \frac{r}{n}\right)B_{n-1} + R$

d. $B_n = \left(1 + \frac{r}{n}\right)B_{n-1} - R$

SUMMARIZE THE MATHEMATICS

In this unit, you investigated key variables, methods, and tools for solving problems involving investments, credit card usage, personal loans, mortgages, auto leases, and annuities.

a. What are the key variables involved in investments and loans that involve compounding of interest?

b. What is the difference between the APR and its corresponding APY? How can this information be used by a consumer?

c. When interest is compounded at regular intervals, how can you determine the future value of an investment and the amount of interest earned?

d. What is the approximate value of the number e and how can that value be approximated by use of a formula?

e. When interest is compounded continuously, how can you determine the future value of an investment and the amount of interest earned?

f. What methods and tools would you use to determine the monthly payment of a loan or home mortgage?

g. What methods and tools would you use to track the balance of a credit card account involving no further charges and minimum monthly payment? A personal loan or mortgage?

h. What is an annuity? Assume deposits of a fixed amount are made monthly and interest is compounded monthly. How would you calculate its future value at the end of 40 years? If the desired annuity value after 40 years is D dollars, how could you determine the monthly deposit amount?

Be prepared to share your responses and reasoning with the class.

✓CHECK YOUR UNDERSTANDING

Write, in outline form, a summary of the important mathematical concepts and methods developed in this unit. Organize your summary so that it can be used as a quick reference in future units.

Binomial Distributions and Statistical Inference

Digital Resources at
ConnectED.mcgraw-hill.com

 Watch
 Tools
 Audio
 eBook

Public opinion polls gather information only from a relatively small sample of the population. Nevertheless, they can estimate proportions, such as the proportion of voters who approve of the job the president is doing, with surprising precision.

Through work on the investigations of this unit, you will learn how polls are conducted and analyzed. You will learn what characteristics make a public opinion poll trustworthy. Finally, you will construct and use binomial distributions to understand how a poll can measure public opinion to within a specified margin of error.

These key ideas will be developed through your work in the following three lessons.

LESSONS

1 Binomial Distributions

Develop the binomial probability formula and construct binomial distributions and compute their expected value. Compute and interpret a *P*-value for the proportion of successes in a sample, deciding whether the result is statistically significant or can reasonably be attributed to chance alone.

2 Sample Surveys

Evaluate and design surveys that satisfy the characteristics of a trustworthy sample survey. Distinguish between random sampling and stratified random sampling. Identify sample selection bias and response bias.

3 Margin of Error: From Sample to Population

Distinguish between point and interval estimates of a parameter. Observe variability in sampling (sampling error) in approximate sampling distributions. Compute and interpret a margin of error and a 95% confidence interval for a proportion. Understand the meaning of *95% confidence*.

Binomial Distributions

The ethnicity, gender, age, and other demographic characteristics of juries have been of great interest in some trials in the United States. When the composition of a jury does not reflect the demographic characteristics of the surrounding community, doubts about fairness of the jury selection process and legal challenges can arise.

Although juries are not selected solely by chance, comparing the actual jury to the composition of juries that would occur if jurors were selected at random can tell lawyers whether there are grounds to investigate the fairness of the jury selection process.

An historic case concerning jury selection, Avery v. Georgia, was brought to the U.S. Supreme Court in 1953. A jury in Fulton County, Georgia had convicted Avery, an African-American, of a serious felony. There were no African-Americans on the jury. At the time, there were 165,814 African-Americans in the Fulton County population of 691,797. The list of 21,624 potential jurors had 1,115 African-Americans. A jury pool of 60 people was selected, supposedly at random, from the list of potential jurors. (However, the names of black and white jurors had been written on different colored slips of paper.) This jury pool, from which the 12 actual jurors were selected, contained no African-Americans. (**Source:** caselaw.lp.findlaw.com/cgi-bin/getcase.pl?court=US&vol=345&invol=559)

Think about the demographics of Fulton County, Georgia and of the jury selected for the trial of James Avery.

a. If 12 jurors were selected at random from the people in Fulton County, how can you compute or estimate the probability that there would be no African-Americans on the jury? Generate as many methods as you can.

b. Do you think that having only 1,115 African-Americans on the list of potential jurors reasonably can be attributed to chance alone or should the lawyers look for another explanation? What strategies could you use to support your choice?

c. The jury pool of 60 people was selected from the list of 21,624 potential jurors. Can getting no African-Americans in a jury pool selected from these potential jurors reasonably be attributed to chance alone or should the lawyers look for another explanation? What strategies could you use to support your choice?

d. The U.S. Supreme Court overturned Avery's conviction. Describe the statistical evidence that you think might have been used by Avery's lawyers.

In this lesson, you will construct the type of *probability distribution* that will enable you to analyze situations such as the selection of jurors for Avery's trial.

INVESTIGATION 1

Rules of Probability and Binomial Situations

Many situations involving probability are called *binomial* because they have two possible outcomes, *success* and *failure*. In a **binomial situation**, you are interested in counting the number of successes that occur in a fixed number of identical, independent *trials*. You flip a coin 5 times and count the number of heads. You roll a pair of dice 12 times and count the number of times you get a sum of 7. You pick 50 students at random from your school and count the number who plan to go to a community college.

As you work on the problems in this investigation, look for an answer to this question:

How can rules of probability help you analyze a binomial situation?

Hero/Corbis/Glow Images

1 This *sample space* shows all possible outcomes of the roll of a pair of dice, one red and one green. There are six ways the red die can land and six ways the green die can land, so there are 36 possible outcomes. These 36 possible outcomes are *equally likely* to occur.

Number on Green Die

	1	2	3	4	5	6
1	1,1	1,2	1,3	1,4	1,5	1,6
2	2,1	2,2	2,3	2,4	2,5	2,6
3	3,1	3,2	3,3	3,4	3,5	3,6
4	4,1	4,2	4,3	4,4	4,5	4,6
5	5,1	5,2	5,3	5,4	5,5	5,6
6	6,1	6,2	6,3	6,4	6,5	6,6

Number on Red Die (row labels, at left)

a. Suppose that you roll a pair of dice, one red and one green. Use the sample space to find the probability that you get doubles (both dice show the same number). What is the probability that you do not get doubles?

b. Now suppose that you roll two dice that are the same color. Why should the sample space remain the same as that above? What is the probability that you get a sum of 7? What is the probability that you do not?

c. What is the probability that you get a sum of 6? What is the probability that you do not?

d. Complete this table that shows the *probability distribution* for the sum of the two dice. Is it best to leave your answers as unreduced fractions, reduced fractions, or decimals?

Sum	Probability
2	
3	
4	
⋮	
12	

e. Which sum is the most likely? Which sums are the least likely?

2 Now suppose you roll a pair of dice two times. Because the results of successive rolls of a pair of dice are independent events, you can compute probabilities using the **Multiplication Rule for Independent Events**:

> If A and B are independent events, the probability, $P(A \text{ and } B)$, that event A occurs and event B occurs is:
>
> $$P(A \text{ and } B) = P(A) \cdot P(B)$$
>
> $P(A)$ denotes the probability that A occurs on the first trial and $P(B)$ the probability that B occurs on the second trial.
>
> Formally, two events, A and B, are **independent** if and only if the rule above holds. An equivalent definition is that events A and B are independent if the occurrence of one of them does not change the probability that the other event occurs.

   a. Show how to use this rule to compute the probability that you get a sum of 7 on both rolls of a pair of dice.

   b. Show how to use this rule to compute the probability that you get a sum of 6 both times.

   c. Show how to use this rule to compute the probability that you get a sum of 7 on the first roll of the dice, but not on the second roll.

   d. Compute the probability that you do not get a sum of 6 on the first roll of the dice, but do on the second roll.

3 The Multiplication Rule for Independent Events can be extended to situations where there are more than two independent trials:

> In n independent trials, the probability $P(A \text{ and } B \text{ and } C \text{ and } \ldots)$ of getting A on the first trial, B on the second trial, C on the third trial, and so on is
>
> $$P(A \text{ and } B \text{ and } C \text{ and } \ldots) = P(A) \cdot P(B) \cdot P(C) \cdot \cdots$$

   a. Use this rule to compute the probability that if you roll a pair of dice five times, you get a sum of 7 on each roll. Then, find the probability that you get a sum of 7 on the first roll, but not on the remaining four rolls.

   b. Use this rule to compute the probability that if you flip a coin ten times, you get a head on each flip. What is the probability that you do not get 10 heads?

   c. Suppose that you roll a pair of dice six times. Find the probability that you never get doubles.

   d. Suppose you roll a pair of dice six times. Recall that the sum of the probabilities of all possible outcomes must add up to 1. Use this fact and the results of Part c to find the probability that you get doubles at least once.

4 The New York Lottery has a game called Quick Draw. In one version of the game, you pick a number from 1, 2, 3, … , 78, 79, 80. The lottery computer then randomly selects 20 of the 80 numbers. If the number you picked is included in the 20 randomly selected numbers, you win.

   a. What is the probability you win if you play the game once?

   b. If you play the game twice, what is the probability that you win both times?

   c. If you play this game 10 times, what is the probability that you win all 10 times?

   d. If you play this game 10 times, what is the probability that you lose all 10 times?

   e. Use the result of Part d to find the probability that if you play this game 10 times, you win at least once.

5 Suppose that you conduct n independent binomial trials. The probability of a success p is the same on each trial.

   a. What is the probability that all n trials are successes?

   b. What is the probability of a failure on a trial? What is the probability that all n trials are failures?

   c. What is the probability that you get at least one success?

6 Another rule of probability that is helpful in binomial situations is called the Addition Rule for Mutually Exclusive Events.

   a. Refer to the sample space in Problem 1. If you roll two dice, what is the probability that you get a sum of 2? What is the probability that you get a sum of 5? What is the probability that you get a sum of 2 or a sum of 5?

   b. In the last question of Part a, to find the probability of getting a sum of 2 or a sum of 5 when rolling a pair of dice once, you may have used the **Addition Rule for Mutually Exclusive Events**:

   If event A and event B are mutually exclusive, $P(A \text{ or } B) = P(A) + P(B)$.

   Mutually exclusive means that the two events cannot happen on the same trial. For example, you cannot get both a sum of 2 and a sum of 5 when you roll a pair of dice just once. Sometimes the term **disjoint** is used instead of mutually exclusive.

   Use the rule to find the probability that if you roll two dice, you get doubles or a sum of 7.

   c. Can you use the rule in Part b to find the probability that, if you roll two dice, you get doubles or a sum of 8? If so, do it. If not, explain why not.

   d. Can you use the rule in Part b to find the probability that if you flip a coin four times, you get either all heads or all tails? If so, do it. If not, explain why not.

In problems involving rolling dice and flipping coins, the trials are identical. For example, when playing a game using dice, the probability of rolling a sum of 7 does not change from one turn to the next. However, when playing a game using a deck of cards, once a card has been drawn, it usually is removed from the deck. This is called **sampling without replacement**. Consequently, the probability of getting, say, a heart on the next draw depends on what cards have been drawn before.

7 In this problem, you will compute a probability two different ways. Suppose a town has a pool of 3,000 eligible voters. Of these, 1,800 are college graduates. A four-person focus group is selected at random from the eligible voters and the number of college graduates is counted.

   a. What percent of the population is the sample size?

   b. What is the probability that the first member of the focus group selected is a college graduate?

   c. If the first member selected was a college graduate and is removed from the pool, what is the probability that the second member selected is a college graduate? Continuing in this way, compute the probability that all four members of the focus group are college graduates.

   d. Now suppose that each name is replaced in the pool after it is drawn. (This is called **sampling with replacement**.) Compute the probability that all four members of the focus group are college graduates.

   e. Is there much difference in the probability from Part c and the probability from Part d that all four members are college graduates? Which was easier to calculate?

Your work in Problem 7 suggests the following guideline used by statisticians to tell whether the probability changes significantly from trial to trial or whether the change is so small you can ignore the fact that you are sampling without replacement when computing probabilities.

Sample Size Guideline

If the size of a random sample is less than 10 percent of the size of the population from which it is taken, then, without much loss of accuracy, you can ignore the fact that the sampling is without replacement when computing a probability.

---

8 Test this guideline using the following situation. Now suppose that there are only 30 eligible voters in a town and that 18 are college graduates. A four-person focus group is selected at random from the eligible voters. Repeat Parts a–e of Problem 7 for this new situation. Does the sample size guideline appear to be reasonable?

SUMMARIZE THE MATHEMATICS

In this investigation, you used rules of probability to compute probabilities in binomial situations.

a. What is the Multiplication Rule for Independent Events? When can it be used?

b. If you conduct n (independent) binomial trials and the probability of a success on each trial is p, what is the probability that all n trials are a success? What is the probability that all n trials are a failure?

c. In a binomial situation, how can you find the probability of getting at least one success?

d. What is the Addition Rule for Mutually Exclusive (Disjoint) Events? When can it be used?

e. When computing probabilities, when can you ignore the fact that the trials are not independent?

Be prepared to share your ideas and reasoning with the class.

✓ CHECK YOUR UNDERSTANDING

Sixty-one percent of 18-year-olds in the United States have a driver's license. (**Source:** *Science Daily,* August 3, 2012, www.sciencedaily.com/releases/2012/08/120803082905.htm) You will select five different 18-year-olds at random from the U.S. and count the number with a driver's license.

a. Is this a case of sampling with or without replacement?

b. If your first 18-year-old has a driver's license, is the probability that the second 18-year-old has a driver's license a bit smaller than 61%, exactly equal to 61%, or a bit larger than 61%? Verify your answer using the fact that there are about 4,389,000 18-year-olds in the U.S.

c. According to the sample size guideline, can you compute the probability that all five 18-year-olds have a driver's license using $p = 0.61$ for each trial, without much loss of accuracy?

d. What is the probability that the first 18-year-old has a driver's license, but the other four do not?

e. What is the probability all five have a driver's license? What is the probability that none of the five 18-year-olds have a driver's license?

f. Find the probability that at least one of the 18-year-olds has a driver's license.

The Binomial Probability Formula

In a binomial situation, you have a population made up of "successes" and "failures," where the proportion of the population that are successes is denoted by p. In this investigation you will use counting methods to compute binomial probabilities.

As you work on the problems in this investigation, look for answers to this question:

How can you find the probability of getting a specified number of successes x in a binomial situation with n trials and probability p of success on each trial?

1 Having a baby and noting the sex is a binomial situation. About 51% of all babies born in the United States are boys. Suppose that a couple is going to have four children.

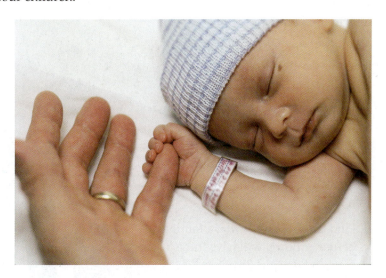

   a. Compute the probability that all four children will be boys.

   b. Name the rule that was used for the computation in Part a. What condition needs to be in place in order to use this rule?

   c. Compute the probability that all four children will be girls.

2 Now consider the following reasoning used by Elena to compute the other possibilities for a family of four children.

To compute the probability of getting one boy and three girls, she calculated

$$(0.51)(0.49)(0.49)(0.49) \approx 0.060.$$

To compute the probability of getting two boys and two girls, she calculated

$$(0.51)(0.51)(0.49)(0.49) \approx 0.062.$$

To compute the probability of getting three boys and one girl, she calculated

$$(0.51)(0.51)(0.51)(0.49) \approx 0.065.$$

a. Place Elena's probabilities and your probabilities from Problem 1 Parts a and c in a copy of the table below.

Number of Boys in a Family of Four Children	Probability
0	
1	
2	
3	
4	
Total	

b. Why can't the probabilities in the table all be correct?

c. What is wrong with Elena's reasoning?

3 You can use careful counting of the possible sequences of births to complete the table in Problem 2 correctly.

a. Imagine all possible families with three boys and one girl.

   i. List all of the sequences of births that would result in a family with three boys and one girl.

   ii. For each of the possible birth sequences, compute the probability that it will occur. What do you notice?

   iii. Find the probability that a family of four children will have three boys and one girl (in any order). What probability rule did you use and why?

b. Imagine all possible families with two boys and two girls.

   i. List all of the birth sequences that would result in a family with two boys and two girls.

   ii. For each of the birth sequences, compute the probability it will occur. What do you notice?

   iii. What is the probability that a family of four children will have two boys and two girls?

c. What is the probability that a family of four children will have one boy and three girls?

d. Using your results from Parts a, b, and c, place the correct probabilities in a copy of the table in Problem 2. Check to be sure that the probabilities in the table add up to 1 (subject to round-off error).

4 In Problem 3, you found the number of possible birth sequences for families of four children by listing them. If you use the idea of combinations from Unit 3, you do not have to list all possibilities.

a. Compute $C(4, 3)$. Why does $C(4, 3)$ give you the number of ways that a family of four children could have three boys and one girl?

b. Use combinations to find the number of ways that a family of four children could have two boys and two girls.

c. Use combinations to find the number of ways that a family of four children could have one boy and three girls.

d. How many ways could a family of n children have exactly x boys?

5 Suppose a softball player has a batting average of .400. In the next game, she expects to be at-bat five times. One model that statisticians have investigated is whether a player's at-bats are independent. Independence would mean that the player has a 0.4 chance of making a hit each time she comes up to bat, no matter what has happened in previous at-bats. In this problem, assume that at-bats are independent; that is, she does not tend to have streaks or slumps that require an explanation other than chance.

a. According to the Multiplication Principle of Counting, how many different sequences of hits and outs are there for five at-bats?

b. Explain why the entries for 4 hits in columns 2 and 3 of the following probability distribution table are correct.

Number of Hits	Number of Possible Sequences	Probability
0	$C(5, 0) = 1$	$1(0.4)^0(0.6)^5 \approx 0.078$
1		
2		
3		
4	$C(5, 4) = 5$	$5(0.4)^4(0.6) \approx 0.077$
5		
Total		

c. Using the same format as in the rows already entered, complete a copy of the table above.

Allan Munsie/Alamy

d. Check your results to be sure you have the right total number of possible sequences and that the probabilities add up to 1.

e. Describe the patterns you see in the table.

f. Make a histogram of the probability distribution similar to the one begun below. Then describe its shape and center.

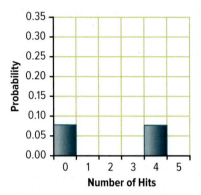

g. Using the graph, what number of hits has the largest probability?

6 In the United States, a person has a right to control inventions, literary and artistic works, and other ideas that they have created. This is called intellectual property rights. About 30% of adults are college graduates. Suppose that this is true in a jury pool where, in a case about intellectual property rights, the jury of 12 people contains only two college graduates. (**Source:** U.S. Census Bureau, 2011 American Community Survey)

Catherine and Houston decided to build a spreadsheet for computing the probabilities for this situation. The first rows of their spreadsheet are shown below.

	A	B	C	D
1	Number of College Graduates	Number of Possible Sequences	Probability of Each Particular Sequence	Probability of That Number of College Graduates
2	0	=COMBIN(12,A2)	=(0.3)^A2*(1−0.3)^(12−A2)	=B2*C2
3	=A2+1	=COMBIN(12,A3)	=(0.3)^A3*(1−0.3)^(12−A3)	=B3*C3

a. Discuss how the entries in rows **2** and **3** will compute the probabilities of 0 college graduates and 1 college graduate.

b. What should be the entries of row 4 of this spreadsheet? Compare your entries with those of your classmates. Resolve any differences.

c. Complete the spreadsheet and display a frequency distribution chart as shown below.

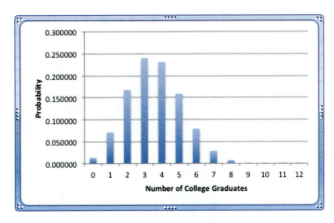

d. Describe the center and spread of the distribution.

e. What is the probability of getting two or fewer college graduates on a jury that is selected at random?

f. How can the probability in Part e be roughly estimated by comparing the sum of the areas represented by the first three bars and the total area represented by the bars?

g. Can you reasonably attribute the composition of the jury in the intellectual property case to chance alone or should the lawyers look for some other explanation? Explain your reasoning.

7 Refer to your work in Problems 5 and 6. Write the formula you have discovered for the probability $P(x)$ of getting exactly x successes in a binomial situation with n trials and probability of success p on each trial. Be sure to define any symbols you use.

8 About 9% of children age 3–17 have Attention Deficit Hyperactivity Disorder (ADHD). Suppose you take a random sample of 50 children age 3–17. (**Source:** Summary Health Statistics for U.S. Children: National Health Interview Survey, 2011, page 1, www.cdc.gov/nchs/data/series/sr_10/sr10_254.pdf)

a. Use your formula from Problem 7 to find the probability that exactly four of them have ADHD.

b. Find the probability that at least one of these children has ADHD.

c. Produce a graph of this probability distribution.

d. What is the probability that four or fewer children have ADHD?

9 The mean of a probability distribution, also called the **expected value**, is the value you would get, on average, in the long run if you repeated the series of trials again and again. The expected value μ of a binomial distribution with probability of success p and n trials may be found using the formula $\mu = np$.

a. Refer to Problem 1 of Investigation 1. You will roll a pair of dice 60 times. What is the expected number of times you will get a sum of 7?

b. Refer to Problem 6. What is the expected number of college graduates on a randomly selected jury of 12 people?

c. Refer to Problem 4 of Investigation 1. It costs $1 to play this lottery game. If you win, you are paid $2. What is your expected net gain (or loss) if you play this game 20 times?

d. Expected value can be used to help price insurance. For example, suppose that a company insures people against being struck by lightning and expects to sell 3,100,000 policies. The probability of being struck by lightning in a year is $\frac{1}{775,000}$. (**Source:** www.lightningsafety.noaa.gov/ odds.htm) If an insured person is struck by lightning, the company would pay them $1,000,000. What is the expected number of insured people who will be struck by lightning? What is the expected total payout to them? What should the company charge each insured person per year (called a *premium*) in order to expect to break even?

SUMMARIZE THE MATHEMATICS

In this investigation, you developed a formula for computing binomial probabilities.

a. Describe your formula for computing binomial probabilities and explain why it works.

b. In what situations can you use your formula to compute binomial probabilities? Explain your reasoning.

c. What is the expected number of successes in a binomial situation with probability of success *p* and *n* trials? Does the expected number of successes have to be a whole number? Explain your reasoning. Give an example to illustrate this computation.

Be prepared to explain your ideas and reasoning to the class.

✓ CHECK YOUR UNDERSTANDING

In the Avery v. Georgia case from the Think About This Situation on page 341, the list of 21,624 potential jurors in Fulton County had 1,115 African-Americans. Suppose that a 12-member jury is randomly selected from the list of potential jurors.

a. What is the probability that the first juror selected is African-American? Is it reasonable to use this same probability for each trial? Explain why or why not.

b. What is the expected number of African-Americans who would be on a jury of 12 people?

c. Complete this probability distribution table for the number of African-Americans on a jury that really was randomly selected from the list of potential jurors.

Number of African-Americans, x	Probability, $P(x)$
0	
1	0.3456
2	0.1033
3	0.0187
4	0.0023
5	
6	
7	0
8	0
9	0
10	0
11	0
12	0
Total	

d. Describe the shape of the probability distribution below. What is the most likely number of African-Americans on a randomly selected jury?

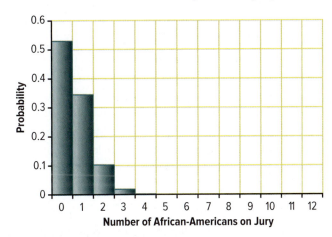

e. Can you reasonably attribute the fact that there were no African-Americans on a jury selected from this list of potential jurors to chance alone or should Avery's lawyers look for some other explanation? Explain your reasoning.

Statistical Significance

Astrologers claim that natal charts, a type of horoscope based on where and when a person was born, can be used to predict personality.

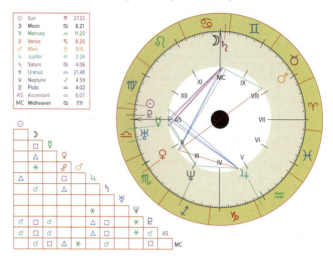

To test this claim, astrologers prepared natal charts for 83 volunteer subjects and wrote a description of the person's personality based only on the chart. Each subject then was given three descriptions (their own and two randomly chosen descriptions that were made for other people). The subject was asked to pick out the one that most correctly described them. Twenty-eight out of 83 (about 34%) selected their own description. (**Source:** Shawn Carlson, "A Double-Blind Test of Astrology," *Nature*, Vol. 318, pages 419–425, December 5, 1985.) If natal charts are complete nonsense, you would expect about 33% of the volunteers to pick their own description just by chance. Getting 34% who pick their own description certainly is not convincing evidence that natal charts can be used to predict personality. So, we say that the result from the experiment is not *statistically significant*. It reasonably can be attributed to chance alone.

As you work on the problems in this investigation, look for answers to this question:

How can you use technology to find out whether the number of successes in a binomial situation is statistically significant?

1 You often will find it best to use technology to compute binomial probabilities. Explore the binomial probability functions of your calculator or *TCMS-Tools* as you answer the following questions.

To use your calculator to compute, for example, the binomial probabilities associated with selecting 12 jurors at random from a large population in which 30% of eligible jurors are college graduates, enter the DISTR menu and select **binompdf(**. (The initials **pdf** stand for *probability density function*.) Then enter the number of trials *n*, the probability of a success *p*, and the number of successes *x*.

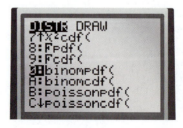

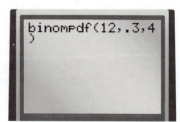

To compute this value in *TCMS-Tools*, open the CAS, select "Auto Numeric" from the Options menu, and type **binompdf(12,.3,4)** in the Command window as shown below.

a. Use the **binompdf(n,p,x)** function of your preferred technology tool to answer these questions. If 30% of people in a large city are college graduates and you select 12 people at random:

    i. what is the probability that exactly three are college graduates?

    ii. what is the probability that three or fewer are college graduates?

    iii. Use your probability from part ii to find the probability that four or more are college graduates.

b. Use a technology-based **binompdf(n,p,x)** function to answer these questions. If 18% of the seniors in a large high school have their own car and you pick 20 seniors at random:

    i. what is the probability that exactly 2 have their own car?

    ii. what is the probability that 2 or fewer have their own car?

    iii. what is the probability that 3 or more have their own car?

c. Use a technology-based **binomcdf(n,p,x)** function to evaluate **binomcdf(12,0.3,3)**. (The initials **cdf** stand for *cumulative distribution function*.) Compare this answer to your answers in Part a.

    i. Explain what the **binomcdf** function does.

    ii. Use the **binomcdf** function to answer the second and third questions in Part b.

d. Find the probability of getting 45% or fewer heads if you flip a coin 400 times. Why would you not want to do this without technology capable of computing binomial probabilities?

e. Suppose you roll a pair of dice 100 times. Find the probability that you will roll doubles on at least 20% of the rolls.

2 Suppose your family has been playing a board game where the player spins a plastic spinner on each turn. The spinner is divided into five equal sections, and one of the sections says, "Go home and start again." You have watched 50 different spins and the player had to "Go home and start again" on 14 of them. You are beginning to think that the spinner is unbalanced so that it is too likely to land on "Go home and start again."

a. If the spinner is balanced, what is the expected number of "Go home and start again" in 50 spins?

b. If the spinner is balanced fairly, what is the probability that the player will have to "Go home and start again" on 14 or more out of 50 spins? Answer this problem two ways.

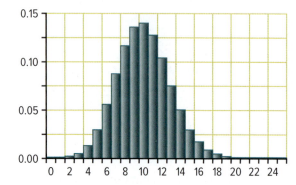

i. Estimate the probability from this graph of the binomial distribution constructed using $n = 50$ and $p = 0.2$.

ii. Compute the probability using the **binomcdf** function.

c. Is it unusual to get "Go home and start again" 14 times out of 50 spins if the spinner is balanced? Explain.

The probability you computed in Part b is called a **P-value**. To compute a P-value, you start with a value of a population proportion p against which you will compare the actual result of x successes out of n trials. Then, using that value of p, you compute the probability of getting, just by chance, x or even more successes (or, in some scenarios, x or even fewer successes) in n trials. When the P-value is less than 0.05, standard statistical practice is to call the result **statistically significant**. You want to look carefully at any actual result that is statistically significant (has low probability of happening just by chance) to see if you can understand why such an unusual event happened.

d. Which of the following is the best conclusion about the spinner?

I You do not have statistically significant evidence that the spinner is unbalanced.

II You do have statistically significant evidence that the spinner is unbalanced.

III The spinner must be balanced.

IV The spinner must be unbalanced.

3 In many sports, the home team is more likely to win a game than is the visiting team. This is called the "home-field advantage." For example, in the 2012 Major League Baseball season, 2,430 games were played. The home team won 1,295 of them. (**Source:** www.baseball-reference.com/games/situational.shtml)

a. If there is no home-field advantage, what proportion of the time do you expect that the team playing at home will win? What was the actual proportion in 2012?

b. What is the *P*-value for this situation? That is, if there is no home-field advantage, what is the probability that the home team will win 1,295 or more of 2,430 games? Answer this problem two ways.

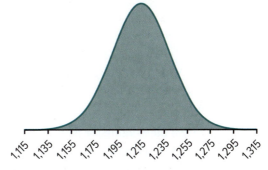

   i. Estimate the probability from the graph of the binomial distribution. This distribution was constructed using $n = 2,430$ and $p = 0.5$.

   ii. Compute the probability using the **binomcdf** function.

c. Is the number of wins by the home team statistically significant? If so, make several conjectures about what the explanation might be.

4 For a project in statistics class, Miguel individually showed 75 adults two different horoscopes taken from yesterday's newspaper. Each adult picked the horoscope that best described what happened to him or her yesterday. The adult did not know it, but one of the horoscopes was the one in the newspaper for his or her sign of the Zodiac and the other was for a different sign. Of the 75 adults, 39 picked the horoscope for their sign of the Zodiac.

a. If horoscopes have no basis in reality, should the *P*-value for this situation be larger or smaller than 0.05? Why?

b. What is the probability that an adult will pick his or her own horoscope? What is the probability that 39 or more out of 75 adults will pick their own horoscope?

c. Which of the probabilities in Part b is the *P*-value for this situation? Is Miguel's result statistically significant?

d. The distribution below illustrates Miguel's situation. Describe how this distribution was constructed. Then describe how to tell from the graph alone (even without a scale on the vertical axis) whether Miguel's result is statistically significant.

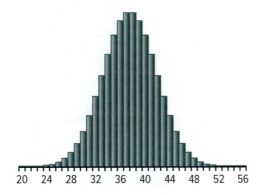

e. What should Miguel conclude?

> Working independently will pay off today. The good life and all that is fine may be what you value just now. You enjoy making your own way and finding solutions to whatever problems you have. Continue to analyze where you may best spend your energies.

5 An important 1977 U.S. Supreme Court case, Castaneda v. Partida, was explicitly decided on statistical evidence. A *grand jury* is responsible for deciding whether a person will have to stand trial. Of the 870 persons who were summoned to serve as grand jurors in Hidalgo County, Texas over an 11-year period, 339, or 39%, were Spanish surnamed. Census figures showed that 79.1% of the county's population had Spanish surnames. Castenada was convicted of a crime in Hidalgo County, Texas and appealed. (**Source:** caselaw.lp.findlaw.com/scripts/getcase.pl?court=US&vol=430&invol=482) While grand jurors were not selected by chance in those days (as they typically are now), comparing the actual jurors to what might happen if jurors were selected randomly can help us decide whether there is any evidence of possible unfairness in the selection process.

   a. Compute the *P*-value for this situation. That is, suppose that 870 people are selected at random from the population of Hidalgo County. Use the **binomcdf** function to find the probability that 339 or even fewer have Spanish surnames.

   b. Is the result from the Hidalgo County sample statistically significant? That is, can getting 339 or even fewer who are Spanish surnamed reasonably be attributed to chance alone or should the court look for another explanation?

   c. Think of an explanation, other than discrimination, that could have resulted in such a small percentage of grand jurors who had Spanish surnames. What data could be collected to provide evidence for or against that explanation?

6 In 2014, eighty men and twenty women served as United States senators. About 51% of the adult population of the U.S. are women.

   a. Suppose that U.S. senators are selected by a process that is totally without regard to sex. The graph below shows the binomial distribution for this situation. What is *n*? What is *p*?

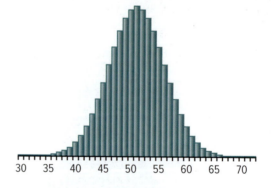

   b. From the graph alone, decide whether the number of women in the Senate in 2014 was statistically significant. What can you conclude?

7 It used to be the case that 30% of the pieces of a well-known chocolate candy had a brown coating. Sam thinks that the proportion is smaller now. To test this, he gets a random sample of 185 candies. Only 41 had a brown coating.

  a. If you take a random sample of 185 chocolate candies from a population where 30% have a brown coating, what is the probability of getting 41 or fewer that have a brown coating?

  b. What should Sam conclude?

SUMMARIZE THE MATHEMATICS

In this investigation, you learned that binomial distributions and the idea of statistical significance are useful not only in examining composition of juries, but also in many other situations.

a. What methods can you use to find binomial probabilities?

b. What is the meaning of a *P*-value? How do you compute it?

c. How can you tell when a result is statistically significant? What does it mean if a result is statistically significant?

Be prepared to share your ideas and reasoning with the class.

✓ CHECK YOUR UNDERSTANDING

In the Think About This Situation on page 341, you read about the Avery v. Georgia case.

a. There were 165,814 African-Americans in the Fulton County population of 691,797. What proportion is this?

b. The list of 21,624 potential jurors in the county had 1,115 African-Americans. What proportion is this?

c. What is the *P*-value? That is, what is the probability of getting 1,115 or even fewer African-Americans if the 21,624 potential jurors were selected at random from the population of Fulton County?

d. Can the composition of the list of potential jurors in Fulton County reasonably be attributed to chance alone? Why or why not?

ON YOUR OWN

APPLICATIONS

1 Cystic fibrosis (CF) is a serious genetic disease. Approximately one in 31 Americans is an unknowing symptomless carrier of the defective gene. A person can be born with CF only if both parents are carriers of the defective gene. In such cases, the child has a 25% chance of being born with CF. (**Source:** www.cff.org/AboutCF/Testing/) Suppose a couple plans to have three children, and both parents are symptomless carriers of the disease.

   a. What is the probability that all of the children are born with CF?

   b. What is the probability that none of the children is born with CF?

   c. What is the probability that at least one of the children is born with CF?

   d. What is the probability that the oldest child is born with CF but the others are not?

   e. Suppose a father and a mother are picked at random. What is the probability that they both are symptomless carriers of CF?

2 According to the U.S. Census Bureau, almost half a million people age 5 or older live in Tucson, Arizona. Of these, 33.5% speak a language other than English at home. (**Source:** quickfacts.census.gov/qfd/states/04/0477000.html) Suppose that you select six people age 5 or older at random from the population of Tucson.

   a. What is the probability that all of them speak a language other than English at home?

   b. What is the probability that none of them speak a language other than English at home?

   c. What is the probability that at least one of them speaks a language other than English at home?

   d. What is the probability that the first two people you pick speak a language other than English at home but the last four speak English?

3 California has more residents than any other state. According to the 2010 U.S. census, there are 308,745,538 residents of the United States and 37,253,956 of them live in California. (**Source:** www.census.gov/2010census/news/releases/operations/cb10-cn93.html) Suppose that you pick five U.S. residents at random. In this task, do not round your answers, but record the entire number given by your calculator.

   a. Use the numbers above to find the probability that the first person selected lives in California.

   b. If the first person selected lives in California, what is the probability that the second person selected lives in California?

c. Are the events *first person selected lives in California* and *second person selected lives in California* independent events? That is, does the result of the first selection change the probability that the second person lives in California?

d. Does the fact that the trials are not independent make much difference in the probabilities in this situation?

4 In the game of Yahtzee®, each player rolls five dice at once. Suppose you are at the end of a game and you need to roll sixes.

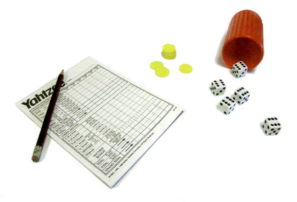

a. Find the probability that none of the five dice will show a six.

b. Find the probability that exactly two of the dice will show sixes.

c. Construct the probability distribution table and its graph for the number of sixes.

5 A softball player bats .350. Assume that every time she comes up to bat, the chance she will get a hit is 0.35. In an upcoming series, she will be at bat 10 times.

a. Compute the missing entries in this probability distribution table for the number of hits she will get in the upcoming series.

Number of Hits, x	Probability, $P(x)$
0	
1	0.0725
2	
3	0.2522
4	0.2377
5	0.1536
6	0.0689
7	
8	0.0043
9	0.0005
10	0.0000
Total	

b. Suppose the player gets only two hits in the upcoming series of games. What is the probability of getting two or fewer hits just by chance?

c. Some people believe that batters have "slumps" and "streaks." What do they mean by this? Do these people believe that hits are independent events? Explain.

6 In 1968, during the Vietnam War, Dr. Benjamin Spock (1903–1998) was on trial in Boston for conspiracy to violate the Military Service Act. He was accused of counseling young men on methods of avoiding the draft. It was thought that women would be more sympathetic to Dr. Spock because a larger percentage of women were opposed to the Vietnam War and because many women had used his book about childcare. After a jury selection process of several stages, Dr. Spock ended up with a jury of twelve men, even though women made up more than half of the eligible jurors.

a. Assume that women make up 50% of the eligible jurors and that jurors are selected at random from those eligible. Complete this probability distribution table and graph for the number of women on a Boston jury consisting of 12 members.

Number of Women	Probability
0	0.0002
1	0.0029
2	0.0161
3	0.0537
4	
5	
6	
7	
8	
9	
10	
11	
12	

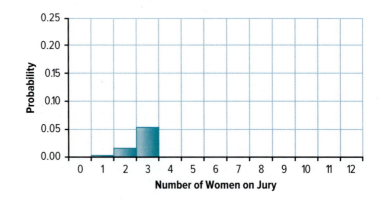

fabulousfaces.com

b. What is the *P*-value for the number of women on Dr. Spock's jury? That is, what is the probability of getting no women on the jury just by chance? Is this a statistically significant result?

c. If jurors are selected at random, what is the expected number of jurors who would be women? How is the value seen in the graph of the probability distribution?

7 In 1995, O.J. Simpson, an African-American former professional football player, was brought to trial for murder. The large jury pool in Los Angeles was 28% African-American. The final jury, which acquitted Simpson, consisted of nine African-Americans. (**Source:** www.law.umkc.edu/faculty/projects/ftrials/Simpson/Jurypage.html)

a. Suppose that jurors were selected at random from the large jury pool without regard to race. The graph below shows the probability distribution for the number of African-Americans on a jury of 12 members. Use the graph to estimate the *P*-value for the actual situation. That is, what is the probability that a jury selected at random would consist of nine or more African-Americans?

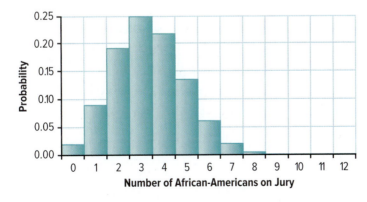

b. What is the expected number of African-Americans on a randomly selected jury of 12 people? How is this value seen in the graph?

c. Can the composition of this jury reasonably be attributed to chance alone? Explain.

8 In 2010, approximately 16% of the population of the United States did not have health insurance. (**Source:** www.census.gov/newsroom/releases/archives/income_wealth/cb11–157.html) To see whether this percentage has changed, this year a polling organization takes a random sample of 1,500 U.S. residents and counts the number without health insurance.

   a. What is n? Is it reasonable to compute probabilities as if the selections were independent? Explain your reasoning.

   b. If the percentage without health insurance remains at 16%, what is the expected number of people in the sample who do not have health insurance?

   c. Suppose that 207 people in this year's random sample of 1,500 U.S. residents do not have health insurance. What percentage is this?

   d. What is the P-value for the situation in Part c? That is, if 16% of the population does not have health insurance, what is the probability that if 1,500 U.S. residents are selected at random that 207 or fewer would have health insurance?

   e. Is the difference from the 2010 percentage statistically significant? What can you conclude?

9 In 2010, approximately 15% of the population of the United States lived below the poverty level. (**Source:** www.census.gov/newsroom/releases/archives/income_wealth/cb11–157.html) To see if the percentage has changed, this year a polling organization takes a random sample of 1,200 U.S. residents.

   a. What is n? Is it reasonable to compute probabilities as if the selections were independent? Explain.

   b. If the percentage living below the poverty level remains at 15%, what is the expected number of people in the sample who live below the poverty level?

   c. Suppose that 166 people in this year's random sample of 1,200 residents live below the poverty level. What percentage is this?

   d. If the percentage has not changed, what is the probability of getting exactly 166 in the sample who live below the poverty level?

   e. If the percentage has not changed, what is the probability of getting 166 or fewer in the sample who live below the poverty level?

   f. Which probability, the one in Part d or the one in Part e, is the P-value?

   g. Is the difference from the 2010 percentage statistically significant? What can you conclude?

CONNECTIONS

10 Look back at the probability distribution table you constructed for the sum of two dice (page 342).

a. Make a graph of the information in the table. Use the horizontal axis for sums and the vertical axis for probabilities.

b. What is the shape of the distribution?

c. What is the mean of the probability distribution? How can the mean be estimated from the histogram?

11 Explore how to compute probabilities using the Multiplication Rule for Independent Events in cases where the events are not identical binomial trials.

a. You roll a pair of dice two times. Compute the probability that you get a sum of 6 on the first roll and a sum of 7 on the second roll.

b. You roll a pair of dice two times. Compute the probability that you get a sum of 7 on the first roll and a sum of 12 on the second roll.

c. You roll a pair of dice two times. Compute the probability that you get a sum of 7 or a sum of 11 on the first roll and doubles on the second roll.

12 Sometimes two events are not mutually exclusive (disjoint). For example, suppose that you pick a number from 1 to 10 at random. What is the probability that it is even or prime? If you use the rule in Investigation 1, Problem 6, you would get $\frac{5}{10} + \frac{4}{10} = \frac{9}{10}$. But only eight numbers—2, 3, 4, 5, 6, 7, 8, and 10—are even or prime, so the correct probability is $\frac{8}{10}$.

a. Where did the rule from Investigation 1, Problem 6 go wrong?

b. When events A and B are not mutually exclusive (disjoint), explain why it is sensible to use the following rule to find $P(A \text{ or } B)$.

$$P(A \text{ or } B) = P(A) + P(B) - P(A \text{ and } B)$$

c. Suppose that 30% of the students in your high school wash their hair in the shower while facing the water, 40% put catsup directly on their fries rather than on the plate, and 15% do both.

If you select one student at random, use the rule to compute the probability that he or she does one or the other (or both). Draw a Venn diagram that illustrates this situation.

d. You roll a pair of dice. Use the rule in Part b to compute the probability that you get a sum of 6 or doubles.

e. Use the rule in Part b to compute the probability that if you draw one card from a deck it is a king or a club.

13 Examine the following tree diagram. Here, p is the probability of a success and $q = 1 - p$ is the probability of a failure in a binomial situation.

a. Examine the diagram below. What patterns do you see? What does the third circle in the bottom row represent?

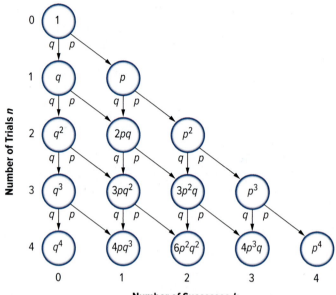

b. Record the next row. What does the third circle in this row represent?

c. Expand $(p + q)^5$.

d. Examine the terms of the expansion. What could they represent?

e. What is the numerical value of $p + q$? Of $(p + q)^5$?

f. What have you just proved?

14 A population of ten items has four "successes" and six "failures."

$$S\ S\ F\ S\ F\ F\ F\ S\ F\ F$$

a. What is the proportion of successes?

b. You can *code* this binomial population by renaming each success as a 1 and each failure as a 0:

$$1\ 1\ 0\ 1\ 0\ 0\ 0\ 1\ 0\ 0$$

Find the mean of these ten numbers.

c. Suppose that you have a population that contains 16 successes and 4 failures.

 i. When this population is coded as above, how many 1s will there be? How many 0s?

 ii. Find the mean of the coded population.

d. Develop a formula for the mean of a coded population of size n with k 1s and $(n - k)$ 0s. Write your formula in a simpler form using the proportion of successes p in the population.

15 This device is called a *binostat*. Balls begin at the top, drop through the pegs, and collect in the nine columns at the bottom. As a ball drops from the top of the binostat to a peg where a branch occurs, the ball is equally likely to go right or left.

a. Can the binostat be considered a binomial situation? Explain.

b. Is the distribution of balls in the columns of the binostat shown to the right typical?

c. What do the number labels on the channels represent? Is this pattern of numbers familiar?

d. Suppose the columns at the bottom are numbered 0, 1, 2, … , 8 from left to right.

   i. How many different routes could a ball take to Column 0? What is the probability of a ball falling into Column 0?

   ii. How many different routes could a ball take to Column 5? What is the probability of a ball falling into Column 5?

16 In Problems 5 and 6 of Investigation 2, you made histograms of binomial distributions. For the softball context, $n = 5$ and $p = 0.4$. For the jury pool situation, $n = 12$ and $p = 0.3$. The graphs below for these two situations were produced using the *TCMS-Tools* "Binomial Distributions" custom app under the Statistics menu. Use this computer software or similar software to explore the following questions.

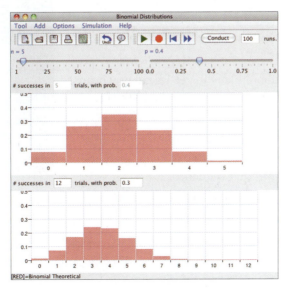

Jim Laser

a. As *n* increases but the probability of a success *p* remains the same, what happens to the shape, center, and spread of the binomial distribution for the number of successes?

b. As *p* increases from 0.01 to 0.99 but *n* remains the same (for example $n = 5$), what happens to the shape, center, and spread of the binomial distribution for the number of successes?

17 According to the United States Census Bureau, the 2010 population of Washington, D.C., was approximately 51% black. (**Source:** www.census.gov/prod/cen2010/briefs/c2010br-06.pdf)

a. Suppose a jury of 12 people was picked at random from the population of Washington, D.C. What is the probability that all of the jurors were black?

b. Suppose that 10 juries were picked at random from the population of Washington, D.C. What is the probability that at least one jury consisted of all black members?

18 According to the United States Census Bureau, the 2010 population of the state of Washington was approximately 11% Hispanic. (**Source:** www.census.gov/prod/cen2010/briefs/c2010br-04.pdf)

a. Suppose a jury of 12 people was picked at random from the population of Washington. What is the probability that none of the jurors were Hispanic?

b. Suppose that 10 juries were picked at random from the population of Washington. What is the probability that at least one jury consisted of all non-Hispanic members?

REFLECTIONS

19 Suppose that you flip a fair coin ten times.

a. Without computing, are you more likely to get 4 heads or 3 heads, or are the probabilities the same? Explain your reasoning.

b. Are you more likely to get 1 head or 9 heads, or are the probabilities the same? Explain your reasoning.

c. List a sequence of five heads and five tails that you could get. What is the probability of this particular sequence?

d. List a sequence of nine heads and one tail that you could get. What is the probability of this particular sequence?

e. If you flip a fair coin ten times, are you more likely to get five heads and five tails or nine heads and one tail? How can you reconcile this with your answers from Parts c and d?

Robert Dant/E+/Getty Images

20 Look back at Applications Task 6. If you used a technology tool in completing that task, what tool did you use and how did you make the decision on tool use? If you did not use technology, reflecting back on the task, what might be an appropriate technology tool to use? Why?

21 About 24% of residents of the United States are under the age of 18. (**Source:** www.census.gov/prod/cen2010/briefs/c2010br-03.pdf)

  a. If you select 10,000 United States residents at random, what is the expected number of them who are under the age of 18?

  b. Use technology to find the probability of getting exactly 2,400 residents who are under the age of 18.

  c. Suppose that you take a random sample of 10,000 residents from your state and exactly 2,400 of them are under age 18. Why doesn't the result of Part b mean that you have a statistically significant result? What is the correct P-value?

22 In New York City, 34% of people over age 24 are college graduates. (**Source:** quickfacts.census.gov/qfd/states/36000.html)

  a. Would it be statistically significant if a jury has 0 college graduates? Only one college graduate?

  b. Suppose a lawyer in New York City keeps track of the number of college graduates on the next 1,000 juries. She finds that 25 of the 1,000 juries have so few college graduates as to qualify as statistically significant. She plans to recommend that the jury selection process be investigated. What would you say to her?

EXTENSIONS

23 This task involves situations where sampling is done without replacement and the sample size is more than 10% of the size of the population.

  The graduating class at Smallville High School has 10 students. Seven of them already have their graduation robes. You plan to randomly select two different students for two "candid" yearbook photos and would like both students to be photographed in their graduation robes.

  a. What is the probability that the first student you select has his or her robe?

  b. Suppose that the first student selected has his or her robe. What is the probability that the second student also has his or her robe?

  c. Compute the probability that both students have their robes.

  d. Now suppose that you will select four students at random. Compute the probability that they all have their robes.

Purestock/SuperStock

24 Chaser is a remarkable border collie who knows the names of more than a thousand objects. To train Chaser, a sample of 20 objects was randomly selected from those Chaser had learned. They were placed randomly around the floor in a different room from where Chaser and the trainer were waiting. The trainer was given a randomized list of the 20 objects and asked Chaser to retrieve the first object listed. When Chaser

Chaser oversees a pile of 1,000 objects, each with a unique proper name.

returned, the trainer continued down the list, without replacing each object, until Chaser had retrieved all 20 items. (**Source:** John W. Pilley and Alliston K. Reid, "Border collie comprehends object names as verbal referents," *Behavioural Processes*, Vol. 86 (2011) pages 184–195.)

a. If Chaser is clueless about the first item and selects one at random to bring back to the trainer, what is the probability she selects the correct item?

b. If Chaser is clueless about each item, what is the probability that she gets all 20 objects correct, just by chance? (She almost always did this.)

c. How would your answer to Part b change if there were still 20 trials but each object was returned to the room before Chaser was asked to retrieve another?

25 Refer to Extensions Task 23. In a large city, 1,465,241 of the 2,000,000 households recycle soft drink cans. If five households are selected at random, what is the probability that all five households recycle:

a. if the households are selected without replacement (all five households must be different)?

b. if the households are selected with replacement (you can select a household more than once)?

c. Does it make much difference in your answers whether the households are selected with replacement or without replacement?

26 Look back at Problem 9 Part d (page 352) where you considered how expected value can be used to help price insurance. Prove that the break-even point for the insurance company occurs at a price of $1.29 per policy *regardless* of the number of policies sold.

27 She thought he was cheating on her. So, Lena Sims Driskell, age 78, shot and killed her boyfriend, age 85. "During jury selection, defense attorney Deborah Poole complained that Mrs. Driskell could not receive a fair trial with a jury of her peers because the juror pool lacked enough older people from which to select." Of the 58 people in the jury pool, all but five appeared to be younger than age 65. (**Source:** www.legalzoom.com/crime-criminals/murder/do-you-have-right and www.gainesville.com/apps/pbcs.dll/article?AID=/20060620/WIRE/60620020/1117/news)

a. Can you use the binomial probability formula to compute the probability that a jury of 12 people that is randomly selected from this pool would contain no one who appeared to be age 65 or older? Explain.

b. Use an appropriate method to find the probability that a jury of 12 people who are randomly selected from this jury pool would contain no one who appeared to be age 65 or older.

c. What can you conclude?

28 **Acceptance sampling** is one method that industry uses to control the quality of the parts it uses or other products for manufacturing. For example, a company that produces organic vegetable juices regularly receives shipments of vegetables from a supplier. To ensure the quality of the vegetables, an employee examines a sample of the vegetables in each shipment. The shipment is accepted if only 5% or fewer of the vegetables in the sample are considered low-quality. Assume that 10% of the vegetables from this supplier are low-quality. Suppose that the employee examines a random sample of 20 vegetables from each shipment.

a. Is this a binomial situation? If so, give the sample size and the probability of a success.

b. Design and carry out a simulation of this situation for 200 random samples.

c. What is your estimate of the probability that a shipment will be accepted?

d. If the company wishes to reduce the probability that they accept a shipment with 10% low-quality vegetables, what should they do?

29 Suppose that a population of size N contains S successes and F failures. You take a random sample of size n from this population.

a. If the sampling is done with replacement, explain why the probability of getting exactly s successes and f failures is given by the following formula:

$$C(n, s)\left(\frac{S}{N}\right)^s\left(\frac{F}{N}\right)^f$$

b. If the sampling is done without replacement, explain why the probability of getting exactly s successes and f failures is given by the **hypergeometric** formula:

$$\frac{C(S, s) \cdot C(F, f)}{C(N, n)}$$

c. Suppose a group of 25 students includes 15 students who play a musical instrument. If the sampling is done with replacement, find the probability that if you select 8 students at random, exactly 6 play a musical instrument. What is the probability if the sampling is done without replacement?

d. A high school football team has 40 players, half of whom are seniors. The coach has eight extra tickets to the homecoming game. He will select eight different players at random to get a ticket. What is the probability that exactly half will be seniors?

e. Refer to Extensions Task 27. Use the appropriate formula to find the probability that a randomly selected jury would have no one on it who appeared to be age 65 or older.

REVIEW

30 In previous studies, you used statistics appropriate to the context and the shape of a distribution to describe the center (median, mean) and the spread (interquartile range, standard deviation) of the distribution. You represented the data with dot plots, histograms, and boxplots. You may have recognized that the mean of a distribution displayed as a dot plot or histogram is the balance point of the distribution. Below is a dot plot of the dissolution times from a chemistry experiment. Students measured the time in seconds for a solute to dissolve. Study the plot to answer the questions below.

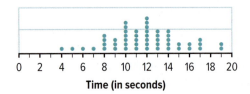

Time (in seconds)

a. How many dissolution times were collected by the science students?

b. Make a quick, rough estimate of the mean of the distribution from the dot plot. Then calculate the mean of the distribution.

c. What percentage of the dissolution times were less than 9 seconds? More than 16 seconds?

d. What percentage of the dissolution times are between 9 and 16 seconds, including 9 and 16?

e. Make a rough estimate from the dot plot of the percentage of dissolution times that are within 2 seconds of 12 seconds. Then calculate that percentage.

31 Rewrite each expression in equivalent form as a single algebraic fraction. Then simplify the result as much as possible.

a. $-a + \dfrac{1}{a}$

b. $\dfrac{a}{b} - 1$

c. $\dfrac{b}{a} + \dfrac{a}{b}$

d. $z\left(z - \dfrac{1}{z}\right)$

e. $\dfrac{x}{y} - \dfrac{1}{xy}$

f. $\dfrac{1}{x-1} - \dfrac{1}{x+1}$

32 Suppose you are asked to figure new prices for items in a music and video store. Show two ways to calculate each of the following price changes—one that involves two operations (either a multiplication and an addition or a multiplication and a subtraction) and another that involves only one operation (multiplication).

a. Reduce the price of a $50 video trilogy set by 20%.

b. Increase the price of a $10 CD by 30%.

c. Reduce the price of a $15 CD by $33\frac{1}{3}$%.

33 The population of a fruit fly colony is given by the function $p(t) = 16(2^t)$, with t representing time in days since the start of the experiment.

a. What is the population of the fruit fly colony after three days?

b. How long will it take for the population to reach 1,000 fruit flies?

34 Rewrite each expression in standard polynomial form.

a. $(3x - 4)^2 - 6$

b. $-3x(2x^2 - 1) - 8x + 7$

c. $8x - 4x^2 + 2x(3x - 4)$

d. $(3 - 4x^2)(3 + 4x^2)$

35 The Clifton Public Library surveyed 1,232 randomly selected adults living in the city about their reading habits. The results of part of their survey are summarized in the table below.

Read an Ebook in the Last Year

		Yes	No	Total
Read a Print Book in the Last Year	Yes	466	495	961
	No	64	207	271
	Total	530	702	1,232

Suppose that you interviewed a randomly selected person from the sample.

a. What is the probability that the person had read both a print book and an ebook in the last year?

b. What is the probability that the person had read an ebook during the last year?

c. What is the probability that the person had read a print book during the last year?

d. Based on your answers to Parts a–c, are reading ebooks and print books independent for the people in this sample?

e. Perform a chi-squared test of independence to determine if there is statistically significant evidence that these two variables are not independent in the population of adults living in Clifton.

36 Write a function rule that matches the transformation of the graph of the function $y = 2^x$.

a. Translated three units up

b. Translated four units to the right

c. Point $A(0, 1)$ is mapped to point $A'(0, 5)$ and point $B(1, 2)$ is mapped to point $B'(1, 10)$.

d. Point $S(0, 1)$ is mapped to point $S'(0, -1)$ and point $T(1, 2)$ is mapped to point $T'(1, -2)$.

37 Solve each equation or inequality. Display solutions for inequalities in interval notation.

a. $3x + 4 = 20$

b. $70 - \frac{3}{4}x < 10$

c. $\frac{x + 1}{3} = \frac{2x - 1}{2}$

d. $45 \leq -5x + 10$

38 Solve each equation for x.

a. $x^2 - x - 6 = 0$

b. $3x^2 - 1 = 47$

c. $16x^3 - x = 0$

d. $5x^2 - 13x = 6$

Sample Surveys

Surveys of samples of people are used often by government agencies, the media, and consumer-oriented businesses to better understand characteristics of the American people, their behavior, and their preferences. Political polls assess public opinion about issues and candidates. A large fraction of these surveys are worthless, but surveys from the U.S. government and reputable companies do a very good job. You may have wondered how they manage to measure public opinion accurately without getting everyone's opinion.

The Gallup company conducts surveys to monitor various characteristics of Americans. For example, it periodically asks a sample of people a series of questions about their lives and then classifies each person as thriving, struggling, or suffering. This survey is conducted by phone with about 28,295 randomly selected American adults. In October 2012, 51% of the Americans surveyed were classified as "thriving," 45% as "struggling," and 4% as "suffering."

(**Source:** www.gallup.com/poll/158543/americans-life-outlook-better-2008-not-best.aspx)

THINK ABOUT THIS SITUATION

Think about the conduct and findings of the Gallup poll described on the previous page.

a. Why do you think this survey is given to a sample of Americans instead of getting information from all Americans?

b. How can these percentages be trusted when only 28,295 Americans were surveyed?

c. What other surveys have you read or heard about?

d. What background information about a survey do you think is important to know in order to decide whether you can trust it?

In this lesson, you will learn how surveys and polls can be conducted so that the results are trustworthy.

INVESTIGATION 1

Trustworthy Surveys

People take surveys for many different reasons. Government surveys are taken to investigate the jobless rate and the cost of living. *Opinion polls* are surveys used to gauge public opinion on political issues. Some of these surveys are completely trustworthy. That is, you can be pretty sure that the percentages reported are fairly close to those that would have resulted had

the entire population been surveyed. On the other hand, the results from some surveys are not worth anything. You will learn about the first kind of survey in this investigation and about worthless surveys in the following investigation.

As you work on the problems in this investigation, look for answers to this question:

What are the characteristics of a trustworthy survey?

1 Individual opinions can be collected by using a census or by using a sample survey. A **census** collects information from every individual in a **population**, the entire set of people or things you would like to describe. In a **sample survey**, questions are asked only of a subset of a population. If the respondents are selected at random, a survey can accurately assess the characteristics of the entire population without contacting every individual. Read the following information from a trustworthy sample survey that estimated the proportion of people age 14–24 in the United States who have been cyberbullied.

Cyberbullying Widespread, MTV/AP Survey Reveals

"More than half of the respondents to an MTV/Associated Press study revealed they have been the targets of mean behavior or fake gossip on social-networking sites or text messages. The pervasiveness of digital abuse and cyberbullying in the study confirms a disturbing trend in which young people are using the Internet and wireless devices to harass each other, but also reveals that more are stepping up and saying something when they see abuse online. According to the study, 76 percent of 14–24 year olds say digital abuse is a serious problem for people their age, with 56 percent reporting that they have experienced abuse through social and digital media."

The study was conducted by Knowledge Networks using 1355 participants, age 14–24, chosen scientifically by a random selection of telephone numbers and residential addresses. Sampling margin of error for a 50% statistic with 95% confidence is ±3.8 for all interviews. The survey has a response rate of about 65%.

Sources: www.mtv.com/news/articles/1671547/cyberbullying-sexting-mtv-ap-survey.jhtml, www.knowledgenetworks.com/knpanel/docs/KnowledgePanel(R)-Design-Summary-Description.pdf, and www.athinline.org/pdfs/2011-MTV-AP_Digital_Abuse_Study_Full.pdf

a. Was this a census or a sample survey?

b. Surveys often are used to estimate a numerical characteristic of the population, called a **parameter**. For example, the percentage of "successes" in a sample can be used as an **estimate** of the percentage of successes in the population (the parameter).

Population of Skittles
Parameter: 20% of all Skittles are red

Sample of 13 Skittles
Estimate: 3/13 ≈ 23.1% of Skittles
in the sample are red

  i. Describe the population that was being studied in the cyberbullying research.

  ii. What parameter is being estimated by 56%?

  iii. Find the statement that gives the margin of error, which tells about how far off that estimate might reasonably be. What is the margin of error?

c. Was the sample selected at random? How large was the sample?

d. What additional information would you want to know in order to decide whether the estimate of the percentage of all people age 14–24 who would say they have been cyberbullied is trustworthy?

2 Describe the population in each case. Then, decide whether you would conduct a census or sample survey to answer each of the questions below.

    a. How do students in your class feel about a particular new movie?

    b. Is a manufacturer producing a high percentage of light bulbs that are defective?

    c. How many students bought a hot lunch at your school today?

    d. Who are people planning to vote for in the next election for governor of your state?

3 The results of a survey may be interesting, but before you trust the results, it is important to evaluate how the survey was constructed and carried out. Here are questions you should ask:

    1. What is the issue of interest or the parameter being estimated?

    2. What is the population?

    3. How was the sample selected?

    4. How large is the sample?

    5. What was the **response rate** (percentage of people contacted to be in the sample who gave a response)?

    6. How were the responses obtained from the sample (personal interview, phone interview, Internet, etc.)?

    7. What were the exact questions asked?

    8. Who sponsored the survey? Did a reputable company conduct the survey?

Refer to the study in Problem 1. Which of the questions above cannot be answered based on the information given?

A trustworthy survey always selects the participants at random. There are two main ways to do this:

- **(Simple) random sampling** is done by a process equivalent to writing the people's names on identical slips of paper, mixing them up, and then selecting slips of paper at random. In a random sample of size n, each possible sample of n people has an equal chance of being the sample.

- In **stratified random sampling**, the population first is divided into non-overlapping **strata**, or groups of similar people, such as men and women or Republicans, Democrats, and other. Then a random sample is taken from each stratum. The sizes of the random samples usually are proportional to the sizes of the strata. For example, suppose you want to take a stratified random sample of 1,500 households from a population in which 42% contain children under the age of 18 and 58% do not. First select a random sample of $0.42(1,500) = 630$ households from those with children. Then select a random sample of $0.58(1,500) = 870$ households from those without children.

4 Consider the following methods of getting a sample of 50 teen drivers from a town.

Method I Go to the high school, ask for a list of students, and select 50 at random. Replace each student who is not a driver with a different randomly selected student until you have 50 drivers.

Method II Contact the Department of Motor Vehicles, get a list of all drivers in the town under age 20. Select 50 at random.

Method III Contact the Department of Motor Vehicles, get a list of all drivers in the town under age 20. Choose a person at random, and then select that person and the next 49 whose names appear on the list.

a. Which method produces a random sample?

b. Why do the other two methods not produce a random sample?

5 In a typical pre-election poll, probable voters are stratified by gender, by political party preference, and by race.

a. Why might polling organizations want to use a stratified random sample rather than a simple random sample?

b. In the United States during the week of September 24th, 2012, about 28% of adults considered themselves Republicans, 32% considered themselves Democrats, 38% considered themselves Independents, and the rest were undecided. (**Source:** www.gallup.com/poll/15370/Party-Affiliation.aspx) Suppose you were taking a political poll that week of 1,200 adults. You wanted to stratify on party preference so that the sizes of the samples were proportional to the sizes of the strata. How many adults should you sample from each group?

6 Suppose you are interested in student opinion about a proposed new school mascot.

a. Describe how you would select a random sample of students from your school.

b. How might you use stratification to be sure that every important group of students is represented in your survey?

SUMMARIZE THE MATHEMATICS

In this investigation, you found that when you read about a survey used to measure public opinion, it is important to evaluate how much you can trust the results.

a. Describe the difference between taking a sample survey of the students in your school and taking a census of the students.

b. What is meant by the "population" when you take a survey? What is a population parameter?

c. What is the best method of getting a sample? Describe the two main methods of doing this.

d. What questions should you ask about the design and conduct of a survey in order to completely understand and accept the results?

Be prepared to share your ideas and reasoning with the class.

Pei has surveyed students about whether they favor the block scheduling that was introduced at their school last fall. To ensure privacy about this contentious issue, he mailed surveys, which could be returned anonymously, to the homes of students in his sample. He could afford to mail the survey to 150 students. Pei believed that the response would vary depending on the student's class year, so he stratified on class year, with the sizes of the random samples proportional to the sizes of the strata. His high school has 450 freshmen, 379 sophomores, 412 juniors, and 352 seniors. Of the 150 surveys he sent out, 135 were returned. Of these, two-thirds favored the new block scheduling.

a. To how many freshmen did Pei send a survey? Sophomores? Juniors? Seniors?

b. Refer to the questions in Problem 3 on page 378. Answer the questions that can be answered based on the information given.

c. How trustworthy do you find Pei's survey?

INVESTIGATION 2

How Surveys Can Go Wrong

When interpreting survey results, it is important to consider how the survey was constructed and carried out. In this investigation, you will examine how bias may occur in a survey.

As you work on the problems in this investigation, look for answers to this question:

What are the characteristics of an untrustworthy survey?

1 As most people realize, the estimate from a sample probably will not be equal to the population parameter. Two possible sources of error are chance and bias.

- **Chance (or sampling) error** results from the fact that a survey based on a random sample does not ask everyone in the population. Thus, the estimate from the sample may not be exactly equal to the population parameter. In Lesson 3, you will learn that a larger sample size tends to reduce sampling error.

- **Bias**, on the other hand, tends to push the estimate to one side of the population parameter. Specifically, in repeated sampling, the estimate from the sample is too big or too small, on average.

In the diagrams below, the center of the target represents the population parameter. The ×s represent estimates from the sample, attempts to hit the center of the target.

a. Which target shows errors due to chance alone? Which shows error due to both chance and bias?

b. Draw a target that shows error due to bias, with very little chance error.

c. Draw a target that shows almost no error due to bias or to chance.

d. Share your targets and explanations with others. Resolve any differences.

2 Shown below is a reduced copy of the set of circles shown on the following page.

Your task in this problem is to estimate the average area of all 115 circles shown on the following page. It certainly would be a lot of work to compute the area of each circle, so this is a situation where sampling might be better. You will be comparing two different methods to get your sample.

a. First use a **judgment sample**, using your best judgment to select five circles for the sample. That is, select five circles you think are fairly typical. Record the radius and area of each of your five circles. Finally, compute the average of the areas of the five circles.

b. Next, use technology to generate five random numbers between 1 and 115 inclusive and locate the corresponding circles. Find the average area of the five randomly selected circles.

c. Now, pool your data with other students in your class, making two lists and two dot plots for the averages generated in Parts a and b. Compare means of the two lists. Are the values about equally spread out in the two plots?

d. Check with your teacher on the actual average area of all 115 of the circles. Compare that area to the means computed in Part c. Was anything surprising in this comparison? Can you make any conjectures about how samples should be selected?

e. A method of selecting a sample is **biased** if repeated samples would yield estimates that are systematically too large (or too small), on average. Types of **sample selection bias** (bias that results from the method of selecting the sample) include

- *size bias*: When using their own judgment, people tend to pick the larger items to be in the sample.

- *voluntary response bias*: Allowing people to decide on their own whether to be in the sample generally results in a sample consisting of people with strong opinions about the issue.

- *convenience sample*: Using a readily available group as the sample typically results in a sample with views that are not representative of the population.

- *wrong population*: The sample was not selected from the population of interest.

Do you think either of the methods of selecting a sample of five circles is biased? If so, which type of bias did it illustrate?

f. Random sampling is used in surveys because it eliminates all but one of the types of bias above. Which one does it not eliminate?

3 The most famous polling mistake in U.S. history changed the way that polls are conducted. This poll had a huge sample, but went wrong anyway. In 1936, the *Literary Digest* conducted a poll of millions of voters and declared that Republican Alfred Landon would win the presidential election. Instead, Franklin Roosevelt won by a landslide. The magazine had sent out ballots printed on more than 10,000,000 postage-paid postcards and over 2,300,000 were returned. The names of people who received the postcards came mostly from *Literary Digest* subscribers, automobile registration lists, and telephone directories. (**Source**: Squire, Peverill. "Why the 1936 *Literary Digest* Poll Failed." *The Public Opinion Quarterly*, Vol. 52, No. 1 (Spring 1988), pages 125–133.)

a. What was the response rate in the *Literary Digest* survey?

b. What types of sample selection bias may have occurred as a result of the sampling procedure used?

c. George Gallup was able to predict the result of the election correctly, with a sample of only 3,000 voters. How do you think he was able to do this?

Surveys also can go wrong because the pollster gets no response or an incorrect response from a perfectly good random sample. Types of **response bias** (bias that results from systematically incorrect responses) include:

- *nonresponse bias*: People who have been selected for the sample cannot be contacted or refuse to answer.

- *questionnaire bias*: The question is worded so as to lead people to one particular answer or worded so that people in the sample misunderstand the question.

- *incorrect response*: People give the wrong answer. For example, maybe they do not remember accurately or they do not want the interviewer to know what they really think.

- *bad timing*: The time the survey was conducted influenced many people's response.

- *measurement error*: A mistake is made in collecting or analyzing the results. For example, a telephone interviewer incorrectly records a person's response or a person accidentally fills in the incorrect response on a written questionnaire.

4 What kind of bias may have occurred in the surveys described below? Do you think that as a result of the bias the estimate from the sample will be larger or smaller than the parameter?

a. Every year, the National Center on Addiction and Substance Abuse at Columbia University surveys a random sample of teens and parents by phone. One question they ask teens is, "Is your school a drug-free school or is it not drug free, meaning some students keep drugs, use drugs or sell drugs on school grounds?" In the 2012 survey, 53% of the teens surveyed answered "Drug free." To check on possible response bias, teens were asked one final question, "As you were speaking with me, was there someone there with you who could overhear your answers?" Twenty-two percent of the teens answered "yes" to the last question. (**Source:** www.casacolumbia.org/templates/publications_reports.aspx)

b. Cali says, "The results of my survey showed that 21% of the boys and 30% of the girls support me for president of the senior class. So, I should win the election with 51% of the vote."

c. A university sent out a survey to all alumni to find out how alumni felt about the usefulness of their education. Twenty-seven percent of the alumni returned the survey. The responses were overwhelmingly positive.

d. Harris Interactive conducted an online survey of 2,331 U.S. adults, asking, "If you had to choose, which one of these sports would you say is your favorite?" The sport chosen most often was professional football (31%). The survey was conducted in the middle of December. (**Source:** www.harrisinteractive.com/vault/HI-Harris-Poll-Favorite-Sport-Football-2011-01-20.pdf)

e. Student.com is an online community for teens and college students. It has a "Poll of the Day" in which members can respond to a question. Recently, the question was "How many piercings do you have?" Of the 363 students who responded, 37.2% said "None." (**Source:** www.student.com/potdmore.php?id=406)

5 The Panama Canal, opened in 1914, was built by the United States.

In 1978, by one vote, the U.S. Senate ratified a treaty that turned over the operation of the canal to the country of Panama. The Senate was influenced by polls taken to gauge public opinion on this issue. The wording of three questions used by various polls is given below. (**Source:** Ted J. Smith III; J. Michael Hogan, *Public Opinion and the Panama Canal Treaties of 1977, The Public Opinion Quarterly*, Vol. 51, No. 1. (Spring, 1987), pages 5–30.)

Question 1: In September, the Presidents of the United States and Panama signed two treaties which would gradually turn the Panama Canal over to the Panamanians but would provide for the continued use and defense of the Canal by the United States. Before these treaties can take effect, the United States Senate must act on it. Do you think the Senate should vote for the new treaties or against them?

Question 2: Do you favor the United States continuing its ownership and control of the Panama Canal or do you favor turning ownership and control of the Panama Canal over to the Republic of Panama?

Question 3: Do you think the time has come for us to modify our Panama Canal treaty or that we should insist on keeping the treaty as originally signed?

a. Which question do you think has the best wording? Explain.

b. For these questions, the percentage of those surveyed who gave an answer that supported the proposed treaty were—not necessarily in order—13%, 30%, and 45%. For which question do you think 45% supported the treaty? Explain your thinking.

c. Which type of response bias does this illustrate?

SUMMARIZE THE MATHEMATICS

In this investigation, you learned that sample selection is an important part of conducting a trustworthy survey. Equally important is getting a good response from each person selected for the sample.

a. Explain why random sampling is a good method for sample selection.

b. What are some types of sample selection bias?

c. What are some types of response bias?

Be prepared to share your ideas and examples with the class.

✔ CHECK YOUR UNDERSTANDING

For each study below, complete Parts a, b, and c to consider whether or not the survey method was trustworthy.

Study 1   A contemporary hit (top 40) radio station wants to estimate the percentage of voters who are in favor of raising the age that people in the state can get a driver's license. It asks its listeners to call in and give their opinion. The sample consists of those who call in on a toll-free number and state their opinion.

Study 2   A psychologist selects a random sample of 60 students from your school for a study about emotion. They are taken to the auditorium where the psychologist asks the students to raise their hand if they have ever cried during a movie.

Study 3   The town council wants to find out what proportion of citizens are bothered by barking dogs. It mails a survey to each citizen asking them to fill it out and return it by mail.

Study 4   An employer personally asks all employees if they think his company's policy about vacation time is fair.

Study 5   The homecoming committee wants to know the percentage of students who plan to buy a ticket to the homecoming dance. They ask all of the students in each of their classes whether they plan to buy a ticket.

a. Identify any possible bias in the method of selecting the sample.

b. Identify any possible response bias.

c. Do you think the estimate is likely to be too high or too low? Explain your reasoning.

APPLICATIONS

1 Read the following summary of a sample survey.

U.S. Religious Knowledge Survey

Researchers from the independent Pew Forum on Religion & Public Life phoned 3,412 Americans and asked them questions about various religions. Atheists and agnostics, Jews, and Mormons scored the highest. This result cannot be attributed to age or educational level. Here is one question that the interviewer read to the respondents. The choices were to be randomized for each respondent.

Please tell me which of the following is NOT one of the Ten Commandments:

- Do not commit adultery.
- Do unto others as you would have them do unto you.
- Do not steal.
- Keep the Sabbath holy.
- All are in the Ten Commandments.

Fifty-five percent answered this question correctly.

Results for this survey are based on telephone interviews conducted under the direction of Social Science Research Solutions (SSRS) among a national sample of 3,412 adults living in the continental United States, 18 years of age or older, from May 19-June 6, 2010 (2,393 respondents were interviewed on a landline telephone, and 1,019 were interviewed on a cell phone, including 444 who had no landline telephone). The survey of the full national population used "random digit dial" (RDD) methodology. Interviews were conducted in English and Spanish. The overall response rate for this study is 17.2%.

Source: www.pewforum.org/2010/09/28/u-s-religious-knowledge-survey/

a. For the question given in the summary, what is the population being studied? Describe the parameter being estimated. What is the estimate of this parameter?

b. What was the method of selecting the sample? How large was the sample? What is the response rate?

c. How were the responses obtained? Is the exact question given?

d. Does there appear to be any reason to find this poll untrustworthy? Explain.

2 Find an example of a survey on TV, in the newspaper, or on the Internet.

a. Write a brief summary of the survey.

b. Referring to the questions in Problem 3 of Investigation 1 (page 378), answer the questions that can be answered based on the information in the media-reported survey.

3 Wareham High School, in Massachusetts, has about 900 students in grades 9–12. Read the following information about a survey to assess student opinion about mandatory school uniforms.

Uniforms May Be Gaining Momentum for Wareham Schools

by Laura Fedak Pedulli

November 9, 2012, 12:00 AM

WAREHAM —The prospect of uniforms at the town's public schools may be gaining momentum.

According to results of a student-wide survey at Wareham High School obtained by The Standard-Times, while many objected to uniforms, 98 of the 184 surveyed expressed a willingness to explore the possibility and 22 students said they would consider working on a uniforms committee or help design the potential new look.

The survey was administered in June by Assistant Principal Debbie Freitas but results weren't tallied until Monday. ...

Both Freitas and Principal Scott Paladino also said uniforms would help with enforcing the current dress code as expected attire would be firmly defined.

Freitas stressed that students' willingness to embrace a uniform policy would be key to whether the idea takes off. ...

Source: www.southcoasttoday.com/apps/pbcs.dll/article?AID=/20121109/NEWS/211090339/-1/NEWS10

a. Referring to the questions in Problem 3 of Investigation 1 (page 378), answer the questions that can be answered based on the information in the abstract above.

b. Does there appear to be any reason to find this survey untrustworthy? Explain.

c. Does the Assistant Principal have statistically significant evidence that a majority of students would have expressed a willingness to explore the possibility of uniforms, if she had asked all of them? Select the best choice in each pair of options below.

Assuming that the proportion of all students who would have expressed this willingness is _exactly/more than_ 50%, we compute $1 -$ **binomcdf(184,0.50,97)** $\approx$ 0.2087. Because this P-value is _more/less_ than 5%, the Assistant Principal _does/does not_ have statistically significant evidence that the proportion of all students who are willing is _exactly/greater than_ 50%. In other words, if the true proportion is 50%, it _would/would not_ be very likely to get _50/53%_ or even more who are willing in a sample of size _98/184_.

4 Yen was quite upset about the way the Senior Class Council was planning graduation. She sent a survey to all members of the senior class. A reporter for the high school newspaper asked the AP Statistics class to evaluate Yen's survey and then wrote the following article based on their report.

Survey About Senior Class Council Found To Be Full of Flaws
by Julie Adalian

Last month's survey of the senior class blasted the work of the Senior Class Council on planning graduation. About three-quarters of the seniors who returned the survey agreed that the Senior Class Council was incompetent.

Following the uproar, Mr. McCune's AP Statistics class analyzed the survey as part of a unit in the design and analysis of surveys. Their review concluded that the survey was "biased," "deeply flawed," and "no valid conclusion could be drawn from it." They said that the survey was a clear attempt to force an election to replace current members of the Senior Class Council.

Yen Chang, a senior who was a candidate for the Senior Class Council, but was not elected, personally handed a copy of the questionnaire to each member of the senior class, asking each person to return it to her. She included a "fact sheet" with the survey that spelled out why she thinks the Senior Class Council is incompetent. She detailed why she thought the planning process for graduation was "amateur, incomplete, wasting our money, and behind schedule."

Mr. McCune's class noted that only 32% of the seniors returned the survey and many of those who did were students in Yen's classes.

Another problem with her survey was lack of confidentiality. Mr. McCune's class suggested that next time such a survey be returned to Mr. Gonzales in the main office, so that students wouldn't be concerned about another student knowing how they responded. Finally, it appears that Yen made some arithmetic mistakes in compiling her results.

Yen defended her survey saying, "Anyone who wanted to could return the survey to me. I took them all, not just those of my friends. I don't think that the fact sheet contained information that most people hadn't heard before, but I thought it important to remind them so they could make an informed decision."

a. Referring to the questions in Problem 3 of Investigation 1 (page 378), answer the questions that can be answered based on the information in the article above.

b. List each source of bias that is mentioned. Classify each source of bias as sample selection bias or response bias.

c. Do you agree that this survey is untrustworthy? Do you think that the result overestimated or underestimated the proportion of seniors who think the Senior Class Council is incompetent? Explain your answers.

5 In the 1948 presidential election, Harry Truman was running against Thomas Dewey. All of the major polling organizations of the day had learned not to make the same mistakes as made by the *Literary Digest* poll of 1936 (see Problem 3 on page 378). Yet, they still were mistaken about who would win the election.

Harry Truman

A good explanation of what went wrong can be found at the Web site of the Truman Library: www.trumanlibrary.org/ whistlestop/study_collections/1948campaign/large/docs/victory_final/what_ happened.htm. Go to the Web site and read the analysis, starting with the paragraph that begins, "Everybody's mistakes, in the end, could be attributed to the polls."

a. List each source of bias that is mentioned.

b. Classify each source of bias as sample selection bias or response bias.

6 In a classic 1940 experiment about the wording of questions, respondents were asked one of the following two questions.

Question 1: Do you think the United States should allow public speeches against democracy?

Question 2: Do you think the United States should forbid public speeches against democracy?

(**Source:** Rugg, D. Experiments in wording questions: II. *Public Opinion Quarterly* 5 (1941) pages 91–92.)

a. Is there any logical difference between the two questions?

b. Which question do you think resulted in a larger proportion of people who did not want to allow speeches against democracy? Explain your reasoning.

7 The three questions below about the scientific evidence for evolution have been asked to large random samples of Americans. (**Source:** pollingreport.com/science.htm)

Question 1: Do you think the scientific theory of evolution is well-supported by evidence and widely accepted within the scientific community?

Question 2: Just your opinion: Do you think that Charles Darwin's theory of evolution is a scientific theory that has been well-supported by evidence, or just one of many theories and one that has not been well-supported by evidence, or do you not know enough about it to say?

Question 3: Last year the National Academy of Sciences recommended that evolution be taught to all public school students as the most convincing theory for how human beings developed. Do you agree or disagree that evolution should be taught in all public schools?

Charles Darwin
developed theory of evolution

a. Rank these questions in order from the one you think resulted in the most support for the scientific evidence for evolution to the one you think had the least support.

b. Which question do you think has the best wording? Explain.

CONNECTIONS

8 Suppose that, in your school, 30% of the students are freshman, 28% are sophomores, 25% are juniors, and 17% are seniors.

You plan to survey a total of 300 students to estimate what percentage of students favor an increase in the size of the student council. If you want a stratified random sample with the sizes of the samples proportional to the sizes of the strata, how many students should you select from each class year?

9 In the United States, there are 2,816,000 high school dropouts, who have not completed high school and are not enrolled in school, between the ages of 16 and 24. Of these, 1,290,000 are employed, 527,000 are unemployed, and 999,000 are not in the labor force. (**Source:** www.census.gov/compendia/statab/2012/tables/12s0274.pdf)

High School Dropouts

a. Suppose a pollster wants to conduct a survey with a stratified random sample of a total of 1,500 dropouts. How many should she select from each group if she wants the sizes of the samples to be proportional to the sizes of the strata?

b. In her sample, the pollster finds that 65% of those employed, 80% of those unemployed, and 32% of those not in the labor force have plans to go back to school. What is her best estimate of the total percentage of dropouts who plan to go back to school?

10 The population of the world is about 33% Christian, the largest percentage of any religion. (**Source:** *CIA Factbook*; https://www.cia.gov/library/publications/the-world-factbook/) Suppose you take a random sample of 1,000 people from Ethiopia. Six hundred of the people in your sample are Christian.

a. Assuming that Ethiopia has the same percentage of Christians as the rest of the world, compute the P-value for this situation. Is this a statistically significant result?

b. What can you conclude?

11 In Problem 2 of Investigation 2 (pages 381–383), you considered the average area of a collection of circles. Think about some characteristics of "average circles."

a. If a circle has a radius equal to the average radius of all circles in a collection, does it have a circumference equal to the average circumference? If so, explain why. If not, give an example of a small collection of circles such that a circle of average radius does not have average circumference.

b. If a circle has a radius equal to the average radius of all circles in a collection, does it have an area equal to the average area? Justify your answer.

c. What explains the difference in your answers to Parts a and b?

12 Below is a group of rectangles.

I

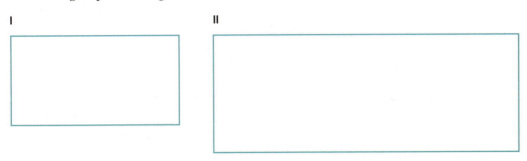

II

III

IV

V

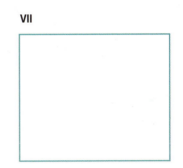

VI

VII

a. Show a copy of rectangles I–VII to 30 people and ask them: "Which rectangle do you prefer?" Write a brief summary of your findings.

b. A **golden rectangle** is a rectangle where the ratio of the length to the width is $\frac{1+\sqrt{5}}{2}$ or approximately 1.618.

   i. Which rectangle is closest to being a golden rectangle?

   ii. It has been claimed that people find the golden rectangle the most pleasing. (See the article at www.maa.org/external_archive/devlin/devlin_05_07.html for a discussion.) What proportion of your sample picked the golden rectangle?

   iii. If, overall, people have no preference among the rectangles, what proportion would pick the golden rectangle?

iv. Use the binomial probability function of your calculator to find the probability of getting as many people as you did in your sample, or even more, who would pick the golden rectangle just by chance.

v. Is your result statistically significant? What can you conclude?

13 The following question was asked of 506 adults nationwide as part of a special end-of-the-millennium poll.

Who do you feel is the greatest figure in the last thousand years, from anywhere in the world, specifically in the field of politics or government?"

Top Responses	%
John F. Kennedy	13
Abraham Lincoln	12
Franklin D. Roosevelt	7
George Washington	6
Thomas Jefferson	5
Bill Clinton	4
Ronald Reagan	4

Source: ABC News Poll. August 16–22, 1999; www.pollingreport.com/20th.htm

a. What is striking about these responses?

b. How could the question have been worded to give better responses?

14 The U.S. Bureau of Labor Statistics (BLS) tracks unemployment in the United States by conducting a monthly survey of 60,000 households. To get a better estimate of the unemployment rate, a question on the survey was reworded as follows.

Former question:

What were you doing most of last week?

- Working or something else?

- Keeping house or something else?

- Going to school or something else?

New question:

Last week, did you do any work for either pay or profit?

a. How might the former question lead to response bias?

b. What effect do you think revising the question will have on the estimate of the unemployment rate? Explain your reasoning.

15 Examine each of the following survey questions. For each survey question, explain how this question might lead to response bias. Then rewrite the question to eliminate the possible bias.

> **Question 1:** This question was asked on an opinion poll about taxes: "Do you agree that the current high tax structure is excessive?"

> **Question 2:** These instructions were given for an opinion poll about a new movie: "Rate the movie 1 to 10, where 1 is best."

> **Question 3:** To determine the number of people looking for jobs, an interviewer asked survey respondents, "Are you unemployed?"

> **Question 4:** Polls that attempt to predict the results of elections need not only to have a random sample, but also want to include only those people who actually are going to vote. Shortly before a presidential election, a pollster attempted to identify people who actually are going to vote by asking each person in the sample, "Are you planning to vote in the upcoming presidential election?"

EXTENSIONS

16 The National Merit Scholarship Corporation conducts two scholarship programs, National Merit Scholars and National Achievement Scholars. All students may compete in the National Merit program. Only Black students may compete in the National Achievement program. In 2012, 8,064 National Merit scholarships and 791 National Achievement scholarships were awarded. You plan to select a sample of these students for in-depth interviews about their college experience.

(**Source:** National Merit Scholarship Corporation, *Allegiance and Support*, 2011–2012 Annual Report; www.nationalmerit.org/annual_report.pdf)

a. You want a sample of 300 students and plan to take a stratified random sample, stratifying on the program in which the student received a scholarship. How many students would you select from each program if you want the number to be proportional to the number who received a scholarship?

b. Sometimes survey experts *oversample* smaller strata by including more people from the smaller strata than they would with proportional sampling. (This allows them to get a good estimate of the parameter from the smaller strata as well as from the larger strata.)

Suppose that you oversample by randomly selecting 100 National Achievement scholars. Of the 200 National Merit scholars in your sample, 8% majored in health sciences. Of the 100 National Achievement Scholars you sampled, 10% majored in health sciences.

  i. How many of the National Merit scholars majored in health sciences? How many of the National Achievement Scholars majored in health sciences?

  ii. What is your estimate of the overall percentage of scholars who majored in the health sciences?

17 Visit the Web site of the American Community Survey at www.census.gov/acs/www/. Then write a report that includes answers to the following questions as well as any other interesting facts you find.

a. Why is this survey taken? What is done with the information collected?

b. What specific questions on the survey would be of interest to students?

c. How is the sample selected? What is the sample size?

18 Visit the Web site of the United States Census Bureau at www.census.gov. Then write a report that includes answers to the following questions as well as any other interesting facts you find.

a. How often is the census taken?

b. When did the census start and why?

c. What is done with the information collected by the census?

d. What questions does the census ask?

e. Is sampling used in the census?

f. How might the census be biased because of nonresponse?

19 Suppose you want to survey your class for some very personal information. If you ask students directly, they might not tell the truth. This would create bias in your survey. Getting honest answers to sensitive questions is a common problem for pollsters. One way to solve this problem is to use a *randomized response* technique.

Using this technique, the survey designer pairs a sensitive question (to be answered "yes" or "no") with a harmless question, to which the interviewer could not possibly know the answer and which has a known proportion of "yes" responses. For example, the harmless question might be "Is the coin you just (secretly) tossed a head?" When the respondent comes to the sensitive question, he or she secretly rolls a die and flips a coin. If the die lands 1 or 2, then the respondent answers the harmless question about the coin flip. If die lands 3, 4, 5, or 6, then the respondent answers the sensitive question. The interviewer records the number of "yes" answers.

a. Explain why only the respondent knows which question is being answered.

b. Suppose in a sample of 60 people there were 32 "yes" responses. Estimate the proportion of "yes" responses to the sensitive question.

c. Suppose in a sample of 40 people, there were 31 "yes" responses. Estimate the proportion of "yes" responses to the sensitive question.

d. Complete the general statement below showing how the proportion of all "yes" responses is related to the proportion of "yes" responses to a sensitive question and to the proportion of "yes" responses to a harmless question.

Proportion of all "yes" responses = … .

20 Select a topic that is of interest to students at your school. Write a question about the topic in a way that you think will cause bias. Then rephrase the question in a way that you think will cause bias in a different direction. Give each question to at least 30 different students. Collect responses and compare the results. Did the difference in wording have the effects you thought it would? Explain.

REVIEW

21 The Highline Ice Cream Shop has five flavors of ice cream and 10 different mix-in choices.

a. How many different possible bowls of ice cream are there that contain only one flavor of ice cream and three different mix-ins?

b. How many different possible bowls of ice cream are there that contain two different flavors of ice cream and one mix-in for each flavor? The mix-ins can be the same or different.

c. How many different possible bowls of ice cream are there that contain three different flavors of ice cream (with no mix-in choices)?

d. Which part of this problem is an example of combinations?

22 Find the coordinates of the vertex, the x-intercepts, and the y-intercept of the graphs of each quadratic function.

a. $y = 10x - x^2$

b. $y = x^2 - 8x + 12$

c. $y = (x - 3)^2 - 4$

23 Solve each equation.

a. $3x - 5(x - 3) = 2x - 9$ **b.** $x(x - 3) = x^2 - 12$

c. $(2x - 1)^2 = 49$ **d.** $2x^2 - 5x - 1 = 2$

24 Rewrite each expression as a product of linear factors.

a. $x^2 + 10x + 24$ **b.** $25 - 16x^2$

c. $x^2 - 12x + 36$ **d.** $x^2 + 23x - 50$

e. $4a^2 - 9b^2$

25 Write each expression in a simpler equivalent exponential form using only positive exponents.

a. $\dfrac{3x^{-1}}{x^2}$ **b.** $\dfrac{x^3(2x^4)^2}{x^{-2}}$

c. $-6x^3y^2(4x^{-2}y^3)$ **d.** $3(21x^2)^{-1}$

26 Find the indicated angle measure and side lengths.

a. Find $m\angle B$.   **b.** Find AC.

 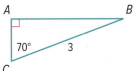

c. Find BC and AC.   **d.** Find AB.

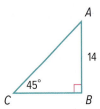 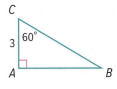

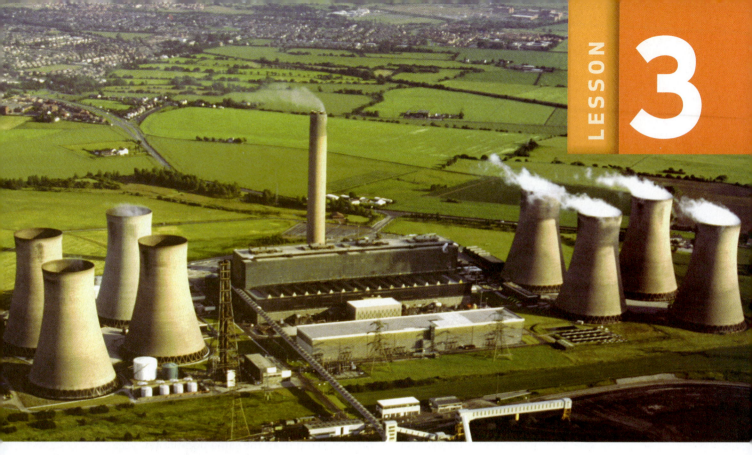

Margin of Error:
From Sample to Population

Perhaps you have wondered how a pollster can make precise claims about the entire population of the United States based on just a relatively small random sample of people.

For example, a Gallup poll asked Americans if they "think nuclear power plants in the United States are safe or not safe." Fifty-seven percent of those surveyed said, "safe." The following technical information about this poll comes from the Gallup Web site.

Survey Methods

Results for this Gallup poll are based on telephone interviews conducted March 8–11, 2012, with a random sample of 1,024 adults aged 18 and older, living in the continental U.S., selected using random-digit-dial sampling. For results based on the total sample of national adults, one can say with 95% confidence that the maximum margin of sampling error is ±4 percentage points. In addition to sampling error, question wording and practical difficulties in conducting surveys can introduce error or bias into the findings of public opinion polls.

Source: www.gallup.com/poll/153452/americans-favor-nuclear-power-year-fukushima.aspx

THINK ABOUT THIS SITUATION

Think about the information about the survey methods used by the Gallup poll.

a. Does this appear to be a trustworthy survey?

b. What sources of possible bias are mentioned?

c. What might be meant by the "margin of sampling error is ±4 percentage points"?

d. What do you think is meant by "with 95% confidence"?

In this lesson, you will develop understanding of the meaning and associated statistical method(s) of each term in the description of the survey methods of the Gallup poll.

INVESTIGATION 1

Variability from Sample to Sample

Public opinion can be measured with a great deal of accuracy by looking at a relatively small random sample from the population. The first step in understanding how this can be done is to understand how much variability there is among different random samples taken from the same population.

As you work on the problems in this investigation, look for answers to this question:

How far is the proportion of successes in a random sample likely to be from the proportion of successes in the population from which it was taken?

1 In Lesson 1, you found the probability of a specified *number* of successes in a binomial situation. When the sample size n is large, people usually find it more useful to know the *proportion* of successes in the sample $\hat{p}$ rather than the number of successes x. (The symbol $\hat{p}$ is read "p hat.") So, typically, when you hear the results of a survey or poll, the proportion of successes is given. For example, a typical report on the President's approval rating will say something like, "From our survey of 1,248 adults, we found that 57% approve of the way the President is handling his job."

a. Suppose that in a sample survey last month, 367 of 1,192 adults approved of the way the President is handling his job. In a similar sample survey this month, 472 of 1,364 adults approved. Was there greater approval for the President among those surveyed this month or among those surveyed last month?

b. The proportion of successes in the sample $\hat{p}$ is a **point estimate** of the *parameter p*, the proportion of successes in the population. Write a formula for computing $\hat{p}$ given the number of successes x in the sample and the sample size n.

2 According to the 2010 U.S. Census, about 49% of all 18-year-olds are female.

Suppose that Nicholas does not know this. To estimate the proportion of all 18-year-olds who are female, he will randomly select 100 18-year-olds and count the number of females. In this problem, you will investigate the *distribution* of all possible outcomes.

a. Nicholas selects his sample and gets 47 females. Compute his point estimate $\hat{p}$ and locate it with a dot on a copy of the plot below.

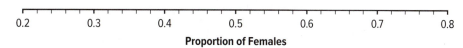

Proportion of Females

b. Nicholas's estimate from the sample of 0.47 is only 0.02 away from the proportion of successes in the population of 0.49. His *sampling error* is 0.02, or 2%.

> **Sampling Error**
>
> **Sampling error** is the absolute value of the difference between the estimate from a sample and the population parameter. When estimating a population proportion p from the proportion of successes $\hat{p}$ in a sample, the sampling error (*S.E.*) is:
>
> $$S.E. = |\hat{p} - p|$$

Usually people can take only one sample. But suppose that Nicholas takes another random sample of 100 18-year-olds and gets 52 females. Add his second point estimate to the plot. What is his sampling error this time?

c. Nicholas continues until he has 1,000 random samples of 100 18-year-olds. The dot plot at the top of the following page shows his 1,000 values of $\hat{p}$. He has created an approximate **sampling distribution**. (The *exact* sampling distribution would have the values of $\hat{p}$ for all possible random samples.)

    i. In how many samples did he get exactly 60 females?

    ii. What was the fewest number of females he got in any sample? The largest number?

    iii. What is the largest sampling error he got?

d. Suppose that you take a random sample of 100 18-year-olds and compute the proportion who are female.

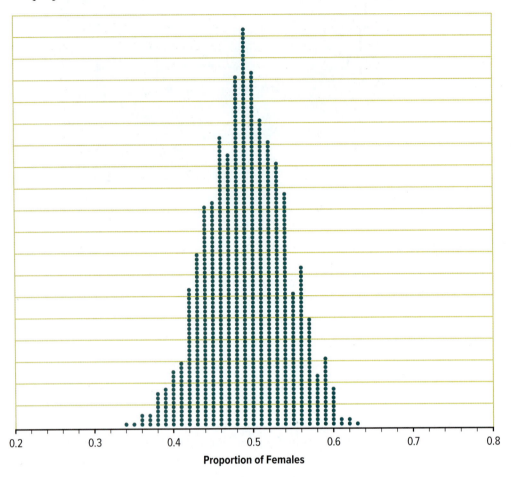

Proportion of Females

    i. Use the approximate sampling distribution above to estimate the probability that you will get a value of $\hat{p}$ that is less than 0.39.

    ii. Estimate the probability that $\hat{p}$ will be more than 0.59.

    iii. Use the approximate sampling distribution to estimate the probability that you get a sampling error of more than 0.10. Of 0.10 or less.

    iv. Make a rough estimate of the probability that you get a sampling error of 0.05 or less.

3 Nicholas's approximate sampling distribution illustrates **variability in sampling**. That is, the point estimate $\hat{p}$ varies from random sample to random sample. Next you will explore the main factor that affects how large the variability in sampling tends to be.

    a. Nicholas notices how spread out the various values of $\hat{p}$ are in the distribution of Problem 2. Often, the sampling error was large. What do you think that pollsters do to minimize the chance of getting a large sampling error?

    b. Nicholas decides to try a larger sample size. Describe how he could make an approximate sampling distribution for samples of size 400 (with p still equal to 0.49).

c. The dot plot below, an approximate sampling distribution, shows the values of $\hat{p}$ from 1,000 random samples of size $n = 400$.

    i. How many 18-year-olds were sampled to get one dot?

    ii. What was the largest sampling error this time?

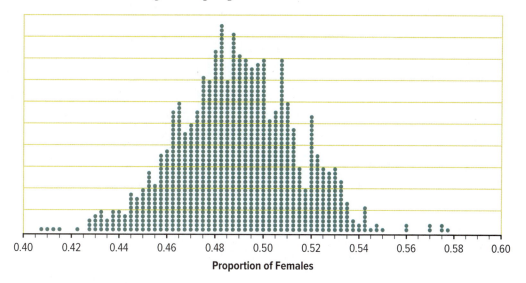

Proportion of Females

d. Use the approximate sampling distribution to estimate the probability of getting a sampling error of more than 0.10. Of 0.10 or less.

e. Use the approximate sampling distribution to estimate the probability of getting a sampling error of 0.05 or less.

f. By increasing the sample size from $n = 100$ to $n = 400$, how does the probability of getting a sampling error of 0.05 or less change?

4 A grade of 3 or better typically is considered "passing" on an Advanced Placement (AP) exam. According to the *AP Report to the Nation* from the College Board, nationwide almost 20% of 2012 high school graduates passed an AP exam during high school. (**Source:** *The 9th Annual AP Report to the Nation*, February 13, 2013, page 37)

Rose does not know this percentage. To estimate it, she will take a random sample of 100 graduates from the class of 2012.

a. In her sample of 100 graduates, Rose finds that 26 have passed an AP exam. What is her point estimate, $\hat{p}$? What is her error attributable to sampling?

b. The dot plot at the top of the following page shows the values of $\hat{p}$ from 500 different random samples of 100 graduates each. Each sample was taken from the class of 2012 where 20% passed an AP exam during high school. What is the name for this type of dot plot? Locate Rose's point estimate on the plot.

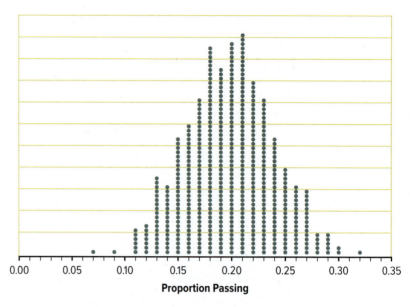

Proportion Passing

c. Using the plot, estimate the proportion of random samples that have a sampling error of 0.10 or less. Estimate the proportion of random samples that have a sampling error of 0.05 or less.

d. Does the sampling error for this situation, with $n = 100$ and $p = 0.20$, tend to be smaller or larger than the one in Nicholas's approximate sampling distribution in Problem 2 for samples of size 100 and $p = 0.49$?

5 As you saw in the previous problems, the size of the sampling error depends on the sample size n. It also depends on the proportion p of successes in the population p. Bigger sample sizes make for smaller errors attributable to sampling. Values of p that are farther from 0.5 make for smaller errors attributable to sampling than values closer to 0.5. The following rule tells you how far apart p and $\hat{p}$ are likely to be.

Rule for Size of Sampling Error

If you take a random sample of size n from a population with proportion of successes p, there is a 95% chance that p and $\hat{p}$ will be no more than $2\sqrt{\dfrac{p(1-p)}{n}}$ apart. Alternatively, there is a 5% chance that the sampling error will be greater than $2\sqrt{\dfrac{p(1-p)}{n}}$.

---

a. Compute the value of $2\sqrt{\dfrac{p(1-p)}{n}}$ for the situation when $n = 100$ and $p = 0.20$. Check whether the rule holds, approximately, for the distribution in Problem 4.

b. Rewrite the first sentence of the rule in the context of Problem 4.

This rule works well only if your sample contains at least 10 successes and at least 10 failures and if the sample size is less than one-tenth of the size of the population.

If the sample size is larger, there is another rule to use, but that is rarely the case in sample surveys.

SUMMARIZE THE MATHEMATICS

In this investigation, you learned to use sampling distributions to estimate the sampling error when taking a random sample from a known population.

a. How would you create an approximate sampling distribution of $\hat{p}$ for the situation of rolling a pair of dice 50 times and computing the proportion of times they land doubles?

b. What is the sampling error? How is it computed? Can people who conduct sample surveys actually compute the sampling error? Why or why not?

c. If you flip a coin repeatedly and use the proportion of heads in your sample as an estimate of the probability that the coin lands heads, will the sampling error tend to be smaller if you use a sample of 100 flips or a sample of 400 flips? Explain your reasoning.

d. Explain in your own words the meaning of the rule:

If you take a random sample of size n from a population with proportion of successes p, there is a 95% chance that p and $\hat{p}$ will be no more than $2\sqrt{\dfrac{p(1-p)}{n}}$ apart. Alternatively, there is a 5% chance that the sampling error will be greater than $2\sqrt{\dfrac{p(1-p)}{n}}$.

Be prepared to share your ideas and reasoning with the class.

✓ CHECK YOUR UNDERSTANDING

According to the 2010 U.S. Census, about 87% of U.S. residents, age 25 and older, are high school graduates. Bernardo does not know this percentage so will take a random sample of 1,500 U.S. residents, age 25 and older to estimate it.

a. What is the value of n? What is the value of p?

b. In his sample of 1,500 U.S. residents age 25 and older, Bernardo finds that 1,300 are high school graduates. What is his point estimate, $\hat{p}$? What is his sampling error?

c. The dot plot below shows the values of $\hat{p}$ from 1,000 different random samples, each of 1,500 U.S. residents age 25 and older. What is the name for this type of dot plot? Locate Bernardo's result on the plot.

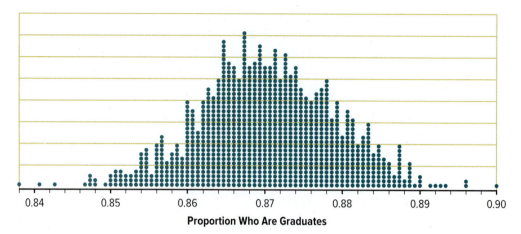

Proportion Who Are Graduates

d. From the plot, estimate the proportion of random samples that have a sampling error of 0.03 or less. Estimate the proportion that have a sampling error of 0.02 or less.

e. Does the sampling error here tend to be smaller or larger than the other ones in this investigation? Why is this the case?

f. Compute the value of $2\sqrt{\dfrac{p(1-p)}{n}}$ for this situation. Write a sentence explaining what the value means in the context of this situation.

INVESTIGATION 2

The Margin of Error

As you worked through the previous investigation, you may have been surprised about how close the proportion $\hat{p}$ of successes in a random sample tends to be to the proportion p of successes in the population even when the sample size is as small as 100 out of a population of millions. Of course, larger samples are better, but "large enough" is smaller than most people think.

In Investigation 1, you were able to compute the **sampling error** $|p - \hat{p}|$ exactly because you knew the value of both $\hat{p}$ and p. But pollsters do not know the value of p, or they would not be taking a sample survey.

As you work on the problems in this investigation, look for answers to this question:

How can you estimate the sampling error when you do not
know the proportion p of successes in the population?

1 The following rule from the previous investigation will make it possible for you to estimate the sampling error.

If you take a random sample of size n from a population with proportion of successes p, there is a 95% chance that the difference between p and $\hat{p}$ will be $2\sqrt{\dfrac{p(1-p)}{n}}$ or less.

Because pollsters do not know the value of p, they substitute $\hat{p}$ into the formula in place of p:

> **Margin of Error**
>
> If you have a random sample of size n from a population with proportion of successes p, then you can be 95% confident that the difference between p and $\hat{p}$ will be
> $2\sqrt{\dfrac{\hat{p}(1-\hat{p})}{n}}$ or less. The value, $2\sqrt{\dfrac{\hat{p}(1-\hat{p})}{n}}$, is called the **margin of error**.

a. To see if the substitution of $\hat{p}$ for p makes much difference, compute both $2\sqrt{\dfrac{p(1-p)}{n}}$ and $2\sqrt{\dfrac{\hat{p}(1-\hat{p})}{n}}$ for the situation of flipping a fair coin 100 times and getting 47 heads.

b. A survey of 727 randomly selected New Yorkers found that 54% thought that stores should not open on Thanksgiving night. The survey gives a margin of error of about 4%. (**Source:** siena.edu/uploadedfiles/home/parents_and_community/ community_page/sri/ independent_research/Hday1112%20Release_Final.pdf)

  i. Verify this margin of error.

  ii. Interpret this margin of error by filling in the blanks in this sentence: If all New Yorkers could have been asked if stores should open on Thanksgiving night, you are 95% confident that the percentage who would have said _____ is no farther than _____ away from the estimate from the sample of _____ .

c. In national polls, major polling organizations typically use a sample size of about 1,200. Suppose that a poll finds that 45% of the 1,200 U.S. adults surveyed approve of the job the President is doing. Compute and interpret the margin of error for this poll.

d. Suppose that Sample X is larger than Sample Y, but the proportion of successes in the samples are the same. How do the margins of error differ? Explain your answer.

e. Suppose that the proportion of successes in Sample P is the same as that in Sample Q. However, the sample size for Sample P is four times as large as that for Sample Q. What is the relationship between their margins of error?

2 The margin of error tells you the farthest apart p and $\hat{p}$ are likely to be. In 95% of all random samples, the difference between p and $\hat{p}$ will be no larger than the margin of error. But that means that in 5% of all random samples, p and $\hat{p}$ are farther apart than the margin of error. Unfortunately, we usually do not know when that is the case. One time that we do is for polls taken right before an election, when the vote on election day tells us how well predictions from polling turned out. In the presidential election of November 2008, Barack Obama got 52.9% of the vote. The table on the next page gives the results from 15 polls taken right before election day. (**Source:** www.realclearpolitics.com/ epolls/2008/president/national.html)

carterdayne/iStockphoto

Name of Poll	Number of Likely Voters Surveyed	Margin of Error	Percent Saying They Would Vote for Obama
Marist	804	4.0	52
Battleground (Lake)	800	3.5	52
Battleground (Tarrance)	800	3.5	50
Rasmussen Reports	3,000	2.0	52
Reuters/C-SPAN/Zogby	1,201	2.9	54
IBD/TIPP	981	3.2	52
FOX News	971	3.0	50
NBC News/Wall St. Journal	1,011	3.1	51
Gallup	2,472	2.0	55
Diageo/Hotline	887	3.3	50
CBS News	714	—	51
ABC News/Wash Post	2,470	2.5	53
Ipsos/McClatchy	760	3.6	53
CNN/Opinion Research	714	3.5	53
Pew Research	2,587	2.0	52

a. Verify the margin of error given for the Battleground (Lake) poll.

b. What was the sampling error for the Battleground (Lake) poll?

c. For the Battleground (Lake) poll, was it the case that the difference between p and $\hat{p}$ was less than or equal to the margin of error?

d. Supply the margin of error for the CBS News poll.

e. For which poll(s) was the difference between p and $\hat{p}$ larger than the margin of error?

f. Suppose that you examine 100 different political polls. What is the expected number of polls where the sampling error is larger than the margin of error?

g. If you examine 15 political polls, what is the expected number where the sampling error is larger than the margin of error? So, is there any reason to be concerned by the result of Part e? Explain.

3 Which two of the following are true statements about the margin of error?

  A. In 95% of all random samples, the absolute value of the difference between p and $\hat{p}$ will be less than or equal to the margin of error.

  B. If you take a random sample, the absolute value of the difference between p and $\hat{p}$ will be less than or equal to the margin of error.

  C. The margin of error tells you exactly how far apart p and $\hat{p}$ are.

  D. The margin of error depends only on the sample size.

  E. If you want a smaller margin of error, you should take a larger random sample.

  F. In a sample survey, the margin of error is equal to the sampling error.

4 Spin a penny 20 times by placing it on edge on the floor and flicking with your finger. Let the penny spin freely until it lands heads up or heads down. Count the number of heads.

a. Do you think that spinning a penny is fair? Explain your reasoning.

b. Combine results with the rest of your class until you have a total of 400 spins. Compute $\hat{p}$, your estimate of the probability a spun penny lands heads up. Test the statistical significance of your result (compared to a fair coin) by computing the P-value.

c. How can you use the margin of error rather than a P-value to help you decide whether you think that spinning a penny is fair?

SUMMARIZE THE MATHEMATICS

In this investigation, you learned how to compute and interpret the margin of error connected with the estimate of the proportion of successes in a population.

a. What does it mean if you are told that the margin of error for a survey is 3%?

b. What is the formula for computing the margin of error? When can you use this formula?

c. What is the difference between sampling error and margin of error?

d. Read the following definition of *margin of error* from a newspaper. Do you think it is a good explanation for the average person? How would you improve it?

> "**Margin of sampling error:** Political polls can never survey all voters. Pollsters use mathematical formulas to determine how much error might be in their results. The more people surveyed, the smaller the margin will be."
>
> (**Source:** Colleen O'Connor, "Science of surveying has margin of error," *Denver Post*, 1/27/2008, www.denverpost.com/nationalpolitics/ci_8088538)

Be prepared to explain your ideas to the class.

The first political poll in the United States was conducted by the *Harrisburg Pennsylvanian* before the 1824 presidential election. The newspaper polled 532 voters in Wilmington, Delaware. Three hundred thirty-five of those polled said they preferred Andrew Jackson.

a. What proportion of those polled preferred Andrew Jackson?

b. What must be true about the method of sampling so that you can compute the margin of error? Is this likely to be the case? Explain your thinking.

c. Regardless of your answer to Part b, compute the margin of error for this poll.

d. What sample size would have cut the margin of error in half? Justify your answer.

INVESTIGATION 3

Interpreting a Confidence Interval

Sometimes news stories, especially those in the medical field, will report a *confidence interval* rather than a margin of error. For example, confidence intervals were reported for a research study that followed 770 people who were newly diagnosed with type II diabetes and had no diabetic retinopathy (damage to the eye caused by diabetes). After six years, 90 of them had diabetic retinopathy. A 95% confidence interval for the proportion of all such diabetics with diabetic retinopathy can be given in this form: (9.4, 14.0)%.

(**Source:** Niels de Find Olivarius et al. "Prevalence and progression of visual impairment in patients newly diagnosed with clinical type 2 diabetes: a 6-year follow up study." *BMC Public Health* 4 February 2011; www.biomedcentral.com/1471-2458/11/80)

As you work on the problems in this investigation, look for answers to this question:

How is a 95% confidence interval computed and how should it be interpreted?

1 Refer to the study about diabetic retinopathy above.

    a. Compute $\hat{p}$, the point estimate, for the proportion of all newly diagnosed type II diabetics who would develop diabetic retinopathy within six years. Write a newspaper headline using this point estimate.

    b. Compute the margin of error.

    c. Examine the 95% confidence interval, (9.4, 14.0)%. Tell how each of the numbers was computed.

d. The endpoints of a **95% confidence interval** (sometimes abbreviated **95% CI**) are given by this formula:

$$\hat{p} \pm 2\sqrt{\frac{\hat{p}(1 - \hat{p})}{n}}$$

   i. What is the name for the first term of this formula? What is the name for the last term of this formula?

   ii. How would you write the 95% CI using interval notation?

e. Here is how a 95% confidence interval should be interpreted.

> **Interpretation of a 95% Confidence Interval**
>
> With random sampling, we are 95% confident that the proportion *p* of successes in the population lies within the confidence interval.

Use the sentence above as a guide to interpret the confidence interval about diabetic retinopathy in the context of the situation.

2 Answer the following questions for each situation below.

   Question 1: What is the population?

   Question 2: What parameter of the population is the survey trying to estimate?

   Question 3: What is the point estimate of that parameter?

   Question 4: What is the 95% confidence interval for that parameter?

   Question 5: What is the interpretation, in context, of this interval?

a. A Gallup poll found that 96% of Americans had a favorable view of Canada; more than for any other country. According to Gallup, the poll was conducted using telephone interviews with a random sample of 1,029 adults living in the United States.

   (**Source:** www.gallup.com/poll/152735/americans-give-record-high-ratings-several-allies.aspx)

b. A survey of the parents of American teenagers who use the Internet found that two-thirds of the parents said they were active on social networking sites themselves. This poll had a margin of error of 4.5%.

   (**Source:** www.pewinternet.org/Reports/2012/Teens-and-Privacy.aspx)

3 To estimate the prevalence of alcohol consumption during pregnancy, a stratified random sample of 12,611 mothers from Maryland who delivered live infants during the years 2001–2008 filled out a questionnaire. The results included this statement, "Nearly 8% (95% confidence interval 7.1–8.4) of mothers from Maryland reported alcohol consumption during the last 3 months of pregnancy." (**Source:** Cheng, Diana; Kettinger, Laurie; Uduhiri, Kelechi; Hurt, Lee, "Alcohol Consumption During Pregnancy: Prevalence and Provider Assessment." *Obstetrics & Gynecology*. 117 (February 2011) pages 212–217.)

a. The actual percentage from the sample was 7.75%. Use this percentage to compute a 95% confidence interval for this situation.

b. Large surveys like this one often use stratified random samples. The formula for a confidence interval for a stratified random sample is a bit different from the formula you are using. Thus, your computation may not always exactly match that reported in the media, but it should be close. Is there much difference between the confidence interval reported in the research article and the one you computed in Part a?

4 You can say that you are 95% confident that the true population proportion lies in your 95% confidence interval. But what, exactly, does the phrase "95% confident" mean? Suppose you use the method of constructing the confidence interval with many different random samples. Then, for some samples, the true population proportion will be inside the confidence interval and for some samples it will not. But, overall, you expect the interval to "capture" the true proportion of successes 95% of the time. The following activity illustrates exactly what this means.

a. On your own, use a calculator or software such as *TCMS-Tools* Simulation software to generate 50 random digits. Count the number of digits that are even. (Remember that 0 is an even digit.) Check your count by also counting the number that are odd.

b. Using your random sample, find a 95% confidence interval for the percentage of all random digits that are odd.

c. Draw a heavy line segment to represent your confidence interval along one of the vertical lines in a copy of the chart below. For example, suppose you are the fifth student to add your confidence interval and you get 24 even digits in your sample of 50, for a sample proportion of 0.48. Then your 95% confidence interval is 0.48 ± 0.14, or (0.34, 0.62). You would draw a line segment from 0.34 to 0.62 along the vertical line above the 5 on the "Student" axis.

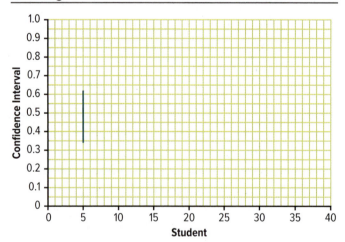

95% Confidence Intervals for the Proportion of Even Digits

d. Are all intervals from the students in your class exactly the same? Should they be?

e. What is the true proportion of random digits that are even? What proportion of the intervals from your class "captured" this true proportion? Is this about what you would expect?

SUMMARIZE THE MATHEMATICS

In this investigation, you learned how to compute and interpret a 95% confidence interval.

a. What is the relationship between a margin of error and a 95% confidence interval?

b. Is $\hat{p}$ always in the confidence interval? Is p always in the confidence interval? Explain.

c. Explain the meaning of "95% confident."

d. Why is it better for pollsters to report a confidence interval (or margin of error) rather than giving just a point estimate?

Be prepared to share your ideas and reasoning with the class.

✓ CHECK YOUR UNDERSTANDING

Suppose you have designed a map that describes a walking tour of your neighborhood. To determine whether it is worth marketing, you would like to know what percentage of people in your neighborhood would be interested in buying such a map.

So, you survey 100 residents of your neighborhood, selected at random. Thirty-four people say they would be interested in buying such a map.

a. What is your population?

b. What percentage are you trying to estimate?

c. What is your point estimate of that percentage?

d. What is the 95% confidence interval for that percentage?

e. Give an interpretation of your 95% confidence interval in the context of this situation.

f. Explain what you mean when you say that you are "95% confident."

1 In the United States, 27% of households with mortgages owe more than their house is worth. This is called being "underwater." Jesse does not know this percentage and picks a random sample of 300 U.S. households with mortgages to estimate the percentage underwater. (**Source:** money.cnn.com/2011/02/09/real_estate/ underwater_mortgages_rising/)

a. What is the value of n? What is the value of p?

b. In Jesse's random sample of 300 households, 92 are underwater. What is Jesse's point estimate $\hat{p}$ of the proportion of all U.S. households that are underwater? What is his sampling error?

c. The dot plot below shows the values of $\hat{p}$ from 1,000 different random samples, each of 300 U.S. households. What is the name for this type of dot plot? Locate Jesse's result on the plot.

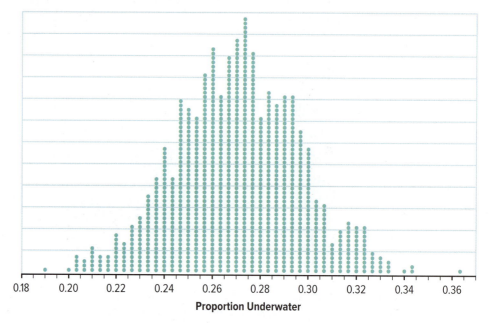

Proportion Underwater

d. From the plot, estimate the proportion of random samples that have a sampling error of 0.05 or less.

e. Compute the value of $2\sqrt{\dfrac{p(1-p)}{n}}$ for this situation. Write a sentence explaining what the value means in the context of this situation.

2 Over a six-month period, only twelve percent of Alaska Airlines flights did not arrive on time. Jasmine does not know this percentage and so takes a random sample of 100 departing flights to try to estimate it. (**Source:** www.rita.dot.gov/bts/sites/rita. dot.gov.bts/files/press_releases/2012/dot089_12/html/dot089_12.html)

a. What is the value of n? What is the value of p?

b. In Jasmine's random sample of 100 flights, 20 were late. What is Jasmine's point estimate $\hat{p}$ of the proportion of all Alaska Airlines flights that are late? What is her sampling error?

c. The approximate sampling distribution below shows the values of $\hat{p}$ from 1,000 different random samples, each of 100 departing flights. From the plot, estimate the proportion of times that a random sample of size 100 will give a proportion of flights that are late as high or even higher than Jasmine got in her sample.

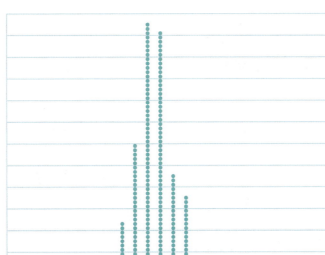

Proportion Late

d. From the plot, estimate the proportion of random samples that have a sampling error of 0.05 or less.

e. Compute the value of $2\sqrt{\dfrac{p(1-p)}{n}}$ for this situation. Write a sentence explaining what the value means in this context.

3 According to the Australian paper, the *Daily Telegraph*, "A Galaxy Research survey of more than 1,200 people aged 18–39 has found 47 per cent have chosen to end it with their partner on or around Valentine's Day after taking stock of their relationship." (**Source:** www.dailytelegraph.com.au/lifestyle/valentines-day-a-popular-time-to-break-up/story-e6frf00i-1226005579514)

a. How many people in the sample of 1,200 people have broken up with someone around Valentine's Day?

b. What is the parameter that Galaxy Research is trying to estimate?

c. What must be true about the method of sampling so that you can compute the margin of error?

d. Assuming that the method in Part c was used, compute the margin of error for this poll.

e. Write an interpretation of the margin of error, in context.

4 Suppose you place a penny on the ground at your school and observe whether the next student to walk by picks it up. After doing this 75 times, you find that 30 students picked up the penny. Assume that the 75 students in your sample can be considered a random sample of students from your school.

a. What parameter are you trying to estimate?

b. What is your point estimate of the proportion of all students at your school who would pick up the penny?

c. Compute the margin of error for this survey.

d. Write an interpretation of the margin of error, in context.

5 A survey of the ethics of high school students reported that 11,781 out of 22,912 students said they had cheated on one or more tests in the last year. (Yet, 96% said that it was important for people to trust them!) (**Source:** charactercounts.org/pdf/reportcard/2012/ReportCard-2012-DataTables-HonestyIntegrityCheating.pdf)

a. What parameter about cheating was the survey trying to estimate?

b. What is the point estimate of that parameter?

c. The researchers did not select the sample randomly, but attempted to acquire responses from a wide variety of high schools. Nevertheless, compute a margin of error.

d. The survey reported, "The survey findings have an error margin of plus or minus less than one percent" Do you agree with their margin of error? Write a more complete sentence reporting the margin of error.

6 At the beginning of this lesson (page 399), a Gallup poll is reported that asked Americans if they "think nuclear power plants in the United States are safe or not safe." Fifty-seven percent of the 1,024 surveyed said, "safe."

   a. What parameter is Gallup trying to estimate?

   b. The margin of error is given as ±4%. Is this close to that given by your formula?

   c. Explain what is meant by "sampling error."

   d. If Gallup had reported a 95% confidence interval instead of the margin of error for this situation, what would that interval be?

   e. Interpret this interval in the context of this situation. Then, explain what is meant by "95% confidence."

7 A study was conducted of children being cared for by a relative (called *kinship care*) after being removed from their family because of maltreatment. Of the 572 children in kinship care, 173 had behavioral problems. These children were not randomly selected, but the sample was selected to be "nationally representative." (**Source:** www.medpagetoday.com/Pediatrics/GeneralPediatrics/24743 based on Sakai S, *et al.* "Health outcomes and family services." *Arch Pediatr Adolesc Med* 2011; 165(2): 159–165.)

   a. What is the population here?

   b. What parameter were the researchers trying to estimate? What is the point estimate of that parameter?

   c. What is the 95% confidence interval for that parameter? What assumption must you make?

   d. What is the interpretation, in context, of this interval?

CONNECTIONS

8 The *approximate* sampling distributions in Investigation 1 were made using *simulation*. You also can make *exact* sampling distributions using the binomial probability formula developed in Lesson 1. In this task, you will construct an exact sampling distribution for the situation of flipping a fair coin 20 times and computing the proportion of times it lands heads. Before beginning, scan the parts of this task to help you make strategic decisions about technology tools that may be helpful in your work.

   a. Use the binomial probability formula to compute the probability that the proportion of heads will be 0.6.

Steve Allen/Brand X Pictures

b. Use the **binompdf(*n,p,x*)** function to compute probabilities for the remaining possible proportions of heads.

c. Make a plot with *proportion of heads* on the horizontal axis and *probability* on the vertical axis.

d. Suppose you flip a coin 20 times and get 14 heads. What is the sampling error in your estimate of the probability that a coin will land heads? Use your exact sampling distribution to find the probability of getting a sampling error no larger than that.

9 Examine the formula for the margin of error, $E = 2\sqrt{\dfrac{\hat{p}(1-\hat{p})}{n}}$.

a. Suppose you thought about taking a sample of size 100, but find that the margin of error is three times as large as you would like. What sample size should you use? Justify your answer.

b. Suppose you thought about taking a sample of size 100, but find that the margin of error is four times as large as you would like. What sample size should you use? Justify your answer.

c. In general, by what factor should you increase the sample size if you want to cut your margin of error by half? By $\dfrac{1}{n}$, where n is an integer greater than 1?

10 Suppose that you expect to get a sample proportion around 0.5 so that the margin of error E is

$$E = 2\sqrt{\dfrac{\hat{p}(1-\hat{p})}{n}} = 2\sqrt{\dfrac{0.5(1-0.5)}{n}}.$$

a. Simplify the expression on the right.

b. Make a graph of the function with sample size n, from 1 to 1,000, on the horizontal axis and margin of error on the vertical axis.

c. Describe how the margin of error changes as the sample size increases.

11 Suppose that you have a sample size of $n = 100$.

a. Consider $E = 2\sqrt{\dfrac{\hat{p}(1-\hat{p})}{100}}$, where E is the margin of error. Make a graph of this function, placing $\hat{p}$ on the horizontal axis and E on the vertical axis.

b. Which value of $\hat{p}$ gives the largest margin of error?

c. By how much does the margin of error change with a change in $\hat{p}$ from 0.5 to 0.6? From 0.8 to 0.9?

James Woodson/Getty Images

REFLECTIONS

12 The dot plots in Investigation 1 were created using simulation software. Below is a screen from three simulations of a binomial distribution with $n = 100$ and $p = 0.5$.

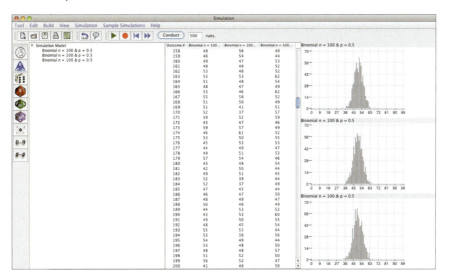

a. How many random samples were generated for each of the three simulations?

b. Explain why the three graphs are not identical.

c. Notice that the labels on the horizontal axis are the number of successes rather than the proportion of successes. How would the shapes of these distributions change if the proportion of successes had been graphed?

d. What are the sampling errors for the 158th sample for each simulation?

13 The following question appeared on the 10th-grade assessment for the state of Wisconsin. (**Source:** oea.dpi.wi.gov/files/oea/pdf/mathrelease10.pdf)

Three candidates are running for mayor in the town of Morganville. The local newspaper conducted two surveys asking potential voters for whom they plan to vote. Survey 1 was taken six months before the election and Survey 2 was taken three months before the election. The results of the two surveys are shown below.

Survey 1

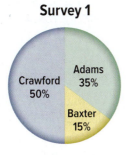

Survey 2

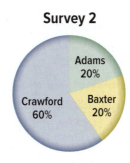

Based on the results of both surveys, which of these statements is true?

Statement 1: Crawford will definitely win the election.

Statement 2: More people will vote for Baxter than Adams in the election.

Statement 3: Crawford is spending the most money of the three candidates.

Statement 4: Baxter gained voter support between Survey 1 and Survey 2.

a. What answer do you think was scored as correct?

b. Why is the question actually impossible to answer? That is, what crucial information is not given that would help you decide whether any of the four statements is necessarily true?

14 A trustworthy sample survey is unbiased (has no systematic tendency to get an estimate that is too large or too small, on average). It also has a small margin of error.

a. How do you design a survey to minimize bias in sampling?

b. How do you design a survey to get more precision?

15 Jeri said that confidence intervals were difficult for her to understand at first. She said it took her a while to realize that what you are 95% confident in is not really a particular interval, but rather a method of generating intervals. Explain what you think she means by this statement.

16 Sometimes confidence intervals (or margins or error) are computed for proportions for two related populations. These then are used to compare the populations.

a. A report on the 2011 mayoral race in Chicago had the following information. Verify that the polls do "overlap." Then explain what is meant by the last sentence of this report.

> **Another Day, Another Debate, Another Poll**
>
> When ABC-7 polled 600 people, Rahm Emanuel stood at 54 percent. Now, the *Chicago Tribune* and WGN have polled 718 people, and it shows Emanuel sitting at 49 percent. Technically, the margin of error means these polls overlap.

Source: www.nbcchicago.com/blogs/ward-room/Race-to-Run-Off-115886274.html

b. The study in Applications Task 7 also examined 736 children placed in foster care, of which 391 had behavioral problems. Compute and interpret a 95% confidence interval for this situation. Are you convinced that the proportion of all children placed in kinship care and the proportion of all children placed in foster care who have behavioral problems are different? Explain.

EXTENSIONS

17 Your calculator will compute confidence intervals when given just the sample size n, the number of successes in the sample x, and the level of confidence desired, which typically is 95%. Find the function **1-PropZInt** in the STAT **TESTS** menu. After entering the number of successes x and the sample size n, enter 0.95 for

a 95% confidence level (**C-Level**), highlight **Calculate** and press ENTER. The calculator uses a factor of 1.96 rather than 2 in the formula for the margin of error, which can make a slight difference in the interval endpoints.

a. Use your calculator to compute a 95% confidence interval for the situation in Applications Task 7 and compare it to the interval computed using the formula for the margin of error.

b. Use your calculator to compute a 95% confidence interval for the situation in Applications Task 6. You must input x the number of successes, as a whole number, so compute that first.

c. How can you use the **1-PropZInt** command to find the margin of error? Use your method to determine the margin of error for the situation in Applications Task 6.

d. Use your calculator to determine the margin of error for this situation: Fifty volunteers were asked by the Massachusetts Public Interest Research Group to phone their credit-card company and ask for a lower interest rate, saying they would switch to another company unless given one. Fifty-six percent of them got a lower rate within 5 minutes. (**Source:** uspirg.org/sites/pirg/files/reports/Deflate_Your_Rate_USPIRG.pdf)

18 Refer to Extensions Task 17 about using your calculator to compute confidence intervals. The confidence intervals you have constructed gave 95% confidence. In some situations people want to have more than 95% confidence that the population proportion is in their interval.

a. If you want to have, say, 99% confidence in your interval, should it be wider or narrower than the 95% confidence interval? Explain your reasoning.

b. A Harris Poll conducted online with 2,016 adult Americans found that 21% said that they had some sort of tattoo. (**Source:** www.harrisinteractive.com/NewsRoom/HarrisPolls/tabid/447/mid/1508/articleId/970/ctl/ReadCustom%20Default/Default.aspx) Compute a 95% confidence interval and a 99% confidence interval for this situation. Compare the widths. Are they consistent with your answer to Part a?

19 The 1948 presidential election is famous because the polling organizations predicted that Dewey would beat Truman. Shortly before the election, *Newsweek* published a poll of 50 of the nation's "leading political writers." All 50 of them predicted a Dewey victory. (**Source:** dailynightly.msnbc.msn.com/archive/2008/01/09/565536.aspx)

a. What population proportion was *Newsweek* trying to estimate?

b. What goes wrong when you try to use the formula for the margin of error?

c. The margin of error rule (page 404) says that "This rule works well only if your sample contains at least 10 successes and at least 10 failures." Is this condition satisfied in the case of the *Newsweek* poll?

20 In Unit 1, *Interpreting Categorical Data*, you learned about relative risk. In the study of kinship versus foster care presented in Applications Task 7 and Reflections Task 16 Part b, of the 572 children in kinship care, 173 had behavioral problems. Of the 736 children in foster care, 391 had behavioral problems.

a. Compute and interpret the relative risk for this situation.

b. Computing a 95% confidence interval for relative risk can be complicated, so a calculator may be found at www.medcalc.org/calc/relative_risk.php, for example. First, compute the relative risk. Note that you must input the number with behavioral problems and the number without for each group (not the sample size *n*). Does the relative risk agree with the one you computed in Part a?

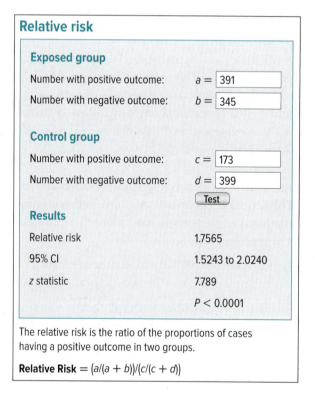

Relative risk

Exposed group

Number with positive outcome:	$a =$	391
Number with negative outcome:	$b =$	345

Control group

Number with positive outcome:	$c =$	173
Number with negative outcome:	$d =$	399

Test

Results

Relative risk	1.7565
95% CI	1.5243 to 2.0240
z statistic	7.789
	$P < 0.0001$

The relative risk is the ratio of the proportions of cases having a positive outcome in two groups.

Relative Risk $= (a/(a + b))/(c/(c + d))$

c. Interpret the confidence interval computed by the online calculator.

d. Compare the relative risk formula used by the online calculator with the representation that you used in Unit 1 (page 9).

REVIEW

21 Write each sum of rational expressions in equivalent form as a single algebraic fraction. Then simplify the result as much as possible.

a. $\dfrac{x}{4} + \dfrac{3x+4}{4}$ **b.** $\dfrac{3x}{5} + \dfrac{2x+1}{2}$ **c.** $\dfrac{4x+5}{x} + 3x$

22 Tonya has a credit card balance of $2,345. Her annual interest rate on any unpaid balance is 21% compounded monthly. Tonya makes a payment of $400 each month.

a. How much interest will be added to Tonya's balance in the first month?

b. What will her balance be after she makes the first payment?

c. Making this payment, and assuming that she does not make any new charges, how long will it take Tonya to pay off the entire credit card balance?

d. How much interest will she have paid by the time she reduces the balance to zero?

23 Solve each inequality and represent the solution using interval notation and a number line.

a. $10 - 8x > 3(6 - 2x)$ **b.** $-12 + 3(x - 5) \leq -9 - 6x$

c. $(x + 5)(x - 6) > 0$ **d.** $-(x + 5)(x - 6) > 0$

24 Suppose that a softball is hit from a height of 1.2 meters with an initial upward velocity of 21 meters per second. Recall that the function rule for the height of a ball is $h(t) = -4.9t^2 + v_0 t + h_0$, where v_0 is the initial upward velocity in meters per second and h_0 is the initial height in meters.

a. Write a function rule that will give the height of the softball after any number of seconds.

b. How high will the ball be after 1 second?

c. At what times will the ball be more than 18 meters above the ground?

d. If the ball is not caught, when will it hit the ground?

25 The table below shows the types of cell phones owned by random samples of U.S. adults who were surveyed in May and December of 2011.

(**Source:** libraries.pewinternet.org/files/2012/04/Topline_for_-e_reading_report_4_surveys.pdf)

	May 2011	December 2011
Smartphone	888	1,404
Basic Cell Phone	1,025	1,194
No Cell Phone	364	388
Total	2,277	2,986

a. For each survey, what percent of the people surveyed owned a smartphone?

b. Perform a chi-square test of homogeneity and state your conclusions.

26 Rewrite each expression as a product of linear factors.

a. $a^2 + 7a + 12$

b. $3x^2 - 13x - 10$

c. $4s^2 + 12s + 9$

d. $t^3 - 8t^2 - 20t$

e. $30x^2 + 3x - 9$

27 Solve each quadratic equation.

a. $x^2 = x + 2$

b. $2x^2 + x - 1 = 0$

c. $2x^2 - 5x = 3$

d. $2x^2 + 2 = 5x$

e. $6x^2 + 5x + 1 = 0$

28 Destiny deposited $1,000 into a savings account that earns 3% interest compounded continuously. Assume that she does not withdraw any money from the account.

a. Write a formula that could be used to find the balance after t years.

b. How much money will be in the account after:

  i. six months?

  ii. one year?

  iii. two years?

c. How much interest will she have earned after 10 years?

d. How long will it take for her money to double?

29 As a fundraiser, the senior class at a high school in Holland, Michigan is considering selling class sweatshirts. They first need to analyze their costs. The printing company will charge $150 to set up the artwork and $8 per sweatshirt.

©Ingram Publishing/Fotosearch

a. Write a formula for the average cost of a sweatshirt if the class purchases n sweatshirts.

b. How many sweatshirts do they need to purchase so that the average cost per sweatshirt is less than $10?

c. Identify any horizontal or vertical asymptotes of the graph of the average cost function. Then explain their meaning in terms of the context of the situation.

30 For each graph, write a function rule that produces the graph.

a.
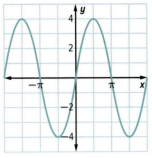

b.
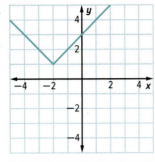

c.
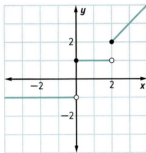

d.
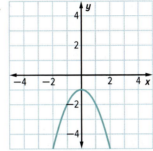

31 Consider the rectangular prism and square pyramid shown below.

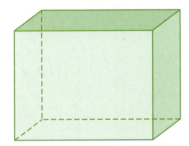
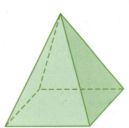

For each shape, complete the following.

a. Without tracing, make a careful sketch of the polyhedron.

b. How many symmetry planes does the shape have? Describe or draw each one.

c. How many axes of symmetry does each shape have? Describe or draw each one and then identify the angles of rotation associated with each axis of symmetry.

d. Imagine a plane that intersects the shape and is parallel to its base. Is the polygon formed by the intersection of the plane and the polyhedron congruent to, similar to, or neither congruent to nor similar to the base of the polyhedron?

Binomial situations often arise in surveys and experiments. In this unit, you have developed and studied statistical methods for making sense of binomial situations. In the first lesson, you reviewed rules of probability, developed the binomial probability formula to compute binomial probabilities exactly, and used the distribution of the proportion of successes to find *P*-values and determine statistical significance. In the second lesson, you learned the characteristics of a well-designed survey and how to identify sources of bias in the method of selecting a sample and in the method of obtaining the response. In the third lesson, you learned what is meant by a margin of error and a confidence interval and how a survey can yield an estimate of the population parameter that is reasonably precise even though only a small portion of the population was in the sample. The tasks in this final lesson will help you review and synthesize these important ideas.

1 In the United States, 8.6% of high school students (in grades 9–12) attend a church-related school. (**Source:** www.census.gov/compendia/statab/2011/tables/11s0234.pdf)

a. Suppose that you pick 2 high school students at random. Demonstrate how to use the Multiplication Rule to compute the probability that neither of them attends a church-related school.

b. Suppose that you select 10 high school students at random. Use the binomial probability formula to compute the probability that exactly 2 of them attend a church-related school.

c. To see if the percentage is larger in the northeastern United States, Carlos takes a random sample of 200 high school students from just the northeast. Twenty-seven of them attend a church-related school. What is the probability of getting 27 or more students who attend a church-related school out of a sample of 200 students if, overall, 8.6% of students in the northeast attend a church-related school? What is the name for the probability you just computed? Is the result from the sample statistically significant? Explain what is meant by "statistically significant."

2 A Gallup poll estimated that the unemployment rate in the U.S. was 9.8%. Read the following excerpts from their explanation of how the poll was conducted.

Survey Methods

Results are based on telephone interviews conducted as part of Gallup Daily tracking Jan. 2 to 31, 2011, with a random sample of 18,778 adults, aged 18 and older, living in all 50 U.S. states and the District of Columbia, selected using random-digit-dial sampling.

For results based on the total sample of national adults, one can say with 95% confidence that the maximum margin of sampling error is ±0.7 percentage points.

In addition to sampling error, question wording and practical difficulties in conducting surveys can introduce error or bias into the findings of public opinion polls.

Source: www.gallup.com/poll/146108/Least-Educated-Likely-Find-Jobs-2010.aspx

a. Describe the parameter being estimated. What is the estimate?

b. The Gallup poll tries to make sure that all regions of the country are proportionally represented. At the time of the survey, about 37% of the population of the United States lived in the South. How many people in the South should have been in the sample?

c. Refer to the questions in Problem 3 of Lesson 2, Investigation 1 (page 378). Answer the questions that can be answered based on the information in the article.

d. Is there any indication that the method of conducting this poll might result in an untrustworthy estimate? What else might you want to know in order to decide?

e. What margin of error is reported? What is the 95% confidence interval?

f. Use the formula to compute the margin of error. Does your computation match the margin of error reported?

g. Can you compute the sampling error in this case? If so, do it. If not, explain why not. If there is sampling error, does that mean that a mistake has been made in the survey?

3 A recent survey of 1,033 American adults found that 53 percent of those polled said that chocolate chip cookies are their favorite kind of cookie. The survey was conducted by a reputable firm, so you may assume the sample was selected at random.

(**Source:** Matthew Reynolds, "Chocolate Chip Remains Cookie King" *Modern Baking*, Vol. 23, No. 13 (October 2009))

a. What is the population?

b. What percentage was the survey trying to estimate?

c. What is their point estimate of that percentage?

d. Compute both the margin of error and the 95% confidence interval.

e. Give an interpretation of the 95% confidence interval in the context of this situation.

f. Why is it important that the sample be selected at random?

g. Suppose that the true percentage of all American adults who would say that chocolate chip cookies are their favorite is 52%. What is the sampling error in this case?

SUMMARIZE THE MATHEMATICS

Many questions about probability can be resolved using the techniques you have learned for analyzing binomial situations.

a. What are the Multiplication Rule for Independent Events and the Addition Rule for Mutually Exclusive Events? When can each be used?

b. Describe how to compute the probability of getting x or fewer successes in a binomial situation with n trials and probability of success p using:

  i. the binomial probability formula.

  ii. a binomial probability function on your calculator or statistical software.

c. Describe the reasoning behind the binomial probability formula.

d. How do you compute a P-value? How can you tell whether a result is statistically significant? What does it mean if a result is statistically significant?

e. What are the characteristics of a trustworthy survey?

f. What is bias in sampling? What is response bias?

g. What is the sampling error and how is it different from the margin of error? Describe the method of computing each of them.

h. What is it that you are 95% confident is in the confidence interval? What is the meaning of "95% confident"?

Be prepared to share your ideas and reasoning with the class.

✔ CHECK YOUR UNDERSTANDING

Write, in outline form, a summary of the important statistical concepts and methods developed in this unit. Organize your summary so that it can be used as a quick reference in future units and courses.

Informatics

Digital Resources at
ConnectED.mcgraw-hill.com

Watch Tools Audio eBook

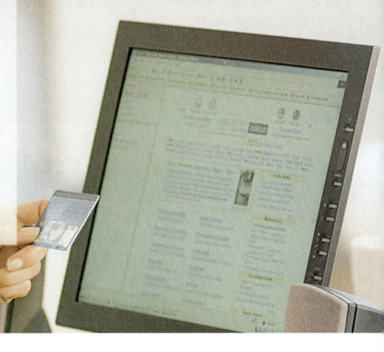

Have you ever wondered how an Internet search engine finds what you are looking for? How a credit card number is sent securely to the online music store where you buy your favorite songs? How the grocery store checkout scanner accurately reads the price and sends product information to the store's inventory computer? How TV shows and other videos are compressed so that many can fit on your portable media player?

Informatics has several different meanings, all related to links between computers and information. In this unit, you will learn some fundamental mathematics of information processing and the Internet that will enable you to answer questions like those above.

In the four lessons of this unit, you will learn key ideas related to issues of access, security, accuracy, and efficiency.

LESSONS

1 Access: Set Theory, Logic, and Searching

Use concepts of set theory and logic, especially in the context of Internet searching, including set representations, set operations, and logical operations.

2 Security: Cryptography

Use concepts and methods of cryptography, specifically symmetric-key cryptography, including the ROT13 substitution cipher; public-key cryptography, including the RSA cryptosystem; digital signatures; and the associated ideas of modular arithmetic.

3 Accuracy: Error-Detecting and -Correcting Codes

Use concepts and methods for detecting and correcting errors when information is electronically read, recorded, or transmitted, specifically error-detecting codes used for ID numbers and error-correcting codes used in data transmission, including the associated ideas of modular arithmetic.

4 Efficiency: Data Compression

Use concepts and methods of data compression, including block codes, variable-length codes, prefix-free codes, and Huffman codes.

Access: Set Theory, Logic, and Searching

The Internet is pervasive in modern life. It has reshaped how companies do business and how people communicate and shop. In this unit, you will learn about the mathematics of information processing and the Internet. You will consider four fundamental issues: access, security, accuracy, and efficiency. Each lesson focuses on one of these issues.

The unit begins with a focus on important mathematical ideas that are useful in accessing information on the Internet. For example, search engines like Google, Bing, and Yahoo! are designed to help you quickly find what you are looking for. Mathematics is at the basis of how these search engines work.

THINK ABOUT THIS SITUATION

Suppose you are looking for some specific old blues music, but you cannot remember the artist's name. You think that part of the singer's name is Holiday.

a. Would you try searching for the music using the search word **Holiday**? Explain your thinking.

b. If you use two or more words in an Internet search, how do you think the search engine processes the words?

c. Suppose you find the blues singer you are looking for and decide to buy a song. How do you think issues of access, security, accuracy, and efficiency are involved in the process of finding the information, buying the song online, and downloading it to your computer?

d. Suppose you want to email your friend a video of the singer, along with a private message. How might issues of access, security, accuracy, and efficiency be important in this process?

In this lesson, you will consider the issue of *access* as it relates to the mathematics of information processing and the Internet. In particular, you will learn some basic ideas of set theory and logic, with applications to Internet searching.

INVESTIGATION 1

Intersection, Union, and Set Difference

There is an immense amount of information on the Internet—data, documents, blogs, images, videos, and so on. Out of all this information, how do you find exactly what you are looking for? **Set theory**, the study of sets and their properties, can be helpful in Internet searches. Roughly speaking, a *set* is a collection of objects. There are many words used to describe collections of objects. We talk about families of people, flocks of birds, forests of trees, bands of musicians, and fleets of ships. We use the word **set** in mathematics to mean a well-defined unordered collection of objects. While a precise treatment of sets requires setting up a system of axioms (assumptions), this informal definition is sufficient for now.

As you complete the problems in this investigation, look for answers to these questions:

How can sets be represented?

How can you combine and operate on sets?

How are operations on sets related to Boolean (or logical) operators?

How are sets and logic used to make Internet searches more efficient?

Representing Sets The individual objects in a set are called its **elements**. Sets are represented by listing or describing the elements within a pair of curly brackets. It does not matter in which order the elements are listed, and the elements can be any "objects." For example, all of the following are sets:

$N = \{2, 3, 5, 7, 11, 13\}$
$P = \{5, \text{exponent, cube, dog}\}$
$S = \{\text{triangle, square, point, prism, icosahedron}\}$
$W = \{\text{www.google.com, www.wmich.edu/tcms/, mathworld.wolfram.com}\}$

1 If an element x is in a set A, you can write $x \in A$, which is read "x is an element of set A" or "x is in A." If x is *not* an element of the set B, write $x \notin B$. Two sets are said to be **equal sets** if and only if they have the same elements.

   a. How many elements are in each of sets N, P, S, and W above?

   b. Is $\{2, 3, 4\} = \{3, 4, 2\}$? Explain.

   c. Give an example of a set A that has the following properties: $5 \in A$, $10 \notin A$, and A has five elements.

2 The elements in a set often share a common property. In such cases, you can define the set in terms of that property. There is specific notation, called *set-builder notation*, that is used to do this. For example, the set of all positive numbers can be written as on the right.

$$\{x \mid x > 0\}$$

the set of · all x · such that · x is greater than zero

   a. Suppose a set A is described in words as follows:

   A is the set containing all numbers x such that x is greater than 2 and less than 10.

   Write A in set-builder notation. (Use the diagram above as a model. The vertical line segment | is read "such that.")

   b. List the elements in each of the sets below.

   i. $S = \{x \mid x \text{ is an integer and } -2 \leq x \leq 5\}$

   ii. $P = \{x \mid x = 2^n, \text{ for } n = 0, 1, 2, \text{ or } 3\}$

   iii. $M = \{x \mid x \text{ is the name of a month that has exactly 30 days}\}$

c. Finish writing each of the following sets using set-builder notation.

    i. $S = \{1, 4, 9, 16, 25\} = \{x \mid x = n^2, \text{ for } n = \ldots\}$

    ii. $E = \{2, 4, 6, 8, 10, \ldots\} = \{x \mid \ldots\}$

    iii. $P = \{\text{triangle, quadrilateral, pentagon, hexagon}\} = \{x \mid \ldots\}$

Boolean (or Logical) Operators When you do an Internet search, you are creating a set—the collection of those Web sites that have the feature defined by your search phrase. For example, at the beginning of this lesson you considered a situation in which you are looking for some blues music by a singer whose name is Holiday.

    A search for **blues** might yield the results shown below—a set of about 223 million Web sites, many of which have nothing to do with blues music.

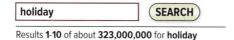

Results **1-10** of about **223,000,000** for **blues**

    A search for **Holiday** may yield even more sites, again with many that do not contain anything related to your desired information.

| holiday | **SEARCH** |

Results **1-10** of about **323,000,000** for **holiday**

    Set theory and logic can be used to narrow an Internet search to a manageable number of sites that are more focused on what you are looking for. For example, the search phrases **blues music** or **blues music holiday** produce fewer and more focused results.

| blues music | **SEARCH** |

Results **1-10** of about **15,400,000** for **blues music**

| blues music holiday | **SEARCH** |

Results **1-10** of about **867,000** for **blues music holiday**

    Search phrases with more than one word are carried out using what are called *Boolean operators* or *logical operators*. Some of these operators are AND, OR, and NOT. In the next problems, you will learn how these operators work.

3 Many Web sites have a "help" page or an "advanced search" page that will help you search more efficiently. For example, consider the help page below adapted from the Library of Congress Web site in 2014.

(**Source:** catalog.loc.gov/help/boolean.htm)

Library of Congress Online Catalog

Boolean Searching

Use Boolean operators and nesting to search for combinations of words or phrases. Enter Boolean operators (i.e., AND, OR, NOT) in either uppercase or lowercase.

Concept	Search Examples	Retrieval Formula
AND	rodgers **AND** hammerstein children **AND** poverty "civil war" **AND** virginia	Retrieves only records containing both terms.
OR	sixties **OR** 60s **OR** 1960s labor **OR** labour email **OR** e-mail **OR** "electronic mail"	Retrieves records containing either one or more terms.
NOT	caribbean **NOT** cuba jockey **NOT** disc "civil war" **NOT** american	Excludes records containing the second term.
NESTING	fruit **AND** (banana **OR** apple) (women **OR** woman) **AND** basketball ((color **OR** colour) **AND** (decorate **OR** decoration)) **NOT** (art **OR** architecture)	Use parentheses ( ) to group portions of Boolean queries for more complex searches.

Based on the information given in the table:

a. describe in words the effect of the operator AND.

b. describe the effect of OR.

c. describe the effect of NOT.

(You will investigate the idea of NESTING later, in Problem 12.)

4 Diagrams like those shown in the Library of Congress table in Problem 3 are called **Venn diagrams** (named for the 19th-century British mathematician John Venn). They are useful in many modern contexts. For example, the social networking site Google+ uses "circles," which are related to the idea of Venn diagrams.

 Circles

A Venn diagram might have three intersecting circles. For example, below is a Venn diagram that represents the search phrase **poverty AND children AND Ecuador**.

a. Describe in words the search results that are in the shaded region of the Venn diagram above.

b. Suppose you have friends in three social groups (three "circles")—math class, Saturday night at the movies, and basketball. Maria and Darius are in each and every one of the three groups. Draw and shade a Venn diagram that represents this situation and these particular friends.

5 In the two-word Google search for blues music, there is no Boolean operator shown. However, one of the operators is in fact being used. Google's Advanced Search Tips states that, "Google's default behavior is to consider all the words in a search. If you want to specifically allow either one of several words, you can use the OR operator…The AND operator, by the way, is the default, so it is not needed." (**Source:** Advanced Search Tips, Google, July 19 2011.)

a. Using Google's policy, draw a Venn diagram representing the search for **blues music**.

b. How does this search differ from a search for **blues OR music**? Draw a Venn diagram for **blues OR music**.

c. Why do you think Google has this policy about the use of the AND operator?

Operations on Sets—Intersection, Union, Set Difference You can think of the circles in a Venn diagram as representing sets. This leads to thinking about *set operations* that correspond to the Boolean operators.

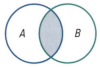 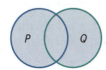 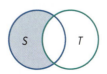

6 AND ↔ Intersection The operator AND corresponds to a set operation called *intersection*. The **intersection** of sets A and B is the set containing all elements that are in *both A and B*. This is denoted $A \cap B$ and is read "*A* intersect *B*." Thus,

$$A \cap B = \{x \mid x \in A \text{ and also } x \in B\}.$$

Using sets and set intersection you can represent a search using the phrase rodgers **AND** hammerstein as follows:

R = $\{x \mid x$ is a Web page containing the word rodgers$\}$

H = $\{x \mid x$ is a Web page containing the word hammerstein$\}$

$R \cap H = \{x \mid x$ is a Web page containing *both* the words rodgers *and* hammerstein$\}$

Suppose you want to do an Internet search for colleges in Boston.

a. Represent this search using set notation.

b. Represent this search using a Venn diagram.

7 Suppose that $C = \{1, 3, 4, 6, 8\}$ and $D = \{2, 3, 4, 8, 10, 12\}$.

a. Find $C \cap D$.

b. Draw and shade a Venn diagram that illustrates this situation. Label the circles and show the elements of the sets in your diagram.

c. Sammi drew the following incorrect Venn diagram for this situation. Explain what is wrong and how to correct it.

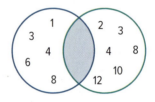

d. Marco correctly described intersection in his own words.

> **I think of intersection as the "common elements."**
> **In a Venn diagram, the intersection is the "overlap" of the circles.**

Explain how Marco's description fits the answers you got in Parts a and b. If his description does not fit your answers, discuss those problems with others.

8 OR ↔ Union The operator OR corresponds to a set operation called *union*. The **union** of sets A and B is the set containing all elements that are in *either A or B or both*. This is denoted $A \cup B$ and is read "*A* union *B*." Thus,

$$A \cup B = \{x \mid x \in A \text{ or } x \in B\}.$$

a. Consider a search for labor **OR** labour.

   i. Represent this search using a Venn diagram.

   ii. Represent this search using set notation.

b. Suppose that you are shopping for a used car. You have entered a used car shopping Web site that has a built-in search engine. You want a used sports car or convertible. Describe the search you would do to find all relevant cars. Describe the search in three ways: using a Boolean expression, using a set operation, and using a Venn diagram.

9 Suppose that $C = \{1, 3, 4, 6, 8\}$ and $D = \{2, 3, 4, 8, 10, 12\}$.

  a. Find $C \cup D$.

  b. Draw and shade a Venn diagram that illustrates this situation. Label the circles and show the elements of the sets in your diagram.

10 You must be very careful using AND and OR. Unfortunately, the mathematical meaning of these words is different than their English meanings.

  a. The mathematical meaning of AND is different than the common English meaning of "and."

   i. Consider what is probably meant in English when someone says, for example, "football team members and baseball team members." Which people are included?

   ii. What is meant when you mathematically state, "football team members AND baseball members"? Which people are included?

  b. Think about the mathematical meaning of OR versus the English meaning of "or."

   i. What is probably meant in English when someone says, "football team members or baseball team members?" Which people are included?

   ii. What is meant when you mathematically state, "football team members OR baseball members"? Which people are included?

11 NOT ↔ Set Difference The operator NOT corresponds to a set operation called *set difference*. The **set difference** of sets A and B is the set containing all elements that are *in A and not in B*. This is denoted $A - B$ and can be read as "A not B." (Another common notation is $A \backslash B$.)

a. Suppose that $X = \{1, 3, 4, 6, 8\}$ and $Y = \{2, 3, 4, 8, 10, 12\}$. Find $X - Y$.

b. Write the definition of $A - B$ using set-builder notation by completing the statement below.

$$A - B = \{x \mid \underline{\hspace{2cm}}\}$$

c. Draw a Venn diagram that illustrates $A - B$.

d. Google uses a minus sign to indicate set difference. According to Google's Advanced Search Tips, "… the query [anti-virus –software] will search for the words 'anti-virus' but exclude references to software." (**Source:** Advanced Search Tips, Google, July 19 2011.)

Consider two searches:

- peanut **NOT** butter

- butter **NOT** peanut

i. Describe how the results of searching for peanut **NOT** butter are different from the results of searching for butter **NOT** peanut.

ii. Draw a Venn diagram that illustrates each search.

iii. Write each search using Google's notation. If you have Internet access, try the searches to see if your description in part i is accurate. Discuss and resolve any differences.

e. Describe how the sets $A - B$ and $B - A$ are related. How is this relationship seen in a Venn diagram of sets A and B?

12 You have been studying operations on sets. In your previous mathematics courses, you studied operations on numbers, for example addition and multiplication. Just as addition and multiplication have certain properties, like the associative property and distributive properties of multiplication over addition (subtraction), so also set operations have similar properties. Consider the following conjectures about set operation properties. (These relate to the NESTING idea shown in the Library of Congress table in Problem 3.)

a. Study Conjecture I and the Venn diagrams below.

Conjecture I: $(A \cup B) \cap C$ = $A \cup (B \cap C)$

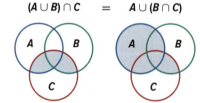

i. Verify that the Venn diagrams correctly represent Conjecture I.

ii. How do these diagrams show that Conjecture I is false?

For Conjectures II–V, use Venn diagrams to show whether the statements are true or false.

Conjecture II: $(A \cup B) \cup C = A \cup (B \cup C)$

Conjecture III: $(A \cap B) \cap C = A \cap (B \cap C)$

Conjecture IV: $(A \cup B) \cap C = (A \cup C) \cap (B \cup C)$

Conjecture V: $(A \cup B) \cap C = (A \cap C) \cup (B \cap C)$

b. For each conjecture that is true, describe the conjecture in words. If you recognize the associative or distributive properties, identify where they occur.

13 A group of friends share an interest in music. Their favorite types of music are rock and blues. They are planning to search for information about these two types of music.

a. Joaquin is particularly interested in crossover artists who play both rock and blues. Think about how to use sets to represent a search in this situation.

   i. Would you use two sets or three sets to represent this situation? Either could be okay. Explain your reasoning.

   ii. Draw a Venn diagram that represents this situation.

b. Bill wants to search for information about either or both of these types of music.

   i. Let R represent the set of all Web pages containing the word **rock**; let B represent the set for **blues**; let M represent the set for **music**. Draw a Venn diagram with the three sets R, B, and M to illustrate Bill's search.

   ii. Use set operations on sets R, B, and M to represent Bill's search.

c. Tionne is a blues purist. She prefers blues music that does not incorporate rock. She does a search for such music on the Internet. Using the three sets R, B, and M from Part b, draw a Venn diagram that represents her search. Also, represent her search using set operations.

SUMMARIZE THE MATHEMATICS

In this investigation, you learned about intersection, union, and set difference and how these relate to the Boolean operators AND, OR, and NOT. You also learned how these ideas are used in Internet searches and on social networking sites.

a. Consider the set of all positive integer factors of 84. Write this set using set-builder notation. Also, write this set by listing the elements in the set.

b. Boolean operators like AND, OR, and NOT are used in many Internet search engines.

    i. Describe the relationship between each of these operators and corresponding set operations.

    ii. For each of the three operators, describe an Internet search that would involve the operator.

c. For arbitrary sets A and B, describe how $A \cap B$ and $A - B$ are related in as many ways as you can.

d. On copies of this Venn diagram, shade the regions that represent each of the following sets.

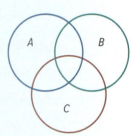

    i. $A \cap (B \cap C)$

    ii. $A \cap (B \cup C)$

    iii. $A \cup (B - C)$

Be prepared to explain your answers to the class.

✔CHECK YOUR UNDERSTANDING

The operations intersection, union, and set difference combine sets to give another set.

a. Describe Internet searches that have not already been discussed that would involve:

    i. the intersection of three sets.

    ii. the operator NOT.

b. Consider the following sets.

$$A = \{2, 4, 6, 8, 3, 17, \pi\} \qquad B = \{1, 3, 5, 7, 2, 8\}$$

    i. Find $A \cap B$, $A \cup B$, $A - B$, and $B - A$. Draw Venn diagrams that illustrate each. Show the elements of the sets in your diagrams.

    ii. Let $E = \{x \mid x \text{ is an even number}\}$. Find $A \cap E$ and $A - E$.

Subsets and Complements

There are other operations and relations involving sets that are important in many situations. As you complete the problems in this investigation, look for answers to these questions:

What is the complement of a set?

What does it mean to be a subset of a set?

What are some properties of complements and subsets?

1 In the last investigation, you studied intersection, union, and set difference. Another useful set operation is *set complement*. The **complement** of a set A is everything in the "universe" that is not in A. This is written A' and is read "A complement" or "not A." In terms of set difference, you can think of A' as "the universe" $- A$. The "universe" is defined by the **universal set**, which is the set of all objects that are under consideration. That is, the universal set defines the overall context.

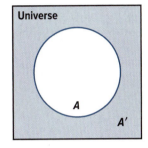

a. Consider the universal set $U = \{5, 10, 15, 20, 25, 30, 35\}$. Suppose $A = \{10, 20, 30\}$ and $B = \{5, 10, 15, 20\}$.

   i. Find A'.

   ii. Find $(A \cap B)'$.

b. Give a general description of A', for any set A, using set-builder notation.

c. Consider the universal set $Z = \{x \mid x \text{ is an integer}\}$. Find the complement of the set $E = \{x \mid x \text{ is an even integer}\}$.

d. You can illustrate the universal set in a Venn diagram by drawing a large rectangle around the circles that represent the sets on which you are operating. Draw Venn diagrams that illustrate the situations in Parts a and c.

2 Use set operations to describe the set indicated by each Venn diagram.

a.
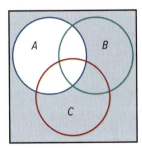

b.
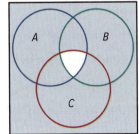

c.
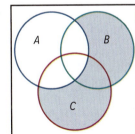

3 Examine the conjectures below relating complements of sets.

   a. Draw Venn diagrams that provide a "proof without words" for whether each conjecture is true or false.

   Conjecture I: $(A \cap B)' = A' \cap B'$

   Conjecture II: $(A \cap B)' = A' \cup B'$

   b. Write and test a conjecture, similar to those above, that begins $(A \cup B)' =$ _____. Compare your conjecture with those of your classmates. Resolve any differences.

   c. For each true conjecture in Parts a and b, write the conjecture in words.

4 A *subset* of a set is, roughly, a set inside the set. Precisely, a set A is a **subset** of set B, denoted $A \subseteq B$, if and only if every element of A is also an element of B. Consider the set $N = \{2, 4, 5, 6, 8, 11\}$.

   a. Is $\{5, 2\}$ a subset of N? Why or why not?

   b. Is $\{4, 6, 8, 10\} \subseteq N$? Why or why not?

   c. Write two other subsets of N.

   d. Is $\{2, 4, 5, 6, 8, 11\}$ a subset of N? Justify your answer by using the definition of subset.

   e. In a search for used Chevrolet trucks on the Internet, think about three possible searching sets:

   $T = \{x \mid x$ is a Web page containing the word truck$\}$

   $U = \{x \mid x$ is a Web page containing the phrase "used truck"$\}$

   $C = \{x \mid x$ is a Web page containing the word Chevrolet$\}$

   Which of these sets is a subset of one of the other sets? Draw a Venn diagram that shows the relationships among these three sets.

5 A subset of a set that does not equal the entire set is called a **proper subset**. To indicate that a set A is a proper subset of a set B, write $A \subset B$. Identify the subsets in Problem 4 that are proper subsets.

6 Now consider relationships between $A \cup B$ and $A \cap B$. Answer each of the questions below, and draw a Venn diagram to illustrate your answer.

   a. Is either $A \cup B$ or $A \cap B$ always a subset of the other?

   b. Under what conditions is one a *proper* subset of the other?

   c. If $A = B$, what can you conclude about $A \cup B$ and $A \cap B$?

   d. If $A \subset B$, what can you conclude about $A \cup B$ and $A \cap B$?

7 The set with no elements is called the **empty set** or **null set**, denoted by $\varnothing$. Thus, $\varnothing = \{\ \}$.

   a. Describe an Internet search whose result would probably be the empty set.

   b. The empty set is considered to be a subset of every set. This may seem a bit strange, but it follows from the definition of a subset given in Problem 4. An argument provided by a student in Legacy High School is shown below. Give reasons for each step in her argument.

   > To prove that $\varnothing$ is a subset of every set, I'll assume the contrary.
   > Assume that there is a set A such that $\varnothing$ is not a subset of A.   (1)
   > There is some element in $\varnothing$ that is not in A.   (2)
   > But this is a contradiction.   (3)
   > Thus, $\varnothing$ must be a subset of every set.   (4)

   c. The above argument is called a *proof by contradiction*. What are the key steps in this proof by contradiction? Explain your reasoning.

8 Two sets that do not share any elements are said to be **disjoint**. That is, sets A and B are disjoint if and only if $A \cap B = \varnothing$. For each pair of sets below:

   - state whether the sets are disjoint or not.

   - draw a Venn diagram that illustrates your answer.

   a. $\{1, 2, 4, 7, 9, 13, 55, 607\}$ and $\{-3, -5, 0, 7\}$

   b. $\{x \mid x$ is an even number$\}$ and $\{x \mid x$ is an odd number$\}$

   c. $\{x \mid x$ is a Web page containing the word mathematics$\}$ and $\{x \mid x$ is a Web page containing the word poetry$\}$

   d. $A - B$ and $B - A$

9 Consider relationships among sets, their complements, and set difference. Answer the questions below. Use Venn diagrams to illustrate your answers.

   a. Is either A or $A - B$ always a subset of the other?

   b. When is $A - B = A$?

   c. When is $A - B = \varnothing$?

   d. Is it true that $A - B = A - (A \cap B)$?

   e. Let U be the universal set. Is it true that $A' = U - A$?

SUMMARIZE THE MATHEMATICS

In this investigation, you learned about sets, set operations, and relationships between sets.

a. Suppose $X = \{1, 2, 3, 4, 5\}$. Which of the following statements are true? Explain.

   i. $3 \in X$ **ii.** $\{3\} \in X$

   iii. $3 \subseteq X$ **iv.** $\{3\} \subseteq X$

b. If P, Q, R, and S are all subsets of a universal set U, what do each of the following expressions tell about the relationships between the sets?

   i. $P \cup R = P$ **ii.** $Q \cap R = R$

   iii. $S \subseteq P'$ **iv.** $S - R = S$

c. Using the ideas of *subset* and *disjoint*, describe the relationship between $A \cap B$ and $A \cup B$. Between $A \cap B$ and $A - B$. Between $A \cup B$ and $A - B$. Draw Venn diagrams to illustrate the relationships.

d. How are the ideas of sets, set operations, and relationships between sets related to making information accessible, especially on the Internet?

Be prepared to share your thinking and Venn diagrams with the class.

✓ CHECK YOUR UNDERSTANDING

Use your understanding of set operations and relationships to complete the following tasks.

a. Use Venn diagrams similar to the one at the right to shade regions representing each of the following sets.

   i. $A \cup B'$

   ii. $A' \cap B$

   iii. $A' \cap B'$

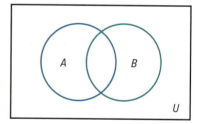

b. Let $A = \{3, 5, 10\}$.

   i. List all subsets of A.

   ii. Which of the subsets that you listed are proper subsets?

   iii. From the subsets that you listed, identify two subsets that are disjoint.

APPLICATIONS

1 Consider some Internet searches using Boolean operators AND, NOT, and OR.

a. Use Boolean operators to design a search that will find information about Irish dancing. Represent the search using sets and set operations. Illustrate the situation with a Venn diagram.

b. Use Boolean operators to design a search that will find information about Irish or English dancing. Represent the search using sets and set operations. Illustrate the situation with a Venn diagram.

c. Describe an Internet search situation that involves the Boolean operator NOT.

2 Consider the following sets:

$$J = \{2, 4, 6, 8, 10\}$$
$$K = \{1, 2, 3, 4, 5, 6\}$$

a. Find $J \cap K$, $J \cup K$, $K - J$, and $J - K$. Draw Venn diagrams that illustrate these four set operations.

b. Write sets J and K using set-builder notation.

3 Examine each conjecture below involving operations on sets A and B.

a. Explain or illustrate why the conjecture is true or false. For example, you might draw Venn diagrams to give a "proof without words" for whether it is true or false.

Conjecture I: $A - B = A \cap B'$

Conjecture II: $A - B' = A \cap B$

b. For each conjecture that is true, describe the statement in words.

4 The inventory system of a used car superstore assigns each vehicle an identification code. Information about each car is included in its code number so a customer or salesperson can search the database to find cars with particular combinations of features.

Suppose that a customer visits the superstore's Web site with preferences for three features in a car: Green color G; Low mileage L; and Two doors T. What types of cars will be found by searches for each of the following?

a. $G \cap (L \cap T)$ **b.** $G \cap (L \cup T)$

c. $L' \cup T'$ **d.** $(L \cup T)'$

5 Suppose that in a probability experiment, you flip a fair coin and then roll a fair die. You record the event of heads or tails on the coin flip and number showing on top of the die.

a. Draw a tree diagram showing all possible outcomes of this two-stage experiment.

b. List the possible outcomes of the experiment as a set S with elements labeled with letter-number pairs like h1 (heads and 1) or t6 (tails and 6).

c. List the elements in the sets $H = \{x \in S \mid x$ shows an outcome with heads$\}$ and $E = \{x \in S \mid x$ shows an outcome with even-number die toss$\}$.

d. Which of the following statements are correct? Explain.

  I $t5 \in E$ **II** $(H \cap E) \subseteq E$ **III** $\{h3\} \subseteq E'$

e. List the elements in the following sets.

  i. $H \cup E$ **ii.** $H \cap E$ **iii.** E'

  iv. $(H \cap E)'$ **v.** $H - E$ **vi.** $H \cup \varnothing$

  vii. $E \cap S$

6 As a project, a senior at Fairfield High School conducted a survey about political opinions in Fairfield, Iowa, by asking 100 randomly selected adults whether they favored (1) public funding of private schools, (2) government control of health care costs, (3) increase in nonmilitary aid to developing countries around the world. Results of the survey showed:

- only 5 people favored all three actions.
- 10 people opposed all three actions.
- 15 favored only the increase in nonmilitary aid to developing countries.
- 13 favored both public funding for private schools and health care cost controls.
- 20 favored health care cost control and aid to developing countries.
- a total of 48 favored health care cost control.
- 25 favored only the public funding for private school action.

Use a Venn diagram similar to the one at the right to model this situation and to help answer the following questions.

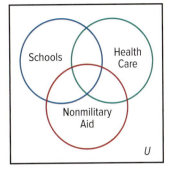

a. In the Fairfield sample, how many people favored at least one of the proposed actions?

b. How many people favored at least two of the proposed actions?

c. How many people favored the health care cost control or aid to developing countries but not the public funding for private schools?

d. Use set notation to describe the set of people in each of Parts a–c, using S for the private school issue, H for the healthcare issue, and A for the foreign aid issue.

CONNECTIONS

7 Set notation and operations are useful in describing geometric figures in a plane. In the diagram below, point X is said to be **between** points A and B since points A and B are collinear and $AX + XB = AB$. This fact is denoted $A-X-B$.

a. Segment AB, denoted $\overline{AB}$, can be described as follows:

$$\overline{AB} = \{A, B\} \cup \{X \mid A-X-B\}$$

Explain why this notation makes sense.

b. Use set notation and operations to describe ray PQ, denoted $\overrightarrow{PQ}$.

c. Use set operations to describe line PQ, denoted $\overleftrightarrow{PQ}$.

d. Use set operations to describe $\angle ABC$, where A, B, and C are noncollinear points.

e. Use set operations to describe $\triangle RST$, where R, S, and T are noncollinear points.

8 Recall that a point in a coordinate plane is represented as an *ordered pair* (x, y) of real numbers. You can think about ordered pairs as elements of a set called the *Cartesian product*. The **Cartesian product** of two sets A and B is the set of all ordered pairs that have first coordinate from A and second coordinate from B. The Cartesian product of A and B is denoted by $A \times B$, which is read as "the Cartesian product of A and B" or "A cross B."

$$A \times B = \{(x, y) \mid x \in A \text{ and } y \in B\}$$

a. Let $A = \{2, 3, 4\}$ and $B = \{50, 60\}$.

   i. Is $(2, 3) \in A \times B$? What about $(2, 50)$? What about $(60, 3)$?

ii. The Cartesian product of A and B is partially written below. Complete the Cartesian product.

$$A \times B = \{(2, 50), (2, 60), (3, 50), \ldots\}$$

b. Let $S = \{y, z\}$ and $T = \{1, 2, 3, 4\}$. Find $S \times T$.

c. How many elements are in $A \times B$ if A has m elements and B has n elements? Explain.

9 The **Inclusion-Exclusion Principle** describes how to count the number of elements in the union of sets. Examine the statements of this principle for the union of two and three sets, below.

Inclusion-Exclusion Principle for the union of two sets:

[# of elements in $A \cup B$]
  = [# of elements in A] + [# of elements in B] − [# of elements in $A \cap B$]

Inclusion-Exclusion Principle for the union of three sets:

[# of elements in $A \cup B \cup C$]
  = [# of elements in A] + [# of elements in B] + [# of elements in C] −
  [# of elements in $A \cap B$] − [# of elements in $A \cap C$] −
  [# of elements in $B \cap C$] + [# of elements in $A \cap B \cap C$]

a. Use the Inclusion-Exclusion Principle to find the number of elements in $A \cup B \cup C$, where $A = \{1, 3, 5, 7, 9\}$, $B = \{1, 4, 9, 16\}$, and $C = \{1, 3, 5, 7, 11, 13\}$.

b. The Venn diagram at the right represents responses of 100 people in Council Bluffs, Iowa, to the public opinion survey described in Applications Task 6.

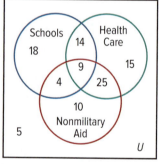

How many people are in the union of all three sets (Schools, Health Care, and Nonmilitary Aid)? Determine the answer by examining the figure and also by applying the Inclusion-Exclusion Principle.

c. Explain why the Inclusion-Exclusion Principle makes sense for the case of two sets. Explain also for the case of three sets.

10 Suppose that Q is the set of all quadrilaterals in a plane, S is the set of all squares, R is the set of all rectangles, P is the set of all parallelograms, and H is the set of all rhombuses. Which of the following statements correctly describes a relationship among these sets? Explain your reasoning in each case. The symbol $\not\subseteq$ means "is not a subset of."

a. $S \subseteq R$

b. $R \cup H = P$

c. $R \not\subseteq P$

d. $R' = P$

e. $S \not\subseteq H$

f. $R \cap H = S$

g. $Q - P' = P$

h. $S \subset P$

11 The following data from the 2014
Chirstina High School graduating
class show numbers of male and
female students who were honors
students and numbers who were
not honors students.

	Honors	Not Honors
Male	20	85
Female	25	70

Suppose that a student at Chirstina High School is chosen at random to be
interviewed on the local television station about his or her experience in local
schools. If M stands for the set of male students, H stands for the set of honors
students, and x represents the chosen student, calculate each of the following
probabilities.

a. $P(x \in M)$

b. $P(x \in M \cap H)$

c. $P(x \in M \cap H')$

d. $P(x \in M' \cup H)$

e. $P(x \in M \mid x \in H)$

f. $\dfrac{P(x \in M \cap H)}{P(x \in H)}$

REFLECTIONS

12 What familiar algebraic tasks are indicated by these examples of set notation?

a. Find $S = \{x \in R \mid 3x + 5 = 12\}$, where R is the set of real numbers.

b. Find $S \cap T$, where $S = \{(x, y) \mid 3x + 2y = 8\}$ and $T = \{(x, y) \mid x - y = 4\}$.

c. Find $S = \{(x, y) \mid x \geq 0, y \geq 0, x + y \leq 2, 3x + 2y \leq 4\}$.

13 As you have seen, set theory is used in Internet searches and social networking.
Find some recent news or examples of this use. Prepare a brief report. For
example, check out the 2010 Search Engine Meeting that was held in Boston
at www.searchenginemeeting.net/2010. Or get more information about using
circles in Google+.

14 The letters W, Z, Q, R, and C are commonly used as labels for the sets of whole
numbers, integers, rational numbers, real numbers, and complex numbers,
respectively. Recall that the whole numbers are the counting numbers and zero,
$\{0, 1, 2, \ldots\}$.

a. Draw a Venn diagram showing the relationships among these sets. Consider
the set of complex numbers as the universal set.

b. Which of the following are correct statements about these number sets? If a statement is not correct, explain why not.

i. $W \in Z$
ii. $Q = \{x \mid x = \frac{a}{b}, a \in Z, \text{ and } b \in Z, b \neq 0\}$
iii. $\sqrt{2} \in Q$
iv. $Q \subseteq R$
v. $\sqrt{-36} \in R'$
vi. $\pi \in (R - Q)$
vii. $W \cup Z = Q$

15 Examine the following statement by Susan Landau, Associate Editor of the *Notices of the American Mathematical Society*, which is from an article entitled, "Internet Time."

> "The information revolution will have as profound an effect on the world as the industrial revolution did. The organization and access of massive amounts of information is a deep and fundamentally mathematical problem."

> **Source:** *Notices of the American Mathematical Society*, Volume 47, Number 3, March 2000, p. 325.

This statement was made over 10 years ago. Do you agree with Dr. Landau? Do you think her prediction came true? Why or why not?

EXTENSIONS

16 The **power set** of a set is the set of all subsets of that set. The power set of a set A is denoted by $P(A)$.

a. Consider the set $A = \{a, b, c\}$. The power set of A is partially written below.

$$P(A) = \{\varnothing, \{a\}, \{b\}, \{c\}, \{a, b\}, \ldots\}$$

List the other elements of $P(A)$.

b. If A has n elements, make a conjecture about the number of elements in $P(A)$. Test your conjecture.

17 Often, the most interesting questions about Cartesian product sets (Connections Task 8) are those in which there is an important relationship between the entries, say x and y, of a subset of the ordered pairs. In those situations, you want to know: "Is y related to x (or x related to y) in a particular way?" Formally, a **relation** between two sets A and B is a subset of $A \times B$. Sometimes a relation is between two sets that are equal. In this case, the relation is called a *relation on A* (or on *B*).

a. Describe the set of points in a coordinate system that is designated by each of the following relations between sets of real numbers.

i. $\{(x, y) \mid y = 0.5x + 1\}$

ii. $\{(x, y) \mid y < x\}$

iii. $\{(x, y) \mid y \geq x^2\}$

iv. $\{(x, y) \mid x^2 + y^2 = 7\}$

b. Consider the "less than" relation on the set of all integers. That is, x is related to y if and only if x and y are integers and $x < y$. Since $3 < 4$, 3 is related to 4, and thus $(3, 4)$ is an element in the relation.

   i. Is $(2, 9)$ in the "less than" relation? Is $(9, 2)$ in this relation? Is $(-4, -3)$?

   ii. Call this relation L. Thus, $L = \{(x, y) \mid x$ and y are integers and $x < y\}$. List three more elements of L. How many elements are in L?

c. Let $A = \{a, b, c, d\}$. Let $T = \{t, u, v, w, z\}$.

   i. Is $R = \{(a, t), (b, w), (c, z), (d, v)\}$ a relation between A and T? Explain.

   ii. Is $K = \{(a, t), (a, u)\}$ a relation between A and T? Explain.

   iii. Is $M = \{(a, t), (z, d)\}$ a relation between A and T? Explain.

   iv. Construct a relation between A and T that has five elements.

d. How many possible relations are there between a set with three elements and a set with two elements? How about between a set with m elements and a set with n elements? Explain.

18 Two *finite sets* have the same number of elements if counting the elements in each set produces the same result. Counting the elements in a set involves tagging each element with a positive integer starting with 1, 2, 3, and so on until all have been tagged.

a. The description of counting given above implies that any two sets with the same number of elements can be put in *one-to-one correspondence*. For example, $B = \{a, b, c, d, e, f\}$ and $E = \{z, y, x, w, v, u\}$ have the same number of elements. Describe or illustrate a one-to-one correspondence between sets B and E.

b. For sets with infinitely many elements, conventional counting is not possible. However, sometimes it is possible to show that two sets with infinitely many elements have the same number of elements by showing that there is a one-to-one correspondence between them. Use this idea to show that the following pairs of sets have the same number of elements.

   i. The positive integers and the negative integers

   ii. The positive integers and the even integers

   iii. The points on $\overline{MM'}$ and the points on $\overline{BC}$, where M and M' are the midpoints of $\overline{AB}$ and $\overline{AC}$, respectively.

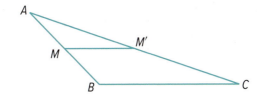

REVIEW

19 Evaluate each of the following using mental computation. Be prepared to explain your method.

a. $\frac{1}{2}\left(2 - \frac{3}{4}\right)$

b. $\frac{5}{8}\left(\frac{2}{3}\right) + \frac{2}{3}\left(\frac{3}{8}\right)$

c. $8 \div (-2 + 10)$

d. $(3.7 - 10.4) + (10.4 - 3.7)$

e. $43(12) - 32(57 - 14)$

20 Fill in each blank with *always*, *sometimes*, or *never* in order to make the statement true. Be prepared to explain your reasoning.

a. A parallelogram is _____ a quadrilateral.

b. A rectangle is _____ a square.

c. A square is _____ a rhombus.

d. A parallelogram is _____ a rectangle.

e. A quadrilateral with exactly two lines of symmetry is _____ a kite.

21 At the beginning of the school year, the 186 seniors at Sterling High School were surveyed and asked about whether they plan to go to college and if they had a summer job. Some of the results are shown in the chart below.

	Had Summer Job	No Summer Job	Total
College Plans			
No College Plans	24		72
Total		56	186

a. Fill in the remaining cells on a copy of the chart.

Suppose one of the seniors at Sterling High School is randomly chosen. Determine each of the following probabilities.

b. *P(college plans)*

c. *P(had summer job)*

d. *P(college plans* and *had summer job)*

e. *P(college plans* or *had summer job)*

f. What is the probability that someone who had college plans also had a summer job? (This can be written in symbols as *P(had summer job | college plans)*. The symbol | means "given that" and restricts the group under consideration. So in this case, rather than determining the probability based on all of the students, we are only considering the 114 students who planned to go to college.)

g. *P(college plans | had summer job)*

22 Consider the following numbers:

$$123 \quad \frac{2}{3} \quad \pi \quad -\frac{8}{4} \quad \sqrt[3]{8} \quad 31.72 \quad -53 \quad 6\frac{3}{8} \quad \sqrt{7}$$

  a. Which of the numbers are whole numbers?

  b. Which of the numbers are integers?

  c. Which of the numbers are rational numbers?

  d. Which of the numbers are irrational numbers?

  e. Order the numbers from smallest to greatest.

23 Carlos deposits $2,875 into a savings account that has an APR of 1.3% compounded monthly.

  a. If he does not deposit any additional money into the account, write a symbolic rule that expresses the account balance as a function of the number of years since he made the deposit.

  b. What is the APY for this account?

  c. Carlos is saving money for a $10,000 downpayment for a new house. If, rather than deciding to not deposit any additional money into the account, he instead adds $350 to the account each month, how long will it take before Carlos has saved enough money for the downpayment?

24 In the diagram below, $\overleftrightarrow{AB} \parallel \overleftrightarrow{DE}$.

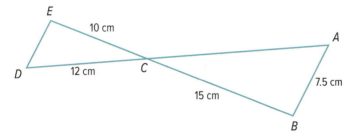

  a. Explain why $\triangle CED \sim \triangle CBA$.

  b. Find the lengths of $\overline{AC}$ and $\overline{DE}$.

25 You may recall that a matrix is a rectangular array of numbers. The tasks below will help you check your recollection or understanding of some key matrix ideas.

a. When you add or subtract two matrices, you add or subtract corresponding entries. Calculate the remaining entries for this matrix sum:

$$\begin{bmatrix} 1 & -6 \\ 0 & 14 \end{bmatrix} + \begin{bmatrix} 0 & 4 \\ -6 & -3 \end{bmatrix} = \begin{bmatrix} 1 & \underline{} \\ \underline{} & 11 \end{bmatrix}$$

b. Find the sum or difference as indicated.

i. $\begin{bmatrix} 7 & 9 \\ 8 & -4 \end{bmatrix} + \begin{bmatrix} -5 & 2 \\ -2 & 16 \end{bmatrix}$

ii. $\begin{bmatrix} 21 & -9 & -12 \\ 18 & 6 & 0 \end{bmatrix} - \begin{bmatrix} 6 & -10 & 1 \\ -8 & 26 & -14 \end{bmatrix}$

c. When you multiply two matrices M and N, you do *not* multiply corresponding entries. Rather, to determine the entry in the first row and second column of the product matrix $M \times N$, you multiply entries in the first row of matrix M by entries in the second column of matrix N and then add awthose products. One example is shown below.

$$\begin{bmatrix} 1 & 3 \\ 0 & 6 \end{bmatrix}\begin{bmatrix} 2 & 4 \\ 5 & 7 \end{bmatrix} = \begin{bmatrix} 1(2) + 3(5) & 1(4) + 3(7) \\ 0(2) + 6(5) & 0(4) + 6(7) \end{bmatrix} = \begin{bmatrix} 17 & 25 \\ 30 & 42 \end{bmatrix}$$

Calculate the following products.

i. $\begin{bmatrix} 5 & 2 \\ 4 & 1 \end{bmatrix}\begin{bmatrix} 11 & 10 \\ 3 & 8 \end{bmatrix}$

ii. $\begin{bmatrix} 1 & 0 \\ 5 & 8 \end{bmatrix}\begin{bmatrix} 2 & 3 & 6 \\ 1 & 4 & 10 \end{bmatrix}$

d. The 2×2 identity matrix I is the matrix $\begin{bmatrix} 1 & 0 \\ 0 & 1 \end{bmatrix}$. You may recall that some matrices have an inverse matrix. When you multiply a matrix by its inverse, the product is the identity matrix. Determine whether the matrices in each pair below are inverses of each other.

i. $\begin{bmatrix} 4 & 7 \\ 1 & 2 \end{bmatrix}$ and $\begin{bmatrix} 2 & -7 \\ -1 & 4 \end{bmatrix}$

ii. $\begin{bmatrix} 0 & 8 \\ 5 & 3 \end{bmatrix}$ and $\begin{bmatrix} 3 & -8 \\ -5 & 0 \end{bmatrix}$

Security: Cryptography

In Lesson 1, you investigated how to organize and access information, particularly in the context of Internet searching. Certainly information is of little use if it is not accessible. The next important issue to consider is security.

There are many situations where information must be secure, that is, private and authentic. For example, secret codes have been part of military and political intrigue for as long as there have been competing groups of people. Information security is important in government, banking, business, and private life, especially in online settings. *Cryptography* helps keep information secure.

THINK ABOUT THIS SITUATION

Think about the issue of information security as it relates to email, e-commerce, and other situations in your own life.

a. Have you ever used secret codes or encryption software? If so, briefly describe the situation and why you wanted to keep the information secure.

b. Examine the email window pictured below. What do you think the padlock and seal icons in the lower-right corner mean? Have you ever used email security features?

c. Consider the Mac Help information just below the email window shown. Have you ever noticed the use of "https" versus "http" in Web site addresses? What, if anything, do you know about "public- and private-key encryption systems" or "digital certificates"?

d. Describe any examples of cryptography or use of secret codes that you have seen in the news, in the movies, in a book, or on a TV show.

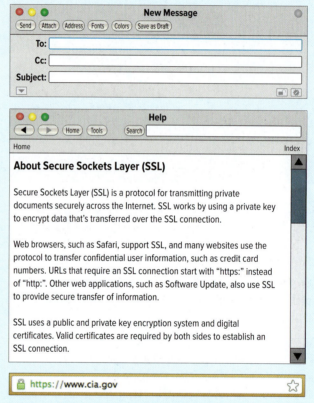

Examples of cryptography used on the Internet

In this lesson, you will investigate symmetric-key (also called private-key) and public-key cryptography. In the process, you will learn some important ideas of number theory related to modular arithmetic.

Symmetric-Key Cryptography

Cryptography is the study and application of mathematical techniques to keep information secure. You will study two fundamental aspects of information security—*confidentiality* and *authentication*. Confidentiality is important when you use a credit card to make an online purchase, or when a spy sends a secret message. Authentication refers to verifying the origin of data, which at the same time can verify the integrity of the data. Authentication is needed to ensure that an email message has not been altered and is from who you think it is from.

As you work on the problems in this investigation, look for answers to the following questions:

What is a symmetric-key cryptosystem?

What is a substitution cipher?

How do the ROT13, Caesar, and Hill ciphers work? How secure are they?

Cryptosystems Cryptography can be used to design *cryptosystems*, which are used to securely transmit confidential information, as illustrated in the diagram below. The original plaintext message is encrypted, so that it is protected and secret. The resulting ciphertext message is decrypted upon receipt. A key is used to encrypt and decrypt the information. A **cryptosystem** is the overall method of encrypting and decrypting using keys.

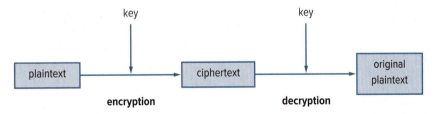

Symmetric-Key Cryptosystems In a **symmetric-key cryptosystem**, the same key is used to encrypt and decrypt. The exact same key might not be used, but at least it is easy to calculate the encryption key from the decryption key, and vice versa. Thus, the security of a symmetric-key cryptosystem depends on the secrecy of the key.

Getty Images

1 An example of a symmetric-key cryptosystem is ROT13. In the **ROT13 cryptosystem**, the encryption method is to "rotate by 13 letters." That is, every letter is replaced by the letter that is 13 places down the alphabet. So, "HI" becomes "UV". This system was popular in the early days of the Internet. For example, it was used as a way to hide punchlines to jokes in discussion forums and as a simple secret code for email messages. It is sometimes used today to hide hints in online games or in Web-based activities like geocaching.

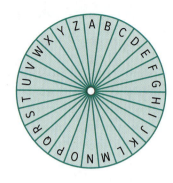

a. Use the ROT13 method to encrypt the message VICTORY IS NEAR.

b. Suppose you receive the following email message, which has been encrypted using ROT13.

> XRRC VG FRPERG, XRRC VG FNSR!

  i. Decrypt this message.

 ii. Explain why this cipher is not very secure. That is, how could you decrypt the message even if you did not know it used a ROT13 cipher?

c. Explain why the exact same ROT13 procedure that encrypts a message will also decrypt it.

2 A fixed-shift, letter-substitution cryptosystem is sometimes called a **Caesar cipher**. Reportedly, Julius Caesar (100–44 B.C.) used a method like this to send secret messages. To encrypt a message using a Caesar cipher, you need to know how many places down the alphabet you must shift to find the letter substitution (that is, you need to know the *key*). Suppose you have just intercepted the secret message below that has been encrypted using a Caesar cipher. Examine the ciphertext carefully to find the shift that has been used, then decrypt the message.

> F EXSB X AOBXJ
>
> JBBQ QLKFDEQ

Julius Caesar used a fixed-shift, letter-substitution cryptosystem.

3 The ciphers in Problems 1 and 2 are called *substitution ciphers* because of the substitution technique used. A **cipher** is an algorithm used for encryption and decryption. Generally, there are two related algorithms, one for encryption and one for decryption. A **substitution cipher** is an algorithm that encrypts a message by replacing each character in the plaintext with another character. The receiver reverses the substitution to decrypt the message. A **key** typically consists of the information needed to carry out the encryption (or decryption) process. In a shift substitution cipher, the key is typically a number. For example, the key in a Caesar cipher is the number of places down the alphabet that you must shift.

a. What is the key for the Caesar cipher in Problem 2?

b. What is the key for the ROT13 cipher?

c. Substitution ciphers like ROT13 and Caesar ciphers are not very secure. That is, it is not very difficult to decrypt the ciphertext even if you do not know the key. Describe some general strategies you think could be used to break substitution ciphers like those above.

d. Create your own substitution cipher. Encrypt a brief message. Then exchange encrypted messages with a classmate and try to decrypt each other's messages. Discuss and resolve any difficulties in the decryption process.

4 Substitution ciphers like those in Problems 1 and 2 are easy to break. A more complex substitution cipher using matrix multiplication was invented by the mathematician Lester S. Hill in the 1920s. It is called a **Hill cipher**.

a. A specific type of Hill cipher is described below. Working with some classmates, complete the steps of the algorithm to see how this Hill cipher encryption works.

Step 1. *Find a 2 × 2 matrix with integer entries that has an inverse with integer entries.*

Suppose you select $A = \begin{bmatrix} 1 & 2 \\ 1 & 3 \end{bmatrix}$. Verify that A^{-1} exists and has integer entries.

Step 2. *Put the plaintext message into a 2 × n matrix, column by column.* Use an underscore (_) for the space between words. Determine the missing matrix entries for the phrase MEET TONIGHT.

$$\begin{bmatrix} M & E & \_ & O & ? & H \\ E & T & T & ? & G & ? \end{bmatrix}$$

Step 3. *Substitute numbers for characters in the matrix message.* Use the following translation table.

_(space)	A	B	C	D	E	F	G	H	I	J	K	L	M
0	1	1	1	4	5	6	7	8	9	10	11	12	13
	N	O	P	Q	R	S	T	U	V	W	X	Y	Z
	14	15	16	17	18	19	20	21	22	23	24	25	26

Complete the following matrix so that it becomes the number version of the MEET TONIGHT matrix above.

$$\begin{bmatrix} ?? & 5 & 0 & 15 & 9 & ?? \\ 5 & 20 & ?? & ?? & 7 & ?? \end{bmatrix}$$

Step 4. *Do a scrambled substitution using matrix multiplication.*
Multiply the numerical message matrix (Step 3) on the left by the matrix *A* (Step 1). Carry out this multiplication and record the result.

Step 5. *Convert numbers back to letters.*
Recall that you have 27 characters, numbered from 0 to 26. So, for example, the number 28 would translate to the letter A. Finally, write the string of letters that is the final ciphertext.

  b. Compare your translation with that of other members of your class. Resolve any differences.

  c. Now that you know how to encrypt a message using a Hill cipher, think about the method of decryption. Starting with the final ciphertext from Part b, work with some classmates to figure out how you could decrypt this ciphertext. Then do the decryption. Check to make sure you end up with the original plaintext: MEET TONIGHT.

  d. Compare the Hill ciphertext for the message MEET TONIGHT to the ROT13 ciphertext for the same message. Which cipher do you think is more secure? Why?

  e. Use the Hill cipher to encrypt a one-word message of your choice. Exchange ciphertext messages with a partner. Decrypt each other's ciphertext message. Discuss and resolve any difficulties in the encryption or decryption process.

Certain symmetric-key cryptosystems have been established as cryptography standards by the U.S. government. For a recent history, you may wish to access "Cryptography Standards" in your book's Resources section on ConnectED.

SUMMARIZE THE MATHEMATICS

In this investigation, you studied symmetric-key cryptography. In particular, you investigated several substitution ciphers—ROT13, Caesar ciphers, and Hill ciphers.

a. The diagram at the beginning of this investigation illustrates the basic process of sending and receiving confidential information using a cryptosystem. Demonstrate each step of this process using the ROT13 cipher and a one-word message.

b. In a symmetric-key cryptosystem, the same key is used to both encrypt and decrypt information, or it is easy to determine the encryption key from the decryption key (and vice versa). Explain why each of the three cryptosystems studied in this investigation is a symmetric-key cryptosystem.

c. Explain why all three ciphers studied in this investigation are examples of substitution ciphers.

d. Explain why the Caesar and ROT13 ciphers are not very secure. Explain why the Hill cipher is more secure than the Caesar or ROT13 ciphers.

Be prepared to share your example and thinking with the class.

Think about the advantages and disadvantages of ROT13, Caesar, and Hill ciphers as you complete these tasks.

a. The following two ciphertext messages were encrypted using a Caesar cipher. By examining the ciphertext, find the substitution key that was used and decrypt the messages.

PSSO FIJSVI CSY PIET

WIMDI XLI HEC

b. Use the ROT13 cipher to encrypt the message SINGING IN THE RAIN.

c. Use a Hill cipher based on the matrix $A = \begin{bmatrix} -4 & -3 \\ 9 & 7 \end{bmatrix}$ to encrypt the message TO BE OR NOT TO BE. Decrypt the message to verify that you get the original plaintext.

INVESTIGATION 2

Modular Arithmetic

Modular arithmetic is often used in cryptography. In fact, it is essential for public-key cryptosystems like those you will study in Investigation 3. In this investigation, you will explore several modular arithmetic systems and examine some general properties of modular arithmetic.

As you complete the problems in this investigation, look for answers to these questions:

> *How do you add, multiply, and find inverses in modular arithmetic?*
>
> *What do "equivalent mod n" and "reduce mod n" mean?*
>
> *What are the "integers mod n (Z_n)" and how do you compute with them?*

1 Suppose that as part of a secret code, you will substitute numbers for letters. You will also need a number to represent a space. Use the following translation table.

_(space)	A	B	C	D	E	F	G	H	I	J	K	L	M
0	1	2	3	4	5	6	7	8	9	10	11	12	13
	N	O	P	Q	R	S	T	U	V	W	X	Y	Z
	14	15	16	17	18	19	20	21	22	23	24	25	26

So, you have 27 characters, numbered from 0 to 26. Think about how you could convert numbers larger than 26 into letters, by starting to count over after you reach 26.

a. Explain why the number 28 translates to the letter A.

b. Translate the numbers 38, 52, and 119 into letters. Compare your translation with those of other students. Resolve any differences.

c. Consider numbers that translate to the letter D. Share your thinking and work with some of your classmates.

   i. Find four other numbers that translate to the letter D. Explain how to find these numbers.

   ii. Describe any patterns you observe for all numbers that translate to the letter D.

d. Think about how to translate negative integers into letters. Test and describe a translation strategy. Compare your strategy with those of other students. Resolve any differences.

2 To translate numbers into characters in Problem 1, you first needed to convert any given integer into an integer between 0 and 26. This is an example of what is called *modular arithmetic*. In particular, you are using a mod 27 system, since you are limited to the 27 integers from 0 to 26. It is possible to do arithmetic in this system. For example, $2 + 3 = 5$ in mod 27, just as in regular arithmetic. But some computations in mod 27 give results that are quite different than those in integer arithmetic.

   Perform the following computations in mod 27. Compare your reasoning and results with other classmates. Resolve any differences.

a. Explain why $25 + 9 = 7$ in mod 27.

b. $18 + 14 = ?$ in mod 27

c. $3 \times 25 = \underline{?}$ in mod 27

d. $7^2 = \underline{?}$ in mod 27

e. $4 \times 5 = \underline{?}$ in mod 27

f. $-5 = \underline{?}$ in mod 27

3 Now explore some other modular arithmetic systems.

a. Consider a mod 33 system. Perform the same computations as in Problem 2, except this time use mod 33.

b. Consider a mod 15 system.

   i. Why does it make sense to say that 23 and 38 are "equivalent mod 15"?

   ii. What is 6^4 mod 15?

Equivalence mod *n* In Problems 1–3, you informally investigated modular arithmetic in particular cases of "mod *n*." The idea of two integers being "equivalent mod *n*" is formally defined as follows.

> *Two integers are **equivalent mod n***
> *if and only if they have the same remainder upon division by n.*

A "three-line equal sign" is used to denote equivalence mod *n*. Thus, $a \equiv b \bmod n$ is read as, "*a* is equivalent to *b* mod *n*."

4 Solve the following problems about equivalence mod n.

   a. Use the definition of equivalence mod n to explain why each statement is true.

      i. $23 \equiv 8 \bmod 5$

      ii. $76 \equiv 4 \bmod 9$

      iii. $-3 \equiv 24 \bmod 27$

   b. Find four integers that are equivalent to 2 mod 27. At least one of the numbers should be a negative integer.

   c. Working with others, brainstorm a possible equivalent, yet different, definition or description of "equivalent mod n."

Reduce mod n As you have seen, many different integers can be equivalent mod n. Finding an equivalent integer mod n that is between 0 and $n - 1$ is called *reducing mod n*.

> To **reduce an integer mod n** means
> to replace the integer by its remainder upon division by n.

For example, reducing 58 mod 7 yields 2, since dividing 58 by 7 leaves a remainder of 2.

5 Reduce each of the integers below, using the indicated modular arithmetic system.

   a. 48 mod 5

   b. 397 mod 10

   c. -24 mod 7

6 What are all the possible results when you reduce integers mod 5? What about when you reduce integers mod 12? When you reduce integers mod n? Explain your reasoning.

Integers mod n (Z_n) In Problem 6, you found that every integer can be reduced mod 5 to 0, 1, 2, 3, or 4, since these are the possible remainders when you divide an integer by 5. You also found that, in general, every integer can be reduced mod n to an integer between 0 and $n - 1$, inclusive. Because of this, a new system of numbers Z_n, called **integers mod n**, can be defined.

$$Z_n = \{0, 1, 2, \ldots, n - 1\},$$
with addition and multiplication mod n

Each "number" in Z_n really represents all the integers that reduce to that number mod n. Even so, you can think of the elements of Z_n as the numbers $0, 1, 2, \ldots, n - 1$. You compute with these numbers using the same modular arithmetic that you have been using throughout this investigation. In particular, you **multiply mod n** by multiplying the integers as usual and then reducing mod n. You **add mod n** by adding the integers as usual and then reducing mod n.

Properties of Z_n The modular arithmetic in Z_n has many interesting properties. Some properties are similar to properties of regular arithmetic with real numbers, while other properties are different.

7 Think about **additive inverses**. Recall the additive inverse of a number is the number that you add to it to get 0. For example, with real numbers, the additive inverse of 3 is −3, since $3 + (−3) = 0$. The number 0 is called the *additive identity*. Find the additive inverse of each of these real numbers: 5, $\frac{3}{4}$, and −1.5.

8 Every real number has an additive inverse; just take the negative of the number. Do you think every number in Z_n has an additive inverse?

   a. Consider Z_{10}.

   i. Find a number in Z_{10} that you can add to 6 to get 0 mod 10. Such a number is the additive inverse of 6 in Z_{10}.

   ii. Find a number in Z_{10} that you can add to 2 to get 0 mod 10. Such a number is the additive inverse of 2 in Z_{10}.

   iii. What is the additive inverse of 3 in Z_{10}? Explain.

   b. Consider some other modular arithmetic systems.

   i. What is the additive inverse of 3 in Z_8? Explain why this answer is different than the answer you got for the additive inverse of 3 in Part aiii.

   ii. What is the additive inverse of 17 in Z_{21}?

   c. Think about Z_n in general.

   i. What is the additive inverse of 1 in Z_n? Why?

   ii. Does every number in Z_n have an additive inverse? Explain your reasoning.

9 Now think about **multiplicative inverses**. Recall the multiplicative inverse of a nonzero number is the number that you multiply it by to get 1. For example, with real numbers, the multiplicative inverse of 3 is $\frac{1}{3}$, since $3 \cdot \frac{1}{3} = 1$. The number 1 is called the *multiplicative identity*. Find the multiplicative inverse of each of these real numbers: 5, $\frac{3}{4}$, and −1.5.

10 Every nonzero real number has a multiplicative inverse; just take the reciprocal of the number. Do you think every number in Z_n has a multiplicative inverse?

   a. Consider Z_7.

   i. Find a number in Z_7 that you can multiply by 3 to get 1 mod 7. Such a number is the multiplicative inverse of 3 in Z_7.

   ii. Find a number in Z_7 that you can multiply by 2 to get 1 mod 7. Such a number is the multiplicative inverse of 2 in Z_7.

   iii. Find the multiplicative inverse of 6 in Z_7. Explain.

b. Next consider Z_{10}.

    i. For each number in Z_{10}, try to find its multiplicative inverse. (Remember that you multiply in Z_{10} using mod 10 modular arithmetic.) To help organize your work, complete a multiplication table like the one below.

Multiplication Table for Z_{10}

×	0	1	2	3	4	5	6	7	8	9
0										
1										
2										
3										
4										
5										
6										
7										
8										
9										

    ii. State any patterns that you see concerning which (nonzero) numbers in Z_{10} do *not* have a multiplicative inverse. How do such numbers relate to 10?

    iii. State any patterns you see for numbers that *do* have a multiplicative inverse.

c. Now consider Z_9.

    i. For each number in Z_9, try to find its multiplicative inverse.

    ii. State any patterns that you see concerning which numbers in Z_9 have a multiplicative inverse and which do not.

d. Does every number in Z_n have a multiplicative inverse, for every n? Explain.

11 In Problem 10, you discovered that not all numbers in a given modular arithmetic system have a multiplicative inverse. Think about when multiplicative inverses exist in Z_n.

a. Make some conjectures about which numbers have multiplicative inverses in Z_n, either for a general n or for particular values of n. For each conjecture, try to prove or disprove it. (Recall, you can disprove it by finding a counterexample.) After making and testing some of your own conjectures, complete and explain the following three statements.

b. When n is _____, then every nonzero integer in Z_n has a multiplicative inverse.

c. In Z_n, how must m and n be related for m *not* to have a multiplicative inverse?

d. Complete this statement: m has a multiplicative inverse in Z_n if and only if _____.

SUMMARIZE THE MATHEMATICS

In this investigation, you studied the modular arithmetic system Z_n and associated ideas of equivalence mod n, reducing mod n, addition and multiplication mod n, and existence of inverses in Z_n.

a. The integers 8 and 5 can be considered elements in many different modular arithmetic systems. Think about the arithmetic in the various modular systems as you complete the following.

  i. $8 + 5 = $ ___ in Z_9

  ii. $8 + 5 = $ ___ in Z_{27}

  iii. $8 \times 5 = $ ___ in Z_9

  iv. $8 \times 5 = $ ___ in Z_{27}

  v. $8^5 = $ ___ in Z_9

  vi. Find the additive inverse of 8 in Z_{12}.

  vii. Find the multiplicative inverse of 5 in Z_{11}.

b. Consider a mod 10 system.

  i. Describe how to determine if two integers are equivalent mod 10.

  ii. Find three integers that are equivalent to -4 mod 10.

  iii. Reduce 346 mod 10.

  iv. Which numbers in Z_{10} have a multiplicative inverse? For each of these numbers, find its multiplicative inverse.

c. Give a general explanation or description of the following.

  i. Two integers are equivalent mod n.

  ii. Reduce an integer mod n.

  iii. Z_n

Be prepared to compare your responses to those of your classmates.

✓CHECK YOUR UNDERSTANDING

Carry out the indicated modular arithmetic procedures.

a. $28 + 16 = $ ___ in Z_{30}

b. $37 + 25 \equiv $ ___ mod 7

c. Reduce 47 mod 12.

d. Reduce 12^8 mod 9.

e. Find the multiplicative inverse of 7 in Z_{12}.

Public-Key Cryptography

In Investigation 1, you studied symmetric-key cryptosystems, in which the same key is used to encrypt and decrypt. A major security problem with symmetric-key cryptosystems is that the key must be confidentially distributed to the sender and receiver. This *key distribution problem* is eliminated in *public-key cryptosystems*.

As you complete the problems in this investigation, look for answers to the following questions:

How does the RSA public-key cryptosystem work?

How can public-key cryptography be used to send confidential messages and provide digital signatures?

How is a public-key cryptosystem different from a symmetric-key cryptosystem?

Public-Key Cryptosystems In a **public-key cryptosystem**, an individual creates a pair of related keys—a **public key**, which can be shared with everyone, and a **private key**, which is kept secret and is known only to the individual. The diagram below illustrates this situation. Alice has created the pair of keys, Bob uses the public key to send an encrypted message to Alice, and Alice uses her private key to decrypt the message.

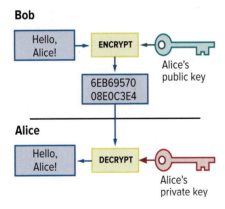

As in all cryptosystems, the sender encrypts and the receiver decrypts. But the key used for each may be different depending on the application. Public-key cryptography has two major applications— secret messages, as illustrated above, and digital signatures. These two applications can be thought of in terms of many people sending confidential messages to one person (secret messages), or one person sending private and authentic messages to many people (digital signatures).

Examples of the first situation—many send secret messages to one—include when many people send their confidential credit card numbers to one online store, or when many embassies abroad send secret messages to their home country. An example of the second situation—one sends authentic messages to many—is when one person sends digitally-signed email messages to many people, so that the recipients know that the message is authentic.

In both situations, one individual creates the keys and uses the private key, while others use the public key.

Both applications of public-key cryptosystems are illustrated by the icons in the lower-right corner of the email message window shown below, from a popular email program in 2014.

The instructions given for the padlock icon are to "click to encrypt this message" (using the recipient's public key). The instructions for the star-shaped seal icon are to "click to digitally sign this message" (using the sender's private key).

In 1977, Ronald L. Rivest, Adi Shamir, and Leonard Adleman at the Massachusetts Institute of Technology developed what is now called the *RSA public-key cryptosystem*. (The name is based on the initials of the last names of the developers.) The idea of a public-key cryptosystem is considered to be one of the most significant developments in the history of cryptography. There are now many such cryptosystems in use.

RSA Public-Key Cryptosystem The keys for the RSA public-key cryptosystem are numbers constructed using large prime numbers and modular arithmetic. The general strategy is shown in the following diagram. The numbers *n* and *e* comprise the *public key*, known to everyone. The numbers *p*, *q*, and *d* are the *private key*. The arrows in the diagram below show which numbers are related to others.

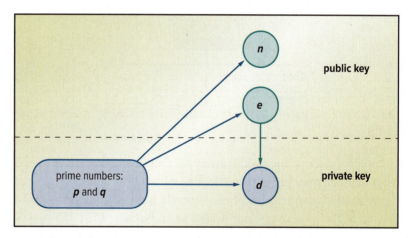

In general, either the public key or the private key can be used for encrypting. Then the other is used for decrypting. For purposes of learning how the system works, assume that the public key is used to encrypt a message. After the message is sent and received, the private key is used to decrypt.

1 Study the steps that follow describing how to use the RSA cryptosystem. Discuss with your classmates any step that you do not understand. (In a later problem you will learn why the procedure works. For now, just make sure you understand what to do at each step.)

Using the RSA Cryptosystem

Construct the Keys

Step 1. Choose two prime numbers, p and q.

Step 2. Compute $n = pq$.

Step 3. Compute $r = (p - 1)(q - 1)$.

Step 4. Choose a number e (for encrypt) such that e has a multiplicative inverse in Z_r.

Step 5. d (for decrypt) is the multiplicative inverse of e in Z_r.

Step 6. Publish e and n in a public directory. This is the public encryption key. It is associated with the individual who constructs the keys, but it can be used by anyone.

Step 7. Keep d secret, along with p and q. This is the private decryption key. It is known only to the individual who constructs the keys.

Encrypt

Step 1. Convert the plaintext message, including spaces, to numbers in $Z_n = \{0, 1, \dots, n - 1\}$.

Step 2. Raise each number to the power e.

Step 3. Reduce each number mod n. The resulting number is called the *ciphertext number*.

Decrypt

Step 1. Raise each ciphertext number to the power d. Reduce mod n.

Step 2. Convert from numbers back to letters.

Step 3. Reduce mod n.

---

2 Use the RSA cryptosystem to encrypt and then decrypt the message FLEE. Discuss and resolve any difficulties with your classmates as you work through the encryption and decryption.

a. Normally, when choosing the two prime numbers p and q, you would use very large prime numbers to keep the system secure. However, for this problem, choose $p = 3$ and $q = 11$.

b. Convert letters to numbers by converting each letter into the integer 1–26 that represents its place in the alphabet.

c. Encrypt one letter at a time. Because RSA encryption is computationally intensive, you may wish to use appropriate features of the *TCMS-Tools* CAS or Wolfram Alpha (www.wolframalpha.com) to support your work.

d. Decrypt one ciphertext number at a time. When you are finished, you should be back to the original message, FLEE.

3 Now use the RSA cryptosystem to send and receive secret messages. Working with your teacher, form teams with your classmates. Choose a name for your team. Your team should choose one of the public encryption keys listed in the public-key directory shown below. Write your team name next to your public key on a copy of the Public-Key Directory. Your teacher will hand each group their corresponding private key. Thus, each team has its own pair of keys—a public key and the associated private key.

Public-Key Directory

Team Name	n	e
	85	13
	55	23
	161	19
	95	31
	91	29
	65	29
	35	13

a. Your team should use the RSA cryptosystem to encrypt and send at least one single-word secret message to another team. As per the cryptosystem rules, *use the receiving team's public key to encrypt the message.*

b. Decrypt any messages you receive. As per the cryptosystem rules, *use your own private key to decrypt the message you receive.*

c. Based on this public-key cryptosystem, discuss and answer the following questions.

   i. Can any team send any other team a secret encrypted message? What information do you need to encrypt a secret message so that only the target team can read it?

   ii. Once you encrypt a secret message to a target team, can any other team decrypt it and read the message? Explain.

   iii. Suppose you receive an encrypted message from another team. What information did that team need to encrypt the message? What information do you need to decrypt it? Can any other team decrypt the message?

Analyze the RSA Public-Key Cryptosystem Now that you know how the RSA cryptosystem works, consider two important questions: *Why does it work?* and *Why is it secure?* These questions will be the focus of the next few problems.

4 For this or any other cryptosystem to work, the decrypting process must undo the encrypting process so that you get back to the original message. In the case of the RSA cryptosystem, you encrypt by raising to the power e (mod n). You decrypt by raising to the power d (mod n), and this gets you back to the original message. So, for a number (message) M, $(M^e)^d \equiv M$ mod n. This works because of a special case of a theorem proven by the Swiss mathematician Leonard Euler (1701–1783).

> **Euler's Theorem (special case)**
>
> $(M^e)^d \equiv M$ mod n, where p and q are prime numbers, $n = pq$,
>
> $r = (p - 1)(q - 1)$, and e and d are multiplicative inverses mod r.

a. Test this theorem by considering the following values:

$$p = 3 \qquad q = 11 \qquad e = 7 \qquad d = 3$$

   i. Verify that these numbers satisfy the conditions of Euler's Theorem.

   ii. Choose several values of M and verify that Euler's Theorem is true in each case.

b. Choose two different (small) values for p and q, and check that Euler's Theorem works for a few values of M.

5 Now that you know *how* the RSA cryptosystem works and *why* it works, examine why it is secure.

a. To be secure, the code must be hard to break. Look back at the diagram illustrating the RSA cryptosystem on page 468. The numbers n and e are the public key. Everyone knows these numbers. To break the code, you must find the private key d. Suppose $n = 55$ and $e = 23$. Find d.

b. Think about the steps needed to find d in Part a. You start with n and e. Then you must:

   - factor n to find p and q.

   - multiply $(p - 1)(q - 1)$ to find r.

   - find the multiplicative inverse of e, mod r.

   Which of these steps was hardest for you in Part a?

c. Look back at Part b. Perhaps surprisingly, the hardest step in general, with very large numbers, is trying to factor n. That is, $n = pq$, but you do not know p and q, so you must factor n to find p and q. This can be very difficult. To help see why this is so, try these factoring problems.

   i. Suppose $pq = 35$. What are p and q?

   ii. Suppose $pq = 77$. Find p and q.

   iii. Suppose $pq = 221$. Find p and q.

   iv. Suppose $pq = 3,431$. Find p and q.

You can see that finding p and q gets more difficult as the product pq gets larger. In real applications of RSA, such as using a credit card to buy a book online, the key may include a product that contains 200 or more digits! *The point is that it is very difficult to factor a large product into its prime factors with or without technology. That is what keeps a secret message secret when you use RSA public-key encryption.*

Cryptographic Protocols As you have seen, a good cryptosystem involves quite a lot of interesting mathematics. In addition, many technical details must be worked out when a cryptosystem is implemented through specific software. Despite all these technical details, a cryptosystem is often initially designed or implemented by thinking in a non-technical way about *protocols*. A **cryptographic protocol** is a step-by-step procedure designed to accomplish a task related to secure communication between two or more parties (people or computers). You will now investigate several non-technical cryptographic protocols.

6 Below is a protocol that describes how to implement a public-key cryptosystem. By carefully modifying the protocol, write a new protocol for implementing a symmetric-key cryptosystem as studied in Investigation 1.

Protocol for Implementing a Public-Key Cryptosystem

Step 1. Alice and Bob agree on a public-key cryptosystem, like RSA.

Step 2. Alice sends Bob her public key.

Step 3. Bob encrypts his message using Alice's public key and sends the ciphertext to Alice.

Step 4. Alice decrypts Bob's message using her (different) private key.

Comparing and Combining Public-Key and Symmetric-Key Cryptosystems Public-key and symmetric-key cryptosystems are not really competitors. Each is better suited for certain purposes since they each have different strengths and weaknesses. One way to compare the two types of cryptosystems is in terms of speed versus key security. Symmetric-key systems are faster, but less reliable in terms of key security. In particular, since the same key is used to encrypt and decrypt, one key must be sent to all authorized users. This makes it more difficult to keep the key secret and thus safeguard the whole system. This is the *key-distribution problem* that was mentioned at the beginning of this investigation. Public-key systems are slower, but they work well for easy and secure key distribution. The public key is freely transmitted as public knowledge, while the private key is kept secret by the one individual who created it and will use it.

7 Given the complementary features of public-key and symmetric-key cryptosystems, sometimes a *hybrid cryptosystem* is used. Study the following protocol for a hybrid cryptosystem. Describe where both symmetric-key and public-key cryptography are used in this protocol.

Protocol for a Hybrid Cryptosystem

Step 1. Alice sends Bob her public key.

Step 2. Bob generates a session key to be used for their communication session. He encrypts this session key using Alice's public key, and sends it to Alice.

Step 3. Alice decrypts the session key using her private key.

Step 4. Bob and Alice encrypt and decrypt their communication using the same session key.

Digital Signatures One of the strengths of public-key cryptography is that it can be used to authenticate data. Think about how you "authenticate data" with old-fashioned pen and paper. You use a signature. When you sign a check or a contract, that provides the certification that it is authentic. This same goal can be accomplished electronically using *digital signatures*.

8 An effective digital signature can be created using public-key cryptography. Study this protocol.

Protocol for a Digital Signature Using Public-Key Cryptography

Step 1. Alice encrypts the document with her private key. This serves as her signature on the document.

Step 2. Alice sends the signed document to Bob.

Step 3. Bob decrypts the document with Alice's public key. This serves to verify the signature.

Suppose you receive an email message from a friend that has been "signed" using the digital signature procedure above.

a. Explain how the digital signature verifies that the message is from your friend.

b. Explain how the digital signature also verifies that the message has not been altered.

For further information on the continuing challenges facing RSA public-key cryptosystems and for related unsolved mathematical problems see "Easy to Encrypt, Hard to Decrypt" in your book's Resources section on ConnectED.

SUMMARIZE THE MATHEMATICS

In this investigation, you studied public-key cryptography. In the process, you learned about the RSA cryptosystem, protocols, and digital signatures.

a. Describe the processes of encryption and decryption used in the RSA cryptosystem.

b. Describe similarities and differences between public-key and symmetric-key cryptosystems.

c. The main theme for this lesson is information security. Two important aspects of information security are confidentiality and authenticity.

    i. Explain how a public-key cryptosystem can be used to provide confidentiality.

    ii. Explain how a public-key cryptosystem can be used to ensure authenticity.

Be prepared to share your descriptions and explanations with the class.

✓ CHECK YOUR UNDERSTANDING

Use your understanding of the RSA public-key cryptosystem to complete the following tasks.

a. Suppose $p = 3$ and $q = 11$. Use these primes and the RSA cryptosystem to encrypt the one-letter message D. Then decrypt to verify that you get back to D.

b. There are at least three characteristics that a digital signature should have.

    (1) The signature must be unforgeable, so that the receiver knows exactly who made the signature.

    (2) The signature must not be reusable, so that the receiver knows the signature was not cut from one document and pasted into a different document.

    (3) The signed document must be unalterable, so that the receiver knows the document was not changed after it was signed.

Explain how the digital signature defined by the protocol on page 473 has each of these characteristics.

1 A variation on a Caesar cipher uses a keyword to shuffle the alphabet, and then you do a letter substitution. Here is how it works: First, choose a keyword that has no duplicate letters, such as, "crypto." Put this keyword in front of the standard alphabet, leaving out the letters in the alphabet that are in the keyword. Thus, in this case you get:

C, R, Y, P, T, O, A, B, D, E, F, G, H, I, J, K, L, M, N, Q, S, U, V, W, X, Z

Now encrypt your plaintext by substituting letters based on this shifted and shuffled alphabet. For example, BE HAPPY becomes RT BCKKX.

a. Use this cryptosystem with the keyword "crypto" to encrypt the message MEET ME AT THE MOVIES. Decrypt the message NTIP BTGK.

b. Make up your own keyword and use it to encrypt a message.

c. Explain why this cipher is a substitution cipher but not a Caesar cipher.

2 Use a Hill cipher based on the matrix $A = \begin{bmatrix} 8 & -5 \\ 5 & -3 \end{bmatrix}$ to encrypt the message

BANANA SPLIT. Then decrypt the message to verify that you get the original plaintext.

3 You can build a device like that shown below for quickly encrypting (or decrypting) messages using a Caesar cipher. The outer wheel represents the plaintext and the inner wheel represents the ciphertext. The inner wheel can be rotated to different positions.

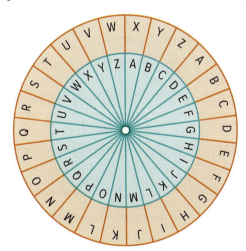

a. Describe the particular cipher shown here in words and in an equation of the form $c = p + n \mod 26$.

b. How many different Caesar ciphers are possible? Explain your reasoning.

c. Experiment with the Caesar Cipher feature of the *TCMS-Tools* "Codes" custom app.

    i. Use the tool to create the cipher in Part a.

    ii. Use the tool to illustrate your answer to Part b.

    iii. Compare the Caesar Cipher feature to the ROT13 Cipher feature of the "Codes" custom app. Explain how they are similar and how they are different.

d. Experiment with the Affine Cipher feature of the *TCMS-Tools* "Codes" custom app.

    i. Explain in words how the Affine Cipher encrypts.

    ii. What settings in the Affine Cipher will create the ROT13 cipher?

    iii. Explain why the Affine cipher is not a Caesar cipher. How is this shown in the slope of the respective linear models?

4 Carry out the indicated modular arithmetic procedures.

a. $3 \times 4 = $ ___ in Z_5

b. Find x if $18 + 36 \equiv x \bmod 7$, where x is in Z_7.

c. Find the additive inverse of 10 in Z_{13}.

d. Find the multiplicative inverse of 8 in Z_{15}.

e. Reduce $6^8 \bmod 14$.

5 Suppose the following RSA public key is published: $n = 77$ and $e = 13$. You intercept a one-letter ciphertext message: "30." What is the plaintext message?

CONNECTIONS

6 A Caesar cipher can be modeled using a linear function with arithmetic mod 26, where A is initially represented as 0, B as 1, and so on, with Z as 25.

a. Encrypt the message YOU HAVE MAIL in two ways:

        (1) using the function $f(x) = x - 8$.

        (2) using the function $g(x) = x + 18$.

    Explain the pattern relating the two resulting ciphertext messages.

b. The message EJDI OCZ XGPW was encrypted with the function $h(x) = x - 5$. What is the original plaintext message?

c. If a message has been encrypted using the function $C(x) = x + b$ with arithmetic mod 26, what general strategy will decrypt the message?

7 The *TCMS-Tools* "Codes" custom app screen below shows algebraic and graphical representations of a cipher.

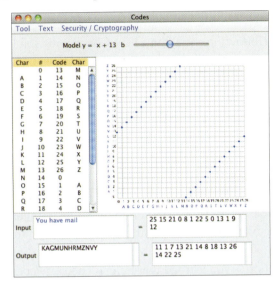

a. What do the dots in the graph represent? Why are there two linear patterns?

b. Describe the display that would be produced when $b = 0$.

c. How do the locations of the dots change as the value for b changes? Explain why this makes sense.

d. Suppose you received the following message coded in a Caesar cipher:

RHNTAUOYTWHKKYWMERTXYWBIAYKYXTMAYTWHXY

Decode this message.

8 The addition and multiplication tables that arise from modular arithmetics have some very interesting patterns.

a. Complete the following addition and multiplication tables for arithmetic mod 7. These tables can be used to investigate properties of addition and multiplication in Z_7.

+	0	1	2	3	4	5	6
0	0	1					
1	1						
2	2						
3							
4			0				
5							
6							5

×	0	1	2	3	4	5	6
0	0	0	0				
1	0	1					
2	0						
3				2			
4						6	
5							
6							

b. Perform these operations in Z_7.

 i. 6^2

 ii. $2 +$ (additive inverse of 6)

 iii. $4 \times$ (multiplicative inverse of 6)

 iv. $2 + (-3)$

c. Solve the following equations in Z_7. Check your solutions.

 i. $3x + 6 = 2$

 ii. $x^2 + 3 = 5$

 iii. $(x + 4)(x - 6) = 0$

 iv. $3x + y = 5$ and $2x - y = 4$

d. Which numbers in the system have additive inverses? Which have multiplicative inverses?

e. Can you solve every linear equation of the form $ax + b = c$ when a, b, and c are integers, $a \neq 0$, in Z_7? Explain.

9 Suppose two integers reduce mod n to the same number. Are these two integers equivalent mod n? Explain, by using the definition in Investigation 2.

10 In Investigation 3, you learned that Euler's Theorem (special case) is one reason why the RSA cryptosystem works. Euler proved this theorem using another famous theorem first proved by the French mathematician Pierre de Fermat (1601–1665), called **Fermat's Little Theorem**:

> *If p is a prime number and a is any integer,*
> *then $a^p \equiv a \bmod p$.*

Check that this statement is true for several values of p and a.

Pierre de Fermat

REFLECTIONS

11 Mathematicians use the symbol Z_n to represent the finite set of numbers $\{0, 1, 2, 3, \ldots, n - 1\}$ because the letter Z is the first letter of the German word "zyklisch" which translates to "cyclic" in English. Why is the word "cyclic" appropriate in connection with modular arithmetic?

12 Public-key cryptography is also sometimes called *asymmetric cryptography*. Explain why the word "asymmetric" is a sensible description of public-key cryptography, especially in comparison to symmetric-key cryptography.

13 In 1999, Sarah Flannery was a 16-year-old high school student in Ireland who made a "breakthrough" in cryptography. In an article written about her, Ronald Rivest (the "R" in RSA) claimed that Flannery knew what she was talking about, but that there was not enough information to evaluate her work. The article also said that "her work must still endure the test of time." Do some research to find out whether her work has endured and write a short report.

(**Source:** www.zdnet.com/rsa-weighs-in-on-teens-breakthrough-3002070539/)

14 The Central Intelligence Agency (CIA) and the National Security Agency (NSA) are two prominent U.S. Government agencies that deal with cryptography. The NSA is said to be the largest employer of mathematicians in the world. Investigate one interesting aspect of cryptography that is pursued by the CIA or the NSA. Write a brief report on what you find. The following topics might be interesting.

- *Kryptos* is an enigmatic sculpture at CIA headquarters that contains an encrypted message in four sections. There has been some competition between the CIA and the NSA to break the code for each section. As of early 2014, the code for three sections has been broken and the messages decrypted, but the puzzle of the fourth section remains unsolved.

- The NSA was formed in 1952. It was so secret that it was jokingly known as "No Such Agency." Nowadays, NSA refers to itself as "America's Codemakers and Codebreakers." While NSA facilities are off limits to the public, you can visit the NSA National Cryptologic Museum. The museum includes an exhibit on Women in American Cryptology, with information on 24 honorees.

Encrypted *Kryptos* sculpture at CIA headquarters

EXTENSIONS

15 To use a Hill cipher as in Investigation 1 (pages 459–460), you need an

encrypting matrix E such that $E = \begin{bmatrix} a & b \\ c & d \end{bmatrix}$ and $E^{-1} = \begin{bmatrix} \dfrac{d}{ad-bc} & \dfrac{-b}{ad-bc} \\ \dfrac{-c}{ad-bc} & \dfrac{a}{ad-bc} \end{bmatrix}$ both have

integer entries.

a. Construct a 2 × 2 matrix with integer entries (different from ones used so far in this lesson) that has an inverse matrix also with integer entries. Use this matrix to encrypt, and then decrypt, the message MATH.

b. The **determinant** of the 2 × 2 matrix E is denoted det(E) and is defined to be $ad - bc$. It turns out that if both E and E^{-1} have only integer entries, then det(E) = 1 or −1. Verify this fact for the matrix you constructed in Part a and the two matrices you used in Investigation 1, reproduced below.

$$\begin{bmatrix} 1 & 2 \\ 1 & 3 \end{bmatrix} \qquad \begin{bmatrix} -4 & -3 \\ 9 & 7 \end{bmatrix}$$

c. Write an argument justifying the fact stated in Part b for all 2 × 2 matrices.

Jim Sanborn

16 When using the mathematics of public-key cryptography, you need to reduce some large numbers mod n. In particular, you need to reduce powers of integers.

a. Let's start small. Consider reducing 10^2 mod 6. Using properties of exponents and what you know about mod n arithmetic, give a plausible explanation for each step in the following three different reduction methods.

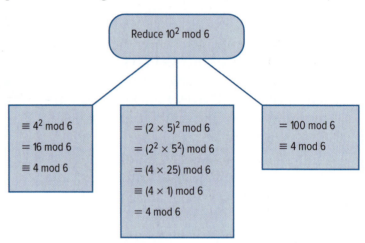

Reduce 10^2 mod 6

$\equiv 4^2$ mod 6
$= 16$ mod 6
$\equiv 4$ mod 6

$= (2 \times 5)^2$ mod 6
$= (2^2 \times 5^2)$ mod 6
$= (4 \times 25)$ mod 6
$\equiv (4 \times 1)$ mod 6
$= 4$ mod 6

$= 100$ mod 6
$\equiv 4$ mod 6

b. Reduce 4^6 mod 12 in two different ways, as described below, to see if they each yield the same answer, and to see if there are any shortcuts.

  i. Reduce 4^6 mod 12 by directly applying the definition of how to reduce mod 12. That is, compute 4^6, divide by 12, and find the remainder. What is the result?

  ii. Using properties of exponents and what you know about mod n arithmetic, give an explanation for each step below.

$$4^6 = (4^2)^3 \text{ mod } 12 \qquad (1)$$
$$= 16^3 \text{ mod } 12 \qquad (2)$$
$$\equiv 4^3 \text{ mod } 12 \qquad (3)$$
$$= (4^2 \times 4) \text{ mod } 12 \qquad (4)$$
$$= (16 \times 4) \text{ mod } 12 \qquad (5)$$
$$\equiv (4 \times 4) \text{ mod } 12 \qquad (6)$$
$$= 16 \text{ mod } 12 \qquad (7)$$
$$\equiv 4 \text{ mod } 12 \qquad (8)$$

c. Based on Parts a and b, you might conjecture the following fact about reducing mod n.

> *When reducing mod n after performing the operations of addition, multiplication, and whole-number powers, you can either perform these operations first and then reduce, or you can reduce first and then perform the operations. You get the same answer in either case.*

Reduce 12^{25} mod 145.

17 In Problem 11 of Investigation 2 (page 465), you may have made the following conjectures.

Conjecture I: *For any modular arithmetic system Z_n and any nonzero element m in Z_n, m has a multiplicative inverse if the greatest common divisor of m and n is 1.*

Conjecture II: *If n is prime, then every nonzero integer in Z_n has a multiplicative inverse.*

Write justifications of these conjectures. In the case of the first conjecture, you will need to use the fact that if the greatest common divisor of x and y is k, then you can find integers a and b so that $ax + by = k$.

18 An important part of the RSA cryptosystem is to find two numbers that are multiplicative inverses mod r. An efficient method for doing this involves what is called the *Euclidean algorithm*. Research and write a brief report on how the method works.

19 Cryptography is featured in many stories, novels, and movies. Read a short story or novel or watch a movie in which cryptography plays a major role. Write a brief report describing how cryptography is featured. Include at least one specific example of how cryptography is used. You might find the following interesting:

- "The Gold Bug," a short story by Edgar Allen Poe

- *Cryptonomicon*, a novel by Neal Stephenson

- *Enigma*, a novel by Robert Harris, made into a movie directed by Michael Apted

- "The Adventure of the Dancing Men," a Sherlock Holmes story by Arthur Conan Doyle

- *The Da Vinci Code*, a novel by Dan Brown, made into a movie directed by Ron Howard (There was a court case related to this book in which the judge, Justice Peter Smith, embedded a secret code in his ruling for the case. This code was discovered and broken in 2006. For information about this code, do an Internet search for the "Smithy Code.")

20 Often it is valuable to rewrite numbers in different but equivalent forms. Doing so may provide insight into properties of the number and it can also allow you to identify relationships between numbers.

a. Rewrite each number as a product of prime numbers.

   i. 18   **ii.** 64

   iii. 140   **iv.** 196

   v. 320

b. Determine the greatest common divisor for each pair of numbers.

   i. 18 and 6   **ii.** 196 and 64

   iii. 64 and 320   **iv.** 140 and 320

   v. 320 and 18

c. How can you use the prime factorization of a pair of numbers to help you determine the greatest common divisor of the numbers?

21 In recent years, the number of students enrolled in colleges and universities in the United States has been increasing linearly. In 2001, there were approximately 15.9 million students enrolled in U.S. colleges and universities (including community colleges). In 2012, total U.S. college and university enrollment had increased to approximately 19.8 million students.

a. Using the information above, write a function rule that can be used to estimate student enrollment in U.S. colleges and universities. Let t represent the number of years since 2000 and $E(t)$ represent the number of students in millions.

b. What is the slope of the graph of this function? Be sure to include units of measure in your answer. What does it tell you about student enrollment in U.S. colleges and universities?

c. What is the y-intercept of the graph of this function? What does it tell you about student enrollment in U.S. colleges and universities?

d. Use your function to estimate college and university enrollment in 2015.

e. Use your function to determine the year during which the enrollment will first be greater than 23 million.

22 Write each expression in a simpler equivalent exponential form using only positive exponents.

a. $(3x^6)^2$ **b.** $(3x^6)(5x^2)$

c. $\dfrac{10x^6}{4x^2}$ **d.** $\dfrac{-2x^2}{8x^6}$

e. $14(7x)^{-1}$ **f.** $(3x^{-6})(2x^4)$

23 In 2013, approximately 90.7% of households in the United States owned an automobile. To see if this percentage has changed, a polling organization takes a random sample of 250 households in the United States and asks if they own an automobile. (**Source**: www.cnbc.com/id/100762511/)

    a. If the percentage has not changed, what is the expected number of households in the sample that will own an automobile?

    b. If the percentage has not changed, what is the probability that 220 or fewer households in the sample will own an automobile?

    c. Is it statistically significant to get only 220 households that own an automobile out of the sample of 250 households?

24 Rewrite each product as a single algebraic fraction. Then simplify the result as much as possible.

    a. $\left(\dfrac{2x^2}{9}\right)\left(\dfrac{12}{8x}\right)$

    b. $\left(\dfrac{3a}{12}\right)\left(\dfrac{a+4}{a^2}\right)$

    c. $\left(\dfrac{y^2-16}{y+1}\right)\left(\dfrac{y}{y+4}\right)$

25 Solve each equation.

    a. $2t + 4(5t + 9) = 3$

    b. $8y(y - 2) + 4(6y) = 4(2y + 1.5y + y)$

    c. $-0.15(p + 3) = p(0.75p - 1.25) - 0.75p^2$

    d. $\dfrac{3}{8}x - \dfrac{1}{4}\left(\dfrac{1}{2}x + 4\right) = \dfrac{3}{2}$

26 Rewrite each expression in standard polynomial form.

    a. $3x(4 - 2x) + 8(x^2 - 5x)$

    b. $(x - 5)(4x^2 + 7x - 9)$

    c. $(5x^3 - 7x^2 + 5) - (x^3 + 10x - 12)$

    d. $\dfrac{4x^3 - 6x^2 + 2x}{2x}$

Accuracy: Error-Detecting and -Correcting Codes

So far in this unit, you have learned how to make information accessible and secure. It is also important for information to be accurate. When you rent a movie on DVD, you want it to play correctly even if the DVD is slightly scratched. When the cashier at the grocery store scans your food purchase, you expect that an accurate price will be sent to the cash register. When you pick up medication at the pharmacy, you want to know for sure that what you receive is the correct prescription. When you check out a book from the library, it is important that the automated barcode reader sends accurate information to the library's database. When NASA sends an unmanned space probe to the far edge of the solar system, and they wait years to receive the information that it gathers, they want the received information to be accurate. However, any time information is electronically read, recorded, or transmitted, there is a chance that errors could occur.

Fuse/Getty Images

THINK ABOUT THIS SITUATION

Think about the issue of accuracy when information is electronically processed.

a. What are some situations in your daily life where information is electronically read or transmitted and accuracy is important? How do you think accuracy is ensured?

b. What information do you think is encoded in each of the two barcodes below? Why is accuracy of information important in these situations?

49008-5201

|.|..||.|..||..||...|.|.|.|.|..|.|||.....||..|||

c. How are your home and school ZIP codes similar? How are they different? Why do you think this is the case?

In this lesson, you will learn some mathematical techniques for detecting and correcting errors when information is electronically processed. In the first investigation, you will study codes used for accurate ID numbers, like UPC and ZIP codes. In the second investigation, you will learn about error correcting codes used to transmit information accurately.

INVESTIGATION 1

ID Numbers and Error-Detecting Codes

Identification (ID) numbers are found on books, driver's licenses, checks, airline tickets, mailing envelopes, grocery products, and in many other places. Many ID numbers are represented as *barcodes*. A **barcode** is a pattern of bars that encodes the ID number. Barcodes allow ID numbers to be scanned into computers using laser scanning devices. Because of this feature, barcodes are sometimes referred to as "keyless data entry."

As you analyze ID numbers and barcodes in this investigation, look for answers to these questions:

What mathematical concepts and methods are used to create ID numbers and barcodes?

How are errors detected and possibly corrected?

To help ensure accuracy, ID numbers consist of two types of digits. Most of the digits, the **information digits**, are used to classify and identify the object. Then one or more extra digits are added to serve as *check digits*. The **check digits** are used to help detect and correct errors that may occur when the ID number is read by or stored in a computer or is electronically transmitted. For example, consider the U.S. Postal Service ZIP code.

ZIP Codes   Zoning Improvement Plan codes (ZIP codes) were first introduced in 1963 as a five-digit code. Nine-digit ZIP codes like the one on the business reply card below were introduced in 1983 and are commonly used today. The ZIP code identifies the delivery location. The first digit identifies a large geographical area, usually a group of states. The next two digits identify a particular mail-distribution center. The fourth and fifth digits indicate the town or local post office. The four extra digits added after the dash identify a specific location within a town.

In March 1993, eleven-digit ZIP codes were introduced. The additional two digits give the last two digits of the street address or box number. These final two digits are often not printed; they only appear in the barcode. Since nine-digit ZIP codes are more common, only nine-digit ZIP codes are analyzed in this investigation.

1 The barcode on the envelope shown above represents the ZIP code. By representing a ZIP code with a barcode, computers are able to read the code and help process the mail. The barcode shown is called a POSTNET (**Post**al **N**umeric **E**ncoding **T**echnique) code. The United States Postal Service is phasing in a new barcode, called an *Intelligent Mail barcode*. As of 2014, both codes are in use. In this problem, you will learn how the POSTNET code works.

   The POSTNET barcode is created by converting the decimal digits 0–9 to binary digits, that is, 0s and 1s, so that a computer can more easily read the numbers. Binary digits are called **bits**. *Each decimal digit 0 to 9 is represented by a five-bit string that contains exactly two 1s.* This is called a **POSTNET 2-out-of-5 code**. The translation of a decimal digit to a binary string is done using the dictionary given at the top of the next page.

POSTNET Binary Code Dictionary

Decimal Digit	2-out-of-5 Binary Code
1	00011
2	00101
3	00110
4	01001
5	01010
6	01100
7	10001
8	10010
9	10100
0	11000

How many different 2-out-of-5 binary strings are possible? Are they all used in the dictionary?

2 Look closely at the series of long and short black bars in the POSTNET barcode at the bottom of the business reply card. Each long bar represents a 1 and each short bar represents a 0. *The first and last long bars are not used in the code; they are only used to define the beginning and end.*

  a. What five-bit string is represented by the first five bars in the code? Is this a 2-out-of-5 string?

  b. Write all the five-bit strings represented by the bars. Separate each five-bit string with a space so that you can easily distinguish them.

  c. Use the POSTNET Binary Code Dictionary in Problem 1 to translate the five-bit strings into decimal digits. Compare the result to the ZIP code printed in the address on the card. Does it match? If not, how is it different?

3 The tenth digit that you found in Problem 2 is the **check digit**. It is used to detect and correct errors that may occur when the ZIP code is read, recorded, or transmitted by a computer. In this problem, you will discover how the check digit works.

  a. Add up all nine digits in the ZIP code on the reply card, and also add the tenth check digit. Then reduce mod 10.

  b. Repeat Part a for the following three additional ZIP codes, with indicated check digits.

    i. 47402-9961, with check digit 8

    ii. 80323-4506, with check digit 9

    iii. 02174-4131, with check digit 7

  c. By examining the results in Parts a and b, state a rule for how the check digit works. Compare your rule with that of others and resolve any differences.

4 Codes and check digits are used to help detect and correct errors. However, not all errors can be detected or corrected.

  a. Suppose that a ZIP code is printed on a business envelope as 55441-6733, with check digit 4. Check if a printing error has been made.

  b. It is good to know that an error has been detected, but it would be even better if the error could be *corrected*. Given that errors are rare, it is reasonable to assume that the error is probably in just one digit. An error in which one digit is incorrect while all others are correct is called a **single-digit substitution error**.

   i. How could you change one digit in the ZIP code in Part a so that the check-digit rule is satisfied?

   ii. Are there other one-digit changes you could make so that the check-digit rule is satisfied? Explain.

  c. Will the ZIP code check-digit system *detect* all single-digit substitution errors? Will it *correct* all single-digit substitution errors? Justify your answers.

  d. Will the ZIP code check-digit system always detect a multiple-digit error, that is an error in which more than one digit has been misprinted? If so, explain how. If not, give an example of a multiple-digit error that is not detected.

5 When a computer-based mail-sorting machine detects a ZIP code error that it cannot correct, it sets the letter aside to be looked at by a human. It would be more efficient if at least some errors could be automatically corrected. In fact, there are some errors that can be corrected if you use the check digit along with information about the 2-out-of-5 binary code. Consider what happens when an error occurs in the bars of the barcode.

  a. Suppose that the first 21 bars of a ZIP-code barcode are printed as below. (As always, the first bar is not part of the code, it only marks the beginning.)

  |.|..||.||....|||.|..

  Find the bar printing error. Explain why it is an error. Is there a unique, single-bar correction that can be made? If so, explain. If not, list all the decimal digits that could be the corrected digit.

  b. Explain why a single-bar error will always be detected when using the 2-out-of-5 binary code. Explain why a two-bar error may or may not be detected.

c. So far in this problem, you have only been using information about the 2-out-of-5 code. With this information, you can detect a single-bar error, but you cannot correct it. Suppose you also use the check-digit system.

Consider the following situation where:

- The first 21 bars of a ZIP code barcode are printed as shown. There is a single-bar error.

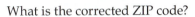

- There are no other errors. The last five correct digits of the ZIP code are 8-3447, and the correct check digit is 5.

What is the corrected ZIP code?

d. Using both the ZIP code check-digit system and the 2-out-of-5 barcode system, explain why you can always detect *and* correct single-bar errors, assuming that there are no other errors.

6 For every code, there must be a process for encoding and a process for decoding. In Problem 2, you worked through the decoding process for the POSTNET code. You started with the barcode, then you decoded to get a series of five-bit binary strings, then you used the dictionary to decode the five-bit strings into decimal digits, and you ended up with the ZIP code. Now reverse that process.

Consider this ZIP code: 22091-1593

Find the check digit, and then encode the ZIP code and check digit into a barcode.

UPC Codes Zip codes use one of the simplest barcode and check digit systems. There are many other codes that work in a similar but more complicated manner. One of the most common such codes is the Universal Product Code (UPC).

The UPC was first used in 1974 to identify grocery items. Now, it is used for almost all retail goods. For example, the UPC from a can of pie cherries is 0-41345-51718-1. A particular brand and style of fly fishing rod has UPC 0-43372-60092-8. The first digit identifies the general type of product. The "0" in both examples here identifies these items as nationally branded products. The next five digits identify the manufacturer, and the following five digits are assigned by the manufacturer to identify the specific product. The last digit is the check digit.

7 Investigate how the UPC check digit works.

a. Consider the fishing rod UPC: 0-43372-60092-8. Carry out the following check-digit algorithm.

Step 1. Add all the digits in the odd positions. (The first digit is 0.)

Step 2. Triple this sum.

Step 3. Add this result to the sum of all the digits in the even positions.

Step 4. Reduce mod 10. This final result should be 0.

b. Carry out the check-digit algorithm for the pie cherries UPC: 0-41345-51718-1. If your final result is not 0, re-examine your work and correct any miscalculations.

8 Suppose the pie-cherries UPC is transmitted and received with a single-digit substitution error in the third position: 0-46345-51718-1.

a. Will this error be detected by the check digit? Explain.

b. Can this error be corrected based on information from the check digit? Explain.

Substitution and Transposition Errors So far, you have considered *substitution errors*, that is, errors in which an incorrect digit is substituted for a correct digit. Now think about another common type of error—transposing two digits, which is called a *transposition error*.

9 Suppose the ZIP code 48195-9822 is incorrectly read by a computer as 48159-9822 (the fourth and fifth digits are transposed).

a. Will this error be detected based on information provided by the check digit? Explain why or why not.

b. Are there any digit transposition errors that will be detected by the ZIP code check-digit system? Explain.

In Extensions Task 13 (page 506), you will explore if the UPC code can detect transposition errors.

SUMMARIZE THE MATHEMATICS

In this investigation, you explored some of the mathematics behind identification numbers, barcodes, and check digits.

a. Explain how modular arithmetic is used to detect errors in a ZIP code. In a UPC code.

b. Suppose you want to explain to a friend how the check digit works in a ZIP code, but your friend has not studied this unit and does not know what mod 10 means. Explain how the check digit works in a way that your friend would understand.

c. An important feature of a code is its ability to detect and correct errors. What types of errors can the ZIP code and the UPC code detect? What types of errors can the ZIP code and the UPC code correct?

Be prepared to share your explanations and summaries with the class.

✓ CHECK YOUR UNDERSTANDING

Apply what you have learned about codes for ID numbers and error checking to complete the following tasks.

a. Find a business reply card that shows a nine-digit ZIP code and the barcode representation. Decode the barcode and verify that you get the ZIP code.

b. Find the check digit for the ZIP code 33664-4162.

c. Although machine scanners are regularly used to read the UPC number from a barcode, sometimes a sales clerk needs to manually key in the number. Unfortunately, the digits can be distorted or otherwise difficult to read. The barcode and UPC number shown at the right are for a particular piece of luggage. The check digit is illegible. What is the correct check digit? (Note that the first digit is the 0 on the far left.)

0   43202 07124

d. Find a retail product that has a 12-digit UPC code. Verify that the check digit given on the code is correct.

Transmit Information Using Error-Correcting Codes

The codes you studied in the last investigation are designed to provide accurate identification numbers. The codes you will now learn about have a different purpose. They are used to transmit data in the presence of "noise" or interference, so that the received data will be accurate.

Whenever information is transmitted, errors can occur. This is particularly true of digital data transmitted electronically, like the electronic signal inside a Blu-ray player or a cell phone signal from a satellite. Error-correcting codes are often used to help provide accuracy. You have already seen that different codes have different error detection and correction capabilities. For example, the ZIP code and UPC code that you studied in the previous investigation can detect some common errors, but neither is very good at correcting errors. In contrast, you will now learn about codes that are able to detect *and* correct certain common errors.

As you work on the problems in this investigation, look for answers to these questions:

> *What mathematical concepts and methods are used to create error-correcting codes?*
>
> *How is a linear code constructed and what are its error-correcting capabilities?*

Binary Codes Sometimes images are sent across millions of miles of space, like the image here of Barred Spiral Galaxy NGC 1672, taken from the Hubble telescope.

Suppose the grid below is one small, enlarged section of that image.

The galaxy image was sent by translating the image into long strings of 0s and 1s that represent all the different shadings of color. In this simplified example, there are only four sections, or *pixels*, and two colors—black and white. As you learned in Investigation 1, the digits 0 and 1 are called binary digits, or bits. A code that translates data into strings of 0s and 1s is called a **binary code**.

1 Consider a very simple binary code. Represent white with a 0 and black with a 1.

    a. Scan the simple grid image at the bottom of the previous page pixel-by-pixel from left to right and top to bottom.

        i. What four-digit string does this scan produce?

        ii. As this string of bits travels through space to Earth (in the form of radio signals), it is possible that some bits will be corrupted by radiation or other factors. This will cause a 1 to be changed to a 0, or vice versa. Suppose the received message is 1101. Sketch the grid that this received message represents.

    b. In Part aii, the received image was quite different from the sent image. An error occurred. Assume that the receiver does not know what the original message is supposed to be. All the receiver can do is analyze the received message. By just examining the received string of 1s and 0s, is there any way to detect the error? Explain your reasoning.

Maximum-Likelihood Decoding In the case of simple codes as in Problem 1, an incorrect received message is decoded as a possible valid message, and thus it is not picked up as an error. Now consider a slightly more complex code, along with a method of decoding.

2 Represent white with 000 and black with 111. The strings 000 and 111 are called the *code words* of this code. In general, **code words** are the specified strings of 1s and 0s that are used to **encode** messages.

    a. Write the string that represents the four-pixel grid image on the previous page. That is, encode the message (image) using the code words 000 and 111. Leave space between code words for easier reading.

    b. Suppose the received message for one pixel is 001. Is this an error? Explain.

    c. What code word should 001 be decoded to? Explain

    d. One natural way to correct the error in Part b is to **decode** the received message into the code word that is most likely to have been transmitted. This is called *maximum likelihood decoding*. You can assume that errors are rare and that it is more likely to have an error in fewer bits than in more bits. Thus, **maximum likelihood decoding** means that you decode each word of the received message to the code word that differs from it in the fewest number of bit places.

    Suppose that the received message for a four-pixel image is:

111   110   000   010

    Decode the received message into valid code words, using maximum likelihood decoding. Sketch and shade the resulting image.

In Problem 2, you used the code words 111 and 000 to encode an image, and you decoded using maximum likelihood decoding. This is an example of the general coding situation illustrated in the diagram below.

The Process of Encoding and Decoding

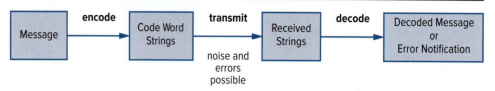

As indicated in the diagram, errors can occur during transmission. The goal of this investigation is to design and analyze codes that can detect and correct such errors. In the remainder of this lesson, you will always use maximum likelihood decoding.

3 Consider the code in Problem 2, along with maximum likelihood decoding.

a. Explain why any error in a single bit can be corrected using this code.

b. Can this code correct double errors? For example, suppose the message 111 is received as 001. That is, a double error occurred in transmission: both of the first two bits were changed. Using maximum likelihood decoding, will this error be corrected? Explain.

4 Now consider some other similar codes.

a. Suppose the code words are 1111 and 0000. Using maximum likelihood decoding, can this code correct single errors? Give an example.

b. Can this code correct double errors? As an example, suppose 1100 is received. Can this double error be corrected? Explain.

c. Consider a code with the following five-bit code words: 11111 and 00000.

   i. How many bit errors can be detected? Explain and give an example.

   ii. How many bit errors can be corrected? Explain and give an example.

5 When trying to correct an error, three things can happen.

   • The error is corrected.

   • A mistake is made when trying to correct the error.

   • The process is indecisive, that is, you cannot choose a single code word to decode to.

a. Give an example from this investigation of each of these three situations.

b. In each of the three situations above Part a, you have *detected* an error, but you cannot always *correct* it. *It is generally true that you can detect more bit errors than you can correct.* Sometimes you must decide whether you would rather just detect and report errors or try to correct the errors. You could set up your system of encoding and decoding so that maximum likelihood decoding is used to try to correct every detected error. Or you could set up the system so that whenever an error is detected an error notification is announced but no attempt is made to correct errors. Think about which system would be best in the following situations.

   i. Suppose you can easily have a message resent if necessary. For maximum accuracy, would you choose a code system that only detects errors, or one that attempts to also correct errors? Explain your reasoning.

   ii. Suppose it is difficult or impossible to have messages resent, like in the case of sending photographs from space. Would you choose a code system that only detects errors, or one that also attempts to correct errors? Explain.

Hamming Distance It is useful to compute a type of distance between code words. The **Hamming distance** between two code words is the number of bit places in which the two code words differ. You may have considered Hamming distance as you answered some of the questions above.

6 There is a strong connection between Hamming distance and error correction, which you may have already noticed.

  a. Consider two codes you studied in previous problems.

    i. The code consists of the two code words 111 and 000. What is the Hamming distance between 111 and 000? How many bits can be corrected using this code?

    ii. The code consists of the two code words 11111 and 00000. What is the Hamming distance between 11111 and 00000? Errors in how many bits can be corrected with this code?

  b. So far, you have only considered codes with two code words. Clearly this limits the kinds of messages you can send! The same connections between Hamming distance and error correction apply to codes with more code words. Suppose there are four code words:

00000 01110 10111 11001

Find the Hamming distance between each pair of code words. What is the smallest Hamming distance between any pair? The smallest Hamming distance between any pair of code words in a code is called the **minimum distance** d for the code.

  c. Using the code in Part b, suppose 11110 is received. Is this an error? To which code word will it be decoded, using maximum likelihood decoding? Assuming it is a one-bit error, did the error get corrected?

  d. The minimum distance for the code in Part b is $d = 3$, and you have been able to correct a one-bit error. This leads to an important conclusion about correcting one-bit errors:

    You can correct all one-bit errors when the minimum distance for a code is 3.

  Explain why this is the case.

7 Now consider how you can generalize the connection between minimum Hamming distance and error correction.

a. Suppose that the minimum Hamming distance for a code is $d = 2m + 1$. Errors in how many bits can be corrected? Explain and give an example. Explain how the diagram below illustrates this situation.

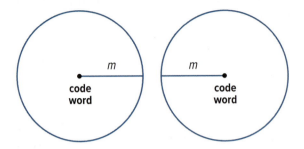

b. Suppose the minimum distance for a code is $d = 2m$. How many bit errors can be corrected? Explain and give an example. Draw a diagram similar to the one above that illustrates this situation.

c. Suppose the minimum distance for a code is $d = n$. How may bit errors can be *detected*? Explain and give an example.

Linear Codes You can determine how many bit errors a code can correct by knowing the minimum Hamming distance. When there are many code words, it can be difficult to find the minimum distance since you must check the Hamming distance between every pair of code words. However, for *linear codes*, finding the minimum distance is easy. A **linear code** is a set of binary code words such that the sum of any two code words in the set (including adding a code word to itself) is also a code word in the set.

8 The first step in understanding linear codes is to understand how to **add** two code words. To **add two code words**, you add them bit-by-bit using mod 2 arithmetic.

a. Consider the code words 011 and 101. Find the sum of these two code words.

$$011 + 101 = ?$$

Compare your code-word sum with those of your classmates. Resolve any differences.

b. Compute the following code-word sums.

  i. 111 + 001

  ii. 11101 + 00000

  iii. 0101 + 0101

9 A set of code words is a linear code if the sum of any two code words is also a code word. For example, {11, 00} is a set of code words. It is a linear code because 11 + 00 = 11, which is a code word.

   a. Is the set {111, 000} a linear code? Explain.

   b. Is {000, 011, 101, 110} a linear code? Explain.

   c. Is {0000, 0101, 1010} a linear code? Explain.

10 Finding Minimum Distance for a Linear Code It is easy to find the minimum Hamming distance for a linear code.

   a. To see why, examine each step of the following argument. Recall that the Hamming distance between any two code words is the number of bit places in which they differ.

   Step 1. If two code words differ in a particular bit place, then when you add them you will get a 1 in this place. If they are the same in a particular bit place, then when you add them you will get a 0 in this bit place.

   Explain and give an example of Step 1.

   Step 2. Thus, to find the Hamming distance between two code words, you can add the two code words and count the number of 1s in the sum.

   Give an example illustrating Step 2.

   Step 3. In a linear code, the sum of two code words is also a code word. Moreover, every code word can be written as the sum of two code words. Thus, to find the minimum Hamming distance for all pairs of code words, you can just find the minimum number of 1s for all the individual code words (do not include the code word that is all 0s). Therefore:

   The minimum distance for a linear code is the minimum number of 1s for all the individual nonzero code words in the code.

   b. Give an example of this generalization by considering this linear code:

   {00000, 01110, 10001, 11111}

   c. Find the minimum distance for this code in a different way than you did in Part b.

Linear codes are desirable codes because the minimum Hamming distance is easy to compute—just find the minimum number of 1s in all nonzero code words. And you know from Problem 6 that if the minimum Hamming distance for a code is 3 or greater, then the maximum likelihood decoding scheme corrects all single errors. Because of these facts, it is useful to create linear codes with minimum Hamming distance 3. To do this requires some work with matrices. Learn how by accessing "Single-Error-Correcting Linear Codes" in your book's Resources section on ConnectED.

SUMMARIZE THE MATHEMATICS

In this investigation, you learned about error-correcting codes, including the ideas of maximum-likelihood decoding, Hamming distance, minimum distance, and linear codes.

a. What is the Hamming distance between two code words?

b. What minimum distance will guarantee that a code can correct all single errors? Explain why this is true.

c. How do you check to see if a set of code words forms a linear code?

d. Why can you find the minimum distance for a linear code by finding the minimum number of 1s for all code words (except the code word with all 0s)?

e. There are three different but related ways to create an error-correcting code.

   (1) Create redundancy by adding bits.

   (2) Create distance between code words.

   (3) Add check bits.

   Give an example of each of these three strategies for creating an error-correcting code.

Be prepared to explain your ideas and examples to the class.

✓CHECK YOUR UNDERSTANDING

As you complete these tasks, think about how they are connected to the entire process of encoding and decoding.

a. Find the Hamming distance between these pairs of code words.

   i. 11101 and 11001

   ii. 1010 and 0001

b. Suppose the grid at the right is a small part of an image sent from a satellite. The image uses four shades of gray (two of which are black and white). Use the linear code {000, 011, 101, 110} to encode this image. You decide how to assign code words to shades of gray.

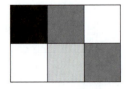

c. Suppose there are four code words: 00000 01110 10111 11001

   i. What is the minimum distance for this code? Explain how you found the minimum distance.

   ii. Suppose you receive the code word 11111. Can this error be detected? Using maximum likelihood decoding, can this error be corrected? If so, explain why and make the correction.

APPLICATIONS

1 In this lesson, you analyzed the typical nine-digit ZIP code. An eleven-digit ZIP code is sometimes used to identify the mailing location even more precisely. The tenth and eleventh digits help sort the mail into the order in which it will be delivered by the carrier. The 11 digits are contained in the barcode. Only nine or five digits are actually printed.

The following return envelope has a nine-digit ZIP code printed, but 12 digits in the barcode, including the check digit.

a. The check digit is chosen so that the sum of all twelve digits will be equivalent to 0 mod 10. Verify that this is the case for the barcode above.

b. Verify that the tenth and eleventh digits give the last two digits of the address or box number.

c. Find a reply card or envelope that has eleven digits plus a check digit in the ZIP-code barcode. Verify that the check digit and tenth and eleventh digits have the properties described in Parts a and b.

2 Each bank in the U.S. has an identification number. This ID number is usually printed at the bottom-left of all checks from the bank. The first eight digits of the ID number are information digits, which identify the particular bank. The ninth digit is a check digit. Suppose the eight information digits are $a_1, a_2, a_3, a_4, a_5, a_6, a_7,$ and a_8. The check digit is chosen so that it is the last digit of $7a_1 + 3a_2 + 9a_3 + 7a_4 + 3a_5 + 9a_6 + 7a_7 + 3a_8$.

a. A bank in Iowa has the following ID number: 073901877. Verify that this number satisfies the check-digit system described above.

b. Explain why choosing the check digit as described above is the same as choosing the check digit so that it is equivalent to the indicated sum mod 10.

c. Examine the bottom of a check from a U.S. bank to find the nine-digit bank ID number, and verify that the ID number satisfies the check-digit system described in this task.

3 A standardized identification number system is used for books around the world. Each book is assigned a unique code—an International Standard Book Number (ISBN). These codes allow books to be more accurately identified and more effectively organized, especially with computerized recordkeeping. From 1970 to 2007, ISBNs had 10 digits. Since January 1, 2007, ISBNs contain 13 digits, and so the numbering system is sometimes called ISBN-13. In this task and in Extensions Task 14, you will consider the older ISBN system. See Extensions Task 15 for an analysis of the ISBN-13 system.

Consider ISBN 0-380-00832-7. A 0 or 1 in the first position indicates that a book was published in an English-speaking country. The next block of digits identifies the publisher. The third block is assigned by the publisher from among the block of numbers the Library of Congress gives the publisher and it identifies the specific book. These first three blocks of digits can be of different lengths, depending on the publisher of the book. The last digit is a check digit. Here is how the check digit works.

> There are ten digits in an ISBN: *abcdefghij*.
> Compute $10a + 9b + 8c + 7d + 6e + 5f + 4g + 3h + 2i + j$.
> If the number is valid, this sum will be equivalent to 0 mod 11.

  a. Verify that the ISBN given above is valid.

  b. Consider this code: 0-88385-423-0. Is this a valid ISBN? Enter the number into the Search feature of a bookseller's Web site. What happens? Calculate to see if the ISBN is valid. (See Extensions Task 14 to explore the error-detecting and correcting capability of the ISBN code.)

4 One of the key results in this lesson is that if the minimum distance for a set of binary code words is 3 or greater, then all single errors can be corrected. When the minimum distance for a code is less than 3, then not all single errors can be corrected, but it may be possible to correct *some* single errors. For example, suppose there are four code words: 00000 01110 11011 11001

  a. Explain why you cannot find the minimum distance by finding the minimum number of 1s in all nonzero code words.

  b. What is the minimum distance for this code?

  c. Suppose you receive 11111. Can this error be detected? Using maximum likelihood decoding, can this error be corrected? If so, explain why and do the correction.

  d. Consider the fourth code word: 11001. Suppose that during transmission an error occurs in the fourth-bit place of this code word. Will this error be detected? Explain.

5 Continue your analysis of the image from Check Your Understanding Part b (page 498). Suppose the following grid is a small part of an image sent from a satellite. The image uses four shades of gray (two of which are black and white).

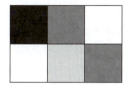

a. Use the linear code {000, 011, 101, 110} to encode this image, based on the following "dictionary." First scan the top row left-to-right, then scan the bottom row left-to-right.

000 – white	011 – light gray
101 – dark gray	110 – black

b. Suppose one of the pixels is received as 111. Will this error be detected? Can it be corrected using maximum likelihood decoding? Explain.

c. Verify that this is a linear code.

d. Compute the minimum distance for this code. Use the minimum distance to explain why single errors cannot be corrected.

e. Find a code that you could use to encode images in four shades of gray that will correct single errors.

CONNECTIONS

6 The ZIP and UPC codes you have studied involve check-digit systems that are based on sums or weighted sums.

This generally creates a stronger error-detecting code than if sums were not used at all. However, not all codes use sums, or perhaps not explicitly. For example, consider U.S. Postal Service money orders. Each U.S. Postal Service money order has a unique serial number, which consists of 10 digits followed by a check digit. As usual, the check digit is included to help ensure accurate identification of the money order. The rule in this case is that the check digit must be what you get when you reduce the 10-digit number mod 9.

a. What is the serial number of the U.S. Postal Service money order shown above? What is the check digit?

b. Suppose a money order serial number begins with the digits 3953988757. What is the check digit?

c. Verify that the 11-digit number 69584712355 satisfies the check-digit criterion for a U.S. Postal Service money order.

d. Illustrate and explain why this check-digit system will not detect the error of transposing two digits in the first ten digits.

7 The Soundex code was established by the National Archives to help locate and accurately identify names in old records such as birth, death, and marriage certificates and passenger lists for the boats that brought immigrants to the United States. This was primarily done to help establish a person's age for Social Security eligibility, since many states did not start recording births until after 1900. Using the Soundex code, surnames (family names) that have the same or similar sound get the same code, even if the  spellings are slightly different. This is done because spelling and pronunciation were not as standardized in the past as they are now. Thus, family names of immigrants may have been spelled in various ways when entered into the records, especially if the names were first transcribed into English from another language, particularly one written in a different alphabet. Soundex codes begin with the first letter of the surname followed by a three-digit code that represents the first three remaining consonants in the surname, as outlined below.

Soundex Code

Step 1. Start with the first letter of the surname (last name).

Step 2. In the remaining letters of the surname, delete A, E, I, O, U, H, W, and Y.

Step 3. Code each of the next three consonants using the numbers below. Ignore all other letters.

1 = B, P, F, V
2 = C, S, G, J, K, Q, X, Z
3 = D, T
4 = L
5 = M, N
6 = R

Step 4. If there are now three digits in the code, you are done. If there are less than three digits in the code, then add 0s at the end as needed to get three digits.

---

a. Examine Step 3 of the Soundex coding scheme. Explain why you think it was designed the way it was.

b. What is the Soundex code for Hart? For Romano? For Williams?

c. When doing historical research, one should consider the possibility that names with the same Soundex code could be the same name. Could the names Hart and Hardy have been the same, but just recorded as different?

d. Suppose that when searching the 1860 census records, you find J. Watkins in North Platte, Nebraska. Then on 1870 census records, you find John Watson also in North Platte. Based on evidence from the Soundex code, is it possible that these are really the same person? Explain your reasoning.

8 Suppose that a computer is reading or sending a code written in binary digits, that is, the only digits used are 0 and 1. Recall that such digits are called bits. Suppose that the probability of an error in one bit place is p. Assume that errors in different bit places are independent of each other. Suppose that you transmit n bits.

a. What kind of probability distribution describes this situation?

b. What is the probability of no errors in any bit place?

c. What is the probability of exactly one error?

d. What is the expected number of errors?

e. The assumption of independence is crucial here, but independence of errors is not always a legitimate assumption. Consider the case of adding two numbers. Is the probability of an error in one decimal place independent of errors in other decimal places? Explain and give an example. (When the errors are not independent, they are said to come in *bursts*.)

REFLECTIONS

9 When the Universal Product Code (UPC) first appeared on products in 1974, it was quite controversial. Think of some plausible reasons why it would be controversial. Then do some research to find out why and write a brief report. One possible resource is the article "Bar Codes: Reading Between the Lines" by Ed Liebowitz, in the February 1999 issue of *Smithsonian* magazine (Vol. 29, No. 11, pages 130–146).

10 ZIP codes have become part of our culture, often appearing in popular media. What are some examples of ZIP codes that are part of TV shows or popular songs? In each example, create the barcode representation of the ZIP code.

11 In a children's story, a mother says to her daughter, "Please put away the dishes that you washed." A little later the mother asks her daughter in astonishment, "Why are there fish in my cupboard?!". "But Mother," the daughter said, "you told me to put away the fish that I watched." How is this an example of errors in information processing? Have you ever misheard something someone said but then quickly corrected the error to know what they really meant? If so, give an example.

12 You have now explored the mathematical structure of ZIP codes, UPC codes, and binary codes. In general, a code is a group of symbols that represents information together with a set of rules for interpreting the symbols. In what sense can Roman numerals and musical scores be thought of as codes?

EXTENSIONS

13 The UPC code for a pair of jeans is 4-90141-17467-6.

a. Suppose that this is incorrectly scanned with the seventh and eighth digits transposed. Will the UPC check-digit system detect this error?

b. Suppose the jeans UPC is incorrectly read with the second and fourth digits transposed. Will this error be detected?

c. Give as complete a description as you can of the types of transposition errors that can be detected by the UPC check-digit system.

14 In Applications Task 3, you considered the 10-digit ISBN code. In this present task, you will see that the ISBN check-digit system will detect any single-digit error or any transposition error. It will not in general correct all errors, although you can often at least narrow down the possible correct numbers.

a. Consider the alleged ISBN from Applications Task 3 Part c: 0-88385-423-0. If you have not already done so in that task, verify that there is an error.

b. In fact, there is a single-digit error in the alleged ISBN in Part a. Can this error be corrected by reasoning about the check-digit system? If so, correct the error. If not, explain why not.

c. Provide an argument justifying that the ISBN check-digit code will detect all single-digit errors, that is, errors in which exactly one digit is incorrect.

d. The ISBN code will also detect all transposition errors. Justify this fact.

Nicole Fonger

15 In Applications Task 3 and Extensions Task 14, you considered the older 10-digit ISBN code, sometimes called ISBN-10. Recall that as of January 1, 2007, ISBNs contain 13 digits, sometimes called ISBN-13. A 13-digit ISBN begins with 978, then the next nine digits are the nine digits of the 10-digit ISBN without its check digit, and the last digit is a new check digit. (When the ISBNs that begin with 978 run out, then additional new numbers will be created that begin with 979.) The first 12 digits provide identifying information for the book. The last digit is a check digit. Below is a procedure to calculate the check digit for 13-digit ISBNs.

Procedure for Calculating an ISBN Check Digit

Step 1. For each of the 12 information digits, alternately multiply by 1 or 3, moving from left to right, and then sum all those products.

Step 2. Reduce this sum mod 10.

Step 3. Subtract this result from 10.

Step 4. Reduce mod 10. This final result is the check digit.

---

a. Consider the ISBN for a book about mathematics teachers: 978-0-415-99010-3. Verify that the check digit is correct.

b. Verify that the ISBN check digit of your *Transition to College Mathematics and Statistics* book is correct.

c. In Extensions Task 14, you found that the 10-digit ISBN code detects all transposition errors. That code uses mod 11, while the newer 13-digit ISBN code uses mod 10. Will the 13-digit ISBN code detect all transposition errors? Provide a justification or a counterexample. Compare to the error-detection capability of the UPC code, which also uses mod 10.

d. Why do you think the 13-digit ISBN code was adopted? What are some advantages and disadvantages compared to the older 10-digit ISBN system?

REVIEW

16 Solve the following systems of equations.

a. $3x + 2y = 4$
$x - 6y = -7$

b. $5x + 9y = 33$
$10x + 3y = 1$

17 Consider the function $f(x) = \frac{32}{x}$.

a. Determine the value of $f(4)$.

b. For what value of x is $f(x) = 160$?

c. If the value of x doubles, what happens to the value of $f(x)$? Explain your reasoning.

d. Which of the following statements are true about the graph of $f(x)$? (Identify all that are true.)

I The y-intercept of the graph is $(0, 32)$.

II The x-axis is a horizontal asymptote of the graph.

III The graph is symmetric across the x-axis.

IV The graph is symmetric across the y-axis.

V The graph has 180° rotational symmetry.

18 A cylindrical silo for storing corn has a diameter of 30 ft and a height of 80 ft.

a. Find the volume of the silo.

b. Find the surface area of the silo including the top and bottom.

19 Use reasoning about square roots to solve each equation without using technology.

a. $\sqrt{x} = 12$

b. $3\sqrt{x} = 12$

c. $\sqrt{x + 20} = 12$

d. $\sqrt{\dfrac{360}{x}} = 12$

20 Factor each of the following expressions into linear factors.

a. $y^2 - 9$

b. $3q^2 - 20q - 7$

c. $ax^2 - 5ax + 4a$

d. $4(p - 1) + k(p - 1)$

e. $6x^2 + 3x - 63$

21 A survey of 648 residents of Summit found that 480 of them thought the town should have a 4th of July parade.

a. What is the parameter that this survey was trying to estimate?

b. What is the point estimate of the proportion of residents of Summit who think the town should have a 4th of July parade?

c. Compute the margin of error for this survey. Then write a sentence explaining what it means in this context.

22 Fifty randomly selected families in a neighborhood in a large city were surveyed and each family was asked how many vehicles were registered at that address. The results of the survey are summarized in the table below.

Number of Vehicles	Number of Families	Percentage of Families
0	17	
1	23	
2	7	
3	2	
4	1	

a. Explain what the 23 in the table indicates.

b. Use the data in the table to compute the average number of vehicles registered per address in this neighborhood.

c. Determine the percentage of families in the neighborhood that had no vehicles registered at their address. Then fill in the rest of that column of the table with percentages. Enter each percentage as a decimal number between 0 and 1. The percentage of times that a data value occurs is also called the *relative frequency*.

d. You can also find the average number of vehicles per address using the relative frequencies. In the formula below, $f(x)$ is the relative frequency of the data value x. Use this formula to find the average number of vehicles registered per address.

$$Mean = \sum (f(x) \cdot x), \text{ where } x \text{ takes on all possible data values.}$$

Your answer here should match the one you got in Part b.

e. What is the median number of vehicles registered per address in this neighborhood?

f. Which measure of center is most appropriate for these data? Explain your reasoning.

Efficiency: Data Compression

In previous lessons, you investigated three important issues related to information processing and the Internet—access, security, and accuracy. You learned how to make information accessible using set theory and logic, how to keep it secure using cryptography, and how to ensure accurate information using codes that detect and correct errors. Another important issue is efficiency. These issues are often related. For example, when you download a photo from the Internet, you would like both accuracy and efficiency. Efficiency in this case means that you want the photo to download quickly. Since a photo consists of a large amount of data, *data compression* is used to provide more efficient transmission of the photo.

There are many other examples of data compression in the digital world of computers and the Internet.

Search Results for Files Related to "Mathematics" Showing Various File Types

☆☆☆☆☆	Mathematics Quiz 2	zip	197.7 KB
☆☆☆☆☆	Mathematics_The Problem_16_Spot Lite (Bonus Track) Fe...	mp3	5,025 KB
☆☆☆☆☆	Telepopmusik - Mathematics	wma	5,514 KB
☆☆☆☆☆	Mathematics_vs_Art_Sample	jpg	2,076 KB
☆☆☆☆☆	Integration (how-to – mathematics)	mpg	197.7 KB
☆☆☆☆☆	Advanced Mathematics Tutorial Video Series #12	3gp	197.7 KB
☆☆☆☆☆	Mos Def feat. DJ Premier - Mathematics	mp3	3,880 KB

THINK ABOUT THIS SITUATION

Think about data compression and some situations where it is useful.

a. To begin, cn u rd ths qustn evn tho mny lttrs r mssng? Rse yr hnd if u cn. Why do you think the previous sentences are readable, even with missing letters? How have the sentences been compressed?

b. The screen shot above shows some of the results when using an Internet file transfer program to search for files related to "mathematics." Examine the column that shows file type. What do you think the file "suffixes" zip, mp3, wma, jpg, mpg, or 3gp mean?

c. Have you ever downloaded or sent what you knew to be a compressed file? Describe the situation and explain why it was useful to compress the data.

In this lesson, you will learn some of the basic concepts and techniques of data compression. Specifically, you will learn about variable-length codes and an especially useful code of this type called a Huffman code.

INVESTIGATION 1

Huffman Codes

Data compression is used every time a fax is sent, every time a photo or video is downloaded from the Internet, and whenever you want to store more data in a limited space on a computer hard drive. Some of the computer applications and file types that involve data compression are Zip and Stuffit (archive utilities and compression formats), JPEG and MPEG (still and moving picture formats), MP3 (music), V.42bis (a modem data compression algorithm), and PPP (an old dial-up protocol to access the Internet).

There are many different types of data compression methods. The original digital data compression scheme, which is still used in most compression algorithms today, including those listed in the previous paragraph, is the *Huffman code*. This code was developed in 1951 by David A. Huffman when he was still a graduate student at the Massachusetts Institute of Technology.

As you complete the problems in this investigation, look for answers to these questions:

What are variable-length codes, and how are they used for data compression?

What are the key characteristics and properties of a Huffman code, and how do you construct such a code?

Variable-Length Codes The first step in understanding how a Huffman code works is to think about the difference between block codes and variable-length codes. A **block code** is a code in which every code word has the same fixed length. A **variable-length code** is a code in which code words have different lengths. All the codes you have used so far in this unit have been block codes.

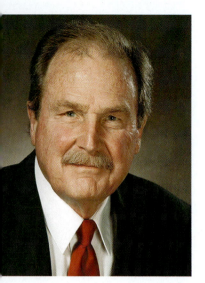

David Huffman

1 The traditional code used to convert text and other characters into 1s and 0s for computer use is called the **ASCII code** (American Standard Code for Information Interchange). This code uses 7-bit blocks to represent each character. Part of the ASCII code structure is shown in the table below.

ASCII Code (A Block Code)

Character	Binary ASCII Code Word
A	100 0001
B	100 0010
C	100 0011
<	011 1100
>	011 1110
=	011 1101
@	100 0000

a. Use this ASCII code to encode the word CAB.

b. Why is the ASCII code a block code?

2 Variable-length codes are constructed by first considering which characters are used more often.

a. Which of the two letters, A or B, do you think occurs more often in the English language? Count the number of As and Bs in the following passage from Lewis Carroll's *Alice in Wonderland*:

> ALICE STARTED TO HER FEET, FOR IT FLASHED ACROSS HER MIND THAT SHE HAD NEVER BEFORE SEEN A RABBIT WITH EITHER A WAIST-COAT POCKET OR A WATCH TO TAKE OUT OF IT …

> **Source:** John C. Winston Company, Publishers, 1923, page 2.

University of California Santa Cruz

b. The focus of this lesson is *efficient* information processing. Keeping this in mind, if you could use code words of different lengths, which of the two letters, A or B, would you encode with a shorter code word? Why?

c. Data compression is the process of recoding source information using fewer bits than used originally, while still maintaining accuracy. In this investigation, you will focus on the method of encoding the most frequently occurring symbols with the shortest code words. Explain why this method of encoding data could be said to *compress* the data.

3 Based on your work in Problem 2, you could compress text data by encoding A with a shorter code word than B. To do so, you need a variable-length code. For example, consider the code below.

Variable-Length Code I

Character	Code Word
A	0
E	1
B	10
C	11

a. Using this code, encode the message CAB.

b. Using this code, decode the string 0110. Compare your decoding with that of others. Resolve any differences.

c. A code is called **uniquely decodable** if every encoded message can be decoded to just one unique original message. Do you think this code is uniquely decodable? Why?

4 In the previous problems, you have seen that two essential features of a data compression code are:

- frequently occurring characters in the message alphabet should have short code words.

- the code should be uniquely decodable.

Consider the code below.

Variable-Length Code II

Character	Code Word
E	0
A	01
C	011
B	111

a. Decode the string 001111101111.

b. Discuss the decoding process that you used in Part a. Describe any disadvantages you see for this code.

c. Do you think this code is uniquely decodable? Why?

5 Ideally, a code should have the property that decoding does not require false starts or backtracking. Consider next the following code.

Variable-Length Code III

Character	Code Word
A	0
C	101
D	111
L	100
V	1101
K	1100

a. Encode the message LACK.

b. Decode the string 1101010101100.

c. Describe the process of decoding that you used in Part b. Did you have any false starts or backtracking? Is there more than one way to decode this string?

6 Certain types of codes have the useful properties that you examined in previous problems. A **prefix-free code** is a code in which no code word is the prefix of any other code word.

a. Verify that the code in Problem 5 is a prefix-free code.

b. Explain why the code in Problem 4 is *not* a prefix-free code.

c. Explain why prefix-free codes have the property that you can decode a string without any false starts or backtracking.

Huffman Codes You have seen in Problem 6 that prefix-free codes are uniquely decodable and do not require any false starts or backtracking when decoding. Now you will learn how to construct an important type of prefix-free code, called a *Huffman code*.

7 A **Huffman code** is a prefix-free code used in many data compression algorithms. It is constructed using the procedure in the following example. Suppose there are six levels of gray in a photo that you want to send to a friend. The table that follows shows the frequency with which each level of gray occurs. You can think of the photo consisting of many pixels, each of which is one of six shades of gray. The table shows the relative frequency (or percentage) for pixels that are colored each shade of gray.

Relative Frequency of Six Levels of Gray

Gray Level	Relative Frequency
white	0.18
black	0.19
gray level 1	0.39
gray level 2	0.14
gray level 3	0.06
gray level 4	0.04

Different levels of gray will be coded with different binary code words. Then the photo can be represented with a string of code words. The photo is sent in this digital form. The code words are created from the relative frequency table using a binary tree, as illustrated below.

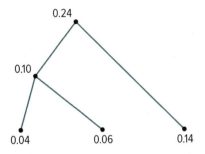

The tree above is the beginning of a Huffman tree. It is constructed according to the **Huffman algorithm** described below.

Huffman Algorithm

Step 1. Draw six vertices in a row at the bottom of a sheet of paper. Label the vertices with the numbers (relative frequencies) in the table. Write them with the numbers in order from smallest to largest, left to right.

Step 2. Draw edges from the two smallest numbers so that the two edges meet above at a new vertex, which is labeled with the sum of the two numbers.

(*Note:* Once two vertices are connected to a new sum vertex, they are not used again. Also, the sum vertex becomes a new candidate vertex, as in Step 3.)

Step 3. Examine the sum vertex and the other vertices that do not yet have an edge drawn up from them. Choose the smallest two, and repeat the process of drawing two edges and creating another sum vertex above.

Step 4. Continue in this way until all six original vertices have been used, and there is one vertex at the top of the tree.

Step 5. Start at the top vertex of the tree. Follow a path from the top vertex to each of the six original vertices at the bottom. As you move down a path, record a 0 at each left branch, and a 1 at each right branch. When you reach each bottom vertex in this way, record the string of 0s and 1s below the vertex. These strings are the code words.

---

a. Carry out Steps 1–4 of this algorithm to complete a copy of the partial tree on the previous page. Why is the top vertex of the tree labeled 1.00?

b. The next step is to assign binary code words to the six original vertices at the bottom of the tree. This is done as indicated in Step 5. The vertex with 0.06 relative frequency should have code word 0001. The vertex with 0.19 relative frequency should have code word 011. Determine and record the remainder of the code words.

8 Suppose a particular text includes the characters and relative frequencies below.

Character	Relative Frequency
B	0.10
C	0.10
D	0.15
F	0.20
G	0.20
H	0.25

a. Construct a Huffman tree for these characters and frequencies. Do not worry about the exact placement of vertices, or if some of the edges cross each other. The key point at each step is simply that the sum vertex should be somewhere above the two vertices that connect to it.

b. Compare your tree to the trees constructed by some of your classmates. Is it possible to construct different Huffman trees for the same characters and frequencies? Explain.

9 The set of code words constructed by the Huffman algorithm is called a **Huffman code**. A Huffman code has several important properties.

a. A Huffman code is a prefix-free code. Verify that the Huffman codes constructed in Problems 7 and 8 are prefix-free codes.

b. In a Huffman code, the most frequently occurring character (or gray level) has the shortest code word. Verify that this is true for the codes in Problems 7 and 8.

Jim Laser

10 The most important property of a Huffman code is that:

A Huffman code has the minimum average code word length among all prefix-free codes.

In order to understand this property, do the following.

a. Compute the average code word length for the Huffman code in Problem 7.

b. Now compare this average code word length to other codes for the same data.

i. Create a non-Huffman code as follows. Take the same six relative frequencies from the table in Problem 7 (page 513). As before, place those six numbers as vertices at the bottom of a piece of paper. Create a tree as you did before, *except* do not choose the two smallest vertices to join together into a sum vertex. Just choose *any two* vertices. Using the same process as before, join the vertices, create the sum vertex, choose another two vertices (any two) and continue until you have completed the tree and the new code words.

ii. For this non-Huffman code, compute the average code word length.

iii. Compare to the average code word length for the Huffman code from Part a.

iv. Compare to the average code word length for codes from different trees that some of your classmates created when they did part i.

v. Does the Huffman code from Part a have the smallest average code word length? If not, re-examine your work with others and make any needed corrections.

11 Code and decode some words with your classmates, as follows.

a. Work with a partner. You take the word "lollapalooza" and your partner takes the word "mississippi." For your word, calculate the letter frequencies and make a table like the one in Problem 8. Construct a Huffman code for your letters. Then encode your word with your Huffman code. Swap encoded words with your partner and decode their word. Discuss and resolve any questions or errors that may occur.

b. For each word in Part a, how many bits are needed to encode the word with the ASCII code (page 510) which uses seven bits for each letter? How many bits are used in your Huffman coding of the word? What percent compression is achieved?

c. Repeat the work in Parts a and b with a secret word of your choice. Choose a word that has some repeated letters. Give your Huffman-encoded (compressed) word to some classmates for them to decode (decompress). Decode the encoded words given to you.

SUMMARIZE THE MATHEMATICS

In this investigation, you learned about efficient information processing through data compression.

a. Describe the difference between a block code and a variable-length code.

b. How are variable-length codes used in data compression?

c. What is a prefix-free code? What are key characteristics and properties of a prefix-free code?

d. Describe how to construct a Huffman code.

e. What are key characteristics and properties of a Huffman code?

Be prepared to share your ideas with the class.

✔ CHECK YOUR UNDERSTANDING

Prefix-free codes are particularly useful in data compression algorithms.

a. Use the code from Problem 5 to decode the string 100110001101100.

b. Construct a Huffman code for these characters and relative frequencies.

Character	Relative Frequency
e	0.58
h	0.20
q	0.12
w	0.10

1 Consider this code.

Character	Code Word
q	1000
z	1001
e	0
@	11
$	101

a. Is this a prefix-free code? Explain why or why not.

b. Decode the following strings.

  i. 01001111000

  ii. 11110

  iii. 10001001101101

2 Consider the characters and frequencies in the following table.

Character	Relative Frequency
d	0.25
m	0.20
r	0.03
c	0.28
z	0.18
g	0.06

a. Construct a Huffman tree and a Huffman code for the characters and frequencies in the table.

b. Find the average code word length for this code.

c. Construct a different binary tree for these characters and frequencies. That is, construct a tree where you do not always choose the smallest two vertices to join together into a sum vertex. Just choose any two vertices at each step to join together into a sum vertex.

  i. Find the code words using this new tree.

  ii. Find the average code word length for this code. Compare to the average code word length for the Huffman code in Part b.

3 In Problem 8 Part b of Investigation 1, you discovered that the same data can generate different Huffman trees. You will now verify that even though the trees, and the associated Huffman codes, can be different, they all have the same average code word length.

  a. Construct at least one Huffman tree for the data in Problem 8 (page 514) that is different from the tree you originally constructed. (You will get different trees when you decide to break ties differently or cross edges differently.)

  b. The tree in Part a and the tree you originally constructed are both Huffman trees, since you followed the Huffman algorithm, but they are different trees and they yield different codes. Compute the average code word length for both codes. Verify that you get the same average for each code.

4 In Problem 10 of the investigation, you learned that a Huffman code has the minimum average code word length among all prefix-free codes. In particular, a Huffman code has smaller (or equal) average code word length compared to block codes.

  a. Suppose you encode the six levels of gray from Problem 7 on page 513 using a binary block code with the smallest possible blocks. What is the length of the smallest blocks you could use?

  b. What is the average code word length for the code in Part a?

  c. Compare to the average code word length for the Huffman code, which you determined in Problem 10. Verify that the Huffman code has a smaller average code word length.

CONNECTIONS

5 The basic version of the ASCII code that you studied in Problem 1 on page 510 was published in 1968 as ANSI X3.4. Today most computers still use character encodings based on this code, although there have been some revisions. Some versions of the ASCII code use 8-bit code words. The eighth bit is used, for example, as a check bit or as a means for encoding languages other than English.

  a. In an 8-bit version of the ASCII code where the eighth bit is a check bit, the check bit works as follows. An eighth bit is added at the beginning of every code word so that the total number of 1s in each code word is even. Use this scheme to find the 8-bit version of the ASCII code words in Problem 1.

  b. Describe the error-detecting and error-correcting capabilities of the ASCII code in Part a.

6 Suppose that you have data about how often records in a database are looked up. That is, you know the lookup frequencies for all records. Suppose that you want to construct a binary search tree to hold all these records. Thus, you will store the records at the *leaves* (the terminal vertices) of a binary tree. You would like to find the most efficient binary tree for doing this.

It is quicker to find a record that is stored at a leaf that is close to the top (*root*) of the tree, than it is to find a record that is stored at a leaf vertex that is many edges down from the top. Consider the table below that gives the lookup frequency for six records.

Record	Lookup Frequency
A	0.10
B	0.10
C	0.15
D	0.20
E	0.20
F	0.25

a. Construct a tree that has the records at the leaves (terminal vertices) and has the most efficient lookup capability.

b. Explain why a Huffman tree will provide the most efficient binary tree in this situation.

c. Construct a Huffman tree for this situation, if you have not already done so in Part a.

d. Compare this task to Problem 8 on page 514. Discuss similarities and differences.

7 In Problem 10 (page 515), you computed the average code word length by multiplying the length of each code word by its relative frequency and then summing all those products. Explain why this procedure works to compute average code word length.

REFLECTIONS

8 The Huffman algorithm is an example of what is called a *lossless compression algorithm*. There is another class of compression algorithms that are called *lossy* algorithms. Why do you think the word "lossless" is used to describe the Huffman algorithm?

9 Explain why the top vertex in a Huffman tree *should* always be labeled 1.0. Does the sum computation always equal 1.0 exactly? Explain.

EXTENSIONS

10 An essential fact about Huffman codes is that, for a given set of characters and relative frequencies, a Huffman code has the minimum average code word length over all prefix-free codes. Research a justification of this fact about Huffman codes in a book or on the Internet. Present the justification in a brief oral or written report. (You might consult *Mathematical Structures for Computer Science*, 3rd edition, by Judith Gersting, published by Computer Science Press, 1993.)

11 The problem of finding a good data compression technique led to the idea of Huffman trees and Huffman codes. As often happens in mathematics, it is possible to use these ideas to solve seemingly unrelated problems. For example, three problems that seem different but can all be solved by Huffman trees are: (a) data compression, which is the focus of this lesson; (b) record lookup using binary search trees, as in Connections Task 6; and (c) merging sorted lists of data, like alphabetized mailing lists of customers or class lists of exam grades. Research and present an example showing how Huffman trees can be used to efficiently merge sorted lists.

REVIEW

12 Consider the three mathematical statements given below.

I $10 = \frac{4}{3}x + 6$

II $10 < \frac{4}{3}x + 6$

III $y < \frac{4}{3}x + 6$

a. Match each statement with the correct graph of its solution.

Graph A

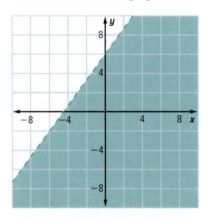

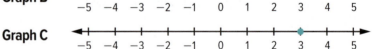

Graph B

Graph C

b. Write another mathematical statement for which Graph A matches the solution.

c. Write another mathematical statement for which Graph B matches the solution.

d. Write another mathematical statement for which Graph C matches the solution.

13 Julia has an unpaid balance of $378 on her credit card from the prior month. The card has an APR of 18% compounded monthly. She has decided that she will pay $40 a month and that she will not make any additional purchases until this balance is paid off.

a. What is the balance after she makes the first payment?

b. How long will it take her to reduce the balance to zero?

c. How much interest will Julia have paid?

14 On the grid below, the line contains the origin and forms a 60° angle with the positive *x*-axis.

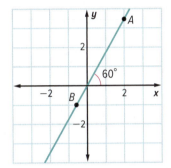

a. The *x*-coordinate of point *A* is 2. Determine the *y*-coordinate of *A*.

b. The *y*-coordinate of point *B* is −1. Determine the *x*-coordinate of *B*.

15 One important property of some geometric shapes is symmetry. Consider the rectangle and square shown below.

a. Sketch each shape and draw all lines of symmetry.

b. Describe the rotational symmetries of each shape by giving the degree measure of each rotational symmetry.

c. How do the numbers of lines of symmetry and rotational symmetries help to explain why a square has more geometric properties than a rectangle?

Looking Back

In this unit, you have learned some of the mathematics of information processing and the Internet. You investigated four fundamental themes of information processing: access, security, accuracy, and efficiency. You learned some basic ideas of set theory, particularly as they apply to making information accessible through Internet searches. You studied cryptography and modular arithmetic as a means to make transmitted information secure. You analyzed error-detecting codes used for ID numbers and error-correcting codes used in data transmission to help ensure accuracy. And you learned about one aspect of efficiency in information processing using data compression, particularly Huffman codes.

In this final lesson, you will review and pull together your understanding of these ideas and apply them in new contexts that draw on connections with geometric transformations and algebraic systems. You will also continue to make use of important mathematical practices such as modeling with mathematics, using technological tools strategically, and making use of structure in mathematical situations.

1 In this unit, you studied several different codes used for ID numbers, such as ZIP, UPC, and ISBN codes. All these codes use a check digit and modular arithmetic. They each have different error-detecting capabilities.

   a. Which of these codes can detect single-digit substitution errors?

   b. Which of these codes can detect all transposition errors? Only some transposition errors? No transposition errors?

Jim Sanborn

2 Now apply the knowledge you gained from this unit to the Verhoeff code, which is a code used for ID numbers that is not based on modular arithmetic. The Verhoeff code detects all single-digit substitution errors and all adjacent transposition errors. This code was used, for example, for the serial numbers on German banknotes. (Adapted from J. A. Gallian "Math on Money," *Math Horizons*, November 1995, 10–11.)

a. The **Verhoeff code** was devised by the Dutch mathematician Jacobus Verhoeff in 1969. It is based on the group of symmetries of a regular pentagon, called the *dihedral group*, D_{10}. The ten elements of D_{10} are the five line reflection symmetries and the five rotation symmetries of a regular pentagon. Consider the regular pentagon *ABCDE* at the right.

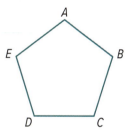

   i. Explain why a regular pentagon can be rotated counterclockwise about its center through angles of 0°, 72°, 144°, 216°, and 288° and coincide with itself (even though the position of the labels may change). These are the five rotation symmetries in D_{10}.

  ii. Illustrate with a sketch the five lines of reflection symmetry in a regular pentagon. These line reflection symmetries are the other five elements in D_{10}.

b. The first step in using the Verhoeff code is to label the elements of D_{10} with the digits 0 through 9, as indicated below.

    0: counterclockwise rotation of 0°

    1: counterclockwise rotation of 72°

    2: counterclockwise rotation of 144°

    3: counterclockwise rotation of 216°

    4: counterclockwise rotation of 288°

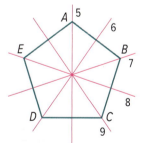

  5–9: reflections across the lines of symmetry as shown in the diagram at the right.

The elements in D_{10} are combined using an operation *, which can be interpreted as "followed by." For example, 2 * 6 means "counterclockwise rotation of 144° followed by reflection across the line labeled 6."

   Describe each of the following computations. Illustrate *each* operation by drawing a diagram or fitting a cut-out pentagon into its outline.

   i. 1 * 3 = 4   **ii.** 2 * 8 = 5   **iii.** 6 * 5 = 1

c. Fill in the five missing entries on a copy of the table below, which shows how to combine all elements in D_{10}.

*	0	1	2	3	4	5	6	7	8	9
0	0	1	2	3	4	5	6	?	8	9
1	1	2	3	4	0	6	7	8	9	5
2	2	3	4	0	1	7	8	9	5	6
3	3	4	?	1	2	8	9	5	6	7
4	4	0	1	2	3	?	5	6	7	8
5	5	9	8	7	6	0	?	3	2	1
6	6	5	9	8	7	1	0	4	3	2
7	7	?	5	9	8	2	1	0	4	3
8	8	7	6	5	9	3	2	1	0	4
9	9	8	7	6	5	4	3	2	1	0

d. The Verhoeff code is similar to some other codes you have studied in that it uses a check-digit system, but one based on a *weighted sum*. First you represent an ID number with the elements of D_{10}. Then you choose a check digit based on the outcome of a weighted sum.

In a Verhoeff code, the "sum" is computed using the operation * from the previous table, and the weights are assigned using the functions w_i that are defined in the table below.

	0	1	2	3	4	5	6	7	8	9
w_1	1	5	7	6	2	8	3	0	9	4
w_2	5	8	0	3	7	9	6	1	4	2
w_3	8	9	1	6	0	4	3	5	2	7
w_4	9	4	5	3	1	2	6	8	7	0
w_5	4	2	8	6	5	7	3	9	0	1
w_6	2	7	9	3	8	0	6	4	1	5
w_7	7	0	4	6	9	1	3	2	5	8
w_8	0	1	2	3	4	5	6	7	8	9
w_9	1	5	7	6	2	8	3	0	9	4
w_{10}	5	8	0	3	7	9	6	1	4	2

For example, refer to the shaded entry in the first row. This entry indicates that the element 4 in D_{10} would be weighted by the function w_1 to yield a value of 2. That is, $w_1(4) = 2$.

Determine the following weights.

   i. $w_1(6)$ ii. $w_3(7)$ iii. $w_9(9)$

e. Now suppose that the serial number on a German banknote is:

AU3630934N7

Note that the banknote serial number includes some letters. Only 10 letters are used in German banknote serial numbers. They are converted to numbers according to the following table.

A	D	G	K	L	N	S	U	Y	Z
0	1	2	3	4	5	6	7	8	9

The last digit of the serial number, 7 in this case, is a check digit used to ensure accuracy. Complete the following steps to see how the check digit is computed.

Step 1. Convert the letters to numbers in the serial number AU3630934N7.

Thus, without the check digit, you now have 10 digits: $d_1, d_2, d_3, d_4, d_5, d_6, d_7, d_8, d_9, d_{10}$.

Step 2. Compute a weighted sum by weighting each digit d_i with the corresponding function w_i and then combine using the operation * as indicated.

$$w_1(d_1) * w_2(d_2) * w_3(d_3) * \cdots * w_{10}(d_{10}).$$

Step 3. Choose the check digit so that when it is included at the end, on the right of the string of digits, the combination equals 0.

Does your check digit agree with the given banknote number?

f. Use the Verhoeff code to find the check digit for the German banknote serial number DA6819403G. Write the complete serial number.

3 Complete the following tasks related to set theory and information processing.

a. Suppose $S = \{2, 3, 9, 4, -2\}$ and $T = \{3, 6, 9, 12\}$.

   i. Find $S \cup T$, $S \cap T$, and $S - T$.

   ii. Draw Venn diagrams illustrating the operations in part i.

   iii. Let $V = \{x \mid x = 6n,$ where n is a positive integer$\}$. Find $T \cap V$.

b. Suppose you are searching on the Internet for information about Republican candidates in Oregon running for either the House of Representatives or the Senate.

   i. Represent this search using Boolean operators.

   ii. Represent this search using sets.

c. State whether each statement is true or false. If it is true, draw a diagram that illustrates the statement. If it is false, give a counterexample.

   i. If $A \subset B$, then $A \cap B = A$.   **ii.** If $3 \in S$, then $\{3\} \subseteq S$.

   iii. If $M \cap N = \emptyset$, then $M - N = M$.   **iv.** $(A \cap B)' = A' \cap B'$

4 Complete the following tasks related to modular arithmetic.

a. Reduce 12^3 mod 6.

b. Compute $6 + 4$ mod 6. In mod 8. In mod 20.

c. Repeat Part b for the product 6×4.

d. Find three numbers that are equivalent to $-3 \bmod 12$.

e. Solve $x + 7 = 2$ in mod 10.

f. Show that in mod 6, if $n(n - 1) = 0$, then it is *not* true that n must be 0 or 1.

5 Use a Hill cipher based on the matrix $E = \begin{bmatrix} 5 & 2 \\ 7 & 3 \end{bmatrix}$ to encode, and then decode, the message SECRET.

6 Use the RSA cryptosystem with $p = 5$ and $q = 17$ along with the *TCMS-Tools* "Codes" custom app to encode the message I UNDERSTAND. Then decode the message.

SUMMARIZE THE MATHEMATICS

In this unit, you studied some important ideas in informatics—the mathematics of information processing and the Internet. You investigated four fundamental issues of information processing: access, security, accuracy, and efficiency. These issues are reflected in the main topics of the unit:

1. Set theory and Boolean operators, especially as connected to Internet searches

2. Symmetric-key cryptosystems

3. Modular arithmetic

4. Public-key cryptosystems

5. Error-detecting codes for ID numbers

6. Error-correcting codes used in data transmission

7. Data compression, especially using Huffman codes

Choose at least one of the topics listed above. Working with a partner, prepare a poster and oral report on the topic, according to the guidelines below.

- Give a brief non-technical overview and summary of the topic.

- Identify which of the four fundamental issues of informatics is addressed by the topic (access, security, accuracy, or efficiency). Explain why and how this issue is addressed. Explain why this issue is important in information processing.

- List the specific mathematical concepts and methods that relate to your topic.

- Develop a thoroughly worked-out example illustrating the topic.

Be prepared to present your report and respond to questions from the class.

✓ CHECK YOUR UNDERSTANDING

Write, in outline form, a summary of the important mathematical concepts and methods developed in this unit. Organize your summary so that it can be used as a quick reference in future units.

Spatial Visualization and Representations

Digital Resources at
ConnectED.mcgraw-hill.com

 Watch
 Tools
 Audio
 eBook

CAT scans and MRI (magnetic resonance imaging) are remarkable technologies that provide two-dimensional renderings of a brain or other organs of the human body. How to represent three-dimensional objects in two dimensions on a computer monitor or on paper is a fundamental problem in geometry and its applications.

In this unit, the focus will be on visualization and reasoning with representations of surfaces of three-dimensional objects such as those of a distant planet and of cross sections of three-dimensional objects such as those that are so useful in medical imaging.

In the following three lessons, you will extend your skills in visualizing surfaces and cross sections and representing those objects, both geometrically and algebraically, through the use of three-dimensional coordinates.

LESSONS

1 Representing Three-Dimensional Objects

Represent three-dimensional surfaces with contour diagrams and identify and sketch cross sections of three-dimensional objects.

2 A Three-Dimensional Coordinate System

Extend ideas of coordinate representation and methods in two dimensions to three dimensions. Visualize, describe, and sketch planes described verbally and by equations. Solve systems of linear equations in three variables.

3 Linear Programming: A Graphical Approach

Recognize problems in which the goal is to find optimum values of an objective function subject to linear constraints on the independent variables. Represent both the objective function and constraints in graphic and algebraic form, and use technology-augmented linear programming techniques to solve the optimization problems.

Representing Three-Dimensional Objects

Hikers regularly use *contour diagrams* of trails to get an overall picture of the terrain, showing where the mountains are and where the flat regions are. The map at the right shows a section of the Presidential Range of the White Mountains National Forest in New Hampshire. Trails are shown by heavier lines that include the name of the trail. The thin lines are *contour lines*. If a hiker walks along a contour line, she stays at the same elevation. On this map, adjacent contour lines differ in elevation by 100 ft.

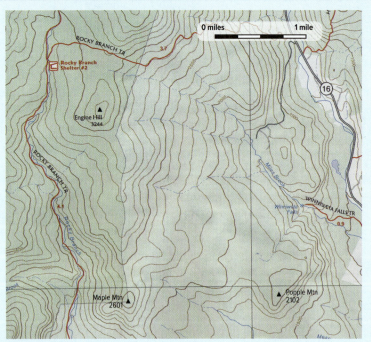

(t) Pixland/AGE fotostock; (b) Courtesy Appalachian Mountain Club

THINK ABOUT THIS SITUATION

Examine the map and think about the information it and similar contour diagrams convey.

a. Why do you think Highway 16 was placed where it is?

b. How do contour lines suggest a mountain peak?

c. How would the arrangement and elevations of contour lines for a long valley be similar to, and different from, the contour lines for a long ridge of mountains?

d. How could a hiker use a map with contour lines to assess the difficulty of a hike?

e. Why will contour lines rarely cross each other? Can you think of a situation where they would cross each other?

Maps showing contour lines are one way to represent three-dimensional objects in two dimensions. Other methods you may have previously studied include orthographic (face-view) drawings and oblique drawings. In this unit, you will learn how to represent and analyze three-dimensional objects using contour diagrams, using cross sections, and using coordinates and equations. You will also review your understanding of linear programming methods involving two variables and extend those ideas to analogous problem situations involving three variables.

INVESTIGATION 1

Using Data to Determine Surfaces

NASA (the National Aeronautics and Space Administration) has sent men to the Moon and spacecrafts to Mars and Venus. In preparing to land on the Moon, a critical concern was identifying a safe location for landing the spacecraft. In the case of our moon, the surface was visible, so relatively flat sites could be located.

However, in seeking a flat landing site on Venus, the problem was more difficult because the surface was obscured by clouds. Orbiting NASA spacecraft used radar to determine the elevations of many points on the surface of Venus. These elevations were then used to make contour diagrams of portions of the surface. The diagrams were used to identify possible landing sites. During the 1978 Pioneer mission to Venus and the 1990 Magellan mission, nearly the entire surface of Venus was mapped using radar altimetry.

As you work on the problems of this investigation, look for answers to these questions:

How can you use altitude data to make and interpret a contour diagram of a surface?

How can you use coordinates to identify locations in space?

Exploration 1 You can simulate the radar altimeter mapping of Venus and other planets by using a partially filled shoe box containing a hand-molded surface inside it and having grid paper on the lid. The molded surface represents a region of the surface of Venus. A bamboo skewer can be used as a radar probe. You will also need grid paper, a centimeter ruler, and a marking pen.

Calibrate your radar probe by making a mark 1 cm from the blunt end. Continue marking and numbering centimeter intervals on the entire skewer. If the box you are using does not have holes in its lid at grid marks, you will need to make holes using the sharp end of the skewer before collecting your data.

1 You are now ready to collect data. Keep the shoe box closed.

> **Step 1.** Keep the radar probe perpendicular to the box top as you insert it into a hole near a corner of the box top. Continue the gentle insertion until the probe touches the surface inside the box. Note the closest centimeter mark on the probe above the box top.

> **Step 2.** On your grid paper, record the number of the centimeter mark determined in Step 1.

> Repeat these steps for each hole in the box top. Your grid paper should represent the grid paper on the box top as if you had been allowed to write directly on the lid.

2 Now, using your data, make a contour diagram of the terrain. Follow the procedure illustrated with the sample data below. Do not open the shoe box and look at the terrain inside.

> **Step 1.** For your data, what number represents the highest peak? For the data below, a cluster of 1s shows the highest peak. It is best to start drawing the contour around either the highest peak or lowest valley. The example below begins with a contour at 1.5 cm, as shown by the loop around the 1s. The open contour drawn around the 10 represents the 9.5-cm contour. It is open because there may be more 10s beyond the edge.

```
10   8   7   5   4   6   5
 8   6   5   5   5   3   3
 2   3   4   5   3   2   2
 2   2   3   3   2   1   3
 2   2   3   3   1   1   3
 6   5   4   3   2   2   4
```

Step 2. Work outward from the chosen peak or valley by drawing contour lines every 2 cm. In this example, working outward from the 1s loop, you need to draw contour lines for 3.5 cm, 5.5 cm, and 7.5 cm. One of the lines is drawn in the diagram below. Why does the contour line at the lower-left corner extend between the 2 and 6?

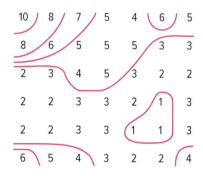

Step 3. Continue drawing the remaining contour lines as illustrated in the diagram below.

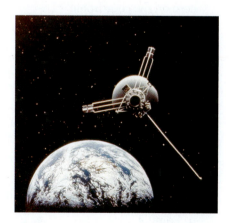

3 Now examine your contour diagram.

a. How is a region of small numbers related to a region of larger numbers? Explain why this makes sense.

b. Should the contour lines drawn for two numbers intersect? Explain your thinking.

c. Based on your data, describe the terrain in the box. Then open the box and compare your description of the terrain to the actual surface.

Exploration 2 The following altitude data were collected by the Pioneer mission. The location is a region near the Venusian equator. For the purpose of this exploration, altitude data are collected at points 90 feet apart and then rounded to the nearest multiple of 10 ft.

4 Examine these data and then, on a copy of the data, draw a contour diagram with contours at 10-ft altitude intervals starting at 725 ft.

720	780	820	830	830	830	830	840	840	840	850	850	850
750	800	820	830	830	830	830	830	840	840	840	850	850
790	820	830	830	830	830	820	830	830	840	840	840	840
820	830	830	820	820	810	810	820	830	840	840	840	840
840	840	830	810	800	800	800	810	830	840	840	840	840
840	840	820	800	790	780	780	800	820	840	850	850	860
830	820	800	780	770	770	770	790	820	850	860	870	890
800	790	780	770	770	770	770	790	820	860	890	910	930
780	770	760	760	770	770	770	790	820	860	910	940	970
760	760	760	760	770	770	780	790	820	860	920	960	980
750	760	760	760	770	780	790	790	810	860	920	970	990

Source: nssdc.gsfc.nasa.gov

a. Using your contour diagram, locate and justify possible landing sites for a space probe.

b. If the captain wishes to land at a high elevation, where would you suggest landing?

c. What can you infer about the terrain by the spacing between the lines?

d. Using the *TCMS-Tools* geometry custom app "Contour," select the NASA Pioneer data set. Compare your contour diagram to the one produced by this software. What does the coloring of the computer-produced diagram show?

5 Discuss the similarities and differences between the two contour diagrams you made in Explorations 1 and 2. For what contexts might a representation such as that in Exploration 1 be most helpful? In Exploration 2?

6 A portion of the altitude data from the bottom-left corner of the surveyed region on Venus is shown here on the coordinate system. Adjacent grid points are 90 ft apart. How could you describe each of the following lettered grid points on the planet using three numbers (coordinates) that give its position in feet, where the third coordinate is altitude?

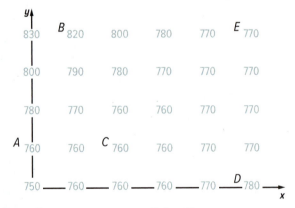

a. Point *A* **b.** Point *B* **c.** Point *C*

d. Point *D* **e.** Point *E*

f. Compare your point coordinates with those of your classmates. Resolve any differences.

7 **A Three-Dimensional Coordinate System** The three-number descriptions of a point in three-dimensional space found in Problem 6 are *rectangular coordinates* of the point. Rectangular coordinates of a point are determined by measuring the directed perpendicular distance from the point to each of three mutually perpendicular planes called **coordinate planes**. The *origin O* is the intersection of the three coordinate planes. The intersections of the pairs of planes are the *coordinate axes*. The axes are usually called the *x*-, *y*-, and *z*-axes. The labels *x*, *y*, and *z* indicate the positive direction on each axis.

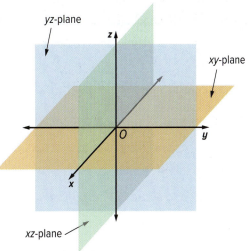

It is customary to represent a three-dimensional rectangular coordinate system with the *y*-axis horizontal and the *z*-axis vertical as shown. The positive *x*-axis appears to come forward out of the page and creates the appearance of depth, or the third dimension. Coordinates of points are given in the order (x, y, z).

a. How would you describe the *xz*-plane? The *xy*-plane?

b. Into how many regions do these three coordinate planes separate space?

c. Describe the location of all points with:

    i. positive *x*-coordinates.

    ii. negative *z*-coordinates.

    iii. positive *y*- and negative *x*-coordinates.

8 Care is needed in plotting and interpreting points in three-dimensional space. When plotting points, it is helpful to show the "path" to the point as indicated in the diagram at the right.

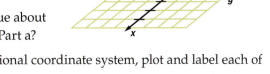

a. Plot and draw the paths to points $P(-2, 3, 4)$ and $Q(2, 3, 4)$ on a copy of the three-dimensional coordinate system shown.

b. What should appear to be true about the plot of the two points in Part a?

c. On a copy of a three-dimensional coordinate system, plot and label each of the following points.

    $A(0, 2, 3)$ $B(3, 2, 4)$ $C(3, -2, -4)$

    $D(-5, 3, 5)$ $E(-4, -7, -2)$

d. Describe in general how you would plot a point $A(a, b, c)$.

SUMMARIZE THE MATHEMATICS

In this investigation, you examined how to draw contour diagrams from altitude data and considered how to use three coordinates to identify locations in space.

a. Describe how contour lines can be drawn to describe a smooth three-dimensional surface.

b. Describe how coordinates can be used to specify the location of a point in three-dimensional space.

c. The altitude data for Venus (Problem 4, page 532) were given every 90 feet. If the x- and y-axes are placed so they intersect at the bottom-left corner of the data set, give the three coordinates of a possible landing site for a spacecraft mission to Venus.

Be prepared to share your group's ideas and reasoning with the class.

✓ CHECK YOUR UNDERSTANDING

The height (in feet) of a surface, represented by the z values in the table below, was measured for the 49 lattice points in a square region: $-3 \leq x \leq 3$, $-3 \leq y \leq 3$.

a. Use these data to plot the positive z values on an xy-coordinate plane. Then construct a contour diagram for the height z of the surface above the xy-plane. Draw contour lines at 1-ft intervals, starting at 9 ft. Use your contour diagram and patterns in the data to describe the shape of the mapped surface.

	−3	**−2**	**−1**	**0**	**1**	**2**	**3**
−3	8.5	10	11	11.5	11	10	8.5
−2	10	12	13.5	14	13.5	12	10
−1	11	13.5	15	15.5	15	13.5	11
0	11.5	14	15.5	16	15.5	14	11.5
1	11	13.5	15	15.5	15	13.5	11
2	10	12	13.5	14	13.5	12	10
3	8.5	10	11	11.5	11	10	8.5

(y is labeled across the top; x is labeled along the left side.)

b. On a three-dimensional coordinate system, plot the point that has z-coordinate of 16 in the contour diagram. Also plot the point that has x-coordinate of −2 and y-coordinate of −3.

Visualizing and Reasoning with Cross Sections

As you saw in Investigation 1, contour diagrams are developed by sampling heights of a terrain at specific intervals. The contour lines represent horizontal cross sections of the terrain. The representation only approximates the true characteristics of the terrain. However, for many people who use contour diagrams, such as hikers, forest rangers, deep sea divers, and ship navigators, the approximate nature of the information is adequate. They want to know generally what the land or sea bottom near them looks like.

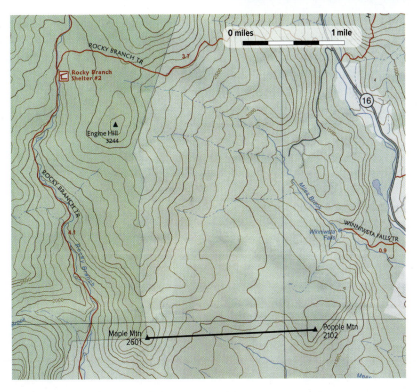

A contour diagram can be used to create a *topographic profile* of a region that also shows the variation in the altitude of the land surface. As you work on the problems of this investigation, look for answers to these questions:

How can you interpret and draw a profile of a region?

How can horizontal and vertical cross sections of a three-dimensional figure be used to determine and sketch the figure?

Topographic Profiles Understanding what contour maps represent can be enhanced through the construction of a topographic profile. A **topographic profile** is a vertical cross-sectional view along a line segment drawn across a portion of a map. In other words, if you could slice vertically through a section of earth, pull away one half, and look at it from the side, the surface would be a topographic profile. Not only does constructing a topographic profile help hikers visualize terrain, it is very useful for geologists when analyzing numerous questions about land formations.

1 Suppose you wanted to hike from the summit of Maple Mountain to the summit of Popple Mountain following a path shown by the line segment drawn near the bottom of the map on the previous page. Consecutive contour lines on this map represent 100-ft elevation differences. Consider the vertical cross section of the region indicated by this segment, called a **relief line**.

   a. Along the relief line, what can you say about the land surface if the contour lines are close together? If they are far apart?

   b. The section of the map with the relief line between Maple Mountain and Popple Mountain is shown below. Using the following procedure, you can make a topographic profile to help visualize the terrain of the hike.

   Step 1. Place a sticky note along the relief line as shown. Make tic marks at the endpoints of the segment. Write the elevation of the starting and ending points next to their tic marks.

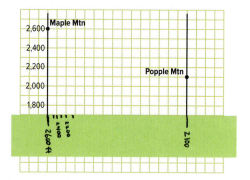

   Step 2. Next, make a tic mark at places the paper crosses a contour line on the map. Write the elevations for the contour lines below their tic marks on the sticky note.

   Step 3. Now, place your marked sticky note along a horizontal axis on grid paper. Draw vertical lines for the starting and ending elevations of the topographic profile as shown at the right. Label the vertical axis with units appropriate for your elevation data.

   Step 4. Plot points representing the elevation data and connect them to visualize the terrain of your hike. Discuss the shape of the terrain with your classmates.

   c. How could you add information to the topographic profile to represent the distance hiked along the trail?

2 A group of divers earning a certification is doing a compass run from a rocky beach to a buoy in Lake Superior. The goal of the exercise is to maintain their heading in spite of terrain and current. Contour lines on this map differ by a depth of 10 feet. Sketch a profile of the lake bottom along their path. What purpose do you think the buoy is serving?

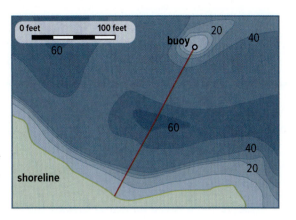

3 Cross Sections Just as a contour diagram can help you visualize portions of the surface of Earth or other planets, **cross sections**—the intersections of planes with physical or mathematical objects—can reveal important features of those objects.

a. The contour diagram at the right shows the series of horizontal cross sections at 2-in. intervals for an object. What polyhedron might this contour diagram represent?

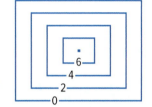

b. Now sketch a circular cone and a triangular pyramid with each shape resting on its base. Assuming that the height of each shape is 8 in. and that contour lines are based on horizontal cross sections at 2-in. intervals, draw the corresponding contour diagrams.

c. How would the contour diagrams for the cone and pyramid change if each shape were resting on the vertex instead of the base?

d. Look back at the map on page 535. Locate Engine Hill and the three closest contour lines to the summit. How is this contour map similar to and different from the contour diagrams formed by horizontal cross sections in this problem?

4 Consider a cylinder with its base in the xy-plane as shown below.

a. Describe the cross sections of the cylinder if the intersecting planes are parallel to the xy-plane.

b. Could any other three-dimensional object have cross sections parallel to its bases exactly like those you described in Part a? If so, describe the object.

c. Describe cross sections of the cylinder if the intersecting planes are parallel to the yz-plane.

d. How could you display or record both the horizontal and vertical cross sections so that they could be used to identify and sketch the cylinder? Compare your method to those of your classmates. Resolve any differences.

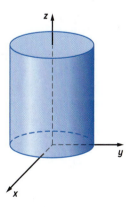

5 The **parabolic surface** at the right is the three-dimensional analog of a parabola. It was formed by rotating a parabola with vertex at the origin about the z-axis.

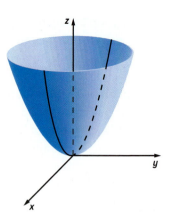

a. Sketch five cross sections of this object determined by equally spaced planes parallel to the xy-plane.

b. Repeat Part a for planes parallel to the yz-plane.

c. How could you label and display the cross sections so that they could be used to identify and sketch the object?

The displays you prepared in Problems 3–5 were made using horizontal or vertical cross sections determined by a series of parallel planes at uniform intervals. It is standard practice to record the series of horizontal cross sections with the one corresponding to the highest point drawn first and subsequent cross sections placed to the right. For the series of vertical cross sections, the first-drawn cross section is the one made by the vertical plane nearest the viewer; subsequent cross sections will be those obtained by intersections with vertical planes at uniform intervals further from the viewer.

6 Consider the following ordered horizontal and vertical cross sections of three-dimensional objects. Use the cross sections to help you describe each object.

a. Horizontal cross sections:

Vertical cross sections:

b. Horizontal cross sections:

Vertical cross sections:

c. Vertical cross sections for another object with the same horizontal cross sections as in Part a:

d. Horizontal cross sections:

Vertical cross sections:

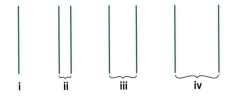

i ii iii iv

SUMMARIZE THE MATHEMATICS

In this investigation, you explored how topographic profiles are created from maps displaying contour lines. You also explored how analysis of both vertical and horizontal cross sections of an object can help you better understand the object.

a. Explain how to make a topographic profile.

b. How is a profile similar to, and different from, a vertical cross section of a surface?

c. Explain how cross sections of a three-dimensional object can be used to figure out what the object looks like.

Be prepared to share your ideas with the class.

✓ CHECK YOUR UNDERSTANDING

Successive horizontal cross sections of a three-dimensional object are shown below. Cross sections are made at 2-cm intervals beginning at the top of the object.

a. Sketch an object that has the given horizontal cross sections.

b. Make sketches of several differently shaped vertical cross sections of the object.

1 Study the portion of a contour map shown below. Describe as precisely as you can the terrain along the 3.5-mile White Ledge Trail including the highest and lowest altitudes. Follow the trail clockwise, beginning at White Ledge Campground.

2 The diagram below shows the contours of the temperature along one wall of a heated room throughout one winter day, with time indicated as on a 24-hour clock. The room has a heater located at the left-most corner of the wall, and there is one window in the wall. The heater is controlled by a thermostat several feet from the window. (**Source:** Adapted from *Multivariable Calculus, Preliminary Edition* by William McCallum, Deborah Hughes-Hallett, Andrew Gleason, et al., New York: Reprinted by permission of John Wiley & Sons, Inc., 1996.)

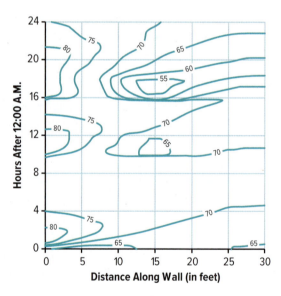

a. Where is the window? When was it open?

b. Why do you think that the temperature at the window at 5 p.m. (17 hours) is less than at 11 a.m. (11 hours)?

c. When was the heat on?

d. To what temperature do you think the thermostat is set? How do you know?

e. How far is the thermostat from the left-most corner of the wall?

3 The altitude data below (in feet) were collected by the Pioneer probe of equatorial Venus, with adjacent probe points 90 feet apart.

a. Use the data to construct a contour diagram of the region. To simplify the contour construction, use one contour line to enclose all values within a 50-ft interval, beginning with 500–549 ft. A contour line that encloses a region of altitudes from 500 to 549 feet has been drawn.

b. Are there possible landing sites in this region? If so, identify and justify your site choice(s). If not, explain why there is no appropriate site.

c. Sketch three vertical cross sections determined by the altitude data in the left, middle, and right columns of the chart.

585	623	631	616	587	572	575	565	533	512	504
625	659	663	658	619	587	587	576	525	507	511
658	666	666	661	636	619	624	617	551	527	531
697	675	672	663	652	655	663	649	591	565	565
760	707	703	673	653	657	659	651	613	584	580
835	744	716	696	671	652	638	634	628	611	591
936	795	736	723	676	651	628	628	642	642	621
998	845	761	747	682	651	633	634	646	661	695
973	856	752	719	682	649	634	633	658	708	753
877	838	740	696	668	630	629	641	701	763	769
834	803	719	686	672	638	631	639	672	729	753
696	671	639	650	648	645	645	627	616	665	724
547	543	566	615	624	637	654	653	674	704	736

Source: nssdc.gsfc.nasa.gov

4 One way to build a sense of points in three dimensions is to consider points plotted on isometric dot paper. Study the three-dimensional coordinate system at the right that shows the location of points *A*, *B*, and *C* in space, each defined by three coordinates (*x*, *y*, *z*). Think of the *x*- and *y*-axes as defining the floor of your room and the *z*-axis pointing straight up. (You might want to compare the diagram at the right to a corner of your room.)

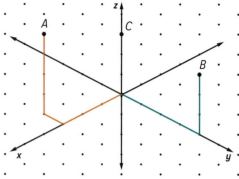

a. Following the marked trails for points *A* and *B*, what are the coordinates for these points?

b. Alanna has proposed that the coordinates for point *C* are (−2, −2, 1). How would you respond to her proposal?

5 An Air Canada plane is flying in a straight line from Thunder Bay, Ontario, to New York, NY. The following contour diagram shows the barometric pressure in millibars (mbar) on the day of the flight. The pressure is assumed to be unchanging throughout the day.

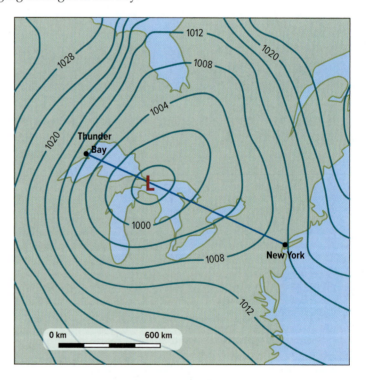

a. Draw a graph of the barometric pressure experienced by the plane as a function of distance from Thunder Bay.

b. Mark a point on your graph representing the location of the "L" (the lowest pressure).

6 A team of engineers at Compudesign is designing an air cooling system for a metal plate in a new model computer. Without the system, the temperatures (in °C) on the plate can be represented by the following contour diagram.

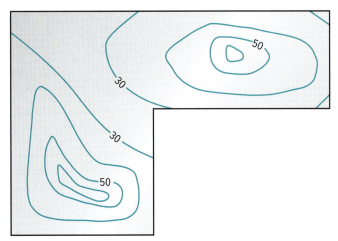

The planned cooling system will consist of two stiff, thin metal fins attached in such a way that they stand perpendicular to the plate. Heat from the plate will travel up into the fins and a fan will blow air across them, cooling the system. The engineers have decided that one fin will be placed along the line shown below.

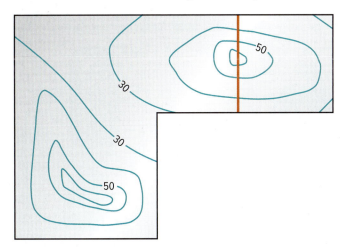

a. Why do you think the engineers chose this location for the fin?

b. Draw a graph of the temperature along the base of the fin as a function of distance from the top of the plate.

c. Choose a location for the second fin and draw a graph of the temperature along the base of the fin similar to that in Part b. What do points on your x-axis represent?

7 Just as rectangles are often replaced by nonrectangular parallelograms to give the impression of depth in two-dimensional drawings, circles are often replaced by ovals or ellipses in three-dimensional drawings. Consider the sphere shown below.

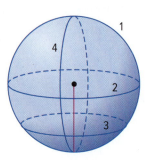

a. Which numbered circular cross sections are drawn as circles? Why?

b. Which numbered circular cross sections on the sphere are congruent? Under what condition will two horizontal or vertical cross sections be congruent?

c. Suppose the diameter of the sphere is 10 cm and that horizontal cross sections are taken at 1-cm intervals. Find the radius of each cross section and sketch the corresponding contour diagram.

d. How would a contour diagram of the bottom hemisphere of the sphere differ from the contour diagram of the sphere? From the contour diagram of the top hemisphere?

e. How does the contour diagram for a sphere differ from that of a cone standing on its base?

8 Assume that the height of each object below is 8 cm. Sketch a diagram for each object that shows horizontal cross sections at 2-centimeter intervals.

a.

b.

c.

d.

CONNECTIONS

9 Draw a three-dimensional coordinate system.

a. Using the origin as a vertex, draw a cube with 4 units on a side with its edges along the positive *x*-, *y*-, and *z*-axes. Label and find the coordinates of each vertex.

b. Draw another 4-unit cube with edges along the negative *x*- and *y*-axes and along the positive *z*-axis. Label and find the coordinates of each vertex.

10 High cholesterol resulting in clogged arteries was once thought to be the major underlying cause of heart attacks. Yet, half of all heart attack victims have cholesterol levels that are normal or even low. Research at Boston's Brigham and Women's Hospital suggests that inflammation, as measured by elevated levels of C-reactive protein, is another important independent trigger.

(**Source:** miriam-english.org/files/Atherosclerosis.pdf)

Study the three-dimensional bar graph at the right that shows the relative risk of cardiovascular problems by levels of cholesterol and C-reactive protein. Explain how each of the following findings is indicated in this plot.

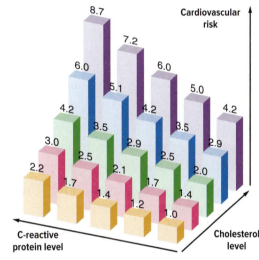

a. A person with the highest combination of cholesterol and C-reactive protein has almost 9 times the risk as someone with the lowest combination.

b. A person with C-reactive protein in the top quintile (fifth) has twice the risk as someone in the lowest quintile with the same cholesterol level.

c. High cholesterol levels seem to contribute more to the risk of cardiovascular problems than do high C-reactive protein levels.

11 Use the "Slicing Polyhedra" custom app to explore different possible cross sections formed by a plane intersecting a rectangular prism. Write a summary, with sketches, of your findings.

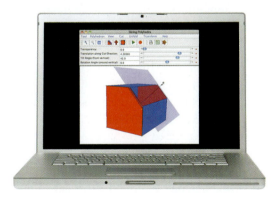

D. Hurst/Alamy

12 Graphs of equations in three variables are *surfaces*. Some of these surfaces can be generated by rotating (or revolving) a curve about a line, sweeping out a **surface of revolution** (see Investigation 2, Problem 5, page 538). Describe, and illustrate with a sketch, how a cone with vertex at the origin can be generated as a surface of revolution from a two-dimensional shape.

13 Sketch a graph of $x^2 + y^2 = 25$ in the *xy*-plane of a three-dimensional coordinate system.

  a. Imagine rotating the circle about the *y*-axis. What kind of surface is formed?

  b. Would you get the same surface if the circle was rotated about the *x*-axis? Explain your reasoning.

14 Consider the segment determined by the points $P(0, 5, 1)$ and $Q(0, 5, 6)$. Imagine rotating the segment about the *z*-axis. What kind of surface is formed? Draw a sketch of the surface.

15 Look back at the map on page 535. Suppose you are standing at Winniweta Falls and looking at the summit of Maple Mountain. What is the angle of elevation of your line of sight?

REFLECTIONS

16 In Investigation 1, you were introduced to a three-dimensional rectangular coordinate system. It is often helpful to picture a three-dimensional coordinate system in terms of a room you are in. Think of the origin of the coordinate system as a corner at floor level where two walls meet.

  a. Describe the *x*-, *y*-, and *z*-axes in this context.

  b. What would points with negative coordinates correspond to in this context?

17 Find an example of how contour diagrams are used to communicate the temperature patterns found across the United States and Canada. How is color sometimes used to indicate the higher temperature regions and the lower temperature regions?

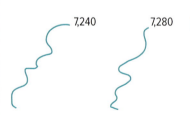

18 Maps displaying contour lines are informative but do not necessarily convey a complete picture of a region. At the left are two contour lines at 40-foot elevation intervals. The horizontal distance between the contours is approximately 300 yards.

  a. Could there be a region between these contours that is higher than 7,280 ft? Explain.

  b. Sketch possible side views (profiles) of the 300 yards between these contours.

  c. How does the existence of a trail across the region between contours eliminate some possible configurations of the terrain?

19 Try to visualize each of the following possibilities. You can use the "Slicing Cones, Cylinders, and Spheres" custom app to aid your thinking as necessary.

   a. If possible, describe a plane whose intersection with a cylinder is:

      i. a point.

      ii. a line segment.

      iii. a rectangle.

   b. If possible, describe a plane whose intersection with a sphere of radius r is:

      i. a point.

      ii. a circle with radius $\frac{1}{2}r$.

      iii. an oval.

EXTENSIONS

20 The Global Positioning System (GPS) locates an object's position on the Earth with three coordinates—latitude, longitude, and altitude. The system is very sophisticated. It can identify a location to within a meter or so. Boaters use GPS devices to locate themselves when they cannot see recognizable landmarks. GPS devices are common in automobiles and cell phones.

   Research and write a short report on the capabilities and underlying mathematical features of GPS. Include information on an application of GPS, other than in autos, boats, and cell phones.

21 Select one of the following projects to learn more about the usefulness of contour diagrams.

   a. Consult your state's Department of Natural Resources or another agency that has responsibility for parks, lakes, and rivers to locate topographical maps of regions in your state. Choose one such map for a land region and one for a lake and describe the region and lake on the basis of the data included in the map. Describe also the contour intervals used in these maps.

   b. If a local weather forecasting service is nearby, contact it and ask for copies of maps describing the atmospheric pressure patterns over the United States on a particular day. Learn how these pressure maps are used to assist in determining wind velocity and direction. Make a sketch of the pressure data. Write a report describing how pressure maps are used in the weather forecasting business.

22 Use the "Slicing Cones, Cylinders, and Spheres" custom app to explore cross sections formed by a plane and a *double cone* as shown at the right.

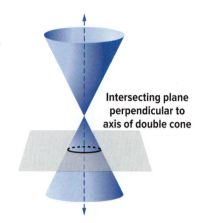

Intersecting plane perpendicular to axis of double cone

    a. Draw sketches of your findings.

    b. Research the technical names for the cross sections you find.

23 Research information on the process of CAT scans and MRIs and the interpretation of their images. Write a short report of your findings, including illustrations. Explain how the ideas are related to the idea of cross sections in Investigation 2.

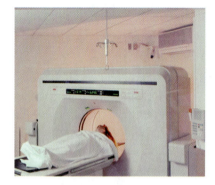

REVIEW

24 When solving equations, it is important to consider carefully the symbolic form of the equations. Consider the two equations below.

$$\text{I. } \sqrt{x} - 4 = 8 \qquad \text{II. } \sqrt{x - 4} = 8$$

    a. Describe how these two equations are similar and how they are different.

    b. Solve each of the equations.

    c. Would the process of solving $\sqrt[3]{x} - 4 = 8$ be more similar to the process of solving equation I or equation II? For the equation that you chose, explain how it would be the same and how it would be different. Then solve the given equation.

25 Determine the length of the line segment connecting each pair of points.

    a. $A(3, 7)$ and $B(-11, 7)$

    b. $X(9, -6)$ and $Y(-3, -6)$

    c. $C(0, 0)$ and $D(4, 10)$

    d. $H(-3, 12)$ and $K(2, 8)$

26 Determine the midpoint of each line segment in Review Task 25.

27 Recall that the equation of a circle with radius r centered at (h, k) has equation $(x - h)^2 + (y - k)^2 = r^2$.

a. What is the equation for the circle shown below?

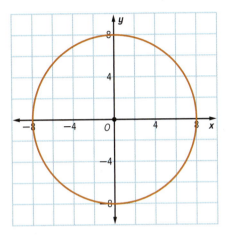

b. What are the y-coordinates of all points on the circle with an x-coordinate of 4?

c. Suppose that point A is on the circle so that $\overline{OA}$ and the positive x-axis form an angle with measure $\frac{2}{5}$ radians. In what quadrant is A located? Find approximate coordinates of point A.

28 A local electronics store is offering a 15%-off sale on all LCD televisions. In addition, the manufacturer is offering a $100 in-store rebate on their televisions.

a. The television that you want to buy is priced at $379. Does it make a difference whether the $100 rebate or the 15%-off sale is applied first? Explain your reasoning.

b. Does your answer in Part a depend on the original price of the television? Explain your reasoning.

29 Consider the line with equation $3x + 2y = 12$.

a. Draw a graph of this line on a coordinate system.

b. Write an equation in $Ax + By = C$ form for a line that does not intersect the given line. Add a graph of your line to the graph in Part a. Label your graph with its equation.

c. Write an equation in $Ax + By = C$ form for a line that intersects the given line in exactly one point. Add a graph of this line to your graph from Part a. Label your graph with its equation.

d. Write an equation for a line that intersects the line with equation $3x + 2y = 12$ at infinitely many points.

30 Solve each equation for x.

    a. $3x + 7y = 12$

    b. $3xy + 5y^2 = 2y$

    c. $5x - b = 3x - t$

    d. $\dfrac{x + a}{y} = 4y$

31 Some students find it difficult to correctly write equations for horizontal and vertical lines and for lines that contain the origin.

    a. Erma says that equations for horizontal lines are always of the form "$x = \ldots$" because the x-axis is horizontal. Is Erma correct or incorrect? Explain your reasoning.

    b. Tucker is confused about finding an equation for line ℓ shown below.

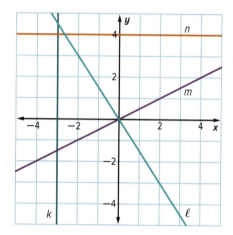

He knows that it contains the point $(0, 0)$ but when he substitutes that point into $y = mx + b$, he gets $0 = m(0) + b$, or $0 = b$. He is not sure what this means or what he should do next. Tucker is correct up to this point.

    i. Explain what $0 = b$ means.

    ii. What else does Tucker need to do to find the equation of this line? Find the correct equation.

    c. Find an equation for each of the remaining lines shown above.

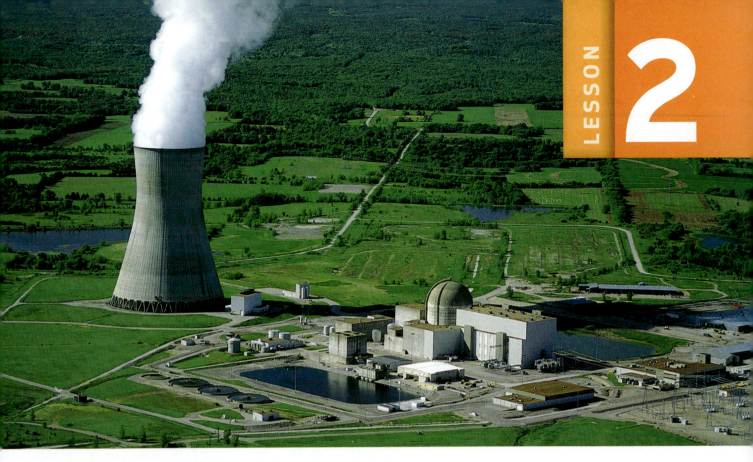

A Three-Dimensional Coordinate System

In the previous lesson, you represented surfaces of planets and familiar objects using contour diagrams and cross sections. You also explored the use of three-dimensional coordinates. Computer models for designing cooling towers like the one above are based on coordinate representations. The images are often produced from information about cross sections derived from the equations of the surfaces of the shapes.

You can use your understanding of coordinate representations of two-dimensional figures to guide your thinking about coordinate representations of related three-dimensional objects.

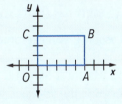
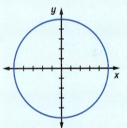
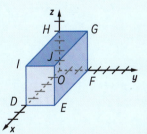
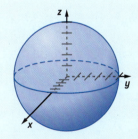

© ImageState/age fotostock

THINK ABOUT THIS SITUATION

Use the diagrams on the previous page to help you think about possible similarities and differences between two-dimensional and three-dimensional objects and their coordinate representations. The scales on the axes are 1 unit.

a. What are the coordinates of point *B*, a vertex of the rectangle? What are the coordinates of point *J*, a vertex of the rectangular prism? How was your reasoning similar in each case?

b. What are the coordinates of the point symmetric to point *B* with respect to the *y*-axis? With respect to the *x*-axis? What do you think are the coordinates of the point symmetric to point *J* with respect to the *xz*-plane? With respect to the *xy*-plane? Explain your reasoning.

c. What is true about all points on the line determined by points *B* and *C*? What is the equation of the line? What is true about all points on the plane determined by points *I*, *D*, and *E*? What do you think is the equation of the plane?

d. How would you determine the length of $\overline{AC}$? How might you determine the length of $\overline{DG}$?

e. What is the equation of the circle? What equation do you think would describe the *sphere*?

In this lesson, you will extend important ideas and reasoning strategies involving coordinates in two dimensions to coordinates in three dimensions.

INVESTIGATION 1

Relations Among Points in Three-Dimensional Space

In Lesson 1, you explored how to represent points in space with *x*-, *y*-, and *z*-coordinates. The *x*-axis and the *y*-axis determine the *xy-plane* as shown below. Similarly, the *x*-axis and the *z*-axis determine the *xz-plane*, and the *y*-axis and *z*-axis determine the *yz-plane*.

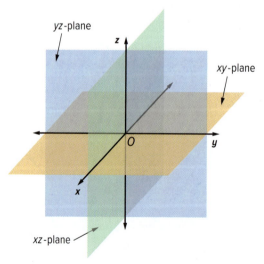

As you work on the problems of this investigation, look for answers to these questions:

How are graphs and equations of planes parallel to a coordinate plane in three dimensions similar to graphs and equations of lines parallel to a coordinate axis in two dimensions?

How are formulas for calculating the distance between two points and finding midpoints of segments in three dimensions similar to the corresponding formulas in two dimensions?

How is the equation of a sphere with center at the origin similar to the equation of a circle with center at the origin?

1 Equations for Special Planes You can determine the equations of the *coordinate planes* and planes parallel to them by reasoning from analogous cases in a two-dimensional coordinate system.

a. Think about the form of equations of horizontal and vertical lines in a two-dimensional (x, y) coordinate system.

  i. What is the equation of the x-axis? Of the y-axis? Why do these equations make sense?

  ii. What are the equations of the lines 5 units away from and parallel to the x-axis?

  iii. What are the equations of the lines 10 units away from and parallel to the y-axis?

b. Now think about the form of equations of horizontal planes in a three-dimensional (x, y, z) coordinate system.

  i. Which coordinate plane is represented by the equation $z = 0$? Why does this make sense?

  ii. What is the equation of the plane 4 units above and parallel to the xy-plane?

  iii. What is the equation of the plane 6 units below and parallel to the xy-plane?

  iv. The diagram at the right shows one of the two planes described above. Reproduce this diagram and then sketch the other plane. Label both planes.

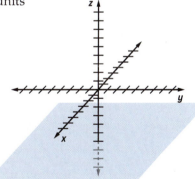

c. What is the equation for the xz-plane? What are the equations of the planes 5 units away from and parallel to the xz-plane? Sketch and label all three planes on the same coordinate system.

d. What is the equation for the yz-plane? What are the equations of the planes 12 units away from and parallel to the yz-plane? Sketch and label those two planes on the same coordinate system.

2 In your previous mathematics coursework, you derived a formula for computing distances between pairs of points in a coordinate plane. If $P(x_1, y_1)$ and $Q(x_2, y_2)$ are points in a coordinate plane, then

$$PQ = \sqrt{(x_1 - x_2)^2 + (y_1 - y_2)^2}.$$

In the Think About This Situation at the beginning of this lesson, you considered how you might compute the distance between points in three-dimensional space. Compare your ideas with the following approach suggested by one class in Winston-Salem, North Carolina. Their approach is based on the rectangular prism at the beginning of this lesson. The scales on the axes are 1 unit.

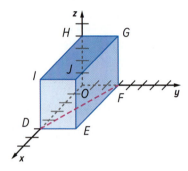

To find the length of $\overline{DG}$, the students suggested drawing $\overline{DF}$ and then using the Pythagorean Theorem twice.

a. Why is $\triangle DOF$ a right triangle? What is the length of $\overline{DF}$?

b. Why is $\triangle DFG$ a right triangle? What is the length of $\overline{DG}$?

c. What is the length of $\overline{OJ}$ in the diagram above?

3 Distance Formula: From Two to Three Dimensions Now use the diagram below to derive a formula for finding the distance PQ if $P(x_1, y_1, z_1)$ and $Q(x_2, y_2, z_2)$ are points in three-dimensional space and the figure shown is a rectangular prism. You may find it helpful to label the coordinates of points S and R.

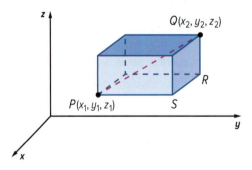

a. In the diagram, point Q is above point P. Does your distance formula hold if point Q is below point P? Explain your reasoning.

b. Use the distance formula you developed to find the distance between the points $A(1, 2, 3)$ and $B(6, -2, -7)$.

c. Does the formula you derived hold for any two points that lie in a plane parallel to a coordinate plane? Explain your reasoning.

4 Equations for Spheres Recall that a circle is the set of points in a plane at a given distance from a fixed point, its center. In three-dimensional space, the set of points at a given distance from a fixed point is a **sphere**. Like a circle, a sphere is determined by its center (the fixed point) and its radius (the given distance).

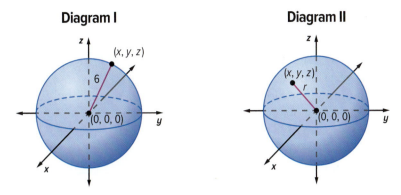

Diagram I   Diagram II

a. Find the equation for a sphere with center at the origin and radius 6 as shown in Diagram I.

b. What are the x-, y-, and z-intercepts of the sphere in Part a? Do the coordinates of these points satisfy your equation?

c. Using Diagram II, derive an equation for a sphere with center at the origin and radius r.

d. How would you modify the equation in Part c if the center of the sphere had coordinates (j, h, k) and radius r? Compare your ideas with those of your classmates. Resolve any differences.

e. The equation derived in Part c represents the surface of the sphere. How would you represent algebraically the solid ball enclosed by the sphere?

5 The pictures below illustrate just a few of the many applications of a solid sphere.

Ball bearings   Bocce balls   Replacement ball-and-socket joints

a. What physical property of spheres makes these ideal for all kinds of uses that involve smooth or rotating movement?

b. The surface areas of a cubic storage tank and of a spherical tank are each 60 m². Which tank has the greater volume? Explain your reasoning.

c. What research question or conjecture does your answer to Part b suggest?

6 Midpoint Formula: From Two to Three Dimensions The distance formula in three-dimensional space is a generalization of the distance formula in two dimensions. Now, investigate how to generalize the formula for the midpoint of a segment in a plane to that of a segment in three dimensions.

a. If $U(x_1, y_1)$ and $V(x_2, y_2)$ are points in a plane, what are the coordinates of the midpoint of $\overline{UV}$?

b. Now consider the points $A(5, 6, 12)$ and $B(1, -4, 4)$.

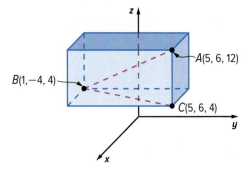

   i. Use your knowledge of computing the midpoint of a segment in a plane to calculate the midpoint of $\overline{BC}$. Then use your result to calculate the midpoint of $\overline{AB}$.

   ii. Use the distance formula to verify that the point you found is the midpoint.

c. Use the diagram at the right to derive a formula for the midpoint M of $\overline{PQ}$. Compare your formula for the midpoint of $\overline{PQ}$ with that of your classmates. Resolve any differences.

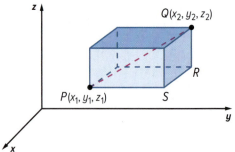

7 Now pull together the ideas you developed in this investigation. Consider the points $S(-3, 1, 5)$ and $T(2, 4, -3)$ in three-dimensional space.

a. Find the equation of the plane containing point S and parallel to the xy-plane.

b. Find the equation of the plane containing point T and parallel to the yz-plane.

c. Find ST.

d. Find the coordinates of the midpoint of $\overline{ST}$.

e. Find the equation of the sphere with center at the origin and containing point T.

f. Find the equation of the sphere with center at point S and containing point T.

SUMMARIZE THE MATHEMATICS

In this investigation, you extended important ideas involving coordinates in two dimensions to coordinates in three dimensions.

a. In three-dimensional space, how are the equations and graphs of planes parallel to a coordinate plane similar to, and different from, the equations and graphs of lines parallel to a two-dimensional coordinate axis?

b. Describe how the formula for the distance between two points in a three-dimensional coordinate system is similar to, and different from, the distance formula for two points in a two-dimensional coordinate system.

c. Describe how to find the coordinates of the midpoint of a segment in three dimensions.

d. How is the equation of a sphere with center at the origin and radius r similar to, and different from, the equation of a circle with center at the origin and radius r?

e. How is the equation of a sphere with center at (a, b, c) and radius r similar to, and different from, the equation of a circle with center (a, b) and radius r?

Be prepared to share your ideas with the entire class.

✓ CHECK YOUR UNDERSTANDING

In the diagram below, vertex T of the rectangular prism has coordinates $(12, 3, 6)$ and O is the origin.

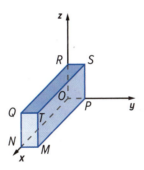

a. Determine the coordinates of the remaining vertices.

b. Write equations for the planes that contain the faces of the prism.

c. Use right triangles to explain why $(PQ)^2 = (PM)^2 + (MN)^2 + (NQ)^2$.

d. Calculate PQ using the equation in Part c. Calculate PQ using the distance formula. Compare your two results.

e. Find the midpoints of $\overline{OT}$ and $\overline{RM}$. What can you conclude?

f. Does the sphere with equation $x^2 + y^2 + z^2 = 200$ contain the prism? Explain.

The Graph of $Ax + By + Cz = D$

In two dimensions, two coordinates locate a point, and equations in two variables can be used to specify lines or curves. For example, the graph of $2x - 3y = 12$ is a line and the graph of $2x^2 + 2y^2 = 24$ is a circle.

In three-dimensional space, three coordinates locate a point, and equations in three variables can be used to specify surfaces. The surface may be flat, as in the case of a plane, or curved, as in the case of a sphere. The challenge, often, is to figure out what the surface looks like given its equation and to make a sketch of it, or to interpret the reasonableness of a technology-produced graph.

As you work on the problems of this investigation, look for answers to this question:

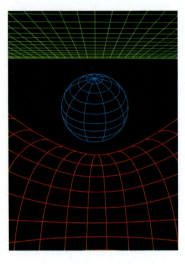

How can you describe and sketch graphs of linear equations of the form $Ax + By + Cz = D$?

1 The equation $Ax + By + Cz = D$ in three dimensions is analogous to the equation $Ax + By = C$ in two dimensions. Based on this analogy, what do you think the graph of $Ax + By + Cz = D$ could be? Explain your reasoning.

2 Now consider a specific instance of the equation $Ax + By + Cz = D$, namely $2x + 4y + 3z = 12$.

  a. Reasoning by analogy from your previous work in a two-dimensional coordinate system, what are the points at which the surface intersects the x-, y-, and z-axes (called the **intercepts**)? Plot these points on a three-dimensional coordinate system.

  b. Cross sections of the surface formed by its intersection with the coordinate planes or with planes parallel to a coordinate plane can provide useful information about the nature of the surface. The first cross sections to check are those made with the coordinate planes. These cross sections are called **traces**.

   i. Explain why you can find the trace of the surface in the xy-plane by setting $z = 0$ and examining the resulting equation.

   ii. Find the equation of the xy-trace for $2x + 4y + 3z = 12$. What is the shape of the trace?

   iii. Find the equation and describe the graph of each of the other two traces.

   iv. Plot the portion of each of the traces between the x- and y-axes, the y- and z-axes, and the x- and z-axes.

  c. Based on your information about the traces of the graph of $2x + 4y + 3z = 12$, what do you think the surface is? Why? Revise your sketch to better display this surface.

d. Additional information about a surface can be obtained by examining the equations of cross sections found by setting x, y, or z to a constant value.

  i. Explain how setting $z = 8$ generates a cross section of $2x + 4y + 3z = 12$ that is in a plane parallel to one of the coordinate planes. Describe the location of the cross section.

  ii. Find the equation of the cross section determined by setting $z = 8$. What is the shape of this cross section? How is this cross section related to the xy-trace?

  iii. Describe the cross section formed by setting $x = 5$. By setting $y = 2$.

  iv. Does the information you found by examining the equations of these cross sections confirm or change your thinking about the surface described by $2x + 4y + 3z = 12$? Explain.

Reasoning like you did in Problem 2 should suggest that a linear equation in x, y, and z is a **plane**.

3 Now consider how you can work backward from information about traces to the equation of a surface. Suppose a plane has traces with equations $x + y = 8$, $x + 4z = 8$, and $y + 4z = 8$. Sketch the plane and find an equation for the plane.

4 Discuss how the graph of the plane shown below is similar to and different from your sketch of $2x + 4y + 3z = 12$. Try to find the equation for this plane using the connection between an equation and points on its graph.

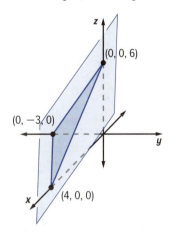

5 Illustrate how you could use the points where the graph of $2x + 3y + 2z = 6$ intersects the x-, y-, and z-axes to quickly sketch the surface on a three-dimensional coordinate system. Explain why your method works.

6 Sketch the graph of $3x + 5z = 15$ on a three-dimensional coordinate system. Describe the surface.

7 Consider the general equation $Ax + By + Cz = D$ for each set of conditions below. Describe the graph of each equation. Identify the intercepts, the traces, and the cross section at $z = 4$.

a. A, B, C, and D are all nonzero.

b. A, B, and D are nonzero, but $C = 0$.

c. $A = B = 0$, while C and D are nonzero.

8 As you have seen, equations of planes in three dimensions are similar to equations of lines in two dimensions. An important characteristic of lines is that they are straight; they have constant slope. Planes have an analogous characteristic: they are flat. But what makes them flat? Consider the portion of the graph of $2x + 3y + 6z = 6$ shown below. Scales are 1 on each axis.

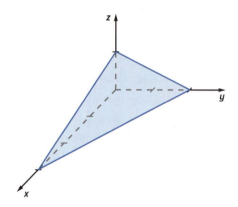

a. How can you quickly check the reasonableness of this graph?

b. Now imagine starting at the point $(0, 0, 1)$ and walking around on this plane. What can you say about the "slope" of your walk if you stay directly above the x-axis? The y-axis?

c. Suppose you walk on the plane along a path with constant y-coordinate. What can you say about the "slope" of your path? Explain your reasoning.

d. Suppose you walk on the plane along a path with constant x-coordinate. What can you say about the "slope" of your path in this case? Why does this make sense?

e. What can you say about the "slope" of walks on the plane along paths with the z-coordinate constant?

9 In your previous work in two dimensions, you saw that two lines could intersect in 0, 1, or infinitely many points. Consider the analogous situation in three dimensions in the case of the graphs of $2x + 3y + 4z = 12$ and $3x + 2y - z = 6$.

a. Sketch the planes and describe their intersection.

b. What are the possible ways the graphs of two nonequivalent equations of the form $Ax + By + Cz = D$ can intersect?

c. Under what conditions would the graphs of $A_1x + B_1y + C_1z = D_1$ and $A_2x + B_2y + C_2z = D_2$ be parallel?

d. Is it possible for the graphs of three equations of the form $Ax + By + Cz = D$ to intersect in 0 points? One point? Infinitely many points? Explain.

SUMMARIZE THE MATHEMATICS

The equation $Ax + By + Cz = D$, where not all A, B, and C are zero, is a first-degree (or linear) equation in x, y, and z.

a. What is the graph of an equation of this form?

b. How can you quickly sketch the graph?

c. What is the nature of each cross section of the graph of $Ax + By + Cz = D$?

d. How are the equations of planes parallel to coordinate planes special cases of $Ax + By + Cz = D$?

Be prepared to explain your ideas to the class.

✓ CHECK YOUR UNDERSTANDING

Sketch the graph of each equation on a three-dimensional coordinate system.

a. $2x - y + 3z = 6$

b. $x + y - 2z = 30$

c. $x - 4y = 8$

INVESTIGATION 3

Systems of Equations

In this lesson, you saw that many geometric ideas—distance between two points, midpoint of a segment, and symmetry in a two-dimensional coordinate model— have analogous representations in a three-dimensional coordinate model. You can also draw parallels with algebraic ideas that involve two- and three-dimensional coordinate systems.

As you work on the problems in this investigation, look for answers to these questions:

> *How can you extend methods for solving systems of linear equations in two variables to systems of linear equations in three variables?*
>
> *How can you examine a system of linear equations in three variables and predict the nature of the solutions?*

1 What are the possible number of solutions for a system of two linear equations in two variables? Describe the solutions geometrically.

2 Consider the following system of linear equations:

$$\begin{cases} -7x + y = 32 \\ 2x + 3y = 27 \end{cases}$$

a. How would you solve this system using the *substitution method*?

b. How would you solve this system using the *elimination method*?

c. Describe another method for solving this system of equations.

d. Choose one of your methods and solve the system. Check your solution.

e. Compare your method with those used by others. Was one method easier than others for this system of equations?

3 The Down Store in northern Michigan is a small family-owned business that manufactures down vests, jackets, and comforters.

	Down (in pounds)	Labor Time (in hours)	Profit (in dollars)
Vests	2	1	6
Jackets	3	2	6
Comforters	4	1	2

One week, the company used a shipment of 600 pounds of down and spent a total of 275 hours of staff time on manufacturing. Their profit for the week was $1,150. With the given information, it is possible to determine how many of each item the company manufactured during that week.

a. The *manufacturing requirement* can be represented by the equation:

$$2x + 3y + 4z = 600$$

   i. What do x, y, and z represent in this equation?

   ii. Explain how this equation represents the manufacturing requirement.

b. What is represented by the following equation?

$$x + 2y + z = 275$$

c. Write an equation representing the *week's profit*.

d. Check that the following system of equations is an accurate model of the problem.

$$\begin{cases} 2x + 3y + 4z = 600 \\ x + 2y + z = 275 \\ 6x + 6y + 2z = 1{,}150 \end{cases}$$

e. For each step below in solving the system of equations, explain what actions were performed on the previous step. Justify the actions using your previous mathematical knowledge.

Step 1. $\begin{cases} x + 2y + z = 275 \\ 2x + 3y + 4z = 600 \\ 6x + 6y + 2z = 1{,}150 \end{cases}$

Step 2. $\begin{aligned} x + 2y + z &= 275 \\ 0x - y + 2z &= 50 \\ 0x - 6y - 4z &= -500 \end{aligned}$

Step 3. $\begin{aligned} x + 2y + z &= 275 \\ -y + 2z &= 50 \\ 0y - 16z &= -800 \end{aligned}$

Step 4. $z = 50$

Step 5. $\begin{aligned} -y + 2(50) &= 50 \\ y &= 50 \end{aligned}$

Step 6. $\begin{aligned} x + 2(50) + 50 &= 275 \\ x &= 125 \end{aligned}$

f. How many vests, jackets, and comforters did the Down Store manufacture to make that week's profit of $1,150?

4 Solve these systems using the **left-to-right elimination method** illustrated above. Check each solution and describe it geometrically. Check your geometric descriptions against a CAS-produced graph of the system.

a. $\begin{cases} x + 2y + 3z = 6 \\ 2x + 3y + 2z = 6 \\ -x + y + z = 4 \end{cases}$

b. $\begin{cases} x + y + z = 6 \\ \phantom{x + {}}y - 2z = -18 \\ x - y - z = 0 \end{cases}$

c. $\begin{cases} x + y + z = 0 \\ 2x - y - z = 0 \\ -x + 2y + 2z = 0 \end{cases}$

d. $\begin{cases} x + 2y + 3z = 1 \\ 2x - y = 3 \\ x + 2y + 3z = 2 \end{cases}$

5 In your prior studies of mathematics, you learned that a system of two linear equations in two variables can be solved using a substitution method. For some systems of three linear equations, the substitution method may be more efficient. Solve each of the following systems of equations using substitution, if possible. Check your solutions.

a. $\begin{cases} x + 2y + z = 2 \\ \quad\quad -y + 3z = 8 \\ \quad\quad\quad\quad 2z = 10 \end{cases}$

b. $\begin{cases} x - y - 8z = 0 \\ \quad\quad y + 4z = 8 \\ \quad\quad 3y + 14z = 10 \end{cases}$

6 Enrique remembered being told in a campus visit that the most commonly used technology tool in college courses across the disciplines is a spreadsheet.

He explored how he could use a spreadsheet to solve the system of linear equations below.

$$\begin{cases} x + y + 2z = 3 \\ 3x + 2y + 2z = 4 \\ x + y + 3z = 5 \end{cases}$$

A start of Enrique's spreadsheet follows. Using *TCMS-Tools*, complete entering the system of three linear equations in three variables into your own version of Enrique's spreadsheet. Then use the built-in Solver functionality to solve the system. Check your solution.

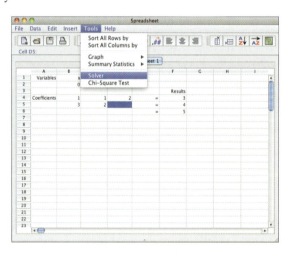

SUMMARIZE THE MATHEMATICS

In this investigation, you learned how to solve systems of three linear equations in three variables.

a. What does it mean to solve a system of three linear equations in three variables? Use the walls, floor, and ceiling in your classroom to geometrically illustrate some possible solutions to such a system.

b. Refer to the system of equations below. Describe in general terms how you could solve this system using the left-to-right elimination method.

$$\begin{cases} 4x - 4y + 4z = -8 \\ x - 2y - 2z = -1 \\ 2x + y + 3z = 1 \end{cases}$$

c. How would you check if $x = 1$, $y = 2$, and $z = -1$ is a solution for the above system?

d. What does it mean to solve a system of two linear equations in two variables by substitution? Explain how the system of three linear equations in Part b can be solved by substitution.

e. How can the system of three linear equations in Part b be solved using a spreadsheet and the Solver tool?

Be prepared to explain your ideas and reasoning to the class.

✓CHECK YOUR UNDERSTANDING

Consider the following system of linear equations in three variables.

$$\begin{cases} 3x - 6y + 9z = 0 \\ 4x - 6y + 8z = -4 \\ -2x - y + z = 7 \end{cases}$$

a. Solve this system by writing and solving an equivalent system in which the coefficient of x in the first equation is 1.

b. Explain why the two systems of equations are equivalent.

c. Check your solution in the original system and interpret your solution geometrically. Check your geometric description against a CAS-produced graph of the system.

1 In this lesson, you saw that geometric ideas such as distance and shape in a two-dimensional coordinate model had analogous representations in a three-dimensional coordinate model.

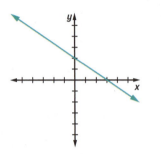 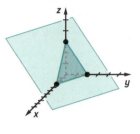

Complete a table like the one below, which summarizes some of the key features of a three-dimensional coordinate model.

Geometric Idea	Two-Dimensional Coordinate Model	Three-Dimensional Coordinate Model
Point	Ordered pair (a, b) of real numbers	
Plane	All possible ordered pairs (x, y) of real numbers	
Distance Between Two Points	For points $A(x_1, y_1)$ and $B(x_2, y_2)$, $AB = \sqrt{(x_1 - x_2)^2 + (y_1 - y_2)^2}$	
Midpoint of a Line Segment	For points $A(x_1, y_1)$ and $B(x_2, y_2)$, midpoint of $\overline{AB}$ is $M\left(\dfrac{x_1 + x_2}{2}, \dfrac{y_1 + y_2}{2}\right)$	
Locus of Points Equidistant from a Fixed Point	Circle $(x - h)^2 + (y - k)^2 = r^2$	Sphere

2 Triangle PQR has vertices $P(1, 2, 3)$, $Q(5, 4, 1)$, and $R(-1, 6, 5)$.

a. Draw $\triangle PQR$ in a three-dimensional coordinate system.

b. What kind of triangle is $\triangle PQR$?

c. Find, plot, and label the coordinates of the midpoints of each side.

d. Is $\triangle PQR$ similar to the triangle with the midpoints as vertices? Justify your response.

3 The table below gives ratings for six laptops by the editors of a magazine. A rating of 5 is the highest rating in each category.

	Hard Drive Size	Cost	Battery Life
Brand A	4	5	4
Brand B	3	5	5
Brand C	3	3	5
Brand D	2	5	5
Brand E	4	2	3
Brand F	3	4	3

a. The three-dimensional plot at the right provides a graphical model for the ratings shown in the above table. What does each axis represent?

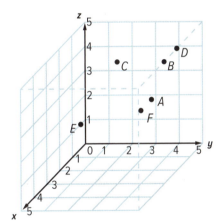

b. Rank the laptops by adding the three ratings for each.

c. Explain how the distance formula might be used to rank the laptops. Then rank them using your described method. Are the rankings the same or different than the ones in Part b? Which method would you recommend and under what conditions?

d. Suppose you use your laptop primarily while sitting at a desk. In this case, its battery life is not as important to you and the cost is very important. How might you modify your method from Part b to take this into consideration?

e. The magazine also included ratings for the laptop weight. How might you modify the distance formula to take this fourth rating into account?

4 Sketch the graph of each of the following equations on separate three-dimensional coordinate systems. Check your sketch using the three-dimensional graphing capabilities of the CAS in *TCMS-Tools*. Be prepared to explain any differences.

a. $x + 2y + z = 8$

b. $x + y + 3z = 3$

c. $4x - 2y + z = 4$

d. $2x + 3y - z = 2$

5 The recession that the United States experienced beginning in 2008 had severe consequences for many citizens including loss of jobs, devaluation of their homes, and even wide-scale foreclosure. The need for social services was high. A social service agency in eastern Arizona specializes in services to clients with these problems.

   The agency can accommodate a total of 500 clients. The service center has a budget of $150,000 for counseling and $100,000 available for emergency food and shelter. Type I clients require an average of $200 for counseling and $300 for emergency assistance. Type II clients require an average of $500 for counseling and $200 for emergencies. Type III clients require an average of $300 for counseling and $100 for emergency assistance. How many clients of each type can the agency serve?

6 Solve each of the following systems using the left-to-right elimination method. Check your solutions.

a. $\begin{cases} x + 2y + 2z = 5 \\ x - 3y + 2z = -5 \\ 2x - y + z = -3 \end{cases}$ **b.** $\begin{cases} x + y + z = 10 \\ 2x + 3y - 3z = -2 \\ 2x + 4y - 2z = 6 \end{cases}$ **c.** $\begin{cases} 2x - 4y + z = 4 \\ x + 3y - z = 5 \\ 4x - 2y + 3z = 6 \end{cases}$

CONNECTIONS

7 In your previous studies, you saw that reflection (or line) symmetry was often an important consideration in architectural design, in art, and in the design of products like automobiles. You likely found reasoning with line symmetry helpful in your study of properties of function graphs, geometric figures, and distributions of data.

   Complete the missing symmetry entries below using visualization or a model of a three-dimensional coordinate system. Add the information to the table you completed in Applications Task 1.

Geometric Idea	Two-Dimensional Coordinate Model	Three-Dimensional Coordinate Model
Reflection Symmetry	Across the x-axis $(x, y) \rightarrow (x, -y)$	Across the xz-plane $(x, y, z) \rightarrow ?$
	Across the y-axis $(x, y) \rightarrow (-x, y)$	Across the yz-plane $(x, y, z) \rightarrow ?$
		Across the xy-plane $(x, y, z) \rightarrow ?$

8 Look back at the table you completed for Applications Task 1. Now consider the similarities and differences between graphs of linear inequalities in two dimensions and in three dimensions.

    a. In a two-dimensional coordinate model, how would you graph $4x + 6y \leq 12$ without using technology? How would you describe the graph?

    b. In a three-dimensional coordinate model, how would you graph $4x + 6y + 3z \leq 12$ without using technology? How would you describe the graph?

    c. Summarize the two- and three-dimensional analogous representation in your table from Applications Task 1.

9 In Unit 2, *Functions Modeling Change*, you saw that changing the rule of a function of a single variable by adding or multiplying by a constant transformed its graph in predictable ways. The graph at the right is a *paraboloid*. Its equation is $z = x^2 + y^2$. Describe the graph of each of the following functions, and describe its relationship to the original paraboloid. Check your predictions using three-dimensional graphing technology.

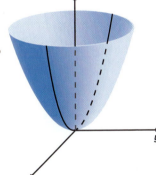

    a. $z = x^2 + y^2 + 5$

    b. $z = 3 - x^2 - y^2$

    c. $z = x^2 + (y - 2)^2$

    d. $z = 5x^2 + 5y^2$

10 The disposable drinking cup shown at the right can be thought of as a surface of revolution with a bottom. The cup has a 5-cm diameter bottom, 7-cm diameter top, and has height 9 cm.

    a. Beginning with an appropriate segment in three-dimensional space, describe how the surface can be generated.

    b. The volume of the cup can be approximated by cutting it horizontally into sections each with a height of 0.5 cm, approximating each section with a cylindrical disk, and then summing the volumes of the disks. Approximate the volume in this manner.

    c. Another way to generate the cup surface is to make the rotating segment part of a line through the origin. In this case, the volume of the cup is the difference of the volumes of two cones. Find the volume in this manner.

    d. Compare the values of the volume found in Parts b and c. Describe how you could improve your approximation in Part b.

Jim Laser/CPMP

11 It can be shown that the volume of any three-dimensional figure bounded by a quadratic surface and two parallel planes (as in the surface shown at the right) can be calculated using the **prismoidal formula**

$$V = \frac{B + 4M + T}{6} \cdot h,$$

where B is the area of the cross section at the base, M is the area of the cross section at the middle, T is the area of the cross section at the top, and h is the height of the figure.

a. Use the prismoidal formula to calculate the volume of the cup in Task 10. Compare your answer to the approximation you calculated in Part b of that task.

b. Use the prismoidal formula to derive formulas for the volumes of the following three-dimensional figures.

I	II	III	IV

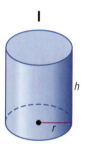

12 Sketch graphs of $z = 0$, $z = 5$, and $z = -2$ on the same three-dimensional coordinate system. Describe how the graphs are related.

13 In two dimensions, a line in a plane separates the plane into two *half-planes*, so that if points P and Q are on opposite sides of the line, $\overline{PQ}$ intersects the line. Write an analogous statement for the case of three dimensions. Illustrate your statement with a sketch.

14 Solve the following system of equations.

$$\begin{cases} x - 7y + z = 3 \\ 2x - 14y + 3z = 4 \end{cases}$$

a. Give one specific point that is a solution.

b. Describe geometrically the solution of this system.

c. Check your description against a CAS-produced graph of the system of equations.

15 In your previous studies, you saw that a system of two linear equations in two variables represents two lines in a plane. The two lines may intersect in no point, one point, or infinitely many points. Rewrite these two statements for the analogous case of a system of three linear equations in three variables.

16 Find values of a, b, c, and d that will guarantee that the system:

$$\begin{cases} x - y - z = -4 \\ x + 2y + 2z = 11 \\ ax + by + cz = d \end{cases}$$

a. has a point as its solution.

b. has infinitely many points as its solution.

c. has no points as its solution.

REFLECTIONS

17 Suppose you are standing at the point $S(4, 5, -3)$. North is in the negative x-direction, east is in the positive y-direction, and up is in the positive z-direction.

a. At what coordinates will you be standing after you move north 6 units, up 5 units, west 8 units, down 3 units, east 10 units, and south 4 units? Call this point A.

b. After completing the trip in Part a, will you be looking up or down at the point $T(-2, 5, 1)$?

c. After completing the trip in Part a, will you be closer to point S or point T?

18 In sketching a plane at the beginning of Investigation 2, why was it important to examine several cross sections, in addition to the traces?

19 Must the graph of a linear equation in three variables have a trace in each of the three coordinate planes? Explain.

20 Below are two systems of linear equations; one in two dimensions, the other in three dimensions.

Two Dimensions

$$\begin{cases} 6x - 4y = 12 \\ ax + by = c \end{cases}$$

Three Dimensions

$$\begin{cases} 6x + 4y - 3z = 12 \\ dx + ey + fz = g \end{cases}$$

a. Find values of a, b, and c for which the first system has no points of intersection. Has one point of intersection.

b. Find values of d, e, f, and g for which the second system has no points of intersection. Has a line as its intersection.

21 Describe at least four different ways of solving a system of three linear equations in three variables. How do you decide which method to use?

22 *Reasoning by analogy* is another important mathematical habit of mind.

a. What is meant by reasoning by analogy?

b. Give two examples where reasoning by analogy was helpful in this lesson.

EXTENSIONS

23 Find the equation of the plane through $P(2, 0, 0)$, $Q(0, 5, 0)$, and $R(0, 0, -8)$.

24 Are points $A(1, 1, 2)$ and $B(2, 4, 3)$ on the same or opposite sides of the plane with equation $x - 3y + 7z = -5$?

25 Solve the following system of three linear equations.

$$\begin{cases} 3x + 2y = 17 \\ 4y + 3z = 22 \\ 2x + 5z = 15 \end{cases}$$

26 In your previous studies, you may have learned how systems of linear equations in two variables can be represented by *matrix equations*.

a. What system of linear equations in three variables is represented by the following matrix equation?

$$\begin{bmatrix} 4 & -2 & 3 \\ 8 & -3 & 5 \\ 7 & -2 & 4 \end{bmatrix} \begin{bmatrix} x \\ y \\ z \end{bmatrix} = \begin{bmatrix} 1 \\ 4 \\ 5 \end{bmatrix}$$

b. The matrix equation in Part a has the form $AX = D$. What does A represent? X? D?

c. If the multiplicative inverse of matrix A exists, the equation $AX = D$ can be solved by multiplying both sides of the equation on the left by the inverse of A, denoted A^{-1}. If you perform this action, what equivalent matrix equation results?

d. Use this **inverse-matrix method** and the matrix capabilities of *TCMS-Tools* to solve the matrix equation in Part a. Check your solution.

e. Use matrices to solve the following system of equations. Check your solution.

$$\begin{cases} 2x - 4y + 7z = 11 \\ x + 3y - 5z = -9 \\ 3x - y + 3z = 7 \end{cases}$$

REVIEW

27 Given the equation $6x + 7y = 10$, determine a second equation so that the system of equations has:

a. no solution.

b. one solution.

c. infinitely many solutions.

28 Recall that the volume formula for both prisms and cylinders is $V = Bh$, where V is the volume, B is the area of a base, and h is the height. The height of these figures is the perpendicular distance between the planes containing their bases. A second related formula, $V = \frac{1}{3}Bh$, gives the volume for pyramids and cones, where h is the perpendicular distance between the vertex and the plane of the base of a pyramid or cone.

a. Sketch and label diagrams illustrating these formulas for a prism, cylinder, pyramid, and cone. At least one sketch should be of an oblique object (for example, a slanted cylinder).

b. Find the volume of a square prism with a height of 10 cm and a base whose sides are 4 cm.

c. Find the volume of a square pyramid with a height of 10 in. and a base whose sides are 4 in.

d. Find the volume of a cone with a height of 5 in. and radius of the base of 10 in.

e. Find the volume of a cylinder with a height of 5 cm and radius of the base of 10 cm.

29 Write an algebraic expression for each situation.

a. At the local grocery store, pears cost \$1.29 per pound, apples cost \$0.98 per pound, and bananas cost \$0.49 per pound. Write an expression for the total cost of buying p pounds of pears, a pounds of apples, and b pounds of bananas.

b. At a garage sale, Carlos bought 4 books and 6 DVDs. If each book cost b cents and each DVD cost v cents, how much did he spend?

c. Srini is preparing postcards and letters in support of his favorite school board candidate. He can prepare 25 postcards per minute and 10 letters per minute. Write an expression for how long it will take Srini to prepare the mailing. Be sure to define any variables that you use in your expression.

30 To encourage students to do their best, all students at Jackson High School who improve their grade from one marking period to the next are eligible to win a coupon for their choice of free ice cream. At the beginning of each marking period, 20 names from the list of eligible students are randomly selected. For the last marking period, 75 of the 500 eligible students were seniors.

a. What is the probability that exactly two seniors are selected?

b. What is the probability that two or fewer seniors are selected?

c. Two of the seniors won a free ice cream coupon. Does this fact give you reason to doubt that students were randomly selected? Explain your reasoning.

31 Match each inequality with the correct graph of its solution set. Be prepared to explain how you made your decisions.

a. $y \leq -3x + 6$

b. $y \geq -3x + 6$

c. $4x + 2y \leq 12$

d. $4x - 2y \geq 12$

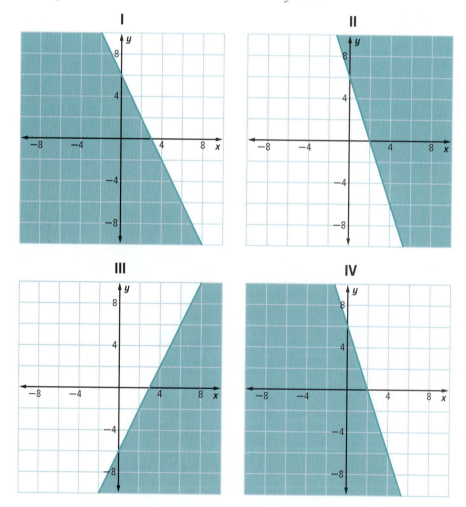

32 It is important to be able to move between different representations of measurements for time and distance.

    a. When might it be easier to work with a decimal representation such as 5.25 hours rather than the equivalent 5 hours and 15 minutes?

    b. When might you prefer to have a measurement of 2 inches rather than 0.167 feet?

    c. It took Maya 2 hours and 24 minutes to complete her chores. How many hours did it take her? How many minutes?

    d. The length of a piece of cloth is 3 feet 9 inches. How many inches long is the cloth? How many feet long is it?

33 In which of the following accounts is it better to invest? Justify your claim.

    Account A: pays 8% annual interest compounded quarterly
    Account B: pays 7.6% interest compounded continuously

34 Let $f(x) = 3(2x - 3)$ and $g(x) = -2x + 7$.

    a. Evaluate $f(-4)$ and $g(-4)$.

    b. Solve $f(x) = g(x)$.

    c. For what values of x is $g(x) < f(x)$? Explain your reasoning.

35 Write each of the following expressions in standard polynomial form.

    a. $(2x^2 - 5x + 7) - (5x^2 + 12x + 18)$

    b. $(x^3 - 5x^2)(10x + 4)$

    c. $5(x - 3) + 6(4x + 8) - 3(7 - 3x)$

    d. $\dfrac{25x^3 - 10x^2 + 5x}{5x}$

Linear Programming: A Graphical Approach

Important decisions in business often involve many variables and relations among those variables. The key to making good decisions is finding a way to organize, represent, and compare options.

For example, suppose that the manager of a small tablet company must plan for and supervise productions of two tablet PCs, the basic TabPC I and the advanced TabPC II. Assume that demand for both tablet PCs is high, so the company will be able to sell whatever is produced. To plan the work schedule, the manager has to consider the following conditions.

- Assembly of each TabPC I model takes 2 hours 24 minutes of technician time, and assembly of each TabPC II model takes 1 hour 12 minutes of technician time. The manager can apply at most 336 hours of technician time to assembly work each day.

- Testing for each TabPC I model takes 30 minutes, and testing of each TabPC II model takes 1 hour. The plant can apply at most 130 hours of technician time each day for testing.

- TabPC I and TabPC II models each take 15 minutes per tablet to package for shipping. The company has limited time available for packaging and can dedicate at most 40 hours of packaging time each day to the task.

- The company makes a profit of $150 on each TabPC I model and $250 on each TabPC II model.

The production planning goal or objective is to maximize daily profit while operating under the constraints of limited technician and packaging time.

THINK ABOUT THIS SITUATION

Suppose that you were the manager of the tablet plant and had to make production plans.

a. How would you decide the time estimates for assembly, testing, and packaging?

b. How would you decide the expected profit for each tablet PC?

c. How might you use all the given information to decide on the number of TabPC I and TabPC II models that should be produced to maximize daily profit for the company?

Many problems like those facing the electronics plant manager are solved by a mathematical strategy called *linear programming*. In this lesson, you will learn how to use this important problem-solving technique and the mathematical ideas and skills on which it depends.

INVESTIGATION 1

Linear Programming with Two Variables

Linear programming problems, like the one faced by managers of the tablet PC company described at the beginning of this lesson, involve finding an optimum choice among many options. As you work on the problems in this investigation, look for an answer to this question:

> *How can algebraic and graphical methods be combined to help solve linear programming problems in two variables?*

Tablet PC Production Planning Look back at the objective and constraints in the tablet PC production-planning problem at the beginning of the lesson.

1 If x represents the number of TabPC I models and y represents the number of TabPC II models produced in a day, what algebraic rule shows how to calculate total profit P for the day? This rule is called the **objective function** for the linear programming problem because it shows how the goal of the problem is a function of, or depends on, the independent variables x and y.

2 The production-scheduling problem at the tablet PC plant requires the manager to find a combination of TabPC I and TabPC II models that will give greatest daily profit. But there are **constraints** or limits on the choice.

a. Explain how the linear inequality $2.4x + 1.2y \leq 336$ represents the assembly-time constraint.

b. Verify that the shaded region on the graph at the right satisfies the constraint inequality.

i. Why does the graph only show points in the first quadrant?

ii. What two additional constraint inequalities are being assumed?

iii. Which point(s) in the shaded region do you think will lead to greatest daily profit for the company? Test your ideas by finding the daily profit at a variety of points in the shaded region.

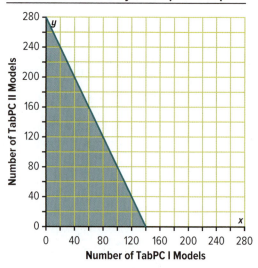

Tablet PCs Assembly Time (in hours)

Number of TabPC II Models (y-axis)
Number of TabPC I Models (x-axis)

3 Next consider the constraint for testing the tablet PCs.

a. Write an inequality that expresses the testing-time constraint in hours. Compare your inequality with that of others in your class. Resolve any differences.

b. On a copy of the graph above, graph the solutions of the testing-time constraint inequality, shading the region with a second color.

c. Describe the region of the graph in Part b that satisfies both the assembly-time constraint and the testing-time constraint.

d. Which point(s) in the double-shaded region do you think will lead to greatest daily profit for the company? Test your ideas by finding the daily profit at a variety of points in that region.

4 Now, recall that the plant can devote at most 40 hours per day to packaging time for the tablets, and that each takes 15 minutes.

a. Write an inequality that expresses this packaging-time constraint. Compare your inequality with your classmates and resolve any differences.

b. On your graph for Problem 3 Part b, graph the solutions of this third inequality, shading the region with a third color.

c. The intersection of the three colored regions is called the **feasible region**. It represents the points with coordinates that meet all the constraints. Which *feasible point(s)* do you think will lead to greatest daily profit for the company? Test your ideas.

5 In Problems 1–4, you developed some ideas about how to maximize profit while satisfying each problem constraint. You can analyze the tablet PC production problem by using a technology tool such as the "Graphical Linear Programming" custom app in the *TCMS-Tools* Geometry menu.

a. Begin by entering each constraint inequality to produce the feasible region. The first constraint inequality is displayed below. (In general, use <= for ≤ and >= for ≥.) Check the reasonableness of each constraint graph.

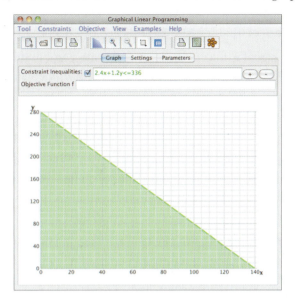

b. Now enter the objective function and use the "Show Obj. Fun. = Value" option under the Objective menu. Move the trace point on the graph of the objective function across the feasible region and observe how the profit value changes as the trace point is moved. Find the maximum profit.

c. How does your solution found in Part b compare with your findings in Problem 4 Part c?

Cattle Diet Planning For some animals, selection of a good diet is a carefully planned scientific process. Dairy farmers want a healthy diet during gestation at minimal cost. Dairy farmers need to consider not only the cost of food, but the way it satisfies a cow and her calf's dietary needs.

6 Suppose alfalfa hay and corn grain are used as feed for milk cows. Both come in 100-pound bags. The farmer is interested in two dietary needs per each 1,300-pound cow: crude protein (CP) and total digestible nutrients (TDN).

- Each bag of alfalfa hay provides 15 pounds of CP and 50 pounds of TDN.

- Each bag of corn grain provides 10 pounds of CP and 80 pounds of TDN.

- Minimum total requirements for the first trimester of the gestation are at least 160 pounds of CP and at least 1,000 pounds of TDN.

- Each bag of alfalfa costs $20 and each bag of corn grain costs $14.

The farmer's goal is to find the least expensive combination of alfalfa hay and corn grain that will fulfill total nutritional requirements of CP and TDN for the first trimester of the gestation.

a. If x represents the number of alfalfa hay bags and y represents the number of corn grain bags, write an objective function C for this situation.

b. Determine each constraint inequality. Then consider each of the following potential choices. Which, if any, of these choices provides at least the minimum of CP and TDN requirements and satisfies the constraint inequalities?

   Choice 1: 12 bags of alfalfa hay and 5 bags of corn grain

   Choice 2: 0 alfalfa hay bags and 16 corn grain bags

   Choice 3: 16 alfalfa hay bags and 2 corn grain bags

   Choice 4: 4 alfalfa hay bags and 10 bags of corn grain

c. Sketch a graph of the constraint inequalities and label the feasible region. Locate the points representing the four choices in Part b on your graph. Do the locations in terms of the feasible region agree with your results from Part b?

d. How does the feasible region for this cow diet-plan problem differ from that found in the tablet PC production problem? Why?

e. Using your objective function from Part a, which of the four choices in Part b minimizes cost while fulfilling the total nutritional requirements of CP and TDN?

7 Now use a technology tool such as the "Graphical Linear Programming" custom app in *TCMS-Tools* to analyze the cow diet-plan situation from Problem 6.

a. Enter the constraint inequalities to find the feasible region. Compare the software-produced graph with your sketch from Problem 6 Part c and resolve any differences.

b. Next, enter the objective function C and use the software to investigate the cost of various feed combinations in the feasible region. Find the combination of alfalfa hay and corn grain that meets the diet constraints and minimizes the cost. Verify that this solution agrees with your answer to Problem 6 Part e.

c. How do you know that there is no other choice that might further reduce the farmer's cost? Carefully explain your response based on graphical reasoning.

SUMMARIZE THE MATHEMATICS

In this investigation, you solved linear programming problems using a combination of algebraic and graphical reasoning.

a. Why are the constraints in the tablet PC production and the cattle diet-plan problems most accurately expressed as inequalities rather than equations? Why are the objective functions most accurately expressed as equations rather than inequalities?

b. What shapes would you expect for the feasible regions in other linear programming problems where the goal is to maximize some objective function? In problems where the goal is to minimize some objective function? How does this relate to the variables of profit and cost?

c. What seems to be the best way to locate feasible points that maximize (or minimize) the objective function in a linear programming problem?

Be prepared to share your ideas and reasoning with the class.

✓ CHECK YOUR UNDERSTANDING

A city recreation department offers Saturday gymnastics classes for beginning and advanced students. Each beginner class enrolls 15 students, and each advanced class enrolls 10 students. Available teachers, space, and time lead to the following constraints.

- There can be at most 7 beginner classes and at most 6 advanced classes.

- The total number of classes can be at most 9.

- The number of beginner classes should be at most twice the number of advanced classes.

a. What are the variables in this situation?

b. Write constraint inequalities for this situation.

c. Find the combination of beginner and advanced classes that will give the most children a chance to participate, assuming that only full classes are allowed.

Linear Programming with Three Variables

Many linear programming problems give rise to an objective function and constraints that involve three or even more variables. Usually, these *variables* are restricted by the situation to non-negative values, often non-negative integer values. These problems can be modeled in a way analogous to that in Investigation 1.

As you complete the problems in this investigation, look for an answer to this question:

How can algebraic and graphical reasoning be combined to help solve linear programming problems in three variables?

Riding the Wave to Maximum Profit There are different types of surfboards for surfing, each of which is individually manufactured. A small surfboard manufacturer in Hawaii produces three different types of surfboards: shortboard, funshape, and longboard. Each board goes through a process of (1) forming and shaping the outer shell; (2) laminating and fin adding; and (3) sanding and final finishing. The number of hours for each surfboard type is summarized in the following table.

Hours Required to Manufacture Surfboards

	Forming and Shaping	Laminating and Fin Adding	Sanding and Final Finishing
Shortboard	2	1	3
Funshape	3	2	3
Longboard	4	1	4

1 During the summer season, the surfboard manufacturer spends at most 600 hours on forming and shaping, 275 hours on laminating and fin adding, and 725 hours on sanding and final finishing. Let x be the number of shortboards, y be the number of funshape surfboards, and z be the number of longboards. Assume non-negativity of these numbers.

a. Explain how the linear inequality $2x + 3y + 4z \leq 600$ represents the forming and shaping constraint for this surfboard manufacturer.

b. Based on your work in Lesson 2 Investigation 3, describe and sketch the region formed by this linear inequality and the inequalities, $x \geq 0$, $y \geq 0$, and $z \geq 0$. Then explore the use of the "Graphical Linear Programming" custom app in *TCMS-Tools* to graph this constraint inequality. Select "3D" from the View menu before entering the constraint inequality. Compare and resolve any differences between your sketch and the technology-produced graph.

c. Write an inequality for the laminating and fin adding time constraint. On a new set of coordinate axes, sketch the region formed by this constraint inequality.

d. Make a conjecture about the shape of the region that satisfies *both* constraint inequalities. Check your conjecture using the "Graphical Linear Programming" custom app.

e. Write an inequality for the sanding and final finishing time constraint. Use the custom app to produce the region that satisfies the three inequalities in Parts b, c, and d. What does this region of space represent in terms of surfboard manufacturing?

2 The small surfboard manufacturer makes a profit on each shortboard of $150, on each funshape surfboard of $160, and on each longboard of $200. Its goal is to maximize its profit.

a. What algebraic rule shows how to calculate total profit P for the season? Enter this objective function into the "Graphical Linear Programming" custom app.

b. Describe the graph of the objective function. Explain why this makes sense.

c. Click and drag the graph of the objective function across the feasible region. What is the greatest profit the surfboard manufacturer can expect? How many of each type of surfboard should be manufactured?

3 **Designing an Exercise Program** As part of promoting a healthy lifestyle, Aimee is designing a monthly exercise program consisting of bicycling, jogging, and swimming. Aimee's schedule permits her to exercise at most 30 hours a month. Her initial monthly routine will involve at most 5 hours of swimming and no more than 15 hours of jogging. She also prefers not to spend more than 20 hours a month bicycling and jogging. In addition, the number of hours swimming and twice the number of hours bicycling combined should not exceed 15 hours. The calories burned per hour for bicycling, jogging, and swimming are 200, 475, and 275, respectively.

a. Explain how the above information can be used to set up a linear programming problem.

b. Based on Aimee's exercise program plan, what is the goal of this problem and what is the related objective function?

c. Determine the inequalities that represent the constraints on Aimee's exercise program. Compare your work with your classmates and resolve any differences.

d. Use the "Graphical Linear Programming" custom app in *TCMS-Tools* to graph the constraint inequalities and the objective function. Print the feasible region and then label key points.

e. What is the optimal exercise program for Aimee? Compare your solution with that of your classmates. Resolve any differences.

4 Maximizing Pizza Profit Pablo's Pizza makes gourmet frozen pizzas to sell to supermarket chains. The company makes three deluxe pizzas—a gluten-free vegetarian, a classic pepperoni, and a deluxe meat.

- Each vegetarian pizza requires 14 minutes of labor, each pepperoni pizza requires 10 minutes of labor, and each deluxe meat pizza requires 12 minutes of labor. The plant has at most 3,800 minutes of labor available each day.

- The plant freezer can handle at most 320 pizzas per day.

- The company can sell at most 250 deluxe meat pizzas each day.

- Sale of each gluten-free vegetarian pizza earns Pablo's Pizza $4.50 profit. Each pepperoni pizza earns $4.00 and each deluxe meat pizza earns $4.25 profit.

Pablo's Pizza would like to plan production to maximize daily profit.

a. What are the variables, objective function, and constraint inequalities?

b. What combination(s) of pizzas do you recommend that will produce maximum profit for Pablo's Pizza?

c. Sales records from the past six months indicate that supermarket chains tend to order about 160 deluxe meat pizzas and about 90 pepperoni pizzas each day. Using these estimates, how many gluten-free vegetarian pizzas should Pablo's Pizza produce each day to obtain a maximum profit?

SUMMARIZE THE MATHEMATICS

In this investigation, you used algebraic and graphical reasoning to explore and solve linear programming problems in three variables.

a. What is the general form for a constraint inequality in three variables? Describe the shape of the region formed by this inequality and provide an illustrative sketch.

b. What is the general form for an objective function in three variables? Describe the shape of its graph.

c. Explain how to identify the feasible region. Describe the general shape of a feasible region.

d. What seems to be the best way to locate feasible points that maximize (or minimize) the objective function in a linear programming problem?

Be prepared to explain your ideas and reasoning to the class.

Examine this computer-produced graph for a linear programming problem.

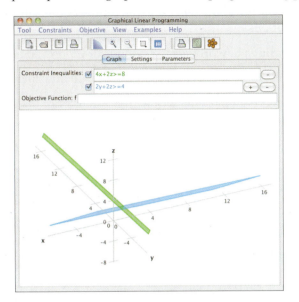

a. The constraint inequalities are $4x + 2z \geq 8$, and $2y + 2z \geq 4$. How would you check the reasonableness of the graph of each inequality boundary?

b. Describe the shape of the feasible region for this *minimization* problem.

c. The objective function is $F = 2x + 4y + 12z$. Identify the coordinates of the point(s) that will minimize this function. What is the minimum value?

INVESTIGATION 3

Linear Programming Using Spreadsheets

In Investigation 2, you solved linear programming problems involving three variables using algebraic and geometric reasoning aided by the "Graphical Linear Programming" custom app in *TCMS-Tools*. The technology tool that is most commonly used in college and careers is a spreadsheet. Spreadsheets have a Solver tool that can be used to efficiently solve linear programming problems in two or more variables.

As you work on the problems in this investigation, look for answers to these questions:

How can spreadsheets be used to solve linear programming problems with two or more variables?

Among available software tools for solving linear programming problems, how does one decide which tool to use?

How are numeric, symbolic, and graphical representations used in solving linear programming problems with software tools?

Revisiting the Cow-Diet Plan Problem Recall that the diet provided for each cow during the first trimester of gestation consists of alfalfa hay and corn grain. Nutrients required are crude protein (CP) and total digestible nutrients (TDN). The farmer's objective is to fulfill the nutritional requirements at minimum cost under the following conditions.

- Each bag of alfalfa hay provides 15 pounds of CP and 50 pounds of TDN.

- Each bag of corn grain provides 10 pounds of CP and 80 pounds of TDN.

- Minimum total requirements for the first trimester of the gestation are at least 160 pounds of CP and at least 1,000 pounds of TDN.

- Each bag of alfalfa costs $20 and each bag of corn grain costs $14.

1 The problem conditions are organized in the spreadsheet shown below. Recall the **$** symbol is used to fix a column letter or row number reference so that when the Fill Down or Fill Across command in the Edit menu is executed, that column letter or row number does not change. Carefully examine the spreadsheet entries by connecting them to problem conditions while completing Parts a–c below.

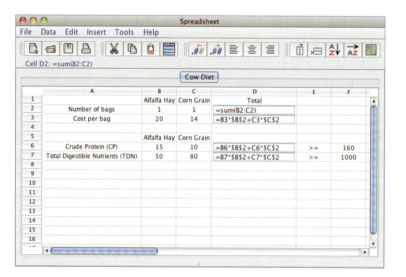

a. What do the formulas in cells **D2** and **D3** represent? What result would you expect when the formulas are evaluated?

b. What do the formulas in cells **D6** and **D7** represent? What result should appear in these cells when the formulas are evaluated?

c. What would be appropriate labels for the cells in columns **E** and **F**?

d. Create a similar spreadsheet with the same headings, quantitative information, and formulas. Compare your spreadsheet with that of others and resolve any differences.

2 Now consider the following choices for the number of 100-pound bags of alfalfa hay and corn grain that will fulfill the CP and TDN nutritional requirements.

    Choice 1: 12 bags of alfalfa hay and 5 bags of corn grain

    Choice 2: 0 alfalfa hay bags and 16 corn grain bags

    Choice 3: 16 alfalfa hay bags and 2 corn grain bags

    Choice 4: 4 alfalfa hay bags and 10 bags of corn grain

a. For each choice, enter the given information in your spreadsheet. Are the constraints satisfied? What is the feed cost for the choice?

b. Which choice(s) will minimize cost to the farmer?

3 In Lesson 2, you saw that some spreadsheet software have a Solver tool. In that lesson, the Solver tool was used to solve a system of three linear equations in three variables. The Solver tool can also be used to solve linear programming problems.

a. Use the spreadsheet Solver tool in *TCMS-Tools* to find the number of 100-pound bags of alfalfa hay and corn grain that will fulfill the CP and TDN requirements. Determine the appropriate optimization type and cell ranges. For example, **B4:C8** is the syntax for identifying the cells in columns **B** and **C**, rows **4** through **8**.

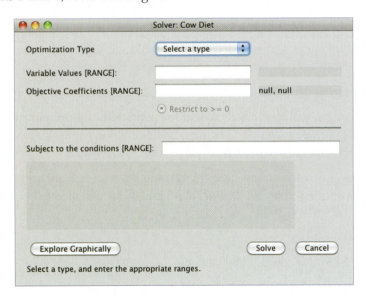

b. What values did the Solver report for the optimal number of alfalfa hay and corn grain bags? Discuss with your classmates whether or not this solution makes sense.

c. Now re-examine the cow-diet plan problem graphically using either the "Explore Graphically" feature of the Solver tool or the "Graphical Linear Programming" custom app in *TCMS-Tools*. How many 100-pound bags of alfalfa hay and corn grain would minimize the farmer's cost? Provide a sketch to illustrate the solution.

4 Suppose that the cost per 100-pound bag of corn grain doubles and the cost of 100-pound bags of alfalfa hay remains the same. How will this affect the setup and solution of the cow-diet plan problem? Test your ideas using the Solver tool and/or the "Graphical Linear Programming" custom app.

5 Allocating Space, Maximizing Sales
Books 'n More have leased space for a new store in Waterfront Mall. Plans call for the leased space to be divided into two sections, one for books and the other for music and videos.

- The store owners can lease up to 10,000 square feet of floor space.

- Furnishings for the two kinds of selling space cost $5 per square foot for books and $7 per square foot for music and videos. The store has a budget of at most $60,000 to spend for furnishings.

- Each square foot of book-selling space will generate an average of $75 per month in sales. Each square foot of music- and video-selling space will generate an average of $95 per month in sales.

The store owners have to decide how to allocate space to the two kinds of merchandise, books and music/video, to maximize monthly sales.

a. Describe the variables, constraints, and objective in this business decision.

b. Write inequalities that represent the constraints. Represent the objective function. Be prepared to explain how you know your algebraic representations are correct.

c. Use *TCMS-Tools* capabilities to find the maximum profit and report the number of square feet of floor space to be allocated to books and to music/videos. Provide screenshots of:

- the spreadsheet setup.

- the Solver setup.

- the graphical representation with solution shown.

6 Strategic Use of Time and Tools Gena is active in afterschool activities and has a part-time job. She wants to maximize her enjoyment from the little free time that she has every day by doing her favorite activities. Her three favorites and their ratings (on a 5-star scale) are as follows:

- Engaging in online activities
  (e.g., social networking or browsing the Web)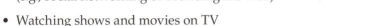

- Watching shows and movies on TV

- Using her cell phone (for calls and texting)

Gena must share her computer and TV time with her siblings, and so can only spend at most 3 hours on those two activities. Gena's parents are concerned that she spends too much time on her cell phone, and so have limited her use to no more than one hour. In addition, they want to spend more time watching TV as a family, and so have stipulated that Gena's TV-watching time should be at least twice as much as she devotes to cell phone use.

a. Is this a two-variable or three-variable linear programming problem? Explain your reasoning. If the problem is represented with a spreadsheet, how many rows and columns would be needed for the numbers and formulas?

b. Think about the computer tools that you have available to you. What tool seems best suited to set up and understand the structure of this linear programming problem? To solve it? Explain your thinking.

c. Divide the work among your classmates so that you are using different tools to help Gena determine the best possible situation for her evening free time.

   i. Clearly indicate the setup for the tool you use, identifying the variables, constraints, and objective as related to the situation.

   ii. Identify the optimal solution(s) and explain what it means for Gena's free time.

d. Compare your work in Part c with a classmate who chose to use a different tool to solve the problem.

SUMMARIZE THE MATHEMATICS

In this investigation, you explored the use of spreadsheets in solving linear programming problems and compared their use to that of other software tools.

a. How are the variables, constraints, and objective in a linear programming problem represented when using a spreadsheet Solver tool? A graphical linear programming app?

b. How does software such as the "Graphical Linear Programming" custom app and the spreadsheet Solver tool help to solve linear programming problems?

c. How is use of the spreadsheet Solver tool similar to use of the "Graphical Linear Programming" app? How is it different?

d. How might you make strategic decisions about which software tool(s) to use when solving a linear programming problem with two variables? With three variables?

Be prepared to explain your ideas and reasoning to the class.

✔ CHECK YOUR UNDERSTANDING

The Aspen Park Committee plans to sell Frisbees, visors, and water bottles to raise money for a new neighborhood park. The Frisbees sell for $4 each, the visors $2.50 each, and the water bottles $2.00 each. Based on the population demographics of the neighborhood, the committee estimates that they will sell at most 25 Frisbees; at most twice as many water bottles as Frisbees; and the total number of visors and Frisbees sold will be at most three times the number of water bottles. The committee wants to maximize income.

a. Identify the variables and write the constraint inequalities and objective function in this situation.

b. What is the maximum income? How many Frisbees, visors, and water bottles are needed?

APPLICATIONS

1 The junior class at Plano Senior High School sells coffee beans at the local businesses to raise funds for their yearly events. The students mix and sell two blends of gourmet coffee, Bold Blend and Jubilant Java.

- Each batch of Bold Blend uses two pounds of Robusta beans and two pounds of Arabica beans.

- Each batch of Jubilant Java uses one pound of Robusta beans and three pounds of Arabica beans.

- The students have an inventory of 120 pounds of Robusta beans and 200 pounds of Arabica beans to use.

- The income per batch is $9.00 for the Bold Blend and $12.00 for the Jubilant Java.

The students would like to plan production to maximize income.

a. Write the constraint inequalities for this situation. Write an expression for the objective function.

b. On a grid like the one at the right, graph the system of constraint inequalities. Shade the feasible region for this linear programming problem. Label each segment of the feasible region boundary with the linear equation that determines it.

c. What is the maximum income to be expected from coffee bean sales? How many batches of Bold Blend and Jubilant Java should they produce?

Coffee Beans Sales Feasibility Options

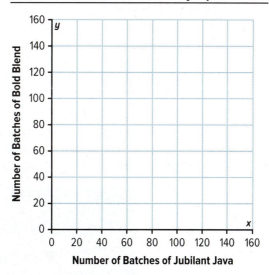

Number of Batches of Bold Blend (y-axis, 0 to 160)

Number of Batches of Jubilant Java (x-axis, 0 to 160)

Plush Studios/Bill Reitzel/Getty Images

2 Without using technology, graph the feasible region and locate the maximum value(s) of the given objective function under the given constraints.

   a. Constraints: $4x + y \leq 16$, $x + 3y \leq 15$, $x \geq 0$, $y \geq 0$
   Objective Function: $F = 6x + 8y$

   b. Constraints: $x \geq 0$, $y \geq 0$, $3x + 4y \leq 24$, $x + 2y \leq 10$
   Objective Function: $F = 4x + 7y$

3 Sunseeker Corporation has solar panel manufacturing plants in Holland and Zeeland, Michigan. Each plant produces three sizes of panels, SK1, SK2, and SK3. The table below summarizes the production capacity for each size panel, the number of current orders for each size panel, and the daily production cost for each of the plants. How many days should each plant operate to complete the orders at minimum cost to Sunseeker?

	Holland Plant	Zeeland Plant	Current Orders
SK1 Panel	80/day	20/day	1,600
SK2 Panel	50/day	20/day	1,900
SK3 Panel	10/day	10/day	500
Daily Production Cost	$20,000	$10,000	

4 Wedel's Flower Bulb Company bags a variety of mixtures of bulbs. There are three customer favorites. The Moonbeam mixture contains 15 daffodils, 15 jonquils, and 32 narcissuses; the Sunshine mixture contains 15 daffodils, 75 jonquils, and 14 narcissuses; and the Starlight mixture contains 30 daffodils, 30 jonquils, and 14 narcissuses. The profit for each bag is $2.30 for the Moonbeam mixture, $2.50 for the Sunshine mixture, and $2.80 for the Starlight mixture.

Wedel's Flower Bulb Company imports 240,000 daffodils, 300,000 jonquils, and 280,000 narcissuses for sale each year. The company wants to maximize its profit.

   a. What are the variables, objective function, and constraint inequalities? Be
   prepared to explain how you know your algebraic representations are correct.

   b. Find the maximum profit and report the number of bags of each mixture.

5 A receptionist for a sports medicine and physical therapy clinic schedules three
different types of appointments: 15-minute walk-in consultation screenings,
45-minute post-surgical rehabilitation sessions, and 20-minute general
evaluation and treatment sessions for injuries. The clinic cannot schedule
more than 9 post-surgical rehabilitation sessions and 14 general evaluation
and treatment sessions per day. In addition, the clinic only has 430 minutes
available for appointments per day. Using the table below, what combination of
appointments will yield maximized earnings for the clinic?

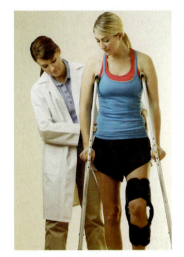

Appointment Type	Earnings per Appointment
walk-in consultation/screening	$90
post-surgical rehabilitation	$200
general evaluation and treatment for injuries	$155

6 A farmer plans to grow and sell two different types of feed for cows—alfalfa
and corn. He has the following constraints that must be taken into consideration.

   • He has allotted $21,000 of his startup funds for buying seed. It costs $300
     to buy seed for each acre of alfalfa and $400 for each acre of corn.

   • He has $2,415 for labor costs. Each acre of alfalfa requires two hours of
     labor to plant and each acre of corn requires five hours. Labor costs are
     $15 per hour.

The expected profit is $565 from each acre of alfalfa and $810 from each acre of
corn. Find a planting plan for the farmer that will provide maximum profit.

7 Malknight's Bakery has created recipes that allow the bakery the option to make
two different types of cookies from the same ingredients by using different
quantities of the ingredients. Using the recipes, total supplies, and number of
servings shown below, how many batches using the original recipe and using
the low-salt recipe will maximize the number of cookies made?

Original Recipe	Low-Salt Recipe	Total Supplies
1 cup softened butter	1 cup softened butter	150 cup butter
1¾ cup sugar	1¾ cup sugar	275 cup sugar
1½ tsp. vanilla	1¾ tsp. vanilla	200 tsp. vanilla
2 eggs	2 eggs	250 eggs
3 cup flour	3 cup flour	450 cup flour
1 tsp. baking powder	1½ tsp. baking powder	160 tsp. baking powder
1 tsp. salt	½ tsp. salt	100 tsp. salt
Servings: 24 large cookies	Servings: 32 small cookies	

8 Maximize the objective function $F = 10x + 5y + 4z$ subject to the indicated constraints:

$$6x + 3y + 3z \leq 150$$
$$4x + 12y \leq 80$$
$$5y + 5z \leq 75$$

CONNECTIONS

9 The diagram at the right shows the feasible region for a linear programming problem. The objective function is $P = 2x + 3y$.

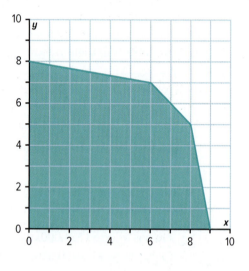

a. On a copy of the diagram, plot the line $2x + 3y = 6$. Explain why the objective function has the value 6 at each point on that line in the feasible region.

b. On the same diagram, plot the line $2x + 3y = 12$. What is the value of the objective function for each point on that line in the feasible region?

c. On the same diagram, plot the line $2x + 3y = 33$. What is the value of the objective function for each point on that line in the feasible region?

d. Explain how the pattern of results in Parts a–c suggests that the lines with equation $2x + 3y = k$ contain no feasible points if $k > 33$.

e. Explain how your work on Parts a–d illustrates the fact that the maximum value of the objective function in a linear programming problem like this will always occur on the boundary of the feasible region.

10 A linear programming problem involving minimizing cost (in dollars) has an objective function $C = 5x + 10y$. The diagram at the right shows the feasible region for the problem and the graph of all combinations of x and y for which $20 = 5x + 10y$.

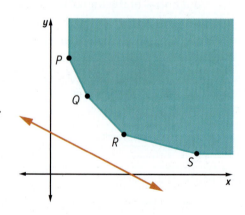

a. Write the equation of this line in slope-intercept form.

b. As the constant cost increases, does the line slide up or down? Explain your reasoning.

c. How are the various lines of constant cost related? Explain.

d. Where does the optimal solution of this linear programming problem occur? Why does this make sense?

11 Consider the feasible region at the right of a linear programming problem. Explain why the objective function $I = ax + by$, with $a, b > 0$, must have its maximum at point Q.

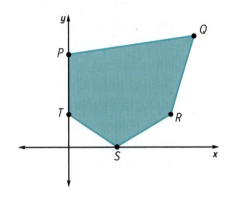

12 Suppose in a two-variable linear programming problem involving cost minimization, another constraint is added. Is it possible for the new minimum cost to be greater than the original minimum cost? Is it possible for the new minimum cost to be less than the original minimum cost? Explain your reasoning.

13 For the feasible region shown at the right:

a. find an objective function of the form $F = ax + by$ that has its maximum at $(14, 4)$.

b. find an objective function of the form $F = ax + by$ that has more than one integer solution.

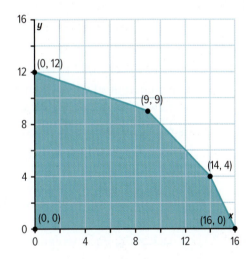

14 How many solutions are possible in a linear programming problem in three dimensions? Provide reasoning to support your conclusion.

REFLECTIONS

15 How are the tablet PC production problem and cow diet-plan problem from Investigation 1 alike? How are they different? Be sure to consider the constraints, objective, and graphical and algebraic representations in your response.

16 Analyze Austin's thinking about the solution to the linear programming problem set up below. Identify any errors in reasoning. If there are errors, decide how they might be corrected.

I know from the given information that the objective function for this situation is F = 5x + 6y. The graph of this function is parallel to one of the boundary lines of the feasible region. This means that the solution to the linear programming problem must be on that line. By looking at the graph, I can tell that the optimal point is (5, 6) because it is an intersection point of that line with another boundary line and it is in the middle of the graph.

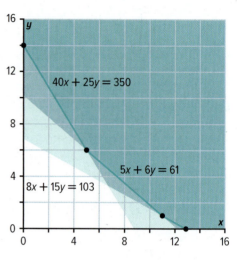

17 Describe the four main steps in solving a linear programming problem graphically.

18 Look back at your solution for Problem 4 (page 584) of Investigation 2. You found that two adjacent vertices of the feasible region gave the maximum value of the objective function.

   a. Will all points on the segment connecting the two vertices also maximize profit for Pablo's Pizza? Explain your reasoning.

   b. Jill noticed that the difference of the two vertices is (125, 125, −250) which has coordinate values that are all divisible by 125, leaving (1, 1, −2). She suggested that by starting at vertex (25, 45, 250), for example, she could produce the other solutions along that segment by adding (1, 1, −2); the next solution is (26, 46, 248). Is Jill correct? If so, why? If not, correct her reasoning.

19 Look back at the linear programming problems in this lesson.

   a. What kind of numerical values make sense for the variables in these problems?

   b. How would you proceed if you determine the optimal solution to a linear programming problem occurs at a point that has non-integer coordinates and the problem requires a solution with integer values?

20 Solving linear programming problems includes finding the boundary of the feasible region. Describe at least three different ways to find the intersection points of the sub-boundaries of the region. Consider both two- and three-dimensional cases in your explanation.

21 Klingman's Furniture Co. specializes in the design and manufacturing of contemporary chairs and sofas. The graph below shows the feasible region for the manufacturing of a particular style of upholstered chair (x) and sofa (y) under constraints in daily labor time available and time required for frame-building, finishing, and upholstering. The objective function $P = 80x + 70y$ gives profit per day.

a. Write the five constraint inequalities and explain the meaning of each.

b. What combination of chairs and sofas produced under those constraints gives maximum daily profit?

c. What is the maximum daily profit?

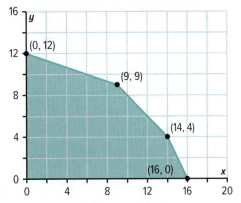

22 Look back at the cow-diet situation and constraints in Investigation 1 (pages 579–580). Suppose instead that the National Research Council indicates that the minimal crude protein (CP) requirement for cows during gestation should be 180 pounds of CP. How will this affect the setup and solution of the cow-diet plan problem? Test your ideas using the Solver tool and the "Graphical Linear Programming" custom app.

23 Natural disaster relief organizations help deliver water, food, and medical supplies around the world. A volunteer leader would like to send at least 180,000 cases of water, 135,000 cases of food, and 90,000 cases of medical supplies for hurricane relief.

The table at the top of page 598 details the contents and cost for each of two relief package options from one organization.

Package Option A	Package Option B
5 cases of water bottles	2 cases of water bottles
3 cases of food	2 cases of food
4 cases of medical supplies	1 case of medical supplies
Cost: $24.00	Cost: $16.00

a. For this particular natural disaster relief organization, what is the total cost for 45,000 packages from Option A and 0 packages from Option B?

b. What number of packages from Option A and Option B do you recommend to the volunteer leader in order to minimize the cost? Is this the only possibility? Explain.

c. What is the minimal cost for sending relief packages?

24 Three main ingredients included in fertilizers include nitrogen, phosphorus, and potassium. Nitrogen promotes overall growth, phosphorus promotes growth of roots, and potash promotes growth of flowers and fruits. A fertilizer company is producing 100-pound bags of lawn, garden, and tree fertilizers  to sell. The following table displays the number of pounds of each ingredient used in a 100-pound bag of each fertilizer type. Other ingredients make up the remaining amount in the bags.

Ingredients (in pounds)

Fertilizer		Nitrogen	Phosphate	Potash
	Lawn	24	2	8
	Garden	10	4	10
	Tree	5	7	7

The fertilizer company has 12,000 pounds of nitrogen, 2,400 pounds of phosphate, and 6,000 pounds of potash. It sells the lawn fertilizer for $60, the garden fertilizer for $35, and the tree fertilizer for $20. The company must produce at least 50 bags of tree fertilizer to meet demand. Their goal is to maximize potential income.

What is the maximum potential income for the fertilizer company? Report the number of bags of each type of fertilizer that would give maximum income.

25 In this lesson, several of the contexts have been related to maximizing profit and minimizing cost. Research some other linear programming problems that have different optimal goals and prepare a brief report of your findings.

26 An online site allows people to search for apartments with certain characteristics. Consider the following three features in an apartment: at least two bedrooms B; monthly rent no more than $1,000 R; and close to public transportation T. Describe the apartments found by searches for each of the following.

a. $B \cap (R \cap T)$

b. $(B \cup R) \cup T$

c. $R' \cap T$

d. $T \cup (B \cap R)$

27 Angelina is buying a pickup truck and has narrowed her choices to two different vehicles. One of them has a hybrid engine and averages 20 miles per gallon of gas. The other has a regular engine and only averages 14 miles per gallon of gas. The hybrid one costs $2,200 more than the one with the regular engine. She expects to drive 8,000 miles per year and estimates that gasoline costs will average about $4.25 per gallon. If she buys the hybrid, how long will it take before Angelina makes up the difference in purchase price by spending less on gasoline?

28 In a recent Presidential race, the following statement was made about the results of a random poll of North Carolina registered voters. "… Barack Obama leads Mitt Romney in a statewide poll, 48 percent to 47 percent. The race is a statistical tie, as the poll of 666 registered voters taken May 10–13 has a margin of error of almost 4 percentage points." (**Source:** www.fayobserver.com/news/local/article_0e65b69d-e24a-5e7a-8626-aacef283fe4c.html)

President Barack Obama

Governor Mitt Romney

a. Verify that the margin of error for this poll is about 4 percent.

b. Explain what the writer means by, "the race is a statistical tie." How is this statement related to the margin of error?

c. The results of another North Carolina poll, which polled likely voters rather than registered voters, indicated that 51% of those polled favored Romney while 43% preferred Obama. How big would this sample have needed to be in order to be able to accurately say that Romney was leading in the polls?

29 Evaluate the following without using a calculator. Then check your answers using a calculator.

a. $9^{\frac{3}{2}}$

b. $-27^{\frac{2}{3}}$

c. $81^{\frac{3}{4}}$

d. $8^{-\frac{2}{3}}$

e. $\left(10^{\frac{1}{2}}\right)\left(40^{\frac{1}{2}}\right)$

f. $\left(\frac{4}{25}\right)^{-\frac{5}{2}}$

30 When Talia got her first apartment, she charged $800 on her credit card. She will not charge anything more until it is completely paid off. After the first month, her credit card company charges 24% annual interest compounded monthly.

a. When she makes the charge, Talia tells herself that she will have it paid off by paying $80 per month for 10 months. Is she correct? Explain your reasoning.

b. Suppose she pays $100 per month. How long will it take her to pay off this $800 charge? Explain your reasoning or show your work.

c. What is the minimum monthly payment she can make in order to pay it off in 10 months?

31 Determine all real number solutions to each equation.

a. $6 - (x + 7) = 2(4x + 10)$

b. $x^2 + 5x - 12 = x^2 - 3x$

c. $x^2 - 7x = -12$

d. $4x^2 - 100 = 0$

e. $x^2 + 5x + 2 = 0$

f. $27^{2x} = 3^{x + 10}$

32 For accounting purposes, businesses often decrease the value of purchased production/distribution-related items by the same amount each year. This is called *straight-line depreciation*. Suppose that a business is using this method of depreciation. One year after an automobile was purchased, it was valued at $21,000. Five years after it was purchased, the value was only $1,000.

a. Write a function rule that expresses the value of the automobile V(t) as a function of t, the number of years elapsed since the automobile was purchased.

b. Sketch a graph of the function from Part a.

c. Identify the slope of the line and explain what it means in this context.

d. Identify the y-intercept of the line and explain what it means in this context.

Looking Back

In this unit, you developed skill in interpreting and representing surfaces in three dimensions. In the process, you investigated the usefulness of contour diagrams and cross sections as means to represent and analyze complex three-dimensional shapes.

The introduction of a three-dimensional coordinate system enabled you to explore and formalize analogs of important geometric ideas in two dimensions including distance, symmetry, and graphs of linear equations and systems of linear equations. Although three-dimensional graphing software enables you to quickly produce graphs of three-dimensional surfaces, being able to visualize the cross sections and intercepts is essential in interpreting the images and judging their reasonableness.

You also learned how to solve linear programming problems in two dimensions and extended those ideas to three dimensions. In the latter case, you had numerous occasions to employ mathematical practices, especially those of modeling with mathematics and of strategically selecting and using technological tools to aid in the solution of problems.

In this final lesson, you will review and consolidate your understanding of these key concepts and skills.

1 The York Pond Trail and the Kilkenny Ridge Trail in the Mahoosuc Range of the White Mountains National Forest meet near Willard Notch. Use the map below to help complete the following tasks. On this map, contour lines represent 100-ft elevation differences.

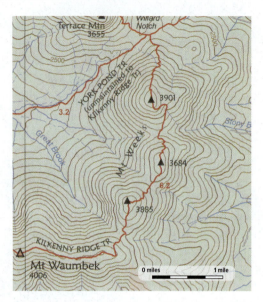

a. What do the three sets of closed curves in the center portion of the map represent?

b. Describe the terrain along the Kilkenny Ridge Trail from Mt. Waumbek to Willard Notch. Include the length of the section of trail and high and low altitudes.

c. Describe the terrain along the York Pond Trail.

d. Visualize a relief line from the Mt. Weeks peak with altitude 3,885 feet to the peak with altitude 3,684 feet. Sketch the vertical cross section along this line.

e. Explain why the contour lines in this map and those in other contour diagrams in this unit never cross each other.

2 Complete the following tasks in the context of a three-dimensional coordinate system.

a. Suppose you are standing above the xy-plane at the point $P(5, 8, 3)$ and looking at the point $Q(1, -3, 5)$. Are you looking up or down? How far is point Q from point P?

b. Describe the set of points whose distance from the y-axis is 4. What equation describes this set of point(s)?

c. Which of the points $A(3, 1, -2)$, $B(2, -3, 0)$, or $C(-10, 0, 0)$ is closest to the yz-plane? Which point(s) are on the xy-plane?

d. Find the equation of a sphere with radius 5 centered at the origin. What is the equation of a congruent sphere with center at $(1, -3, 2)$?

e. For the sphere in Part d with center at $(1, -3, 2)$, is the point $(-2, 2, -2)$ on the sphere, in its interior, or in its exterior?

3 Consider the plane with equation $6x + 4y + 3z = 24$.

   a. Sketch the plane.

   b. How would you describe the horizontal cross sections of this plane?

   c. Write the equation of a plane parallel to the given plane.

   d. Write the equation of a plane that intersects the given plane at its *yz*-trace.

4 Solve the following system of equations *by hand* using a method of your choice. Then check the solution by solving the system using a technology tool. Correct any mistakes.

$$\begin{cases} -x + y + z = 3 \\ 3x \quad\;\; - z = 1 \\ 2x - 3y - 4z = -2 \end{cases}$$

5 Walt Armstrong is one of two candidates in the fall election for sheriff of Lenawee County. His election committee plans to promote the candidate through television, radio, and the local online newspaper. Electorate estimates, media costs, and maximum media ads are subject to the conditions given below.

Conditions	Television	Radio	Online Newspaper
Expected number of electorate reached per ad	100,000	18,000	40,000
Cost per ad	$2,000	$300	$600
Maximum media ads	10	20	10

The campaign media budget is $18,200. To ensure a balanced media campaign, radio ads are not to exceed 50% of the total number of authorized ads. Also, television is to account for at least 10% of the total number of authorized ads.

   a. The election committee wants to maximize the number of electorate reached by the ad campaign. Write the objective function and *six* constraint inequalities described above.

   b. How many candidate ads should run on each medium to maximize the electorate reached? What is the total electorate reached?

SUMMARIZE THE MATHEMATICS

In this unit, you strengthened your visualization, geometric reasoning, and algebraic skills and learned new methods for representing and analyzing surfaces. This included extending coordinate methods of a plane to three-dimensional space. The added dimension enabled you to solve more complex linear programming problems.

a. Describe how altitude or depth data can be used to draw a map displaying contours of the surface of a region. What other types of data can be displayed using contour diagrams?

b. Compare the manner in which the following ideas are represented or determined in a two-dimensional coordinate system and in a three-dimensional coordinate system.

  i. point

  ii. distance between two points

  iii. midpoint of a segment

  iv. plane

  v. graph of a linear equation

  vi. symmetry with respect to the x- and y-axes (with respect to the xy-, yz-, and xz-planes)

c. What strategies would you use to sketch a graph or check the reasonableness of a technology-produced graph of a three-dimensional surface represented by a linear equation in three variables? In two variables?

d. Consider systems of linear equations like:
$$\begin{cases} ax + by + cz = d \\ ex + fy + gz = h \\ px + qy + rz = s \end{cases}$$

  i. What are the possible numbers of solutions for such systems? How are those possibilities illustrated in graphs of the systems of equations?

  ii. What methods can be used to find the solution(s)?

  iii. What are the key steps in the application of each method?

  iv. How do you decide which method to use in a particular case?

e. Describe the roles played by the following elements of linear programming problems.

  i. constraints   **ii.** feasible region   **iii.** objective function

f. In a linear programming problem represented in a two-dimensional coordinate system, where in the feasible region will the objective function have its optimum (maximum or minimum) value? Why does this make sense? What if the problem required a three-dimensional model?

Be prepared to share your descriptions and thinking with the class.

✓ CHECK YOUR UNDERSTANDING

Write, in outline form, a summary of the important mathematical concepts and methods developed in this unit. Organize your summary so that it can be used as a quick reference in your future work.

Mathematics and statistics help you in making intelligent decisions in daily life and in careers. You have likely seen many examples of this in your previous mathematical studies. You used exponential functions in making investment decisions. You used statistics and probability in making data-based decisions. You used ideas of geometry in making design decisions.

This unit focuses on mathematical ideas that are helpful for group decision-making in daily life, business, and politics in a democratic society.

In the two lessons of this unit, you will develop important mathematical methods related to voting and fair division.

LESSONS

1 Social Choice and Voting

Understand and apply the mathematics of voting—different methods of voting, including ranked-choice voting, approval voting, and weighted voting and different methods for analyzing the results of voting, including plurality, majority, runoff, pairwise-comparison, points-for-preferences, and instant runoff voting with attention to Arrow's Theorem and the Banzhaf power index.

2 Fair Division

Understand and apply basic concepts and methods of fair division for apportioning seats in the U.S. House of Representatives, for partitioning a continuous object like a region of land, and for allocating indivisible items and assets in an estate that must be fairly shared among the heirs.

Social Choice and Voting

The basic question of *social choice* is how groups such as citizens of a community, state, or country, legislative bodies, and even coaches of athletic teams best arrive at group decisions. Social choice involves methods for turning individual preferences into a single group choice. This process is at the heart of a democratic society.

One of the hallmarks of a democratic society is that its leaders are chosen "by the people" through a process of voting. The purpose of voting is to make fair decisions based on the opinions of the voters. Mathematics is used to design voting methods and analyze the results so that the best decisions are made.

THINK ABOUT THIS SITUATION

Think about situations involving voting that you are familiar with or have heard about.

a. What are some recent situations in the news that involve voting? Briefly describe the situation, the voting process, and how the winner was chosen.

b. Think about the last time you voted in some sort of an election. Do you think the final results were fair? Did the results clearly reflect "the will of the people?" Why or why not?

c. In voting for state governor, does every person's vote count the same? What about in presidential voting? Explain.

d. Voting is straightforward when there are two choices. Think about voting situations in which there are more than two choices. In such situations, can you think about different ways to vote, or different methods of analyzing the results? Describe any voting methods that you know about, perhaps in sports, politics, business, or in your school.

There are in fact several different methods of voting, along with different ways to analyze the data from voting. In this lesson, you will learn about *ranked-choice voting* and *weighted voting*. You will also learn some common ways to analyze voting data, including plurality, majority, runoff, pairwise-comparison, points-for-preferences, and instant runoff voting.

INVESTIGATION 1

Ranked-Choice Voting

The voting method with which you are probably most familiar is the common method of voting for your single favorite candidate. This works well when there are just two candidates. Why could this method be problematic when there are more than two candidates?

To counter some of the disadvantages of voting for a single favorite candidate, some organizations are using a ranked-choice voting method as described below.

An Electoral Experiment in North Carolina
By McKay Coppins, *Newsweek*, October 24, 2010

The logic of general elections is simple: winner takes all. This, of course, can encourage nasty campaigning—and at the end of a race with more than two candidates, the victor often wins with only a plurality (not a majority) of support. Searching for a solution, a handful of cities have experimented with an alternative approach, in which voters rank candidates in order of preference.

Source: www.newsweek.com/2010/10/24/north-carolina-tries-instant-runoff-voting.html

Ranked-choice voting was used for the first time in a statewide election in 2010 in North Carolina. It is currently used throughout the country in many local elections, and since 2000 more than 20 states have considered moving to this voting system.

As you work on the problems in this investigation, look for answers to the following questions:

What is ranked-choice voting, and what are the advantages of ranked-choice voting over the method of voting for your single favorite candidate?

How can you analyze the data from ranked-choice voting, and what are advantages and disadvantages of each vote-analysis method?

1 Suppose your class is going on a field trip to a nearby math and science museum. Everyone will be on the same bus, and you will all eat at the same restaurant. The choices are KFC, McDonald's, and SUBWAY. To best decide where everyone will eat lunch, in a way that most accurately reflects everyone's opinion, you might try **ranked-choice voting** as described below.

a. Everyone in your class should vote using a ballot like the one at the right. Vote by ranking the restaurants. Rank the restaurants according to your preference by writing a 1 next to your favorite restaurant and a 2 and 3 next to your 2nd and 3rd preferences.

b. Discuss some ideas for how to analyze the voting results. Which restaurant should be chosen for lunch? Compare your method and answer with those of your classmates.

> **BALLOT**
>
> Rank these restaurants in order of preference.
>
> ☐ KFC
>
> ☐ McDonald's
>
> ☐ SUBWAY

2 One group of students in the class organized the individual rankings of restaurants in a **preference table** as shown below.

Restaurant Preferences

	Rankings					
KFC	1	1	2	2	3	3
McDonald's	2	3	1	3	1	2
SUBWAY	3	2	3	1	2	1
Number of Voters	6 voters	4 voters	6 voters	7 voters	5 voters	5 voters

a. How many students are in this class? How can you tell from the preference table?

b. Explain what the entries in the first column mean.

c. How many students ranked SUBWAY as their 3rd preference? How many ranked SUBWAY as their 1st preference?

d. How many different rankings were made by this class? Do you think the same number of rankings will appear in all preference tables involving 3 choices? Explain your reasoning.

3 Examine each of the opinions below about which restaurant is the winner.

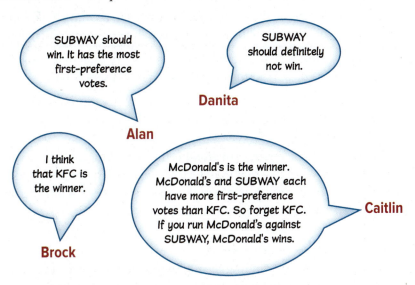

a. With which of these students do you agree? Why?

b. How could Danita explain to Alan that SUBWAY should not win?

c. Verify Caitlin's claim that McDonald's and SUBWAY each have more 1st-preference votes than KFC. Explain why McDonald's is the winner using Caitlin's method.

d. Give a reasonable explanation for Brock's thinking.

4 Look back at Problem 2 and consider a different type of voting.

a. Suppose everyone only voted for their favorite restaurant, and they did not rank the restaurants by preference. In this case, which restaurant is the winner? Explain.

b. Do you see any drawbacks to the voting method where you only vote for your favorite? If so, describe.

Vote-Analysis Methods You have seen that there are different ways to analyze the results of ranked-choice voting. Now you will systematically investigate the most commonly used vote-analysis methods. Some of these methods may be ones that you have already considered.

For Problems 5–9, consider the following preference table, which summarizes voting on preferred athletic shoe brands in a class at Elmwood Park High School.

Athletic Shoe Preferences

	Rankings			
Adidas	1	1	3	3
Nike	3	2	1	2
Reebok	2	3	2	1
Number of Voters	4 voters	6 voters	7 voters	8 voters

5 You are probably most familiar with *plurality* and *majority* methods.

   a. A **plurality** winner is the candidate that receives the most 1st-preference votes. Find the plurality winner for athletic shoe preferences, if there is one.

   b. A **majority** winner is the candidate that gets more than half of the 1st-preference votes. Find the majority winner, if there is one.

   c. Is a majority winner always a plurality winner? Is a plurality winner always a majority winner? Be prepared to explain your reasoning.

6 Another common vote-analysis method is the **runoff method**. Since there are several different ways that you might carry out a runoff, the runoff method used in this problem and throughout this lesson is carried out using the specific method described here.

Runoff Method

This method works by finding the top two candidates based on 1st-preference votes, and then running those two against each other to find the winner. Here are the steps:

   Step 1. Count the 1st-preference votes to find the top two candidates.

   Step 2. Eliminate all the other candidates. So, now you have just two candidates.

   Step 3. Some voters have had their 1st-preference candidate eliminated. So, reassign their votes based on their rankings.

   a. Apply the runoff method to find the winner in the athletic shoe voting. You might use these hints:

     • Which are the top two shoe brands, based on 1st-preference votes?

     • Which brand gets eliminated? Cross out the row for that brand.

     • How many voters voted for the eliminated brand as their 1st preference? Which shoe brand will they now vote for as 1st preference?

     • Now which brand has the most 1st-preference votes?

   b. Compare your winner with that determined by some of your classmates. Discuss and resolve any differences.

   c. Describe some situations you know where some kind of runoff method is used to decide on a group choice. Discuss your answer with those of your classmates.

   d. Look back at the vote-analysis strategies used by Brock, Caitlin, and Alan in Problem 3.

     i. Which student used the runoff method?

     ii. Who used the plurality method?

7 Using the **pairwise-comparison method**, a runoff is computed for each pair of candidates, and the winner (if there is one) is the candidate who beats all the others. This method was developed in the eighteenth century by the philosopher and social scientist Marie Jean Antoine Nicolas Caritat, the Marquis de Condorcet. It is sometimes called the *Condorcet method*. (Condorcet is pronounced "con-door-say".)

Pairwise-Comparison Method

Step 1. Find *all* pairs of candidates.

Step 2. For *each* pair, eliminate all other candidates and run off the two against each other. (Remember to reassign the eliminated candidates' 1st-preference votes to one of the two candidates in the runoff.)

Step 3. Identify the winner as the candidate who beats every other candidate in such a head-to-head runoff.

a. Use the pairwise-comparison method on the athletic shoe preference voting results (page 609) to find the pairwise-comparison winner, if there is one.

b. Apply the pairwise-comparison method to the restaurant preference voting data (page 608). Which restaurant is the pairwise-comparison winner?

8 Another common vote-analysis method is the **points-for-preferences method**. This method was first proposed by an amateur mathematician, Jean-Charles de Borda, who was a French cavalry officer and naval captain in the 18th century. It is sometimes called the *Borda method*. Using this method, points are assigned to each preference and the winner is the candidate that gets the most total points. For example, 3 points could be assigned to 1st preference, 2 points to 2nd preference, and 1 point for 3rd preference. There can be many other different point assignments.

a. Refer back to the ranked athletic shoe preference voting data. Using the point assignments above, how many total points does Reebok get?

b. Find the points-for-preferences winner using the point assignments above.

c. What are some other situations where the points-for-preferences method is used to choose a winner? Compare your answer with those of your classmates.

d. Apply the points-for-preferences method to the restaurant voting data in Problem 2. Which restaurant is the points-for-preferences winner?

9 In Problems 5–8, you used different methods to analyze the results of ranked-choice voting. Consider the most common type of non-ranked-choice voting—voters vote for their one favorite candidate and the candidate with the most votes wins. This is just like the plurality method in Problem 5, except that voters only mark the ballot showing their favorite candidate.

    a. Which shoe brand is the winner under this non-ranked-choice voting method using the plurality analysis method?

    b. Compare the plurality winner to the winners using the runoff, pairwise-comparison, and points-for-preferences methods. Do you think the plurality winner is a fair winner? Describe some disadvantages of using the plurality method to choose a winner.

    c. As you probably concluded in Problem 5, the plurality winner may or may not be the majority winner.

       i. If the plurality winner gets a majority of the votes, do you think the plurality winner is a fair winner?

      ii. Do you know of any elections in which the winner did not get a majority of the votes? Think about local elections, state elections, and U.S. presidential elections.

    d. Assume an election has 5 candidates and that voters' favorite candidate is found using the plurality method.

       i. Determine if a candidate could win by receiving 30% of the vote.

      ii. What is the smallest percentage of the vote a candidate could receive to be declared the winner?

10 In Problems 5–9, you used different vote-analysis methods to determine the favorite athletic shoe brand based on voting data given in a preference table. Compare your results with your classmates. Which shoe brand do you think should be considered the overall winner?

Most people agree that a majority winner, based on all 1st-preference votes, is a fair winner. Since a plurality winner is not always a majority winner, some election laws require that in this situation another runoff election be held to determine a majority winner. Holding a new election is expensive and often has low voter turnout. Because of these disadvantages and others, the use of ranked-choice voting is growing. For example, countries such as Australia and Ireland use ranked-choice voting. Many cities in the United States use ranked-choice voting, including Minneapolis, MN, Memphis, TN, San Francisco, CA, and Tacoma Park, MD.

Another vote-analysis method for ranked-choice voting is gaining popularity, called *Instant Runoff Voting* (*IRV*). The word "instant" highlights the important feature that (as in all the methods you have studied in this lesson) the results are computed instantly based on one election, avoiding any need to hold a later follow-up election.

11 **Instant Runoff Voting** (**IRV**) is another method for analyzing ranked-choice voting data. IRV is very similar to the runoff method you studied in Problem 6. In the runoff method, you eliminate all candidates except the two who get the most 1st-preference votes, then you redistribute the 1st-preference votes for the eliminated candidates. The winner is the candidate with the most 1st-preference votes after the redistribution, which will be a majority of votes (or a tie) since there are only two candidates in the runoff. In the IRV method, you eliminate the candidates with the fewest 1st-preference votes *one at a time*, each time *redistributing* the eliminated candidate's 1st-preference votes until one candidate remains with a majority of votes. This method is advocated, for example, by the non-profit, non-partisan group FairVote. They describe IRV as follows.

Instant Runoff Voting Method: Instant runoff voting allows voters to rank candidates in order of preference (i.e. first, second, third, fourth and so on). Voters have the option to rank as many or as few candidates as they wish, but can vote without fear that ranking less favored candidates will harm the chances of their most preferred candidates. First choices are then tabulated. If more than two candidates receive votes, a series of runoffs are simulated, using voters' preferences as indicated on their ballot.

The candidate who receives the fewest first place choices is eliminated. All ballots are then retabulated, with each ballot counting as one vote for each voter's highest ranked candidate who has not been eliminated. Specifically, voters who chose the now-eliminated candidate will now have their ballots added to the totals of their second ranked candidate—just as if they were voting in a traditional two-round runoff election—but all other voters get to continue supporting their top candidate who remains in the race. The weakest candidates are successively eliminated and their voters' ballots are added to the totals of their next choices until two candidates remain. At this point, the candidate with a majority of votes is declared the winner. (Some jurisdictions choose to end the count as soon as one candidate has a majority of votes, as this cannot be defeated.)

Source: FairVote, www.fairvote.org/How-Instant-Runoff-Voting-Works, 2/11/2014

a. The preference table below shows the results of student voting in an election for Senior Class President at Cooper High School. Based on the description given above, determine the IRV winner.

Senior Class President Preferences

	Rankings					
Charnell	1	5	5	5	5	5
Amarjit	5	1	2	4	2	4
Lamar	4	4	1	2	4	2
Rodene	2	3	4	1	3	3
Richard	3	2	3	3	1	1
Number of Voters	36 voters	24 voters	20 voters	18 voters	8 voters	4 voters

b. Using the "Ranked-Choice Voting" custom app in *TCMS-Tools* or similar software, determine the winner under the other methods you have learned in this lesson. Who do you think should be declared the winner?

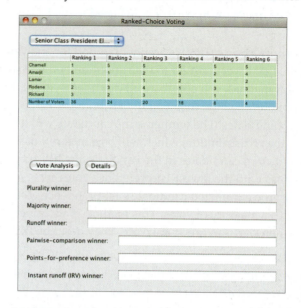

c. Explain why the IRV method described on page 613 and the runoff method described on page 610 are identical when there are exactly three candidates. Are these two methods always the same?

SUMMARIZE THE MATHEMATICS

In this investigation, you learned some of the mathematics of voting. In particular, you learned about ranked-choice voting and six analysis methods.

a. Describe how ranked-choice voting works, and how it differs from voting only for your favorite choice.

b. Describe each of the six ranked-choice, vote-analysis methods below.

  i. plurality   **ii.** majority

  iii. runoff   **iv.** pairwise-comparison

  v. points-for-preferences   **vi.** instant runoff voting

c. Did you use any vote-analysis methods other than the six listed above as you worked on this investigation? If so, describe them.

d. You have probably noticed some drawbacks to the voting methods you have been studying. For each of the vote-analysis methods used in this investigation, describe one advantage and one disadvantage of the method.

Be prepared to share your ideas and thinking with the class.

Ranked-choice voting is only relevant when there are more than two candidates. Not many Presidential elections in the United States have had more than two viable candidates. The strongest showing of a third-party candidate in the last 100 years was in the 1992 Presidential election. In this case, there were three major candidates: Bill Clinton, George Bush, and Ross Perot. Although Clinton won, both of the other two candidates received a significant share of the popular vote.

Of course, the President is elected through the Electoral College, but think about how the election might have turned out if some of the voting methods you have been studying were used instead. The final popular vote tallies from the election were as follows.

Clinton:	43,728,375
Bush:	38,167,416
Perot:	19,237,247

Since voters did not vote by ranking the candidates, there is no exact preference table for this election. But opinion polls can be used to construct an approximate preference table. Opinion polls at the time suggest these three assumptions.

- If a person voted for Clinton, then Perot would have been the second choice.

- If a person voted for Bush, then Perot would have been the second choice.

- For the people who voted for Perot, half would have chosen Clinton as their second choice and half would have chosen Bush.

a. These assumptions, though broad and open to some debate, provide a basis for constructing a reasonable preference table. Use these assumptions to complete the preference table below for the 1992 Presidential election.

Estimated Ranked 1992 Presidential Candidates

	Rankings			
Clinton	1			
Bush	3			
Perot	2			
Number of Voters	43,728,375	38,167,416	9,618,623	9,618,623

b. Is there a majority winner?

c. What percent of the total votes does the plurality winner get?

d. Determine the winner using runoff, points-for-preferences, and pairwise-comparison. Any surprises? Explain.

Fair Is Fair, Isn't It?

You have probably noticed that each of the voting methods you have studied so far has some drawback. You might be wondering if any voting method is always fair. Or, even more fundamental, you might be wondering just what "fair" means. These are very important and difficult questions.

The mathematical approach to answering these questions is to formulate a mathematical definition of fairness and then see if any voting method satisfies the definition. In 1949, the Nobel Prize-winning economist Kenneth Arrow did just this. His result, called *Arrow's Impossibility Theorem*, is one of the most surprising and famous theorems of the 20th century. You will now learn about Arrow's Impossibility Theorem.

As you work through the following problems, look for answers to these questions:

Kenneth Arrow, Professor Emeritus, Stanford University

What are essential characteristics of fairness?

Are there situations in which any voting method is not fair?

1 Consider the following preference table for three candidates for chairperson of a local planning commission.

Planning Commissioner Preferences

	Rankings			
Candidate A	1	2	3	2
Candidate B	2	3	1	1
Candidate C	3	1	2	3
Number of Voters	12 voters	10 voters	8 voters	4 voters

  a. Determine the runoff winner.

  b. Suppose the four voters represented in the last column change their ranking to 1-2-3, so that they now give more support to candidate A. Modify the preference table and determine the new runoff winner.

  c. What seems unfair about this situation?

Arrow's Impossibility Theorem Arrow proposed the following six fairness conditions that should be satisfied by any fair voting method. Then he proved that *no* voting method can always satisfy all six conditions when there are more than two candidates.

2 Carefully study the fairness conditions below. Which are violated in the situation in Problem 1?

Arrow's Fairness Conditions

- **Agree with Unanimous** — A fair voting method should produce results that agree with the unanimous will of the voters. That is, if all voters prefer candidate X over candidate Y, then X should be ranked above Y in the results.

- **Decisive** — A fair voting method should be decisive. That is, it should decide for any two candidates X and Y whether X beats Y, Y beats X, or there is a tie.

- **Ordered** — A fair voting method should produce results that establish a clear order among the candidates, so that if X is ranked over Y and Y is ranked over Z, then X is ranked over Z.

- **Consistent** — A fair voting method should produce results that are consistent with preference trends among the voters. That is, if voter preferences change so that X is raised in some of the voters' rankings, then if X beat Y before the change, X should still beat Y after the change.

- **Relevant** — A fair voting method should ignore irrelevant alternatives. The decision about whether X beats Y should depend only on how the voters rank X versus Y, not on how they rank other candidates. That is, for any two candidates X and Y, if some voter preferences change but the relative preference for X versus Y remains the same for all voters, then if X beats Y before the change, X should still beat Y after the change.

- **Non-Dictatorial** — There is no dictator. That is, there is no voter whose preferences completely determine the outcome of an election no matter what the preferences of other voters may be.

3 The preference table below summarizes voter preferences in an election for class president.

Class President Preferences

	Rankings			
Jill	1	1	4	4
Sammi	4	3	1	3
Amir	3	4	2	1
Orlando	2	2	3	2
Number of Voters	45 voters	27 voters	45 voters	36 voters

a. Who is the runoff winner?

b. Suppose Jill's family moves just after the election. The winner in Part a claims this should have no effect on who wins the election. However, the other candidates demand that Jill be removed from the preference table, and the results recomputed.

> **i.** Study the ranking of the 27 voters in the second column who had Jill as their 1st preference. If Jill is removed, who is their new 1st preference? Their new 2nd preference? Their new 3rd preference? Update the preference table with your answers.

Modified Class President Preference Table

		Rankings		
Sammi			1	3
Amir			2	1
Orlando			3	2
Number of Voters	45 voters	27 voters	45 voters	36 voters

> **ii.** Similarly, modify the table to reassign the votes of the block of 45 voters who had Jill as their 1st preference.

c. According to your new preference table in Part b, who is the runoff winner after Jill drops out?

d. Analyze this situation with respect to Sammi and Orlando.

> **i.** In the results from the election in Part a, who ranks higher, Sammi or Orlando? Who ranks higher in the results of the election in Part c?

> **ii.** Did any voters change their relative preference for Sammi versus Orlando from the election in Part a to the election in Part c?

> **iii.** Do you think this situation involving Sammi and Orlando is fair? Explain.

e. Which of Arrow's fairness conditions is violated in this situation? Explain.

4 You have seen that there are drawbacks to all the voting methods you have studied. Arrow's Impossibility Theorem proves that when there are more than two candidates, every possible voting method violates at least one of his fairness conditions. In this sense, there is no voting or vote-analysis method that is perfectly fair in all situations. Thus, you should apply your knowledge of different voting methods to each particular decision-making situation, and then decide on the best voting method to use. Think about which voting method you would recommend in each of the situations below.

a. There are two choices for a new school mascot. All students will vote to decide which mascot to adopt. Which voting method would you recommend? Why?

b. There are three candidates for president of the junior class. Two candidates have some differences, but their views are generally similar and they are both popular. The remaining candidate has views that are very different from the other two, and those views are shared by a significant group of students. Which voting methods would you *not* recommend for this situation? Which method would you recommend? Explain your reasoning.

c. There are six finalists in the school talent show. All students will vote to choose the overall winner. Choose one voting method that you would recommend. Choose one method other than majority that you would not recommend. In each case, defend your answer.

Even though there is no perfect voting method, some are better than others. The plurality method is arguably the worst (when there are three or more candidates). Experts often recommend the *points-for-preferences method* (Borda), *Instant Runoff Voting* (IRV), *approval voting* (see Applications Task 5), or, if a winner is produced, the *pairwise-comparison method* (Condorcet).

SUMMARIZE THE MATHEMATICS

Fairness and fair voting are important issues in a democratic society.

a. How is fairness of voting methods analyzed mathematically?

b. Arrow's conditions are widely accepted as a good way to characterize a fair voting method (although mathematicians continue to look for characterizations that may be even better). Discuss why each of Arrow's conditions should be satisfied by any fair voting method.

c. Summarize what Arrow's Impossibility Theorem says about fair voting.

Be prepared to share your ideas and thinking with the entire class.

✓ CHECK YOUR UNDERSTANDING

Examine the voting results in the table below.

Candidate Preferences

	Rankings		
Candidate A	1	3	3
Candidate B	2	1	2
Candidate C	3	2	1
Number of Voters	40 voters	36 voters	14 voters

a. Who is the plurality winner?

b. Suppose candidate C drops out. Now who is the plurality winner?

c. Explain using these voting results why plurality violates Arrow's Relevant condition.

Weighted Voting and Voting Power

Ranked-choice voting, which you studied in Investigation 1, is based on the principle of one-person/one-vote. Each person votes once, by ranking all the candidates. It is fraudulent for one person to cast more than one vote. However, there are situations in which it is sensible and legal for one person to have many votes. This type of voting is called *weighted voting*. For example, each shareholder of a corporation has a number of votes equal to the number of shares owned.

In the U.S. Electoral College, each state has a different number of votes depending on the state's population. Thus, each state has a *weighted vote*.

USA 2012 Electoral Votes by State

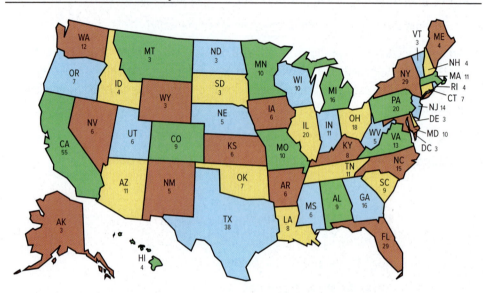

As you complete the problems in this investigation, look for answers to the following questions:

What is the difference between the weight and the power of a person's vote?

How is voting power calculated and interpreted?

Weight versus Power Consider the common situation in which a yes or no decision will be made by voting. For example, this occurs when voting on a resolution in a government committee, or a proposal for one of two homecoming dinner menu items, or when there are two suggested locations for the senior class trip. In all these situations, there is an agreed-upon quota of yes votes needed to win, which is usually at least a majority of votes.

1 Consider a small local company with three shareholders and a total of 6,000 shares owned. Wai owns 3,000 shares and thus has 3,000 votes, Morales owns 2,000 shares with 2,000 votes, and Harper owns 1,000 shares with 1,000 votes. Suppose that corporate regulations require that 5,000 or more votes are needed to pass any proposal. Thus, the quota is 5,000. Shorthand notation that summarizes all this information is:

$$[5{,}000: 3{,}000, 2{,}000, 1{,}000]$$

Think about how voting must happen in order for a proposal to pass.

a. Give a complete description of who must vote for a proposal in order for it to pass. In your description, be sure to address these questions.

- Can any of these shareholders determine the outcome of the voting just by their votes alone?

- Can two shareholders form a winning coalition? Explain.

- Does Harper have any impact on the voting?

b. Based on your analysis in Part a, give a reasonable description of the voting power of each shareholder. Explain your reasoning.

c. The **weight** of a person's vote is the number of votes that person has. Think about weight versus power.

    i. If someone has greater voting weight, does he or she necessarily have greater voting power?

    ii. Can someone have voting weight and yet no power? Explain your reasoning.

    iii. Which is more important, the *weight* or the *power* of a person's vote? Why?

d. Using the shorthand notation on the previous page, give an example of a weighted voting situation where there is a *dictator*, that is, someone whose vote completely determines the outcome regardless of how anyone else votes.

Measuring Voting Power   You have seen that the *weight* of a person's vote can be different than the *power* of their vote. This can result in voting situations that are unfair. To address this problem, several methods have been devised to measure voting power. A commonly-used method is the *Banzhaf power index*, named for John F. Banzhaf III, a law professor at George Washington University who devised the method as part of a legal case in 1965 involving alleged voting inequities on the Nassau County Board in New York.

The Banzhaf power index measures the power of a person's vote by counting the number of possible winning coalitions in which their vote is *critical*. A **critical vote** (sometimes called a "swing vote") is a vote that is necessary to win; that is, if the person leaves a winning coalition, then the coalition no longer has enough votes to win. To see precisely how this works, consider the following example.

John F. Banzhaf III

2 A weighted voting situation is represented as:

$$[5: 4, 3, 2]$$

This could represent a corporation with 9,000 shares, three shareholders, A, B, and C, with 4,000, 3,000, and 2,000 shares, respectively, and a majority is required to pass; thus, the quota is 5,000 (all numbers represented in thousands).

Since the Banzhaf power index is based on whether a voter's vote is critical in winning coalitions of voters, the first step is to list all possible winning coalitions. Then, for each voter in each winning coalition, decide if their vote is critical. The **Banzhaf power index** (**BPI**) for each voter is the ratio:

$$\frac{\textit{number of critical votes for that voter}}{\textit{total number of critical votes for all voters}}$$

This method is organized and illustrated in the partially completed tables below.

Winning Coalitions	Voters Critical to the Winning Coalition (Voters who have a "swing vote" in the coalition)
[4, 3]	A, B
[4, 2]	
[3, 2]	
[4, 3, 2]	none

Critical Votes for Voter X (Number of winning coalitions that become losing coalitions if X leaves the coalition)	Banzhaf Power Index (BPI) # of critical votes for voter X / total # of critical votes for all voters
A: 2	For A: $\frac{2}{6}$
B:	For B:
C:	For C:

a. Recall that the voting situation described in the tables is [5: 4, 3, 2], with three shareholders, A, B, and C, respectively. Make sure you understand the tables by discussing the items below.

  i. Explain why [4, 3] represents a winning coalition of voters A and B.

  ii. Explain why A is a critical voter for the coalition represented by [4, 3].

  iii. Explain why no voter is critical in the coalition represented by [4, 3, 2].

  iv. Explain why all the winning coalitions are shown, there are no others.

b. Complete the tables.

c. What is the *weight* of each voter, A, B, and C? What is the *power* of each voter, as measured by the BPI?

d. Describe at least one interesting comparison between weight and power for these voters. Compare your observation with those of others.

3 Consider what happens in a similar voting situation, [6: 4, 3, 2]. The only difference between this situation and that in Problem 2 is the quota has changed from 5 to 6.

a. Find the BPI for each of the three voters in this situation.

b. Describe at least one interesting comparison between weight and power for these voters. Compare your finding with those of others.

4 The Banzhaf power index can be thought of as a measure of the probability that a voter will be a critical voter.

    a. Under this interpretation, what is the probability that the second voter represented in this situation will be a critical voter?

    b. Why can the Banzhaf power index be thought of as the probability that a voter will be a critical voter? Compare your answer with your classmates. Resolve any differences.

SUMMARIZE THE MATHEMATICS

In this investigation, you learned about weighted voting.

a. What is weighted voting? How does it differ from ranked-choice voting? What are some examples where weighted voting is used?

b. Briefly describe how the Banzhaf power index is computed.

c. Describe some differences, with examples, between the weight of a person's vote and the power of the person's vote.

Be prepared to share your responses and thinking with the entire class.

✔CHECK YOUR UNDERSTANDING

In addition to the "Weighted Voting" custom app in *TCMS-Tools*, there are Web sites that provide customized calculators for computing the Banzhaf power index. Consider a weighted voting situation represented by [9: 6, 4, 2, 1].

a. Find the Banzhaf power index for each voter. Compute by hand, without using any Web site or technology.

b. Use the "Weighted Voting" custom app or a BPI calculator from a Web site to compute the Banzhaf power index for each voter in the same weighted voting situation as in Part a. Compare results. Resolve any differences. (One useful calculator can be found at cow.math.temple.edu/bpi.html.)

APPLICATIONS

1 Consider the following ballots from a vote to determine the evening entertainment for a local summer festival.

Concert	2		Concert	1		Concert	3		Concert	3		Concert	2
Ball game	3		Ball game	2		Ball game	2		Ball game	2		Ball game	1
Dance	1		Dance	3		Dance	1		Dance	1		Dance	3

Concert	2		Concert	2		Concert	2		Concert	2		Concert	2
Ball game	1		Ball game	3		Ball game	1		Ball game	1		Ball game	1
Dance	3		Dance	1		Dance	3		Dance	3		Dance	3

Concert	3		Concert	1		Concert	2		Concert	3		Concert	2
Ball game	2		Ball game	2		Ball game	3		Ball game	2		Ball game	1
Dance	1		Dance	3		Dance	1		Dance	1		Dance	3

Concert	2		Concert	2		Concert	1		Concert	2		Concert	2
Ball game	1		Ball game	1		Ball game	2		Ball game	3		Ball game	3
Dance	3		Dance	3		Dance	3		Dance	1		Dance	1

Concert	1		Concert	3		Concert	2		Concert	1		Concert	2
Ball game	2		Ball game	2		Ball game	1		Ball game	2		Ball game	3
Dance	3		Dance	1		Dance	3		Dance	3		Dance	1

a. Construct a preference table summarizing the results of the voting.

b. Use the preference table you constructed in Part a to find a winner using two different methods you learned in this lesson. Describe the methods you used.

2 Some friends and family are planning a weekend picnic. As part of the day's activities, they want to play a group ball game. To decide which game to play, they order their preferences from 1st choice to 5th (last) choice. The results are shown in the preference table below.

Ball Game Preferences

	Rankings		
Baseball	1	5	5
Soccer	2	1	2
Basketball	3	4	1
Football	4	3	4
Volleyball	5	2	3
Number of Voters	18 voters	16 voters	3 voters

a. Which game is the plurality winner? Do you think the group should choose this game to play? Why or why not?

b. Which is the pairwise-comparison winner?

c. Check out the winners under at least two other vote-analysis methods.

d. A decision must be made. Which vote-analysis method would you recommend to the group? Defend your answer.

3 The New Hampshire Primary is the first primary in every U.S. Presidential election year. (The Iowa caucuses are earlier, but these are not in the format of a primary election.) Public opinion throughout the nation can be greatly influenced by how a candidate performs in the New Hampshire Primary. In the 2012 Republican primary, five candidates received a significant percentage of the votes. The results for these five candidates were as follows.

Candidate	Votes	Percentage
Mitt Romney	97,591	39.28%
Ron Paul	56,872	22.89%
Jon Huntsman	41,964	16.89%
Rick Santorum	23,432	9.43%
Newt Gingrich	23,421	9.43%

Source: www.cnn.com/election/2012/primaries/state/nh

a. Mitt Romney was declared the winner, with 39.28% of the votes. What vote-analysis method is used to get this result?

b. Think about one possibility if ranked-choice voting was used. Suppose that:

- all the voters who voted for Huntsman had Paul as their second choice.

- all the voters who voted for Santorum had Paul as their second choice.

- all the voters who voted for Gingrich had Romney as their second choice.

In this situation, who is the runoff winner?

4 The IRV method that you studied in Problem 11 of Investigation 1 is also called the **sequential-elimination method**. Consider the preference table on the following page, showing the results of ranking of preferences for different types of energy sources. Copy the preference table onto your own paper. You need to do this so that you will have plenty of room to cross things out and make changes.

Energy Source Preferences

	Rankings				
Oil	1	4	4	4	2
Solar	2	3	1	2	3
Coal	3	1	3	3	4
Nuclear	4	2	2	1	1
Number of Voters	37 voters	32 voters	30 voters	21 voters	7 voters

a. Which energy source received the fewest 1st-preference votes? Eliminate this choice by crossing out its row.

b. Which groups of voters had the energy source you eliminated as their 1st preference?

c. For each group of voters in Part b, assign a new 1st preference. Change their most-preferred remaining choice into their new 1st preference by changing the lowest remaining preference number in the column for that group to a "1".

d. Repeat Parts a–c for the modified schedule.

e. Continue in this way until there is only one choice left. This remaining choice is the **sequential-elimination** winner. Using this sequential-elimination method, which energy source is the voters' preferred energy source?

f. Compare the process of sequential elimination described in Parts a–e to the description of the IRV method in Problem 11 on page 613. Explain why the IRV method is the same as the sequential-elimination method.

5 Another common voting method is **approval voting**. Since 1987 the Mathematical Association of America (MAA) has elected its officers using approval voting. The United Nations Security Council also uses approval voting. Using this method, every voter selects *all* the candidates he or she approves of. You can still only cast one vote per candidate, but you can vote for as many candidates as you like. For example, if there are five candidates and you approve of three of them, then you can cast a vote for each of the three. The winner in such an election is the candidate that receives the most votes. The 2013 MAA election candidates for president-elect and first vice-president are shown at the right.

**The Mathematical Association of America
2013 Ballot**

Balloting is by approval voting. For each office,
you are advised to vote for one or two candidates.
One will be elected.

President-Elect (2014)
☐ Deanna Haunsperger
☐ Frank Farris
☐ Francis Su
☐ _____

First Vice-President (2014–15)
☐ Jenna Carpenter
☐ Michael Dorff
☐ James Sellers
☐ _____

a. Why do you think the MAA advises members to vote for one or two candidates?

b. When using approval voting, you do not rank the candidates; you simply cast a vote for all candidates of whom you approve. Thus, a preference table is not needed. However, to get an idea how approval voting works, consider again the Senior Class President preference table reproduced below.

Senior Class President Preferences

	Rankings					
Charnell	1	5	5	5	5	5
Amarjit	5	1	2	4	2	4
Lamar	4	4	1	2	4	2
Rodene	2	3	4	1	3	3
Richard	3	2	3	3	1	1
Number of Voters	36 voters	24 voters	20 voters	18 voters	8 voters	4 voters

   i. For each group of voters in the table, place a check mark next to the candidates you think they will approve. For example, you might assume that voters will approve of their top two choices or maybe their top three choices. Use whatever approval criterion you think is reasonable.

   ii. Compute the approval winner and explain your method.

6 The preference table below summarizes the results of asking all seniors at Escondido High School to rank the importance of four environmental protection policies. You may wish to use the *TCMS-Tools* "Ranked-Choice Voting" custom app or similar software to help you explore "what-if" scenarios for this situation.

Environmental Policy Preferences

	Rankings				
Recycle	1	4	4	4	2
Plant Trees	2	3	1	2	3
Conserve Electricity	3	1	3	3	4
Carpool	4	2	2	1	1
Number of Voters	56 voters	48 voters	41 voters	35 voters	29 voters

a. Using the points-for-preferences method, where 1st preference gets 3 points, 2nd preference gets 2 points, 3rd preference gets 1 point, and 4th preference gets 0 points, which policy wins?

b. Do you think the point scheme in Part a is reasonable? Why or why not?

c. Does the winner change if you double all the point allocations in Part a? How about if you square all the point allocations in Part a?

d. One student suggested a combination of a modified approval method (see Applications Task 5) and the runoff method. Here is how it works. Based on the information in the preference table, everyone casts a vote for their two most preferred choices, then the two choices that get the most votes undergo a runoff. Find the winner under this method.

7 There were many close races in the November 2010 U.S. midterm elections. A key factor in many races was a viable third-party candidate, in addition to the Democrat and Republican, often a candidate from the Libertarian Party, Green Party, or Tea Party. For example, the Indiana 2nd District race for the U.S. House of Representatives was won by Democrat Joe Donnelly, but only by a narrow margin over Republican Jackie Walorski, with a strong showing from the Libertarian candidate Mark Vogel. Here are the vote totals, as reported in a local newspaper.

Indiana 2nd District Election Results for the U.S. House of Representatives

Joe Donnelly (Democrat)	91,330
Jackie Walorski (Republican)	88,787
Mark Vogel (Libertarian)	9,445

Source: "Joe Donnelly's narrow win was built on a typical Democratic power base," *Elkhart Truth*, by Josh Weinhold, 11/4/2010, www.elkharttruth.com/news/2010/11/04/Joe-Donnelly-s-narrow-win-was-built-on-a-typical-Democratic-power-base.html

Some analysts claim that Vogel was the "spoiler" in this election, which Walorski otherwise would have won. As reported in the news story, "Any third-party candidate is always charged with playing the role of the spoiler," according to Elizabeth Bennion, a political science professor at Indiana University South Bend. Bennion points out that this may not always be true, since the third-party candidate might take votes equally from both of the other candidates. However, in this case, the article states that, "Vogel claimed a larger portion of votes in more rural areas, … suggesting he had greater appeal in the district's more conservative territory."

a. Suppose that ranked-choice voting had been used in this election. Based on the news story above, explain why the preference table below is plausible.

Indiana 2nd District Hypothetical Preference Table

	Rankings			
Donnelly	1	3	3	2
Walorski	3	1	2	3
Vogel	2	2	1	1
Number of Voters	91,330	88,787	6,000	3,445

b. Using this hypothetical preference table, give an argument for how the election might have been won by Walorski.

c. According to the article, "the Indiana Democratic Party sent a controversial districtwide mailing last weekend, promoting Vogel as the only 'true conservative' in the race." Why do you think the Democratic Party would do this? Explain how using ranked-choice voting might have eliminated any effect or even the idea of this mailing.

8 Examine the voting results in the preference table below.

Candidate Preferences

	Rankings		
Candidate A	1	2	3
Candidate B	3	1	2
Candidate C	2	3	1
Number of Voters	8 voters	6 voters	4 voters

a. Determine the pairwise-comparison winners.

b. Which of Arrow's Fairness Conditions is violated by pairwise-comparison?

9 Find the Banzhaf power index for each voter in the weighted voting situation represented by [7: 6, 4, 2, 1].

CONNECTIONS

10 There are different ways to represent the results of ranked-choice voting. Below are three different representations of the same voting data. The first is the preference table format used in this lesson. The other two are alternative formats.

I.

	Rankings			
Candidate A	1	3	3	2
Candidate B	3	1	2	1
Candidate C	2	2	1	3
Number of Voters	15 voters	13 voters	11 voters	10 voters

II.

	15 Voters	13 Voters	11 Voters	10 Voters
1st Preference	A	B	C	B
2nd Preference	C	C	B	A
3rd Preference	B	A	A	C

III.

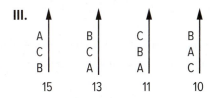

a. Explain how the information shown in Format I is the same as in Format II.

   b. Explain how the information shown in Format III is the same as in Format II.

   c. One of the reasons the preference table in Format I is used in this lesson is because it matches the fixed listing of candidates on a ballot. Describe some advantages and disadvantages of each of the three formats.

11 In this task, you will count the total number of rankings possible (without ties) in a preference table.

   a. With three choices, A, B, and C, there are six possible rankings:

   ABC, ACB, BAC, BCA, CAB, CBA

   i. Explain why there are 720 different rankings possible with 6 choices.

   ii. Explain how to use the fact in part i to find the number of possible rankings of 7 choices.

   b. If $f(n)$ is the number of rankings possible for a certain number of choices n, and $f(n + 1)$ is the number of rankings possible with one more choice, write a recursive formula showing the relationship between $f(n)$ and $f(n + 1)$.

   c. You may have previously studied permutations and combinations in your previous coursework. If so, is counting the number of possible rankings in a preference table a problem of counting permutations or combinations? Explain.

12 When using the pairwise-comparison method, you must run off all possible pairs of choices. Think about how many possible pairs there could be.

   a. List all the pairs that need to be run off if there are 3 choices: A, B, and C. How many pairs are there?

   b. You may have previously studied permutations and combinations. If so, is counting the number of possible pairs in the pairwise-comparison method a problem of counting permutations or combinations? Explain.

   c. Make a table with the number of choices in one column and the number of possible pairs in another column. Record your results from Part a. Continue the table for the cases of 4, 5, 6, and 7 choices.

   d. Describe any patterns you see in your table from Part c. Write a recursive formula for the sequence of numbers of pairs.

13 The pairwise-comparison method can be modeled with a digraph (directed graph), as follows. The vertices represent the candidates, and two vertices are connected by a directed edge (arrow) from one to the other if the one candidate beats the other in their pairwise runoff. For example, if A and B are two of the candidates and A beats B in the A-B runoff, then there is a directed edge from A to B.

Digraph

a. Draw a digraph representing a pairwise-comparison analysis of each of the following preference tables.

Table I

	Rankings			
Candidate A	1	3	3	2
Candidate B	3	1	2	3
Candidate C	2	2	1	1
Number of Voters	8 voters	6 voters	4 voters	4 voters

Table II

	Rankings		
Candidate A	1	2	2
Candidate B	3	1	3
Candidate C	2	3	1
Number of Voters	8 voters	6 voters	4 voters

b. Can you tell just by looking at the digraph model whether or not there is a pairwise-comparison winner? Explain and illustrate your answer.

14 Suppose that a group of 12 students decides to vote to play volleyball, softball, or soccer at their picnic. Each student votes only for his or her most preferred choice. If the results of the voting are represented in a bar graph, describe the possible shape of the bar graph if:

a. there is a majority winner.

b. there is a plurality winner, but no majority winner.

c. there is no plurality winner.

15 As you learned in Investigation 3 of this lesson, the Banzhaf power index is computed with reference to winning coalitions of voters. Think about how many coalitions of voters are possible, both winning and losing. Suppose you have a weighted voting system with n voters. Explain why the total number of coalitions of voters, both winning and losing coalitions, is $2^n - 1$. (Recall that in the *Informatics* unit, you may have discovered that a set with n elements has 2^n subsets.)

16 The city of Oakland, CA, used ranked-choice voting for the first time in November 2010. There were 10 candidates in the race. Read the excerpted story below from the *California Beat*.

"Perata concedes Oakland mayoral race, faults ranked-choice voting for loss"

California Beat, Jennifer Courtney, 11/11/2010

On election night, November 2, Perata led Quan by a formidable 35 to 24 percent in first-choice votes. But Quan edged past him when she picked up more second and third-choice votes from those who ranked Rebecca Kaplan as their first choice. Kaplan, Oakland's at-large city councilperson, came in third place.

But while Perata said he believes the electorate overwhelmingly supported him, he added that because ranked-choice voting was new to Oakland voters this November, many did not understand how to fill out their ballots correctly.

"I think people were confused by it," he said. "Most of us are used to voting for one person. It will take awhile for voters to get used to that not being the case."

Perata added that he has had reservations about ranked-choice voting since he first learned about the process 12 years ago while chairing the state senate's election committee, and still does not know how its algorithms work to eliminate candidates. Perata evaded answers to questions posed by reporters who pressed him on how it is that he was confused.

Ranked-choice voting was passed by Oakland, San Leandro and Berkeley voters in an effort to save money by avoiding separate run-off elections.

---

a. In the article, losing candidate Perata criticizes ranked-choice voting. He says, "I think people were confused by it." Do you find ranked-choice voting confusing? Why or why not?

b. To avoid possible voter confusion about ranked-choice voting, the Registrar of Voters in Alameda County, which includes Oakland, released a how-to video about ranked-choice voting. Do an Internet search to find this video. (You might find it at www.acgov.org/rov/rcv/video.htm.) Do you think the video provides a good explanation of ranked-choice voting? Why or why not?

c. The article states at the very end that, "Ranked-choice voting was passed by Oakland, San Leandro, and Berkeley voters in an effort to save money by avoiding separate run-off elections." Explain how ranked-choice voting helps avoid "separate run-off elections."

d. This election made national news on November 19, 2010 on PBS NEWSHOUR: "Oakland Election System Allows Mayoral Runner-Up to Make a Surprise Win." View the video of this newscast and write a brief report summarizing the report. (You may find the video at: www.pbs.org/newshour/bb/politics/july-dec10/oakland_11-19.html)

17 The article at the beginning of Investigation 1 reported on the use of ranked-choice voting in North Carolina. Do an Internet search to find out about the results and reaction to the voting. Prepare a brief report on what you find.

18 In the United Kingdom, ranked-choice voting is sometimes called *Alternative Vote*. This voting method has been used in local elections for some time. A national referendum was held on May 5, 2011, "to decide if MPs [Members of Parliament] should be elected by the alternative vote, where voters' choices are ranked." Do an Internet search to find out about the results of this referendum. Write a brief report summarizing the results and public reaction. (**Source:** "Ed Miliband joins campaign for voting reform," by Patrick Wintour, theguardian.com, December 7, 2010.)

19 Consider the Zimbabwean presidential election in 2008, which took place in two rounds separated by three months. The first round did not produce a majority winner and so a runoff was required by law. Between the voting rounds, there was political unrest and violence in the country. The second round was controversial. Below is a summary of the election results in both rounds.

Summary of the March 29 and June 27, 2008 Zimbabwean Presidential Election Results

Party	Candidate	1st Round		2nd Round	
		Votes	%	Votes	%
Movement for Democratic Change	Morgan Tsvangirai	1,195,562	47.9	233,000	9.3
Zimbabwe African National—Patriotic Front	Robert Mugabe	1,079,730	43.2	2,150,269	85.5
Mavambo/Kusile/Dawn	Simba Makoni	207,470	8.3		
Independent	Langton Towungana	14,503	0.6		
Invalidated				131,481	5.2
Totals		2,497,265	100.0	2,514,750	100.0

Source: www.electionguide.org/elections/id/2071/; "Mugabe sworn in after Zimbabwe's one-man election," Agence France-Presse, 6/30/2008

   a. Examine the results shown in the table above. Describe any interesting patterns or observations.

   b. Robert Mugabe is the leader of the ZANU-PF Party. Simba Makoni was expelled from this party, and the ZANU-PF Party announced that anyone supporting Makoni would also be expelled. (**Source:** "Mugabe rival expelled from party," BBC News, February 12, 2008.)

Steve Allen/Brand X Pictures

Morgan Tsvangirai is the leader of one of the factions of the MDC Party. The leader of the other faction of the MDC, Arthur Mutambara, decided not to run and declared that he would back Makoni. (**Source:** Fikile Mapala, "Mutambara withdraws from race, backs Makoni," newzimbabwe.com, February 15, 2008.) Based on this information and the first-round results shown in the table, who do you think would have won the election using the runoff method from this lesson? Explain.

c. In the case of the 2008 Zimbabwean presidential election, do you see any advantage to using one of the runoff methods you learned in this lesson over the two-round runoff method actually used? Explain.

20 Physical and social scientists, medical researchers, economists, actors, authors, athletes, and others are often recognized for outstanding contributions or achievements in their field. Selection of these awardees is done by voting of a well-defined group. Complete one of Parts a, b, or c below and be prepared to share your findings with the class and respond to any questions they may have about the method of voting.

a. Research and prepare a short report on how voting is used in the selection of a Nobel Prize winner.

b. Research and prepare a short report on how voting is used in the annual selection of the Heisman Trophy winner.

c. Research and prepare a short report on how voting is used to choose the site for the Olympics games.

21 Sometimes in an election there are several candidates, but only two seem to have much of a chance at winning. The others are "long-shots." For example, this is often the case when there are third-party candidates in a presidential election. In such situations, you sometimes hear voters complain that they would like to vote for the perceived long-shot candidate, but they feel that doing so would be throwing their vote away. One advantage of ranked-choice voting is that you can vote for a long-shot candidate and still vote for one of the front-runners as well.

Ask some adults you know who have voted in presidential elections if they ever felt that there was a candidate they wanted to vote for, but did not because it seemed like they would just be throwing their vote away. Explain to these adults how ranked-choice voting would solve this problem. Write a paragraph describing who you talked to, what they said, and how they reacted to your explanation of ranked-choice voting.

22 How do you and your friends make group decisions? Are they always fair decisions? Give an example. Discuss how what you have learned about voting in this lesson could help you and your friends perhaps make fairer decisions.

23 Voting and social decision-making are topics that have intrigued philosophers as well as mathematicians for ages. For example, read the three quotations below. Choose the one that you find most interesting, explain what you think the author meant, and briefly discuss whether you agree or not.

a. "The principle of majority rule must be taken ethically as a means of ascertaining a real 'general will,' not as a mechanism by which one set of interests is made subservient to another set. Political discussion must be assumed to represent a quest for an objectively ideal or 'best' policy, not a contest between interests."

(**Source:** Rousseau, in *The Social Contract*, English Translation, New York and London: G.P. Putnam's Sons, second edition, revised, 1906, pp. 165–166.)

b. "The idealist doctrine then may be summed up by saying that each individual has two orderings, one which governs him in his everyday actions and one which would be relevant under some ideal conditions and which is in some sense truer than the first ordering. It is the latter which is considered relevant to social choice, and it is assumed that there is complete unanimity with regard to the truer individual ordering." (**Source:** Kenneth Arrow, in *Social Choice and Individual Values*, New Haven and London: Yale University Press, second edition, 1963, pp. 82–83.)

c. When writing about individuals being subject to the laws of a society, and whether this subjection can be the basis of a free society, T. H. Green claims, "What is certain is that a habit of subjection founded upon fear could not be a basis of political or free society to which it is necessary, not indeed that everyone subject to the laws should take part in voting them, still less that he should consent to their application to himself, but that it should represent an idea of common good, which each member of the society can make his own so far as he is rational, or capable of the conception of common good, however much particular passions may lead him to ignore it … ."

(**Source:** T. H. Green, *Lectures on the Principles of Political Obligation*, New York and London: Longmans, Green and Co., 1895, p. 89.)

ON YOUR OWN

EXTENSIONS

24 The presidential election in France in 2012 was a close contest between the political left and right. No candidate received a majority of votes in the first round of voting, so by French law, a runoff was required. The runoff winner was not determined using the methods you studied in this lesson. Instead, a second round of voting took place two weeks after the first round. According to one popular source, the first-round voting results were as follows, showing candidates who received more than 1% of the votes.

Candidates	Political Identification	Percent of Vote
François Hollande	Socialist Party (left)	28.63%
Nicolas Sarkozy (incumbent)	Union for a Popular Movement (center-right)	27.18%
Marine Le Pen	National Front (far-right)	17.90%
Jean-Luc Mélenchon	Left Front (far-left)	11.10%
François Bayrou	Democratic Movement (centrist)	9.13%
Eva Joly	Europe Écologie – The Greens (green)	2.31%
Nicolas Dupont-Aignan	Arise the Republic (right)	1.79%
Philippe Poutou	New Anticapitalist Party (far-left)	1.15%

Source: IEMed. Mediterranean Yearbook

a. Who is the winner using the plurality method?

b. Using the runoff method that you learned about in this lesson, which two candidates will run against each other? Which of these two candidates do you think will win a runoff? Explain, using the information about other candidates shown in the table.

c. Suppose just the top five candidates are running for president. Construct a reasonable preference table for the election. To construct the preference table, use your own reasonable assumptions about how people might rank the candidates based on the information in the table above, and you might also do some research about the election. For example, the radio station rfi reported, "Surprise backing for Hollande from centrist Bayrou in presidential run-off with Sarkozy." (**Source:** www.english.rfi.fr/france/20120504-centrist-bayrou-gives-support-hollande-ahead-sundays-presidential-run; May 4, 2012). Using your preference table, find the winner of the election using at least two different vote-analysis methods.

d. The winner of the 2012 French presidential election, after two rounds of voting, was François Hollande. How does this compare to the winner(s) you found in Parts b and c? What are some comments you have about the efficiency and outcome of this election?

25 Sometimes voters vote insincerely to try to change the outcome of an election. Such voting is called **insincere** or **strategic voting**. For example, consider the preference table for Senior Class President that you analyzed in Investigation 1, Problem 11 (page 613) reproduced below. You may wish to use the "Voting" custom app in *TCMS-Tools* or similar software to help you explore "what-if" scenarios.

Original Senior Class President Preference Table

	Rankings					
Charnell	1	5	5	5	5	5
Amarjit	5	1	2	4	2	4
Lamar	4	4	1	2	4	2
Rodene	2	3	4	1	3	3
Richard	3	2	3	3	1	1
Number of Voters	36 voters	24 voters	20 voters	18 voters	8 voters	4 voters

Suppose the 20 voters in the third ranking in the preference table decide to vote strategically (and insincerely) and switch their 1st and 2nd preferences. The modified preference table is shown below.

Modified Senior Class President Preference Table

	Rankings					
Charnell	1	5	5	5	5	5
Amarjit	5	1	1	4	2	4
Lamar	4	4	2	2	4	2
Rodene	2	3	4	1	3	3
Richard	3	2	3	3	1	1
Number of Voters	36 voters	24 voters	20 voters	18 voters	8 voters	4 voters

a. Determine the plurality, majority, runoff, pairwise-comparison, points-for-preferences, and IRV winners based on the original preference table. Do the same for the modified preference table.

b. Did this strategic switch in preference produce different winners?

c. Give a reason why the 20 voters who switched preferences might have gotten together to plan the switch.

d. Suppose the voters know the election will be analyzed using the runoff method. Find an instance of strategic voting that will benefit one of the groups of voters.

26 As you have seen, sometimes two vote-analysis methods produce the same winner, and sometimes they do not. Consider the IRV and plurality methods.

 a. Construct a preference table where the plurality winner is different than the winner using the IRV method.

 b. Construct a preference table where the plurality winner is the same as the IRV winner.

27 You now have experience in finding winners with several voting methods. Think about whether those methods always produce winners.

 a. Is there a plurality winner for all possible preference tables? If not, construct a simple preference table where there is no plurality winner.

 b. Is there a majority winner for all possible preference tables? If not, construct a simple preference table where there is no majority winner.

 c. Using the preference table below, find a ranking by the group of 4 voters so that there is no pairwise-comparison winner.

	Rankings		
Choice A	1	2	
Choice B	3	1	
Choice C	2	3	
Number of Voters	8 voters	6 voters	4 voters

28 Describe the method of voting used in your school for electing class officers. Discuss the method with some other students who have learned about different methods of voting and ways to analyze the results. Work with them to prepare a report for the student government proposing a fairer voting strategy to use in the next election. Make a strong case for adopting your strategy over the system currently in use.

29 In Investigation 2, you learned how the runoff method can violate some of Arrow's fairness conditions. Do a search for "Arrow's Impossibility Theorem" on the Internet to find an example showing how a different voting or analysis method can violate a fairness condition. Prepare a written or oral report describing the example and showing how it violates the condition.

30 In Investigation 3, you learned about the Banzhaf power index. Another well-known index for measuring voting power in a weighted voting system is the *Shapley-Shubik power index*, named for the developers Lloyd Stowell Shapley at the University of California, Los Angeles, and Martin S. Shubik at Yale University. Both the Banzhaf index and the Shapley-Shubik index are based on the idea of a voter who is *critical* for a coalition.

Lloyd Shapley receiving the Sveriges Riksbank Prize in Economic Sciences

However, for the Shapley-Shubik index, you consider so-called *sequential coalitions* and define critical in terms of those coalitions. A sequential coalition is not just a coalition of voters, rather it is an ordered sequence of all voters. A voter is considered critical for a particular sequential coalition if he is the first voter in the sequence such that he and all the previous voters comprise a winning coalition. The Shapley-Shubik index for a voter is the ratio of the number of critical votes for the voter to the total number of sequential coalitions, which is $n!$ if there are n total voters.

Use the Shapley-Shubik index to determine voting power for each voter in one of the weighted voting systems in Investigation 3. Compare the Shapley-Shubik indices to the power indices yielded by the Banzhaf power index.

REVIEW

31 Consider the function $r(x) = x^2 + 2x - 24$.

   a. Evaluate $r(3)$ and $r(-5)$.

   b. Find the zeroes of $r(x)$.

   c. For what values of x is $r(x) > 0$?

   d. For what values of x is $r(x) = 12$?

32 A rectangular prism has one vertex at the point $(0, 0, 0)$. Three faces of the prism are contained in the planes $x = 4$, $y = 10$, and $z = 2$.

   a. Sketch the prism on a three-dimensional coordinate system.

   b. What are the coordinates of the eight vertices of the prism?

   c. What is the volume of the prism?

   d. One diagonal of the prism joins the point $(0, 0, 0)$ with the point $(4, 10, 2)$. What is the length of this diagonal?

33 Write an equation for the line satisfying each set of conditions.

　a. Contains the points (2, −1) and (−2, 5)

　b. Contains the points (4, 0) and (0, −5)

　c. Is perpendicular to the line with equation $y = \frac{4}{3}x - 4$ and contains the point (6, 0)

34 Karla has a credit card bill of $4,750. The interest on this account is 24% annual interest compounded monthly. Suppose that she does not charge anything more and that she makes $200 monthly payments.

　a. What will her credit card balance be after one month? After two months?

　b. If her current balance is in cell **A3** on a spreadsheet, what formula should you put in cell **A4** so that it contains the balance after the next payment?

　c. How long will it take Karla to pay off the $4,750 balance?

　d. How much interest will she have paid when her balance is reduced to zero?

35 Sketch a graph of each function. Then state the domain and range of the function.

　a. $f(x) = -\frac{2}{3}x - 5$

　b. $g(x) = 2(3^x)$

　c. $h(x) = \frac{9}{x}$

　d. $p(x) = |x + 3|$

36 For the functions in Task 35, for which functions could you determine its domain and range just by examining the symbolic rule? Explain.

37 In the figure at the right, $\overline{BE} \parallel \overline{CD}$. If $AB = 10$, $BC = 2$, $BE = 8$, and $ED = 3$: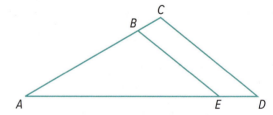

　a. find CD.

　b. find AE.

38 Solve each equation or inequality. Display solutions for inequalities on a number line.

　a. $3 + x + x^2 = 12 + x$

　b. $85 - \frac{2}{3}x < 15$

　c. $\frac{3 - x}{2} = \frac{2x + 1}{3}$

　d. $32 \leq -12x - 16$

39 Consider the function $f(x) = 3x^2 + x - 2$. Determine the rule for the function $f'(x)$ whose graph is the image of the graph of $f(x)$ under each transformation.

   a. Reflection across the y-axis

   b. Reflection across the x-axis

   c. Translation up 5 units

   d. Translation to the left 4 units

40 Find the indicated angle measure and side lengths.

   a. Find $m\angle B$.

   b. Find exact values for BC and AC.

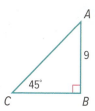

   c. Find exact values for AB and BC.

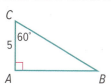

Fair Division

Fairness is a basic tenet in any democratic society, and in life more generally. In Lesson 1, you investigated fairness in the context of voting. Another important aspect of fairness is fair division, for example, when two companies merge or in the case of estate assets. A key feature of objects that you want to fairly divide is whether they are *divisible* or *indivisible* (sometimes referred to as continuous or discrete, respectively). For example, you might want to fairly divide a divisible object like a region of land. Or you might fairly divide a collection of indivisible objects like seats in the U.S. House of Representatives.

THINK ABOUT THIS SITUATION

There are many situations that involve fair division. Think about a few of those.

a. Suppose you and a classmate want to share a chocolate chip cookie at lunch. What method would you suggest using to ensure fairness? Explain your thinking.

b. Suppose you and two friends want to fairly share a medium pizza after a movie. What method could you use to ensure fairness? Explain your thinking.

c. What are some other examples in your own life where you might want to fairly divide something? What is divided? How would you divide it fairly?

d. The U.S. Senate has 100 members. Compared to the other 49 states, do you think your state has fair representation in the Senate? Explain your thinking.

e. The U.S. House of Representatives has 435 members. Do you think your state has fair representation in the House of Representatives? Explain your thinking.

In this lesson, you will learn some classical and recent methods to fairly divide both divisible and indivisible objects. To begin, you will study historical and current methods for apportioning the indivisible seats in the U.S. House of Representatives among the fifty states. You will then learn some established and some newly discovered methods for other types of fair division situations.

INVESTIGATION 1

Political Apportionment

One of the most important tasks for demographers is conducting census counts of populations to determine fair representation in government. The constitution of the United States requires a complete national census every 10 years. On the basis of the census results, there are often changes in the number of representatives to Congress from the various states. These changes can have many repercussions. For example when a state loses a representative, the state, and often one of the political parties, loses influence, which may affect economic and social policies.

White House photo by Paul Morse

In this investigation, you will analyze mathematical methods, both historical and current, for apportioning seats in the U.S. House of Representatives. As you work through the following problems, look for answers to these questions:

What methods can be used to allot representatives to states based on population?

How are mathematics, logic, and technology helpful in apportionment methods?

The United States Congress consists of the Senate and the House of Representatives. Every state has two senators. The House of Representatives has 435 members with different numbers of representatives for different states based on the census counts. The following spreadsheet contains data about apportionment populations and political representation in the U.S. House of Representatives for the 50 states from the censuses conducted in 2000 and 2010.

U.S. House of Representatives Apportionment

	A	B	C	D	E
1	State	2000 Population	2010 Population	Representatives in 2000	Representatives in 2010
2	Alabama	4,461,130	4,802,982	7	7
3	Alaska	628,933	721,523	1	1
4	Arizona	5,140,683	6,412,700	8	9
5	Arkansas	2,679,733	2,926,229	4	4
6	California	33,930,798	37,341,989	53	53
7	Colorado	4,311,882	5,044,930	7	7
8	Connecticut	3,409,535	3,581,628	5	5
9	Delaware	785,068	900,877	1	1
10	Florida	16,028,890	18,900,773	25	27
11	Georgia	8,206,975	9,727,566	13	14
12	Hawaii	1,216,642	1,366,862	2	2
13	Idaho	1,297,274	1,573,499	2	2
14	Illinois	12,439,042	12,864,380	19	18
15	Indiana	6,090,782	6,501,582	9	9
16	Iowa	2,931,923	3,053,787	5	4
17	Kansas	2,693,824	2,863,813	4	4
18	Kentucky	4,049,431	4,350,606	6	6
19	Louisiana	4,480,271	4,553,962	7	6
20	Maine	1,277,731	1,333,074	2	2
21	Maryland	5,307,886	5,789,929	8	8
22	Massachusetts	6,355,568	6,559,644	10	9
23	Michigan	9,955,829	9,911,626	15	14
24	Minnesota	4,925,670	5,314,879	8	8
25	Mississippi	2,852,927	2,978,240	4	4
26	Missouri	5,606,260	6,011,478	9	8
27	Montana	905,316	994,416	1	1
28	Nebraska	1,715,369	1,831,825	3	3

	A	B	C	D	E
29	Nevada	2,002,032	2,709,432	3	4
30	New Hampshire	1,238,415	1,321,445	2	2
31	New Jersey	8,424,354	8,807,501	13	12
32	New Mexico	1,823,821	2,067,273	3	3
33	New York	19,004,973	19,421,055	29	27
34	North Carolina	8,067,673	9,565,781	13	13
35	North Dakota	643,756	675,905	1	1
36	Ohio	11,374,540	11,568,495	18	16
37	Oklahoma	3,458,819	3,764,882	5	5
38	Oregon	3,428,543	3,848,606	5	5
39	Pennsylvania	12,300,670	12,734,905	19	18
40	Rhode Island	1,049,662	1,055,247	2	2
41	South Carolina	4,025,061	4,645,975	6	7
42	South Dakota	756,874	819,761	1	1
43	Tennessee	5,700,037	6,375,431	9	9
44	Texas	20,903,994	25,268,418	32	36
45	Utah	2,236,714	2,770,765	3	4
46	Vermont	609,890	630,337	1	1
47	Virginia	7,100,702	8,037,736	11	11
48	Washington	5,908,684	6,753,369	9	10
49	West Virginia	1,813,077	1,859,815	3	3
50	Wisconsin	5,371,210	5,698,230	8	8
51	Wyoming	495,304	568,300	1	1
52	Totals	=SUM(B2:B51)	=SUM(C2:C51)	=SUM(D2:D51)	=SUM(E2:E51)

Sources: 2010.census.gov/2010census/data/ and www.census.gov/population/apportionment/data/2010_apportionment_results.html

1 Examine the spreadsheet of the apportionment data.

   a. What do you notice about the population changes in states that gained or lost a representative? Do the new numbers of representatives seem fair to you?

   b. What does the formula in cell **D52** mean? What number should appear in this cell when the Enter key is pressed?

2 In your discussion in Problem 1, you may have wondered how the number of representatives is decided. The theoretical solution is that each state should get a proportion of the total number of representatives (435) according to its proportion of the total population. The U.S. Constitution mandates that each state have at least one representative.

a. You may recall from your use of spreadsheets in Unit 4, *Mathematics of Financial Decision-Making*, that the **$** is used to fix a column letter or row number reference so that when the Fill Down or Fill Across command in the Edit menu is used, that column letter or row number does not change. Explain how the formula in cell **F2** computes the exact proportional representation for Alabama in the sample spreadsheet below.

Exact Proportional Representation

	A	B	C	D	E	F
1	State	2000 Population	2010 Population	Representatives in 2000	Representatives in 2010	2010 Exact Proportion
2	Alabama	4,461,130	4,802,982	7	7	=(C2/$C$52)*435
3	Alaska	628,933	721,523	1	1	1.02
4	Arizona	5,140,683	6,412,700	8	9	9.02
5	Arkansas	2,679,733	2,926,229	4	4	4.12
⋮						

b. Access the "U.S. House of Representatives Apportionment" data set in the *TCMS-Tools* spreadsheet. Save a new spreadsheet that shows the exact proportional representation for 2010. Use two-decimal-place accuracy in the spreadsheet settings. Check that your values for 2010 Exact Proportion agree with the three values in column **F** above.

c. Compare the values in column **F** to the actual whole numbers in column **E** for the 2010 apportionment. Can you determine the method used to allot representatives to states?

3 You will now do some spreadsheet calculations that help you analyze how population growth affects the allotment of representatives to each of the states. Begin with the original spreadsheet. Create and save a new spreadsheet with two new columns to show the growth for each state from 2000 to 2010 as shown below.

Effect of Population Growth

	A	B	C	D	E	F	G
1	State	2000 Population	2010 Population	Representatives in 2000	Representatives in 2010	Population Growth	% Population Growth
2	Alabama	4,461,130	4,802,982	7	7	341,852	7.66
3	Alaska	628,933	721,523	1	1	92,590	14.72
4	Arizona	5,140,683	6,412,700	8	9	1,272,017	24.74
5	Arkansas	2,679,733	2,926,229	4	4	246,496	9.20
⋮							

a. How well does growth in *number* of people predict which states gained or lost representatives from 2000 to 2010? Give examples that illustrate your answer.

b. How well does *percent* growth predict which states gained or lost representatives from 2000 to 2010? Give examples that illustrate your answer.

c. Which of the growth trends shown in the spreadsheet are most consistent with what you know about population change in the U.S. Which are surprising?

Apportionment Methods to Address Rounding As you explored the relationship between state populations and numbers of representatives in the United States Congress, you probably discovered a variety of complications caused by the fractions that arise from proportional calculations. For example, it appears that in 2010, Minnesota was entitled to 7.48 representatives and Missouri to 8.46 representatives. What number of representatives would be assigned to these states using standard rounding rules?

Notice from your spreadsheet that the number of representatives for Missouri was decreased, while Minnesota was increased. Also, if you apply standard rounding rules to all the states, you get a total of 433 representatives, instead of the required 435.

Since change in the number of representatives from a state can mean a loss of political power for a state and possible loss of jobs and money, fractions and rounding have become very important issues in this context. All of the apportionment methods that you will now study address the issue of rounding fractions in different ways.

4 Several different apportionment methods have been devised over the course of the history of the United States. The standard starting point in all methods is the people-per-representative rate. Based on results from the 2010 census, that figure is $\dfrac{309,183,463}{435} \approx 710,767$.

a. Explain why the fraction $\dfrac{309,183,463}{435}$ can sensibly be called the "people-per-representative rate."

b. Explain why the people-per-representative rate given in Part a entitles California to 52.54 representatives.

c. The number (52.54) of California representatives is the same number of representatives you computed using the exact proportional method in Problem 2. Provide an argument that the computation used for the exact proportional method is equivalent to the computation used for the people-per-representative method.

d. Since California is entitled to 52.54 representatives, the question is: "What should be done with the fraction?" Brainstorm with classmates on possible ways to deal with the fractions in assigning representatives.

5 Statesmen and mathematicians have been proposing apportionment methods since the beginning of the United States. Alexander Hamilton, Secretary of the Treasury during the presidency of George Washington, proposed the following method (explained below using the current 435 seat House):

Step 1. Calculate the theoretical exact number of representatives to which each state is entitled, including fractional parts.

Step 2. Round the exact calculations *down* to the nearest lower integer value.

Step 3. Add these 50 integer values and see how far short of 435 the total is.

Step 4. Add single representatives to the states with highest fractional parts from the exact calculation stage until a total of 435 is achieved.

The **Hamilton method of apportionment** was used from 1850 to 1900. Create a spreadsheet to test Hamilton's method using the 2010 census data. You can do the rounding with cell formulas like **=INT(CELL)**. You can scan the fractional parts to see which are larger, or you can create a column containing only the fractional parts and then sort those values from largest to smallest. The commands for ordering data in columns or rows are usually listed in the Data menu of a spreadsheet with the key word **SORT**.

a. Apply Hamilton's method to the 2010 census data. How do the assigned numbers of representatives under Hamilton's method compare with the actual numbers of representatives allotted after the 2010 census?

b. Modify your spreadsheet from Part a to apply Hamilton's method to the 2000 census data. How do the assigned numbers of representatives under Hamilton's method compare with the actual numbers of representatives allotted after the 2000 census?

c. What advantages and disadvantages can you see for Hamilton's method?

6 Thomas Jefferson, the third President of the United States, proposed a different procedure for assigning numbers of representatives. The **Jefferson method of apportionment** was used from 1790 to 1830. This method works as follows:

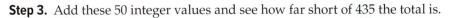
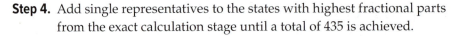

Step 1. Calculate exact proportional values and round them *down* to the nearest integer values.

Step 2. Instead of adding single representatives to selected states, revise the people-per-representative rate until the sum of whole numbers of representatives matches the total number of U.S. representatives.

a. Apply Jefferson's method to the 2010 census data. To experiment to find a suitable rate, it might help to put the proposed people-per-representative number in a particular cell of a spreadsheet. Then refer to the cell with a fixed reference. Begin with the exact people-per-representative rate, $\frac{309{,}183{,}463}{435} \approx 710{,}767$, using zero-decimal-place accuracy in the spreadsheet settings.

Step 1. Strategically choose a different people-per-representative rate to compute an "exact" number of representatives.

Step 2. Round your answer *down* to get an actual number of representatives.

Step 3. Add all the whole numbers to check if the sum is 435.

Step 4. If the sum is not 435, adjust the people-per-representative rate and recalculate until the sum is 435.

b. Compare the results of using Jefferson's method with the actual allotment of representatives used after the 2010 census. For which states does the method give a different number of representatives?

c. What advantages or disadvantages do you see of Jefferson's method?

7 John Quincy Adams, another early United States president, proposed a mirror image of Jefferson's procedure. Instead of rounding down at the start, he proposed rounding up and then adjusting the people-per representative rate. The **Adams method of apportionment** was never actually used.

a. Explain why the formula =INT(CELL)+1 will round the non-integer entry in that cell up to the nearest integer.

b. Apply Adams's method to the data from the 2010 census. Then compare the results with the actual allotment of representatives after the 2010 census.

8 Construct a table showing the numbers of representatives allotted by each of the three procedures—Hamilton's, Jefferson's, and Adams's—along with the actual results from the 2010 census. Describe any patterns in the data suggesting which method would be preferred by large states. By small states. Explain your reasoning.

9 The **Huntington-Hill method of apportionment** used today was devised by mathematicians Edward Huntington and Joseph Hill. It is based on the **geometric mean** of two numbers a and b, which is equal to the square root of their product. The Huntington-Hill method works as follows:

Step 1. Calculate the exact proportional representation for each state and round those values both down and up to the nearest integer values.

Step 2. Calculate the geometric mean of the two rounded values.

Step 3. If the proportional representation for a state is greater than the geometric mean, choose the higher integer value; if it is less than or equal to the geometric mean, choose the lower integer value.

Step 4. Adjust the people-per-representative rate until a total of 435 representatives is obtained.

For example, in 2010, the proportional representation for Washington is the state population divided by the people-per-representative rate, which is:

$$6{,}753{,}369 \div \frac{309{,}183{,}463}{435} \approx \frac{6{,}753{,}369}{710{,}767} \approx 9.502$$

This rounds down to 9 and up to 10. The geometric mean of those two numbers is $\sqrt{(9)(10)} \approx 9.49$. The proportional representation of 9.502 is greater than the geometric mean of 9.49, so Washington is allotted the higher number of representatives, 10.

a. Create a spreadsheet to calculate the Huntington-Hill allotments. The syntax for the square root of a number and the logical statement IF-THEN-ELSE (if **C2** > **F2**, then **E2**, else **D2**) are shown below.

Huntington-Hill Method

	A	B	C	D	E	F	G
1	State	2010 Population	Proportional Representation	Proportional Representation Round Down	Proportional Representation Round Up	Geometric Means	Apportionment
2	Alabama	4,802,982	6.757	6	7	=SQRT(D2*E2)	=IF(C2>F2,E2,D2)
3	Alaska	721,523	1.015	1	2		
4	Arizona	6,412,700	9.022	9	10		
5	Arkansas	2,926,229	4.117	4	5		
⋮							

i. How does the formula in cell **G2** work to represent Alabama's apportionment? (You may wish to experiment with this formula in your spreadsheet in order to formulate your response.)

ii. What people-per-representative rate results in 435 representatives being assigned using the Huntington-Hill method?

b. Compare the apportionment of representatives by the Huntington-Hill method to those of the Hamilton, Jefferson, and Adams methods and to the actual values from the 2010 census. Discuss with others any patterns you see. Share your observed patterns with the rest of the class.

SUMMARIZE THE MATHEMATICS

In this investigation, you studied a variety of methods for apportionment of political power in the United States Congress. In the process, you applied some elementary mathematics, used some formal logic, and learned to use new spreadsheet commands.

a. Each of the apportionment methods you studied in this lesson can be characterized by how fractions are rounded and how adjustments are made to meet the requirement of 435 seats in the House of Representatives. Give a brief "round-and-adjust" description for each of the apportionment methods.

b. You used a spreadsheet to implement some mathematical and logical operations. Describe and explain the spreadsheet instructions you used to do the following.

  i. Round values up or down to the nearest integer.

  ii. Determine entries based on making IF-THEN-ELSE logical decisions.

c. As you have seen in other units, a spreadsheet can be a useful tool. Describe two ways you used a spreadsheet in this lesson (different from Part b) to help in your analysis and learning of apportionment methods.

d. What kinds of mathematics that you previously studied are most closely related to the apportionment methods you learned in this lesson?

Be prepared to share your ideas and thinking with the entire class.

✓ CHECK YOUR UNDERSTANDING

Consider the state in which you live. Suppose every other state's population remains the same, but your state population increases. Experiment using a spreadsheet to see what increase in your state's population would be required to increase the number of representatives by 1. What state would lose that representative? Use the present-day apportionment method (the Huntington-Hill method) to carry out this analysis. Summarize your findings.

Getting Your Fair Share

Getting your fair share is a basic expectation in a democratic society. In the last investigation, you learned about fair division in the context of apportioning seats of Congress. In this investigation, you will study two more fundamental fair division problems: dividing a cake and dividing an estate.

Each of these two problems has many applications outside the context in which it is classically studied. For example, dividing a cake relates to dividing land as in the division of Germany after World War II, or dividing a parcel of land among heirs. Dividing an estate relates to, for example, the problem of negotiating the Panama Canal treaty in 1974, reaching a fair settlement in a divorce, or any situation in which you want to fairly divide a contested set of goods.

As you complete the problems in this investigation, make note of answers to these questions:

How can you fairly divide a continuous object like a parcel of land?

How can you fairly divide a discrete object or a set of discrete objects as in an estate?

What are some criteria to use in judging fair division methods?

Cake Cutting: Fairly Dividing a Continuous (Divisible) Heterogeneous Object

A cake can be thought of as representing an object that is *continuous*, in the sense that it is *divisible*, as opposed to a *discrete* object like a painting or a family heirloom, which is *indivisible*. A cake is also *heterogeneous* since it is not uniformly the same throughout, for example there might be different flavored layers, thicknesses of frosting, or decorations. Similarly, a parcel of land that is hilly and flat with trees and ponds is also a continuous, divisible, heterogeneous object. Mathematicians, economists, and political scientists have worked hard to figure out good ways to fairly divide a cake, because it is an interesting problem that has many applications.

1 Consider the situation of dividing a cake between two people.

   a. How would you fairly divide a cake between yourself and a friend? Describe a fair division procedure. Explain why you think your procedure is fair.

   b. One common method of fair division in this case is the **cut-and-choose procedure**, whereby one person cuts the cake into two pieces that are equally desirable for the cutter, and then the other person chooses one of the two pieces. If you did not already do so in Part a, explain why this procedure could be considered fair.

You confronted the question of fairness in Lesson 1 with respect to voting. In that context, Arrow's fairness conditions (page 617) provide a good answer. The following criteria are often used to judge fairness in division procedures.

Fairness Criteria for Fair Division

- **Proportionality** If there are n people, then each person thinks he or she is getting at least $\frac{1}{n}$ of the total value.

- **Envy-Freeness** No person envies another person's portion; no person would be happier with another person's portion; no person wants to give up his or her portion in exchange for the portion someone else receives. That is, each person receives a portion he or she considers at least tied for largest or most valuable or most desirable.

- **Equitability** Each person's subjective valuation of his or her portion is the same as every other person's subjective valuation of their portion. Thus, in a two-person situation, the value that person A gives to his portion is the same as the value that person B gives to her portion.

Two additional fairness criteria are often applied: **Pareto-Optimality** (also called *efficiency*), where there is no other allocation that is better for one person and at least as good for everyone else, and **Strategy Proof**, where a person cannot misrepresent his or her valuation and assuredly do better. You will explore these two criteria in Extensions Task 19.

2 Are the fairness criteria—Proportionality, Envy-Freeness, and Equitability—met by the cut-and-choose procedure for dividing a cake between two people? Explain.

3 Think about the cut-and-choose procedure in a slightly more complicated situation. Suppose a rectangular-shaped cake is 8 inches long and frosted with chocolate frosting on one half and butterscotch frosting on the other half. You want to fairly divide the cake between you and your friend, using the cut-and-choose procedure. For convenience, assume that the cake is positioned with the 8-inch side parallel to the table edge, you are sitting at the table, and the chocolate half is on your left. Both you and your friend want to get the best piece possible, based on your own subjective valuations of the cake. Suppose that you value chocolate-frosted cake twice as much as butterscotch-frosted cake, so you want a large piece for sure, but one that includes as much of the chocolate-frosted part as possible. Your friend values both flavors just the same, so her goal is to get the largest piece possible regardless of the frosting. Neither of you knows the other's valuations in advance.

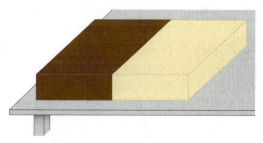

a. If you are the cutter, where would you cut the cake, with a cut perpendicular to the 8-inch side, so that each piece has the same value to you?

b. If your friend is the cutter, where would she cut the cake, with a cut perpendicular to the 8-inch side, so that she values both pieces the same?

c. For each of the situations in Parts a and b, explain how the Equitability criterion is not satisfied.

Steven J. Brams, a professor of politics at New York University, is a leading researcher in the area of fair division. In 2006, he and two colleagues, Michael A. Jones, a mathematician at Montclair State University, and Christian Klamler, an economist at the University of Graz, Austria, proposed a new two-person, cake-cutting procedure called the **surplus procedure**.

4 Using the surplus procedure, each person cuts the cake at a point where he or she thinks the cake is fairly divided. Generally, these cut points will be at different places. So, there is a surplus—the portion of the cake between the cut points. This surplus is then divided between the two people. Thus, each person receives what they think is a 50% share based on their own cut point, plus they receive some surplus. What a good deal!

Study the description of the surplus procedure below. Clarify any questions you may have with your classmates.

Surplus Procedure (SP)

A fair division procedure for dividing a cake between two people

Assumptions: Only parallel, vertical cuts, perpendicular to the horizontal axis, are made. A referee gathers valuation information independently from each person and uses that information to carry out the procedure. The steps of the procedure are then as follows.

Step 1. A mark is made, labeled point a, indicating person A's opinion of a 50-50 cut of the cake. Another mark is made, labeled point b, indicating person B's opinion of a 50-50 cut.

Step 2. If a and b coincide, then the cake is cut at $a = b$, and each person gets one of two pieces.

Step 3. If a and b are not the same point, then assume that a is to the left of b. Person A receives the piece to the left of a, and person B receives the piece to the right of b. So far, each person feels they have received a 50% share of the cake, according to their own respective valuations.

Step 4. The portion between a and b is the portion of the cake that remains, called the *surplus*. Let c be the cut point between a and b that divides this surplus so that each person receives the same proportion of the value of the surplus, relative to their respective valuations of the cake in the surplus. Then person A receives the surplus portion to the left of c, and person B receives the surplus portion to the right of c. Thus, person A receives in total everything to the left of c, and person B receives everything to the right of c.

5 Suppose two friends want to divide a cake that has fudge frosting on one half and buttercream frosting on the other half. The cake is rectangular-shaped, 10 inches long. They will divide the cake by making one vertical cut, perpendicular to the 10-inch side. A vertical line at 5 inches separates the fudge from the buttercream frosting. One friend, AJ, strongly prefers fudge frosting, while the other friend, Beth, slightly prefers buttercream frosting. To quantify these preferences, consider vertical segments of cake one-inch wide. Suppose AJ values a one-inch-wide segment of fudge-frosted cake at $4 and a one-inch-wide segment of buttercream-frosted cake at $1. Beth's valuations are $2 and $3 for fudge and buttercream, respectively. Use the surplus procedure to fairly divide the cake, as follows.

a. Find point *a* in Step 1 of the surplus procedure. That is, find the point indicating AJ's opinion of a 50–50 cut of the cake.

b. Find point *b* in Step 1 of the procedure.

c. Points *a* and *b* are different, so there is a surplus. Fairly divide the surplus. That is, find point c in Step 4 of the surplus procedure.

d. With regard to fairness criteria, explain why your division of the cake satisfies Proportionality and Envy-Freeness, but not Equitability.

6 You have studied two cake-cutting procedures—cut-and-choose, which is probably thousands of years old, and the surplus procedure, which was invented in 2006. You have determined that both procedures satisfy Proportionality and Envy-Freeness, but not Equitability. So, why use the new surplus procedure? An answer to that question is given by the inventors of the procedure in their 2006 article.

> "This procedure, which we call the *surplus procedure* (SP), gives each person at least 50% of the entire cake and generally more. By contrast, cut-and-choose limits the cutter to exactly 50% if he or she is ignorant of the chooser's preferences." (**Source:** "Better Ways to Cut a Cake," Brams, Jones, and Klamler, *Notices of the American Mathematical Society*, December, 2006.)

Explain the claims made about how much of the cake each person gets using each method.

Estate Division: Fairly Dividing a Contested Set of Goods So far, you have learned about cake-cutting fair division strategies. Now consider the second fundamental question in this investigation—how to fairly divide an estate among heirs or, more generally, how to fairly divide a contested set of goods.

In a manner similar to how you studied cake cutting, you will first learn a common method, called *Knaster's procedure*, which satisfies some of the fairness criteria, and then you will learn a newer method, called the *Adjusted Winner algorithm*, which is stronger.

Use Knaster's procedure on the following page to verify the given entries in the table and complete the missing entries.

7 In the 1940s, several mathematicians worked on the estate division problem, including Hugo Steinhaus and Bronislaw Knaster. Based on their work, a common method used today is often called **Knaster's procedure**.

Suppose you wish to fairly divide an estate among three heirs, Nora, Luis, and Pilar. The items to be divided are a house, a car, and some jewelry.

Use Knaster's procedure on the following page to verify the given entries in the table and complete the missing entries. Would you agree that this seems to provide a fair division of the estate?

	Nora	Luis	Pilar
Bid for House	$180,000	$190,000	$175,000
Bid for Car	$20,000	$22,000	$25,000
Bid for Jewelry	$20,000	$18,000	$10,000
Result of Bids (highest bid wins)		House	
Sum of Bids	$220,000		
1/n Share of Total Bids	$73,333.33		
Remaining Claim Based on Bids	$53,333.33	−$113,333.33	
Surplus from Bids		$15,000	
1/n Share of Surplus	$5,000		
Cash Value	$78,333.33		
Settlement	Jewelry + $58,333.33	House − $108,333.33	

Knaster's Procedure

A fair division procedure for allocating several items among n people

Step 1. *Bid*—Each person makes a sealed confidential bid for each item. These bids are the values each person assigns to each item.

Step 2. *Award*—Each item is awarded to the highest bidder. If there is a tie, break the tie randomly. It may happen that someone does not receive any item, or someone receives more than one item.

Step 3. *Sum of Bids*—Sum the bids for each person. This is each person's total valuation for all items.

Step 4. *Proportional Shares*—Compute the proportional share for each person based on that person's sum of bids. That is, compute $\frac{sum}{n}$ for each person.

Step 5. *Remaining Claims*—For each person, compare the proportional share from Step 4 to the value for the item(s) won in the bidding. Subtract the item(s) value from the proportional share. This is the remaining claim.

Step 6. *Surplus from Bids*—People with negative remaining claims (from Step 5) pay their claim into a temporary account. People with positive remaining claims receive payment for those claims from this temporary account. The balance of the account is the surplus.

Step 7. *Share of Surplus*—Distribute the surplus evenly to each of the n people. That is, each person receives $\frac{1}{n}$ of the surplus.

Step 8. *Cash Value*—For each person, compute the cash value of everything that person has received, both items and cash.

Step 9. *Settlement*—State the final settlement for each person, by describing the item(s) received and the cash either paid out or received.

---

8 Analyze Knaster's procedure with respect to the fairness criteria.

a. Does Knaster's procedure satisfy the Proportionality criterion? Explain.

b. Does Knaster's procedure satisfy the Equitability criterion? Explain.

c. Now think about whether Knaster's procedure satisfies proportional equitability. That is, "Does each person receive the same *proportional value*, with respect to each person's valuation?" As you answer this question, consider the following questions.

   i. Consider each person's share before the surplus from the bids is distributed. (See Step 4 of the algorithm.) At this point, what proportion of each person's valuation of the estate (sum of bids) does each person have?

   ii. Each person then receives the same cash share of the surplus. However, is each person's share of the surplus the same proportion of his or her total valuation of the estate?

    iii. Consider each person's final cash value (Step 8 of the algorithm). For each person, the final cash value is what percentage of the total valuation of the estate?

    So, does Knaster's procedure satisfy proportional equitability?

d. Do you think Knaster's procedure satisfies the Envy-Freeness criterion? Why or why not?

Knaster's procedure works fairly well, but it does not satisfy equitability. Also, when there are more than two parties, it is not envy-free. (See Extensions Task 17.) In the case of two parties, a stronger method has been developed by Steven J. Brams, a political scientist at New York University, and Alan D. Taylor, a mathematician at Union College in New York. They call their method the **Adjusted Winner (AW) method**, which they patented in 1999. The AW method is both equitable and envy-free.

9 Using the Adjusted Winner method, each person is initially given 100 points. They bid on the contested items by distributing their 100 points across all items according to their preferences. Then, as in Knaster's procedure, the highest bidder wins each item. But now the method proceeds very differently from Knaster's procedure. In the Adjusted Winner method, items are now transferred based on some proportional calculations until everyone has the same total points. To achieve the same point total for both people, one item may need to be divided. Read the step-by-step description of the Adjusted Winner method on the following page. For now, examine Step 4 and answer these questions.

Steven J. Brams and
Alan D. Taylor

a. Why is the proportion computed in Step 4 always greater than or equal to 1?

b. You could say that the item with smallest proportion in Step 4 is the item for which the two bids are closest. Explain why.

Adjusted Winner (AW) Method

A fair division method for allocating several items between two people

Step 1. *Bid*—Each person is given 100 points. Each person makes a sealed confidential bid for each item, by distributing 100 points among the items.

Step 2. *Initial allocation based on winning bids*—For each item for which one person's bid is higher than the other person's bid, award the item to the person with the higher bid. Reallocate the points so that the winner receives all of his or her points for the item awarded and the loser receives 0 points.

Step 3. *Initial allocation of tied items*—If there is a tie on an item, award the item to the person with the lowest point sum from the items awarded so far. If two or more items are tied, then allocate the tied items one at a time, each time awarding the tied item to the person with the lowest point total so far. As before, once an item is awarded, the points are reallocated so that the person to whom the item is awarded gets all of his or her points for that item and the loser receives 0 points.

Step 4. *Transfer items to achieve equal point totals*—Suppose person A has the higher point total for items awarded so far, and person B has the lower point total. Sequentially transfer items for which bids are proportionally closest from A to B until point totals are equal, as follows.

- For each item currently awarded to A, compute the following proportion:

$$\frac{\#\text{of points initially bid on the item by A}}{\#\text{of points initially bid on the item by B}}$$

- For the item with smallest such proportion, transfer this item from A to B. Find the condition below that matches the new point allocation to complete Step 4.

   (i) If this reallocation results in equal point totals, then stop.

   (ii) If after this reallocation A's point total is still higher than B's point total, then transfer the item with next smallest proportion to B. Continue in this way as long as A's total is greater than B's total.

   (iii) If after this reallocation B's point total becomes greater than A's total, you must divide the item and distribute a proportional amount of points to A and B so that their point totals become equal.

Step 5. *Settlement*—State the final settlement for each person, by showing the final point allocation, which items each person receives, and, if there is an item that must be divided, what proportion of that item each person receives.

10 Now suppose Nora and Luis must divide four items. They have agreed to use the Adjusted Winner method. Their confidential bids on the items are shown in the table below. Carry out the AW method and show the final settlement by working through Parts a–c below.

Bids (based on 100 points)

	Nora	Luis
Car	25	20
Boat	35	30
Jewelry	5	5
House	35	45

a. Find the point allocation after assigning items to the higher bidder and resolving any ties.

Point Allocation After Ties Resolved

	Nora	Luis
Car	25	0
Boat		
Jewelry		
House		
Point Total		

b. After items have been awarded based on higher bids and ties, Nora has more points than Luis (60 points compared to 50 points). To equalize the point totals, something must be transferred from Nora to Luis. Assume that the parties have agreed in advance that any item can be divided if necessary.

   i. According to Step 4 of the AW method, the boat must be divided and points shared between Nora and Luis. Explain why the boat is the item chosen in Step 4 to be divided and shared. How might the boat be divided and points shared? Compare your ideas with those of others.

   ii. Explain why simply transferring 5 points from Nora to Luis is not equitable.

   iii. Let p represent the proportion that makes the point totals equal. What does $60 - 35p$ represent? What does $50 + 30p$ represent?

   iv. What proportion of the boat should be transferred from Nora to Luis?

c. Show the final point allocation.

Final Point Allocation

	Nora	Luis
Car	25	0
Boat	29.615	
Jewelry	0	5
House	0	45
Point Total		

d. Describe the final settlement, including who receives which items and what proportion of the divided item each person receives. If the boat is sold for $142,000, how much does Nora receive? How much does Luis receive?

SUMMARIZE THE MATHEMATICS

In this investigation, you analyzed two classes of fair division problems: cake cutting and estate division.

a. Describe how to use each of the following fair division methods.

- cut-and-choose procedure

- the surplus procedure (SP)

- Knaster's procedure

- the Adjusted Winner method (AW)

b. What are some real-world situations in which the fair division methods in Part a could be used?

c. You analyzed fair division methods in terms of certain fairness criteria. In particular, you considered the criteria of Proportionality, Envy-Freeness, and Equitability.

i. Explain what is meant by each of these three criteria, and how each relates to fairness.

ii. For each of the fair division methods from Part a, which of the three fairness criteria are satisfied by the method?

Be prepared to explain your ideas and reasoning to the class.

✓CHECK YOUR UNDERSTANDING

Suppose Camille and her sister inherit their mother's diamond necklace and car. Camille bids $24,000 for the car and $13,000 for the necklace. Her sister bids $28,000 for the car and $10,000 for the necklace. Describe how they might reach a fair division of this inheritance.

APPLICATIONS

1 Another U.S. House of Representatives historical apportionment method was formulated by Daniel Webster. Webster served in the House of Representatives and the Senate during the first half of the 19th century, and he was also Secretary of State under President Millard Fillmore. His apportionment method was used in 1840, 1910, and 1930. The "Webster method," adapted to the 435-member House, works as follows:

Step 1. Find the exact proportional representation for each state using the people-per-representative rate based on 435 representatives.

Step 2. Round these calculations to the closest whole number (round up if *greater than* 0.5). This will likely give you a total number of representatives that is not 435.

Step 3. Revise the people-per-representative ratio until the total of whole numbers of representatives sums to the desired 435.

a. Using the data from the 2010 census given in the table at the beginning of this lesson, use Webster's method to determine the allocation of Congressional seats by state. What people-per-representative rate did you use?

b. Compare the apportionment from Webster's method to the actual apportionment shown in the 2010 census data table. Describe any patterns that you see.

c. Compare the apportionment from Webster's method to the apportionments from the Jefferson, Adams, and Hamilton methods. Describe differences and similarities in the results.

2 Consider again the situation from Investigation 2 Problem 5 (page 656), but this time with regard to the *cut-and-choose procedure.* Two friends want to divide a cake that has

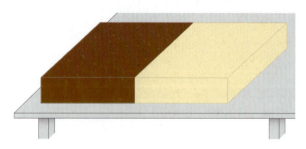

fudge frosting on one half and buttercream frosting on the other half. The cake is rectangular-shaped, 10 inches long. They will divide the cake by making one vertical cut, perpendicular to the 10-inch side. A vertical line at 5 inches separates the fudge from the buttercream frosting.

One friend, AJ, strongly prefers fudge frosting, while the other friend, Beth, slightly prefers buttercream frosting. To quantify these preferences, consider vertical segments of cake one-inch wide. AJ values a one-inch-wide segment of fudge-frosted cake at $4 and a one-inch-wide segment of buttercream-frosted cake at $1. Beth's valuations are $2 and $3 for fudge and buttercream, respectively.

a. Using the cut-and-choose procedure, where will AJ cut the cake? If Beth is the cutter, where would she cut the cake?

b. With regard to fairness criteria, explain why each division in Part a satisfies Proportionality and Envy-Freeness, but not Equitability.

3 Consider a situation similar to that in Investigation 2, Problem 3 (page 653), but this time with regard to the *surplus procedure*. An 8-inch long rectangular cake has chocolate frosting on the left half, white frosting with sprinkles on the right half, Malcolm likes chocolate frosting twice as much as the white sprinkled frosting, and Mark has no frosting preference. As you determined in Problem 3, Malcolm's mark for cutting the cake would be at $a = 3$ inches from the left edge of the cake, and Mark's mark for cutting the cake would be at $b = 4$ inches.

The key condition in Step 4 of the SP is that each person receives the same proportion of the value of the surplus relative to their respective valuations of the surplus. Find the cut point c that satisfies this key condition. Be prepared to explain your reasoning.

4 Knaster's procedure can be used to resolve a dispute over a single item between two people. Suppose Tomas and Steve enter a contest together and they win a mountain bike. They decide that one person will keep the bicycle and give the other a certain amount of cash. They submit confidential bids on the value of the bike. Tomas bids $325 and Steve bids $275. Use Knaster's procedure to decide who will keep the bike and how much cash he will give to the other person.

5 Suppose three heirs must fairly divide a condominium, a sailboat, and a classic car. They submit confidential bids as shown in the table below.

	A	B	C
Condominium	$150,000	$175,000	$130,000
Sailboat	$60,000	$70,000	$75,000
Classic Car	$100,000	$85,000	$75,000

a. Use Knaster's procedure to fairly divide these three items among the three heirs.

b. With regard to fairness criteria, explain why the division in Part a satisfies Proportionality but not Equitability.

6 Cake-cutting and estate-division are the classic contexts in which fair division may be studied, but there are many other applicable situations. For example, instead of fairly dividing contested property in an estate, we might be interested in finding a fair resolution of contested issues in a negotiation. Consider the Panama Canal treaty negotiation in 1974 between the United States and Panama. Bids, or weighted measures of importance, on ten contested issues are shown in the following table. Use the Adjusted Winner method to find a fair resolution.

	United States	Panama
U.S. Defense Rights	22	9
Use Rights	22	15
Land and Water	15	15
Expansion Rights	14	3
Duration	11	15
Expansion Routes	6	5
Compensation	4	11
Jurisdiction	2	7
U.S. Military Rights	2	7
Defense Role of Panama	2	13

Source: *Fair Division: From Cake-Cutting to Dispute Resolution*, by Steven Brams and Alan Taylor, Cambridge University Press, 1996.

CONNECTIONS

7 You rounded fractions using the *geometric mean* in the Huntington-Hill apportionment method. You are probably more accustomed to using the **arithmetic mean**, often simply called the *mean*, which is computed by adding the designated numbers and dividing the sum by the number of numbers. For example, the arithmetic mean of 200 and 300 is $\frac{200 + 300}{2} = \frac{500}{2} = 250$, while the geometric mean is $\sqrt{200 \cdot 300} = \sqrt{60,000} \approx 244.9$.

a. Examine the first three steps of the Huntington-Hill apportionment method on page 650. Rewrite those steps using arithmetic mean instead of geometric mean. Then explain how the rewritten procedure is the same as rounding to the closest whole number.

Paul Katz/Getty Images

b. If a and b are positive numbers, when is the geometric mean of a and b equal to the arithmetic mean of a and b?

c. Provide arguments justifying the following two properties of arithmetic and geometric means.

    i. The arithmetic mean of two consecutive integers is halfway between the two integers.

    ii. For any two consecutive positive integers, the geometric mean of the integers is less than their arithmetic mean.

8 Suppose a and b are any two distinct positive numbers. Without loss of generality, assume that $a > b$.

a. Explain why $\triangle PQR$ is a right triangle.

b. Based on your work in Part a, state a general relationship between the arithmetic mean and the geometric mean of any two distinct positive numbers.

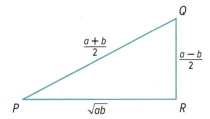

c. Using your result in Part b, explain as precisely as you can why the Huntington-Hill apportionment method favors less-populated states more than does the Webster method (Applications Task 1).

9 At the beginning of Investigation 2, you considered three fairness criteria: Proportionality (P), Envy-Freeness (EF), and Equitability (E). There are several relationships between these three criteria, as listed below.

    I. P does not imply EF.   **II.** P does not imply E.

    III. E does not imply EF.   **IV.** EF does not imply E.

a. Choose at least one of these relationships and prove it.

b. After thinking hard about the three fairness criteria, Emily claimed that there is another relationship: *P is the same as EF in two-person situations.* Study Emily's argument below. Justify or correct each statement in her argument.

> **Prove: P is the same as EF in two-person situations.**
>
> If P, then each person thinks he or she is getting at least $\frac{1}{2}$ of the total value. Thus, neither person envies the other person's portion, neither person wants to give up his or her portion in exchange for the other person's portion, each person receives a portion he or she considers at least tied for largest or most valuable or most desirable. Thus, EF.
>
>   Conversely, if EF, then each of the 2 people receives a portion he or she considers at least tied for most valuable. Since there are exactly 2 people, this must mean that each person thinks he or she is getting at least $\frac{1}{2}$ of the total value. Thus, P.

Are you convinced of the truth of Emily's claim? Explain.

ON YOUR OWN

10 In this lesson, you learned some of the history associated with apportionment methods for the U.S. House of Representatives. Write a report that includes what you have learned about apportionment and also provides a more complete and systematic historical account of apportionment methods. Useful information can be found at the following Web sites:

- www.census.gov/population/apportionment/about/history.html
- www.ams.org/samplings/feature-column/fcarc-apportion2/

11 For the Equitability fairness criterion, it is important that equitability is attained for the most appropriate measure. For example, the Adjusted Winner method may not equalize money but it does equalize points. Explain why points is the appropriate measure to be equalized.

12 The Adjusted Winner method, which applies to two-person situations, is proportional, equitable, and envy-free. If Knaster's procedure is used with two people, and if the total of the bids for each person is the same, then in this case Knaster's procedure is also proportional, equitable, and envy-free. So, for two-person fair division situations in which both methods could be applied, which should be used? Brams and Taylor, the inventors of the Adjusted Winner (AW) method, state that, "Compared with Knaster's procedure, AW is, we think, somewhat easier for the players to understand and apply." Do you agree? Explain. (**Source:** *Fair Division: From Cake-Cutting to Dispute Resolution*, by Steven Brams and Alan Taylor, Cambridge University Press, 1996, p. 111.)

13 Fair division is an issue that occurs in literature quite often. For example, consider the following story adapted from *Aesop's Fables*. What criteria of "fair" are applied in the fable?

A Lion, a Fox, and a Donkey went out hunting together. After a successful hunt, the Lion requested the Donkey to divide the food between them. The Donkey divided it all into three equal parts, and modestly begged the others to take their choice; at which the Lion, bursting with fury, sprang upon the Donkey and tore him to pieces. Then, glaring at the Fox, he bade him make a fresh division. The Fox gathered almost the whole in one great heap for the Lion's share, leaving only the smallest possible morsel for himself. "My dear friend," said the Lion, "how did you get the knack of it so well?" The Fox replied, "Me? Oh, I took a lesson from the Donkey."

14 Read the following Indian folktale about fair division. Explain Maulvi's solution to this fair division problem. Why were both men satisfied that it was a fair division? (**Sources:** Original article in *Highlights for Children*, October 2003; paraphrasing from www.edb.utexas.edu/empson/?p=47.)

> Two farmers, Ram and Shyam were eating chapatis. Ram had 3 pieces of the flat, round bread and Shyam had 5. A traveller who looked hungry and tired rode up to the two men. Ram and Shyam decided to share their chapatis with him. The 3 men stacked the 8 chapatis (like pancakes) and cut the stack into 3 equal parts. They shared the pieces equally and ate until nothing was left. The traveller, who was a nobleman, was so grateful that he gave the two farmers 8 gold coins for his share of the food.
>
> After the traveller left, Ram and Shyam wondered how they should share the 8 gold coins. Ram said that there were 8 coins and only 2 people, so each person should get an equalshare of 4 coins. "But that's not fair," said Shyam, "since I had 5 chapatis to begin with." Ram could see his point, but he didn't really want to give 5 of the coins to Shyam. So he suggested they go see Maulvi, who was very wise. Shyam agreed.
>
> Ram and Shyam told the whole story to Maulvi. After thinking for a long time, he said that the fairest way to share the coins was to give Shyam 7 coins and Ram only 1 coin. Both men were surprised. But when they asked Maulvi to explain his reasoning, they were satisfied that it was a fair division of the 8 coins.

EXTENSIONS

15 In Lesson 1 on voting, you learned about Arrow's Impossibility Theorem. Roughly, it says that when there are more than two candidates, no voting or vote-analysis method is perfectly fair in all situations. A similar result applies to apportionment methods. In 1982, Michel Balinski and H. Peyton Young published a book that includes what is known as *Balinski and Young's Impossibility Theorem*. This theorem states, roughly, that when the number of seats is fixed, there is no apportionment method that is perfectly fair in all situations. More precisely, the theorem states (similarly to Arrow's Theorem) that any apportionment method will produce paradoxes in some situations. Some of the paradoxes that can be produced are the following.

- Violation of the "Quota Rule"—The quota rule says that the allotted number of seats for a state should be either the whole number just above the exact proportional representation or the whole number just below.

- The "Alabama Paradox"—This paradox occurs if adding one more seat to the total number of seats causes a state to lose one of its seats, even though the population does not change.

- The "New States Paradox"—This paradox occurs if the addition of a new state with new seats according to its population (along with a corresponding increase in the total number of seats) changes the apportionment of seats for other states.

  a. Do some research to find out more about Balinski and Young's Impossibility Theorem and possible paradoxes. Describe the "Population Paradox."

  b. Choose one paradox and construct an example to show how the paradox can be produced by one of the apportionment methods from this lesson.

16 Suppose that in the decade between the 2010 Census and the 2020 Census, the major population trends in the U.S. are captured in the following two hypothetical newspaper headlines: "Auto Industry Makes Big Comeback" and "Retired Baby-Boomers Move West." Based on population trends indicated by these two headlines, make predictions for state populations in the 2020 Census. You should change all state populations according to systematic computations that are consistent with the headlines. When you have your predicted 2020 Census complete, use the Huntington-Hill method to find the numbers of representatives to which the states will be entitled in 2020. Then write a short report in the form of a news article commenting on the big gainers and big losers in the adjustment of political influence in the U.S. House of Representatives.

17 In Investigation 2, you learned two envy-free procedures for fairly dividing a cake between two people. Finding an envy-free, cake-cutting procedure that works for any number of people is much harder. In fact, this was a famous unsolved problem for much of the latter half of the 20th century. It was solved by Steven Brams and Alan Taylor in 1995.

From the 1960s through the 1990s, several proportional cake-cutting procedures were devised for fairly dividing a cake among three or four people, some of which are also envy-free. Several of these procedures are listed below.

Do some research on one of the procedures, or find an interesting procedure not on this list. Write a brief report about it, explaining how it works, describing some of its properties, and illustrating with an example. (One good resource is *For All Practical Purposes*, 8th Edition, by COMAP, W. H. Freeman, 2009.)

- Lone Divider Procedure (Steinhaus; 3 people; proportional)

- Last Diminisher Procedure (Banach and Knaster; 4 or more people; proportional)

- Selfridge-Conway Procedure (3 people; envy-free)

- Stromquist Moving-Knife Procedure (3 people; proportional)

- Webb Moving-Knife Procedure (3 people; envy-free)

18 In Investigation 2, after Problem 8, it was stated that Knaster's procedure is not envy-free. To see why, consider the following example. Three roommates are sharing the rent for a house with three bedrooms. The total rent is $1,500. They have agreed that their shares of the rent will depend on which room they get. They will assign rooms using Knaster's procedure. Note that in this context, exactly one item (one room) must be awarded to each person. Thus, if one person is the highest bidder on more than one room, then a new condition must be used to make awards. This condition is that rooms

shall be awarded in a way that maximizes the total winning bids. The bids, in dollars, for the rooms are shown in the following table.

	Isabel	Rayna	Ginny
Large Room	500	530	600
Smaller Corner Room	500	520	550
Small Room	500	450	350

a. Carry out Knaster's procedure to assign rooms and rent, with the modified condition stated above that each person gets one room and the total of the winning bids is maximized.

b. Explain why Ginny envies Rayna's room assignment and rent.

19 In Investigation 2, you focused on three fairness criteria—Proportionality, Envy-Freeness, and Equitability. Two other fairness criteria that are often used to judge fair division procedures are *Pareto-Optimality* and *Strategy Proof*.

a. **Strategy Proof** means that a person cannot misrepresent his or her valuation and assuredly do better. Explain why the two-person, cut-and-choose, cake-cutting procedure is strategy proof.

b. **Pareto-Optimality** (also called **efficiency**) means that there is no other allocation that is better for one person and at least as good for everyone else. In other words, any allocation that is better for one person is worse for someone else. There is some disagreement about the meaning and application of this condition. Consider the two-person, cut-and-choose, cake-cutting procedure. Try to give a plausible explanation for why this procedure satisfies Pareto-Optimality. Then try to give a plausible explanation for why this procedure does *not* satisfy Pareto-Optimality. Your explanations may use different interpretations of the Pareto-Optimality condition.

20 Write each expression in a simpler equivalent exponential form using only positive exponents.

a. $xy^{-3}(x^{-2}y)$

b. $-6x^2y(2xy^2)^2$

c. $\dfrac{5a^4b^3}{a^2b^5}$

d. $\dfrac{x^3(x^{-1})}{(2x^2)^{-3}}$

21 Charter buses to parks, museums, and other attractions have special rates for groups. Suppose that two groups paid the following for charter buses from Capital Tours:

Group A: paid $7,096 for 342 people

Group B: paid $2,295 for 85 people

a. Which group paid less per person? How much less did they pay per person?

b. If Capital Tours charges the same price per person for groups of 250 to 500 people, how much will charter buses cost for a group of 435 people?

22 Three vertices of a parallelogram are located at $P(2, 1)$, $Q(3, 6)$, and $R(10, 3)$. The fourth vertex S is also located in the first quadrant.

a. Find the coordinates of vertex S.

b. Find the lengths of each side of the parallelogram.

c. At what point do the diagonals intersect?

23 If possible, factor each expression as a product of linear factors.

a. $x^2 + 3x - 18$

b. $9 - 16x^2$

c. $x^3 + x^2 + 5x$

d. $x^2 - 10x + 24$

e. $8x^2 + 18xy - 5y^2$

f. $12t^2 + 54t + 42$

24 Steve deposits $600 in a special savings account that earns 4% interest compounded annually. He does not withdraw or deposit any additional money.

   a. Write an exponential function rule that can be used to calculate the amount of money in the account after t years.

   b. How much money will be in the account after 15 years?

   c. How long will it take Steve's money to triple in value? Explain your reasoning.

25 A random survey of 2,013 U.S. adults found that 1,046 of them had taken a multivitamin in the last year.

   (**Source:** www.naturalproductsinsider.com/news/2013/09/crn-survey-most-adults-take-supplements.aspx)

   a. What is the population of interest for this survey?

   b. What percentage were they trying to estimate?

   c. What is the point estimate of that percentage?

   d. What is the margin of error for this result?

   e. What is the 95% confidence interval for that percentage?

   f. Interpret the meaning of the 95% confidence interval in this situation.

26 Solve the following system of equations. Check your answer.

$$2x + 3y - z = 16$$
$$x + y + z = 3$$
$$x - y + 2z = -7$$

27 Shown below is the feasible region for a linear programming problem.

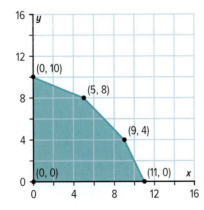

   a. Write the constraint inequalities for this feasible region.

   b. Where would you search for the optimal value of the objective function for this problem? Explain.

28 Each of the function graphs below is a transformation of the graph of $f(x) = \frac{1}{x^2}$. Describe each transformation in words and write a function rule for the image graph.

a.

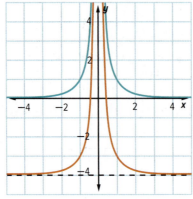

b.

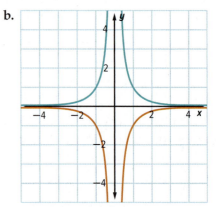

c.

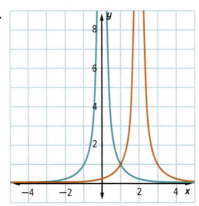

d.

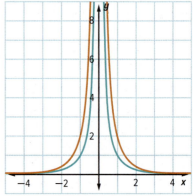

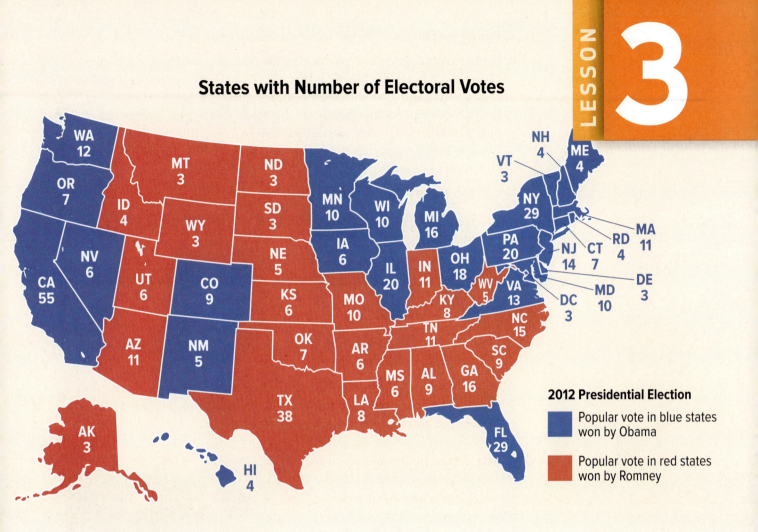

States with Number of Electoral Votes

2012 Presidential Election

■ Popular vote in blue states won by Obama

■ Popular vote in red states won by Romney

Looking Back

In this unit, you studied some mathematics of strategic decision-making in situations involving social choice, particularly in cases of voting and fair division. To help make the best decisions in these situations, you applied a mathematical analysis of ranked-choice voting, weighted voting, and fair division of both divisible objects (like a cake) and indivisible objects (like family heirlooms or seats in the U.S. House of Representatives). The tasks in this final lesson give you a chance to review and pull together the key ideas and methods of the unit.

1 Waterloo, Iowa, adopted an ordinance in 2002 requiring candidates for city office to get more than 50% of the vote. This first affected the election for Mayor in 2003. On November 4, 2003, the mayoral election results were as follows.

Candidate	Votes
Miller	4,554
Hurley	3,885
Rooff	3,421
Abebe	1,547

Source: www.ourcampaigns.com/RaceDetail.html?RaceID=674698

a. Who is the plurality winner? Is there a majority winner?

b. In the runoff election four weeks after the first election, Hurley defeated Miller and was elected Mayor of Waterloo. Consider a hypothetical ranked-choice voting situation with the top three candidates, Miller, Hurley, and Rooff. Newspaper accounts indicate that Rooff and Hurley had similar policy positions that were markedly different from Miller's positions. Assume Miller supporters are evenly divided in their opinions of Hurley and Rooff. A possible preference table is given below.

Top Three Mayoral Candidate Preferences

	Rankings			
Miller	1	1	3	3
Hurley	2	3	2	1
Rooff	3	2	1	2
Number of Voters	2,277 voters	2,277 voters	3,421 voters	3,885 voters

Explain why this preference table could be used as a possible representation of the election description above.

c. Do you think that the new ordinance in Waterloo, which required a runoff for this election instead of a plurality winner, resulted in a fairer outcome? Justify your answer by analyzing the preference table.

2 As you learned in this unit, there are several different methods for fairly dividing goods between people. Some methods are designed to work for just two people, like the Adjusted Winner method, while other methods work for any number of people, like Knaster's procedure. You may be able to apply more than one method in a given situation.

a. The table below shows the confidential point bids from two heirs for five items in an estate. Use the Adjusted Winner method to find a fair allocation of the items to the two heirs.

Bids (based on 100 points)

	Alanna	Bilal
Car	20	15
Pickup Truck	10	20
Jewelry	5	5
House	45	35
Summer Cabin	20	25

b. Suppose that instead of making a points bid, Alanna and Bilal submit confidential bids for the cash value of each item, as shown in the table below. Use Knaster's procedure to fairly divide the items.

Bids (cash value in dollars)

	Alanna	Bilal
Bid for Car	20,000	15,000
Bid for Truck	10,000	20,000
Bid for Jewelry	15,000	15,000
Bid for House	450,000	350,000
Bid for Cabin	40,000	50,000

SUMMARIZE THE MATHEMATICS

There are many situations in politics, business, and everyday life in which you need to make decisions that are optimal in specific ways. In this unit, you have learned some mathematical concepts and methods that help you make good decisions. You have mathematically analyzed voting and fair division. In particular, you have studied the following topics.

1. Ranked-choice voting and related vote-analysis methods

2. Arrow's theorem

3. Weighted voting

4. Political apportionment (methods of apportioning seats in Congress)

5. Fair division of divisible and indivisible objects

Choose at least one of the topics listed above. Working with a partner, prepare a written and oral report on the topic using the guidelines below.

- Give a brief non-technical overview of the topic.

- Outline the sub-topics or details that you learned for your chosen main topic.

- Develop a thoroughly worked-out example illustrating one detail or sub-topic for your chosen main topic.

Be prepared to present your report to the class and respond to questions.

✓ CHECK YOUR UNDERSTANDING

Write, in outline form, a summary of the important mathematical concepts and methods developed in this unit. Organize your summary so that it can be used as a quick reference in your future work.

English

A

Español

Absolute risk (p. 3) The proportion or percentage of people in the group of interest for whom an undesirable event occurs.

Riesgo absoluto (pág. 3) Proporción o porcentaje de personas en el grupo de interés para quienes ocurre un suceso indeseado.

Absolute risk reduction (p. 9) The difference in the absolute risk for two groups.

Reducción del riesgo absoluto (pág. 9) Diferencia entre el riesgo absoluto de dos grupos.

Absolute value function (p. 146)

$$f(x) = |x| = \begin{cases} x, & \text{for } x \geq 0 \\ -x, & \text{for } x < 0 \end{cases}$$

Función valor absoluto (pág. 146)

$$f(x) = |x| = \begin{cases} x, & \text{por } x \geq 0 \\ -x, & \text{por } x < 0 \end{cases}$$

Add mod n (p. 463) Add integers as usual as real numbers, then reduce mod n.

Adición mod n (pág. 463) Sumar enteros al igual que se suman números reales, luego reducir a mod n.

Add two binary code words (p. 496) Add bit-by-bit using mod 2 arithmetic.

Sumar dos palabras de código binario (pág. 496) Sumar bit por bit usando aritmética mod 2.

Addition Principle of Counting (p. 210) The total number of outcomes from two tasks is the sum of the number of outcomes from each task minus the number of outcomes that are common to both tasks, if there are any.

Principio de conteo de la adición (pág. 210) El número total de resultados de dos tareas es la suma del número de los resultados de cada tarea menos el número de resultados que son comunes a ambas tareas, si los hay.

Addition Rule for Mutually Exclusive Events (p. 344) If A and B are mutually exclusive events, the probability that one or the other occurs is $P(A \text{ or } B) = P(A) + P(B)$.

Regla de la adición para sucesos mutuamente exclusivos (pág. 344) Si A y B son sucesos mutuamente exclusivos, la probabilidad de que ocurra uno o el otro es $P(A \text{ o } B) = P(A) + P(B)$.

Ammortization (p. 313) The process of repaying a loan.

Amortización (pág. 313) Proceso de reembolsar un préstamo.

Amplitude of sine and cosine functions (p. 118) One-half the difference of the (*maximum value* − *minimum value*) of the function in one cycle of its graph.

Amplitud de las funciones seno y coseno (pág. 118) Mitad de la diferencia del (*valor máximo* − *valor mínimo*) de la función en un ciclo de su gráfica.

Anecdotal evidence (p. 14) Evidence based on observation of a few non-randomly-selected cases.

Evidencia anecdótica (pág. 14) Evidencia basada en la observación de unos pocos casos no elegidos al azar.

Annual percentage rate or **APR** (p. 272) Stated rate of annual interest on which any compounding is based.

Tasa de porcentaje anual o **TPA** (pág. 272) Tasa nominal de interés anual en la cual se basa cualquier interés compuesto.

Annual percentage yield or **APY** (p. 274) The effective annual interest rate on an investment, taking into account the effect of *compounding interest*.

Rendimiento anual porcentual o **RAP** (pág. 274) Tasa anual efectiva de interés de una inversión teniendo en cuenta el efecto del *interés compuesto*.

Annuity (p. 292) A financial plan involving making an initial deposit and then equal deposits of the same amount at regular intervals.

Anualidad (pág. 292) Plan financiero en el que se realiza un depósito inicial y luego depósitos iguales de la misma cantidad a intervalos regulares.

Apportionment methods (p. 646) Mathematical methods used to apportion seats in the U.S. House of Representatives based on state populations.

Métodos de distribución (pág. 646) Métodos matemáticos que se usan para distribuir los escaños en la Cámara de Representantes de los Estados Unidos, basándose en las poblaciones de los estados.

English

Arithmetic sequence (p. 136) A sequence of numbers in which the difference between any two consecutive terms is a fixed nonzero constant d. Symbolically, $a_n - a_{n-1} = d$. Recursively defined by $a_n = a_{n-1} + d$.

Arrow's Impossibility Theorem (p. 616) A theorem about fair voting, proved by Kenneth Arrow in 1949, in which Arrow proposed six fairness criteria and then proved that no voting method can always satisfy all six conditions when there are more than two candidates.

Banzhaf power index (BPI) (p. 622) A method used in weighted voting situations to measure the power of a person's vote (as opposed to the weight, which is the number of votes the person has), by counting the number of winning coalitions in which the person's vote is critical. A person's vote is critical if by leaving a winning coalition the coalition no longer has enough votes to win. In particular, the Banzhaf power index (BPI) for a voter is the ratio:

$$\frac{\textit{number of critical votes for the voter}}{\textit{total number of critical votes for all voters}}$$

Base 10 (common) logarithmic function (p. 34) A function with rule of correspondence that can be expressed in the form $f(x) = \log_{10} x$, where $\log_{10} x = y$ if and only if $10^y = x$.

Base e (natural) logarithmic function (p. 294) A function with rule of correspondence that can be expressed in the form $f(x) = \log_e x$, where $\log_e x = y$ if and only if $e^y = x$.

Bias (p. 380) A method of estimating a population parameter that makes the estimate from repeated sampling too large or too small, on average.

Binary code (p. 487) A code that translates data into strings of 0s and 1s.

Binomial distribution (p. 342) A probability distribution where the outcomes are the possible numbers of successes in a fixed number of binomial trials and the probability of each outcome is given by the binomial probability formula.

Español

Sucesión aritmética (pág. 136) Serie de números en la cual la diferencia entre dos términos consecutivos cualesquiera es una constante d fija distinta de cero. Simbólicamente, $a_n - a_{n-1} = d$. De otra forma, como $a_n = a_{n-1} + d$.

Teorema de la imposibilidad de Arrow (pág. 616) Teorema sobre la votación imparcial, demostrado por Kenneth Arrow en 1949 en el cual se proponían seis criterios de imparcialidad y luego se demostraba que ningún método de votación podía satisfacer siempre todas las seis condiciones cuando hay más de dos candidatos.

B

Índice de poder de Banzhaf (IPB) (pág. 622) Método que se usa en situaciones de votación ponderadas para medir el poder del voto de una persona (a diferencia del peso, que es el número de votos que tiene la persona), contando el número de coaliciones ganadoras en las cuales el voto de la persona es crítico. El voto de una persona es crítico si al dejar una coalición ganadora, dicha coalición deja de tener suficientes votos para ganar. Particularmente, el índice de poder de Banzhaf (IPB) para un votante es la proporción:

$$\frac{\textit{número de votos críticos para el votante}}{\textit{número total de votos críticos para todos los votantes}}$$

Función logarítmica de base 10 (común) (pág. 34) Función con regla de correspondencia que se puede expresar de la forma $f(x) = \log_{10} x$, donde $\log_{10} x = y$ si y solo si $10^y = x$.

Función logarítmica de base e (natural) (pág. 294) Función con regla de correspondencia que se puede expresar de la forma $f(x) = \log_e x$, donde $\log_e x = y$ si y solo si $e^y = x$.

Sesgo (pág. 380) Método para estimar el parámetro de una población que hace que la estimación a partir del muestreo repetitivo sea, en promedio, muy grande o muy pequeña.

Código binario (pág. 487) Código que convierte datos en series de 0 y 1.

Distribución binomial (pág. 342) Distribución probabilística donde los resultados son los números posibles de éxitos en un número fijo de ensayos binomiales y la probabilidad de cada resultado viene dada por la fórmula de la probabilidad binomial.

English

Binomial probability formula (p. 347) The probability, $P(x)$, of getting exactly x successes in n binomial trials with probability of success p on each trial is $P(x) = C(n, x)p^x(1 - p)^{n-x}$.

Binomial situation (p. 341) A chance situation involving binomial trials.

Binomial Theorem (p. 250) For any real numbers a and b and positive integer n, $(a + b)^n = C(n, 0)a^n + C(n, 1)a^{n-1}b + C(n, 2)a^{n-2}b^2 + \cdots + C(n, k)a^{n-k}b^k + \cdots + C(n, n-2)a^2b^{n-2} + C(n, n-1)ab^{n-1} + C(n, n)b^n$.

C

Cartesian product of two sets (p. 447) The Cartesian product of two sets A and B is the set of all ordered pairs that have first coordinate from A and second coordinate from B, denoted $A \times B$. That is, $A \times B = \{(x, y) \mid x \in A \text{ and } y \in B\}$.

Categorical data (or, **categorical variable**) (p. 3) Data that can be summarized by counting the number of individuals in the group of interest that fall into each of a set of disjoint (mutually exclusive) categories.

Census (p. 376) A study that collects information from every individual in a population.

Chance error (p. 380) See *Sampling error*.

Check digits (p. 486) The digits in an ID number for an object that are used to help detect and correct errors that may occur when the ID number is read by or stored in a computer or is electronically transmitted.

Chi-square statistic χ^2 (p. 44) A measure of how far two groups are from being homogeneous, or a measure of how far two variables counted on a single group are from being independent:

$$\chi^2 = \sum \frac{(O - E)^2}{E}$$

Español

Fórmula de la probabilidad binomial (pág. 347) La probabilidad, $P(x)$, de obtener exactamente x resultados favorables en n ensayos binomiales con probabilidad de éxito p en cada ensayo es $P(x) = C(n, x)p^x(1 - p)^{n-x}$.

Situación binomial (pág. 341) Situación aleatoria que involucra ensayos binomiales.

Teorema binomial (pág. 250) Para todo número real a y b y un entero positivo n, $(a + b)^n = C(n, 0)a^n + C(n, 1)a^{n-1}b + C(n, 2)a^{n-2}b^2 + \cdots + C(n, k)a^{n-k}b^k + \cdots + C(n, n-2)a^2b^{n-2} + C(n, n-1)ab^{n-1} + C(n, n)b^n$.

Producto cartesiano de dos conjuntos (pág. 447) El producto cartesiano de dos conjuntos A y B es el conjunto de todos los pares ordenados que tienen la primera coordenada de A y la segunda coordenada de B, denotado como $A \times B$. Es decir, $A \times B = \{(x, y) \mid x \in A \text{ y } y \in B\}$.

Datos categóricos (o **variable categórica**) (pág. 3) Datos que se pueden resumir contando el número de individuos en el grupo de interés que pertenecen en cada uno de los conjuntos de categorías disjuntas (mutuamente exclusivas).

Censo (pág. 376) Estudio que reúne información sobre cada individuo en una población.

Error aleatorio (pág. 380) Ver *Error de muestreo*.

Dígitos de control (pág. 486) Dígitos en el número de identificación de un objeto que se usan para detectar y corregir errores que pueden ocurrir cuando dicho número es leído o almacenado en una computadora o transmitido electrónicamente.

Estadístico chi-cuadrado χ^2 (pág. 44) Medida de la distancia a la que están dos grupos de ser homogéneos o medida de la distancia a la que están dos variables, contadas en un solo grupo, de ser independientes.

$$\chi^2 = \sum \frac{(O - E)^2}{E}$$

English

Chi-square test of homogeneity (p. 49) A statistical test that uses the size of the chi-square statistic computed using two random samples classified on the same categorical variable to decide whether it is plausible that the two samples were taken from populations that are homogeneous or whether the difference in the samples is statistically significant.

Chi-square test of independence (p. 81) A statistical test that uses the size of the chi-square statistic computed from a random sample classified on two categorical variables to decide whether it is plausible that the two variables are independent in the population from which the random sample was taken.

Circle (p. 166) The set of points in a plane that are a fixed distance r, called the *radius*, from a given point O, called the *center*. A circle with center at the origin and radius r can be represented by an equation of the form $x^2 + y^2 = r^2$, where r is a positive number.

Circular functions (p. 68) Functions (sine and cosine) pairing points on the real number line with coordinates of points on a unit circle by "wrapping" the real number line around the unit circle—non-negative real axis is wrapped *counterclockwise* starting at $(1, 0)$; negative real axis is wrapped in *clockwise* direction. For any real number r, if the wrapping procedure pairs r with (x, y), then $\cos r = x$ and $\sin r = y$.

Combination (p. 224) An arrangement of objects or numbers in which order does not matter, that is, different orderings are not counted as different possibilities, and repetitions are not allowed. Equivalently, a combination is a subset of a set.

Common logarithm (p. 34) If $10^x = y$, then x is called the base 10 logarithm of y; it is often denoted $x = \log y$.

Complement of a set (p. 441) The complement of set A is the set of all elements in the universal set that are not in A, denoted A'. In terms of set difference, $A' = U - A$.

Español

Prueba de homogeneidad de chi-cuadrado (pág. 49) Prueba estadística que usa el tamaño del estadístico chi-cuadrado calculado usando dos muestras aleatorias clasificadas en la misma variable categórica para decidir si es verosímil que las dos muestras se hayan tomado de poblaciones homogéneas o si la diferencia en las muestras es estadísticamente significativa.

Prueba de independencia de chi-cuadrado (pág. 81) Prueba estadística que usa el tamaño del estadístico chi-cuadrado calculado a partir de una muestra aleatoria clasificada en dos variables categóricas para decidir si es verosímil que las dos variables sean independientes en la población de la cual se tomó la muestra aleatoria.

Círculo (pág. 166) Conjunto de puntos en un plano que están a una distancia fija r, llamada *radio*, de un punto dado O, llamado *centro*. Un círculo con centro en el origen y radio r se puede representar por una ecuación de la forma $x^2 + y^2 = r^2$, donde r es un número positivo.

Funciones circulares (pág. 68) Funciones (seno y coseno) que relacionan puntos en la recta numérica real con coordenadas de puntos en un círculo unitario "envolviendo" la recta numérica real alrededor del círculo unitario, el eje real no negativo está envuelto *en sentido contrario a las manecillas del reloj* comenzado en $(1, 0)$; el eje real negativo está envuelto *en el sentido de las manecillas del reloj*. Para cualquier número real r, si el procedimiento de envoltura relaciona a r con (x, y), entonces $\cos r = x$ y sen $r = y$.

Combinación (pág. 224) Arreglo de objetos o números en que el orden no es importante. Es decir, las disposiciones diferentes no cuentan como posibilidades diferentes y no se permiten repeticiones. Equivalentemente, una combinación es un subconjunto de un conjunto.

Logaritmo común (pág. 34) Si $10^x = y$, entonces x es el logaritmo en base 10 de y; a menudo se denota como $x = \log y$.

Complemento de un conjunto (pág. 441) El complemento del conjunto A es el conjunto de todos los elementos en el conjunto universal que no están en A, denotado como A'. En términos de la diferencia de conjuntos, $A' = U - A$.

English

Compound interest (p. 272) Interest that is applied to previous interest as well as to the original amount of money borrowed or invested.

Conditional probability (pp. 93, 254) The probability that event A happens, when it is known that event B happens.

Confidence interval (95%) (p. 411) An interval constructed using the estimate from a random sample by a method that gives us 95% confidence that the population parameter lies within the interval. That is, if we construct 95% confidence intervals using 100 different random samples, the expected number of intervals that contain the population parameter is 95.

Congruent figures Two figures are congruent if and only if one figure is the image of the other under a rigid transformation (line reflection, translation, rotation, glide reflection) or a composite of rigid transformations. Congruent figures have the same shape and size, regardless of position or orientation.

Constraint (p. 577) A limitation on values that variables may assume in a problem situation. For example, the linear constraint $3x + 5y < 21$ expresses a condition for acceptable combinations of values for the variables x and y.

Contour diagram (p. 528) By analogy with a map, a diagram that illustrates how the value of one quantity, such as elevation, varies with changes in the values of two variables, such as those representing position.

Control group (p. 14) In an experiment, a group that does not get the treatment of interest, so that its response can be compared to that of a group receiving the treatment of interest. May receive a placebo or an older treatment.

Convenience sampling (p. 383) Using a readily available group as the sample. Often results in sample selection bias.

Converse of an if-then statement Reverses the order of the hypothesis and conclusion of the if-then statement. Given the statement $p \Rightarrow q$, its converse is $q \Rightarrow p$.

Español

Interés compuesto (pág. 272) Interés que se aplica al interés anterior, así como también al monto inicial de dinero prestado o invertido.

Probabilidad condicional (pág. 93, 254) Probabilidad de que suceda el suceso A cuando se sabe que sucede el suceso B.

Intervalo de confianza (95%) (pág. 411) Intervalo construido con la estimación de una muestra aleatoria por un método que nos proporciona 95% de confianza de que los parámetros de la población yacen dentro del intervalo. Es decir, si construimos intervalos de confianza de 95% usando 100 muestras aleatorias diferentes, el número esperado de intervalos que contienen el parámetro de la población es 95.

Figuras congruentes Dos figuras son congruentes si y solo si una figura es la imagen de la otra bajo una transformación rígida (reflexión sobre una recta, traslación, rotación, deslizamiento, reflexión) o una composición de transformaciones rígidas. Las figuras congruentes tienen la misma forma y tamaño, sin importar la posición u orientación.

Restricción (pág. 577) Limitación en los valores que una variable puede asumir en un problema. Por ejemplo, la restricción lineal $3x + 5y < 21$ expresa una condición sobre las combinaciones de valores aceptables para las variables x y y.

Diagrama de curvas de nivel (pág. 528) Por analogía con un mapa, un diagrama que muestra cómo el valor de una cantidad, como la altitud, varía con los cambios en los valores de dos variables, como las que representan la posición.

Grupo control (pág. 14) En un experimento, un grupo que no recibe el tratamiento de interés, de manera que sus respuestas se puedan comparar con las de un grupo que sí recibe el tratamiento de interés. Puede recibir un placebo o un tratamiento antiguo.

Muestreo a conveniencia (pág. 383) Usar un grupo fácilmente accesible como muestra. A menudo resulta en sesgo en la selección de la muestra.

Recíproco de un enunciado "si-entonces" Invierte el orden de la hipótesis y la conclusión del enunciado "si-entonces". Dado el enunciado $p \Rightarrow q$, su recíproco es $q \Rightarrow p$.

English

Critical value (p. 49) The largest value of χ^2 for which it is plausible, in a test of homogeneity, that the populations are homogeneous or, in a test of independence, that the variables are independent in the population.

Cross section (p. 537) The intersection of a plane with a three-dimensional object.

Cryptography (p. 457) The study and application of mathematical techniques to keep information secure.

Cryptosystem (p. 457) A system based on cryptography used to securely send confidential information by encrypting and decrypting the information using keys.

D

Data compression (p. 511) The process of recoding source information using fewer bits than used originally, while still maintaining accuracy.

Determinant of a 2 × 2 matrix (p. 480) The determinant of the 2 × 2 matrix $E = \begin{bmatrix} a & b \\ c & d \end{bmatrix}$, denoted $\det(E)$ or $|E|$, is $ad - bc$.

Digital signature (p. 473) An electronic method for authenticating data, analogous to signing a document with pen and paper, often done using a public-key cryptosystem.

Disjoint events (p. 344) See *Mutually exclusive events.*

Disjoint sets (p. 443) Sets that do not share any elements, that is, their intersection is empty.

Distance formula in three dimensions (p. 554) Formula for calculating the distance between two points (x_1, y_1, z_1) and (x_2, y_2, z_2)

Distribution A collection of values, typically summarized in a table or plotted so that the number or proportion of times that each value occurs can be observed.

Double blind (p. 15) An experiment where neither the subject nor the person who evaluates how well the treatment works knows which treatment the subject received.

Español

Valor crítico (pág. 49) El mayor valor de χ^2 para el cual es verosímil que las poblaciones sean homogéneas, en una prueba de homogeneidad o, en una prueba de independencia, que las variables sean independientes en la población.

Sección transversal (pág. 537) Intersección de un plano con un objeto tridimensional.

Criptografía (pág. 457) Estudio y aplicación de técnicas matemáticas para mantener la información segura.

Sistema criptográfico [Criptosistema] (pág. 457) Sistema basado en la criptografía usado para enviar de manera segura información confidencial, cifrando y descifrando la información con el uso de claves.

Compresión de datos (pág. 511) Proceso de recodificar información original usando menos bits que los que se usaron originalmente mientras se mantiene la fidelidad.

Determinante de una matriz 2 × 2 (pág. 480) El determinante de la matriz 2 × 2 $E = \begin{bmatrix} a & b \\ c & d \end{bmatrix}$, denotado $\det(E)$ o $|E|$, es $ad - bc$.

Firma digital (pág. 473) Método electrónico para autenticar datos, análogo a firmar un documento con bolígrafo y papel. A menudo se hace con un sistema de encriptado de clave pública.

Sucesos disjuntos (pág. 344) Ver *Sucesos mutuamente exclusivos.*

Conjuntos disjuntos (pág. 443) Conjuntos que no comparten elementos, es decir, su intersección es vacía.

Fórmula de distancia en tres dimensiones (pág. 554) Fórmula para calcular la distancia entre dos puntos (x_1, y_1, z_1) y (x_2, y_2, z_2)

Distribución Colección de valores, normalmente resumidos en una tabla o graficados, de manera que se pueda observar el número o la proporción de veces que ocurre cada valor.

Doble ciego (pág. 15) Experimento en el cual ni los sujetos ni la persona que evalúa cómo funciona el tratamiento saben qué tratamiento recibió el sujeto.

English		Español

Doubling time for investment (p. 273) The time it takes for an investment to double in value.

Tiempo para duplicar una inversión (pág. 273) Tiempo que le toma a una inversión duplicar su valor.

E

---

e (p. 283) *e* is an irrational number whose value is approximately 2.71828. It is the limit of $\left(1+\frac{1}{n}\right)^n$ as n becomes very large. Formally, $\lim_{n\to\infty}\left(1+\frac{1}{n}\right)^n = e$.

e (pág. 283) *e* es un número irracional cuyo valor es aproximadamente 2,71828. Es el límite de $\left(1+\frac{1}{n}\right)^n$ a medida que n tiende a ser muy grande. Formalmente, $\lim_{n\to\infty}\left(1+\frac{1}{n}\right)^n = e$.

Element of a set (p. 432) An individual object in a set.

Elemento de un conjunto (pág. 432) Objeto individual en un conjunto.

Equivalent mod *n* (p. 462) Two integers, *a* and *b*, are equivalent mod *n* if and only if they have the same remainder upon division by *n*, denoted $a \equiv b$ mod *n*.

Equivalentes mod *n* (pág. 462) Dos enteros, *a* y *b*, son equivalentes mod *n* si y solo si tienen el mismo residuo cuando se dividen entre *n*; se denota como $a \equiv b$ mod *n*.

Error attributable to sampling (p. 401) See *Sampling error*.

Error atribuible al muestreo (pág. 401) Ver *Error de muestreo*.

Estimate (or, **point estimate**) (pp. 377, 400) A numerical characteristic of a sample that is used to estimate the population parameter.

Estimación (o **estimación puntual**) (pág. 377, 400) Característica numérica de una muestra que se usa para estimar el parámetro de la población.

Euler's Theorem (special case) (p. 471) $(M^e)^d \equiv M$ mod *n*, where $n = pq$, *p* and *q* are prime numbers, *e* and *d* are multiplicative inverses mod *r*, with $r = (p-1)(q-1)$.

Teorema de Euler (caso especial) (pág. 471) $(M^e)^d \equiv M$ mod *n*, donde $n = pq$, *p* y *q* son números primos, *e* y *d* son inversos multiplicativos mod *r*, con $r = (p-1)(q-1)$.

Expected frequency (or, **expected count**) *F* (p. 43) For a chi-square test of homogeneity, the number that would be in the cell of a frequency table if the groups were homogeneous and the marginal totals were unchanged. For a chi-square test of independence, the number that would be in a cell if the variables were independent and the marginal totals were unchanged. In both cases, computed using:

$$\frac{(row\ total)(column\ total)}{(grand\ total)}$$

Frecuencia esperada (o **conteo esperado**) *F* (pág. 43) En una prueba de homogeneidad de chi-cuadrado, el número que estaría en la casilla de una tabla de frecuencia si los grupos fuesen homogéneos y los totales marginales inalterados. En una prueba de independencia de chi-cuadrado, el número que estaría en una casilla si las variables fuesen independientes y los totales marginales inalterados. En ambos casos se calculan con:

$$\frac{(total\ de\ la\ fila)(total\ de\ la\ columna)}{(gran\ total)}$$

Expected number (or, **expected value**) (p. 39) The product of the size of the sample and the probability or proportion of times the specified outcome occurs overall.

Número esperado (o **valor esperado**) (pág. 39) Producto del tamaño de la muestra por la probabilidad o proporción de veces que en general ocurre el resultado particular.

Expected value of a probability distribution (p. 351) The mean μ of a probability distribution. For a binomial distribution, the product of the number of trials *n* and the probability *p* of a success on each trial: $\mu = np$.

Valor esperado de una distribución de probabilidad (pág. 351) La media μ de una distribución de probabilidad. Para una distribución binomial el producto del número de ensayos *n* por la probabilidad *p* de éxito en cada ensayo: $\mu = np$.

English	Español

F

Experiment (p. 14) A study designed to compare the difference in the effect of two or more treatments, which are randomly assigned to an available group of subjects.

Explanatory variable (p. 5) The groups, conditions, or treatments to be compared in an experiment or observational study.

Exponential function (p. 124) A function with rule of correspondence that can be expressed in the algebraic form $f(x) = a(b^x)$.

Experimento (pág. 14) Estudio diseñado para comparar la diferencia del efecto de dos o más tratamientos asignados aleatoriamente a un grupo de sujetos disponibles.

Variable explicativa (pág. 5) Grupos, condiciones o tratamientos a ser comparados en un experimento o en un estudio por observación.

Función exponencial (pág. 124) Función con regla de correspondencia que se puede expresar en la forma algebraica $f(x) = a(b^x)$.

Factorial notation (p. 224) A compact way of writing certain products of consecutive non-negative integers; specifically, when n is a positive integer, "n factorial" is written $n!$ and is computed as $n! = n \times (n-1) \times \cdots \times 2 \times 1$. For convenience, $0!$ is defined to be 1.

Fair division procedures for cutting a cake and dividing an estate (pp. 652, 655) Mathematical procedures used for fair division when the object to be divided is continuous, divisible, and heterogeneous, like a cake, or discrete, indivisible, and non-identical, like pieces of furniture in an estate.

False negative (p. 72) A negative result on a screening test when the person does have the condition.

False positive (p. 72) A positive result on a screening test when the person does not have the condition.

Feasible region (p. 578) In a linear programming problem, the feasible region is the set of all points whose coordinates satisfy all given constraints.

Fermat's Little Theorem (p. 478) If p is a prime number and a is any integer, then $a^p \equiv a \bmod p$.

Frequency A count of the number in the group that fall into a given category.

Function A relationship between two variables in which each value of the independent variable x corresponds to exactly one value of the dependent variable y. The notation $y = f(x)$ is often used to denote that y is a function of x.

Notación factorial (pág. 224) Forma compacta de escribir ciertos productos de enteros consecutivos distintos de cero; específicamente, cuando n es un entero positivo, "n factorial" se escribe como $n!$ y se calcula como $n! = n \times (n-1) \times \cdots \times 2 \times 1$. Por conveniencia, $0!$ es igual a 1.

Procedimientos de división justa para cortar un pastel o dividir una propiedad (pág. 652, 655) Procedimientos matemáticos que se usan para divisiones justas cuando el objeto a dividir es continuo, divisible y heterogéneo, como un pastel, o discreto, indivisible y no idéntico como los muebles de una casa.

Falso negativo (pág. 72) Resultado negativo a un examen de detección cuando la persona sí padece de la condición.

Falso positivo (pág. 72) Resultado positivo a un examen de detección cuando la persona no padece de la condición.

Región factible (pág. 578) En problemas de programación lineal, la región factible es el conjunto de todos los puntos cuyas coordenadas satisfacen todas las restricciones dadas.

Pequeño teorema de Fermat (pág. 478) Si p es un número primo y a es cualquier entero, entonces $a^p \equiv a \bmod p$.

Frecuencia Conteo del número en el grupo que yace en una categoría dada.

Función Relación entre dos variables donde a cada valor de la variable independiente x le corresponde exactamente un valor de la variable dependiente y. La notación $y = f(x)$ se usa mucho para expresar que y es una función de x.

English	Español

Future value of an investment, _F_ (pp. 278, 282)
Can be calculated using the formula

$$F = I\left(1 + \frac{r}{n}\right)^{nt} + A\left(\frac{\left(1 + \frac{r}{n}\right)^{nt} - 1}{\frac{r}{n}}\right),$$ where _I_ represents

the initial deposit amount, _A_ represents the fixed regular deposit amount, _r_ represents the annual interest rate (APR, expressed as a decimal), _n_ represents the number of investment (compounding) periods per year, and _t_ represents the number of years.

Valor futuro de una inversión, _F_ (pág. 278, 282)
Se puede calcular con la fórmula

$$F = I\left(1 + \frac{r}{n}\right)^{nt} + A\left(\frac{\left(1 + \frac{r}{n}\right)^{nt} - 1}{\frac{r}{n}}\right),$$ donde _I_ representa

la cantidad inicial depositada, _A_ representa la cantidad fija que se deposita, _r_ representa la tasa de interés anual (expresada en forma decimal), _n_ representa el número de períodos (compuestos) por año y _t_ representa el número de años.

G

Geometric mean (p. 650) The geometric mean of two positive numbers _a_ and _b_ is the positive number _x_ such that $\frac{a}{x} = \frac{x}{b}$ or $x = \sqrt{ab}$.

Media geométrica (pág. 650) La media geométrica de dos números positivos _a_ y _b_ es el número positivo _x_ tal que $\frac{a}{x} = \frac{x}{b}$ o $x = \sqrt{ab}$.

Geometric sequence (p. 136) A sequence of numbers in which the ratio of any two consecutive terms is a fixed nonzero constant _r_. Symbolically, $a_n = r \cdot a_{n-1}$.

Sucesión geométrica (pág. 136) Sucesión de números donde la razón de cualesquiera dos términos consecutivos es una constante _r_ distinta de cero. Simbólicamente, $a_n = r \cdot a_{n-1}$.

H

Half-turn symmetry In three dimensions, a figure has half-turn symmetry if there is a line (called the _axis of symmetry_) about which the figure can be turned 180° in such a way that the rotated figure appears to be in exactly the same position as the original figure.

Simetría de media vuelta En tres dimensiones, una figura tiene simetría de media vuelta si hay una recta (llamada _eje de simetría_) sobre la cual la figura se puede girar 180° de tal manera que la figura girada pareciera estar exactamente en la misma posición que la figura original.

Hamming distance (p. 495) The number of bit places in which two binary code words differ.

Distancia de Hamming (pág. 495) Número de posiciones de bits en el cual difieren dos palabras en código binario.

Homogeneous groups or populations (p. 37) When two groups are sorted into the same categories, the proportion of the first group that falls into the first category is equal to the proportion of the second group that falls into that same category. And so on for all of the categories.

Poblaciones o grupos homogéneos (pág. 37) Cuando dos grupos se clasifican en las mismas categorías, la proporción del primer grupo que cae en la primera categoría es igual a la proporción del segundo grupo que cae en esa misma categoría. Y así sucesivamente para todas las categorías.

Horizontal asymptote (p. 125) A function _f_ has a horizontal asymptote at _y_ = _b_ if and only if _f_(_x_) approaches _b_ as |_x_| becomes very large.

Asíntota horizontal (pág. 125) Una función _f_ tiene una asíntota horizontal en _y_ = _b_ si y solo si _f_(_x_) se aproxima a _b_ a medida que |_x_| se hace muy grande.

Horizontal stretch or compression (p. 178) Let _f_ be a function and _k_ a constant. The graph of _y_ = _f_(_kx_) can be obtained from the graph of _f_ by a horizontal stretch if |_k_| > 1 or a horizontal compression if |_k_| < 1. If _k_ < 0, then the graph of _f_ must also be reflected across the _y_-axis to obtain the graph of _y_ = _f_(_kx_).

Compresión o extensión horizontal (pág. 178) Sea _f_ una función y _k_ una constante. La gráfica _y_ = _f_(_kx_) se puede obtener de la gráfica de _f_ por una extensión horizontal si |_k_| > 1 o una compresión horizontal si |_k_| < 1. Si _k_ < 0, entonces la gráfica de _f_ también debe ser reflejada sobre el eje _y_ para obtener la gráfica de _y_ = _f_(_kx_).

English

Horizontal translation (p. 154) A transformation that "slides" all points in the plane the same horizontal distance.

Huffman code (p. 512) A code used in many data compression systems, named for its developer David A. Huffman, generated by an algorithm using a binary tree in which the most frequently occurring character is coded with the shortest code word, it has the minimum average code word length among all prefix-free codes.

If and only if statement A combination of an if-then statement and its converse. In symbols, "p if and only if q" is written as $p \Leftrightarrow q$, and is understood to mean $p \Rightarrow q$ and $q \Rightarrow p$.

If-then statement Frequently used in deductive arguments because if the hypothesis is satisfied then the conclusion follows. If-then statements can be represented symbolically as $p \Rightarrow q$ (read "if p, then q" or "p implies q"), where p represents the hypothesis and q represents the conclusion.

Inclusion-Exclusion principle (p. 448) A principle that describes how to count the number of elements in the union of sets. For example, for three sets,
[# of elements in $A \cup B \cup C$] =
   [# of elements in A] + [# of elements in B]
   + [# of elements in C] − [# of elements in $A \cap B$]
   − [# of elements in $A \cap C$]
   − [# of elements in $B \cap C$]
   + [# of elements in $A \cap B \cap C$].

Incorrect response bias (p. 384) A type of response bias where people do not supply the correct answer because, for example, they do not remember accurately or they do not want the interviewer to know what they really think.

Independent events (pp. 78, 343) Events A and B are independent if $P(A \text{ and } B) = P(A) \cdot P(B)$. Equivalent definitions use the idea that the occurrence of one event does not change the probability that the other occurs:
$P(A) = P(A \mid B)$ if $P(B) \neq 0$ or
$P(B) = P(B \mid A)$ if $P(A) \neq 0$.

Español

Traslación horizontal (pág. 154) Transformación que "deliza" todos los puntos en el plano a la misma distancia horizontal.

Código Huffman (pág. 512) Código que se usa en muchos sistemas de compresión de datos llamado como su desarrollador David A Huffman. El código se genera por un algoritmo que usa un árbol binario en el cual el carácter con la mayor frecuencia de ocurrencia se codifica con la palabra clave más corta. Tiene la menor longitud promedio de palabras clave entre todos los códigos sin prefijo.

Enunciado si y solo si Combinación de un enunciado si-entonces y su recíproco. En símbolos, "p si y solo si q" se escribe $p \Leftrightarrow q$, y se entiende que significa $p \Rightarrow q$ y $q \Rightarrow p$.

Enunciado si-entonces Frecuentemente se usa en los argumentos deductivos porque si se satisface la hipótesis, entonces se llega a la conclusión. Los enunciados si-entonces pueden representarse simbólicamente como $p \Rightarrow q$ (se lee "si p, entonces q" o "p implica q"), donde p representa la hipótesis y q representa la conclusión.

Principio de inclusión-exclusión (pág. 448) Principio que describe cómo contar el número de elementos en la unión de conjuntos. Por ejemplo, para tres conjuntos,
[# de elementos en $A \cup B \cup C$] =
   [# de elementos en A] + [# de elementos en B]
   + [# de elementos en C]
   − [# de elementos en $A \cap B$]
   − [# de elementos en $A \cap C$]
   − [# de elementos en $B \cap C$]
   + [# de elementos en $A \cap B \cap C$].

Sesgo de respuesta incorrecta (pág. 384) Tipo de repuesta sesgada donde las personas no suministran la respuesta correcta porque, por ejemplo, no recuerdan precisamente o no quieren que el entrevistador sepa lo que realmente piensan.

Sucesos independientes (pág. 78, 343) Los sucesos A y B son independientes si $P(A \text{ y } B) = P(A) \cdot P(B)$. Definiciones equivalentes se basan en la idea de que la ocurrencia de uno de los sucesos no cambia la probabilidad de que ocurra el otro:
$P(A) = P(A \mid B)$ si $P(B) \neq 0$ o
$P(B) = P(B \mid A)$ si $P(A) \neq 0$.

English

Independent variables (p. 78) Two categorical variables are independent if $P(A \text{ and } B) = P(A) \cdot P(B)$ for all categories A that make up the first variable and all categories B that make up the second variable. Equivalently, knowing which category an individual falls into on one variable does not help you better predict which category it falls into on a second variable.

Inequality A statement like $3x + 5y < 9$ or $t^2 + 2 \geq 4$ comprised of numbers or algebraic expressions connected by an inequality symbol ($<, \leq, >, \geq$). The solution of an inequality is all values of the variable(s) for which the statement is true.

Information digits (p. 486) The digits in an ID number for an object that are used to classify and identify the object.

Integers mod n (Z_n) (p. 463) The set $Z_n = \{0, 1, 2, \dots, n-1\}$, with addition and multiplication mod n. (Each "number" in Z_n really represents all the integers that reduce to that number mod n. Even so, you can think of the elements of Z_n as the numbers $0, 1, 2, \dots, n-1$.)

Intersection (AND) (p. 436) The Boolean operator AND corresponds to the set operation intersection. The intersection of sets A and B is the set containing all elements that are common to both A and B, denoted $A \cap B$.

Inverse function For a given function f, if there is a function g with domain equal to the range of f, with range equal to the domain of f, and the graph of g is the reflection image of the graph of f across the line $y = x$, then g is the inverse of f, denoted f^{-1}.

Español

Variables independientes (pág. 78) Dos variables categóricas son independientes si $P(A \text{ y } B) = P(A) \cdot P(B)$ para todas las categorías A que componen la primera variable y todas las categorías B que componen la segunda variable. Igualmente, saber en qué categoría cae un individuo en una variable, no ayuda a predecir mejor en qué categoría cae en una segunda variable.

Desigualdad Enunciado como $3x + 5y < 9$ o $t^2 + 2 \geq 4$ compuesto por números o expresiones algebraicas conectadas por un símbolo de desigualdad ($<, \leq, >, \geq$). La solución de una desigualdad son todos los valores de la(s) variable(s) para los cuales se cumple el enunciado.

Dígitos de información (pág. 486) Dígitos en el número de identificación de un objeto que se usan para clasificar o identificar el objeto.

Enteros mod n (Z_n) (pág. 463) Conjunto $Z_n = \{0, 1, 2, \dots, n-1\}$, con adición y multiplicación mod n. (Cada "número" en Z_n realmente representa todos los enteros que se reducen a ese número mod n. Aun así, puedes pensar en los elementos de Z_n como los números $0, 1, 2, \dots, n-1$.)

Intersección (Y) (pág. 436) El operador lógico Y corresponde a la operación de conjuntos intersección. La intersección de los conjuntos A y B es el conjunto que contiene todos los elementos comunes tanto a A como a B, denotado como $A \cap B$.

Función inversa Para cualquier función f, si hay una función g con dominio igual al rango de f, rango igual al dominio de f y la gráfica de g es la imagen reflejada de la gráfica de f sobre la recta $y = x$, entonces g es la inversa de f, denotada como f^{-1}.

J

Judgment sampling (p. 381) A method of obtaining a sample by having a person select individuals that seem fairly typical or that seem to best represent the population. This method often results in sample selection bias. See also size bias.

Muestreo por juicio (pág. 381) Método de obtención de muestras en el cual una persona elige individuos que parecen razonablemente típicos o que parecen representar mejor a la población. Este método a menudo produce un sesgo en la selección de la muestra. Ver también sesgo por tamaño.

English | L | Español

Lease (p. 316) An agreement whereby a person uses (borrows) an object from the owner of the object for a fixed period of time and agreed-upon amount of money.

Level One of the values or categories making up a categorical variable.

Linear code (p. 496) A set of binary code words such that the sum of any two code words in the set (including adding a code word to itself) is also a code word in the set.

Linear function (p. 124) A function with rule of correspondence that can be expressed in the algebraic form $f(x) = mx + b$.

Linear programming (p. 577) A mathematical procedure to find values of variables that satisfy a set of linear *constraints* and optimize the value of a linear *objective function*.

Logarithm, natural (p. 294) If $a = e^b$, then b is the natural or base-e logarithm of a. This relationship is often indicated by the notation $\log_e a = b$ or $\ln a = b$.

Lurking variable (p. 17) A variable that helps to explain an association between the treatments and the response, but is not the explanation that the study was designed to test.

Alquiler (pág. 316) Acuerdo mediante el cual una persona usa (pide prestado) un objeto durante un período fijo de tiempo y le paga al dueño una cantidad preestablecida de dinero.

Nivel Uno de los valores o categorías que conforman una variable categórica.

Código lineal (pág. 496) Conjunto de palabras clave binarias, tales que la suma de cualesquiera dos palabras clave en el conjunto (incluyendo sumar una palabra clave a sí misma) también es una palabra clave en el conjunto.

Función lineal (pág. 124) Función con regla de correspondencia que se puede expresar de la forma algebraica $f(x) = mx + b$.

Programación lineal (pág. 577) Procedimiento matemático para hallar valores de variables que satisfacen un grupo de *restricciones* lineales y optimizan el valor de una *función lineal objetivo*.

Logaritmo natural (pág. 294) Si $a = e^b$, entonces b es el logaritmo natural o base-e de a. Esta relación se expresa frecuentemente con la notación $\log_e a = b$ o $\ln a = b$.

Variable latente (pág. 17) Variable que ayuda a explicar una asociación entre los tratamientos y las respuestas, pero no es la explicación para la cual se diseñó el estudio.

M

Margin of error (p. 407) A number that tells you how large the sampling error is likely to be. For example, in 95% of all random samples of size n, the difference between the sample proportion $\hat{p}$ and the population proportion p will be no more than the margin of error of $2\sqrt{\dfrac{\hat{p}(1-\hat{p})}{n}}$.

Marginal total (p. 42) For a frequency table, a row or column sum.

Mathematical modeling (p. 164) The process of formulating and using an appropriate mathematical representation of a real-world problem situation, phenomenon, or data pattern to analyze it, to better understand it, and to help answer related questions. The mathematical representation of the situation is called a *mathematical model*.

Margen de error (pág. 407) Número que indica qué tan grande es posible que sea el error de muestreo. Por ejemplo, en 95% de todas las muestras aleatorias de tamaño n, la diferencia entre la proporción de la muestra $\hat{p}$ y la proporción de la población p, no será más que el margen de error de $2\sqrt{\dfrac{\hat{p}(1-\hat{p})}{n}}$.

Total marginal (pág. 42) En una tabla de frecuencias, la suma de una fila o una columna.

Modelado matemático (pág. 164) Proceso de formular y usar una representación matemática apropiada de una situación de la vida real, un fenómeno o un patrón de datos, para analizarlo, entenderlo mejor y ayudar a contestar preguntas relacionadas. La representación matemática de la situación se llama *modelo matemático*.

English	Español

Maximum likelihood decoding (p. 493) The method of decoding used when transmitting data with a binary code whereby a received code word is decoded to the code word that is most likely to have been transmitted, which is taken to be the code word that differs from it in the fewest number of bit places.

Decodificación de máxima verosimilitud (pág. 493) Método de descifrado que se usa cuando se transmiten datos con un código binario gracias al cual una palabra clave recibida se descifra y se convierte en la palabra clave que más probablemente se haya transmitido, la cual se considera como la que difiere de ella en el menor número de posiciones bits.

Measurement error (p. 384) A type of response bias where a mistake is made in recording or analyzing the results of a survey.

Error de medición (pág. 384) Tipo de sesgo de respuestas donde se comete un error al registrar o analizar los resultados de una encuesta.

Midpoint formula in three dimensions (p. 556) Formula for calculating the midpoint of a segment with endpoints (x_1, y_1, z_1) and (x_2, y_2, z_2)

Fórmula del punto medio en tres dimensiones (pág. 556) Fórmula para calcular el punto medio de un segmento con extremos (x_1, y_1, z_1) y (x_2, y_2, z_2)

Minimum payment balance formula for credit cards (p. 307) *balance after n minimum payments =* *initial balance* $\times \left(\left(1 + \frac{r}{12}\right)(1 - m) \right)^n$

Fórmula para el saldo de pagos mínimos para tarjetas de crédito (pág. 307) *saldo después de n pagos mínimos* *= saldo inicial* $\times \left(\left(1 + \frac{r}{12}\right)(1 - m) \right)^n$

Mortgage (p. 310) A legal agreement in which a person borrows money to buy property (such as a house) and pays back the money with interest over a period of years.

Hipoteca (pág. 310) Acuerdo legal en el cual una persona pide dinero prestado para comprar una propiedad (como una casa) y paga el dinero con intereses durante cierto período de años.

Multiplication Principle of Counting (p. 208) To count all the outcomes from a sequence of tasks, count how many outcomes there are from each task and multiply those numbers together. More precisely, if there are distinct combined outcomes from a finite sequence of tasks, and if the number of outcomes from each task is finite and the same no matter which outcomes happened in previous tasks, and if the first task in the sequence has n_1 outcomes and for each of these the second task has n_2 outcomes and for each of these the third task has n_3 outcomes and so on, then the number of possible combined outcomes from the whole sequence of tasks is $n_1 \times n_2 \times n_3 \times \cdots$.

Principio de conteo de la multiplicación (pág. 208) Para contar todos los resultados de una sucesión de tareas, se cuenta cuántos resultados hay de cada tarea y se multiplican esos números. Con más detalle, si hay resultados combinados diferentes de una sucesión finita de tareas y si el número de resultados de cada tarea es finito e igual sin importar qué resultados sucedan en tareas anteriores, y si la primera tarea en la sucesión tiene n_1 resultados y para cada uno de estos la segunda tarea tiene n_2 resultados y para cada uno de estos la tercera tarea tiene n_3 resultados y así sucesivamente, el número de resultados posibles combinados para toda la sucesión de tareas es $n_1 \times n_2 \times n_3 \times \cdots$.

Multiplication Rule for Independent Events (pp. 244, 343) If A and B are independent events, then, by definition, the probability that both occur is $P(A \text{ and } B) = P(A) \times P(B)$.

Regla de multiplicación para sucesos independientes (pág. 244, 343) Si A y B son sucesos independientes, entonces, por definición, la probabilidad de que ambos ocurran es $P(A \text{ y } B) = P(A) \times P(B)$.

Multiplication Rule for Probability (general) (p. 245) If A and B are two events, $P(A \text{ and } B) = P(A) \times (B \mid A)$.

Regla de multiplicación para probabilidad (general) (pág. 245) Si A y B son dos sucesos, $P(A \text{ y } B) = P(A) \times P(B \mid A)$.

English	Español
Multiply mod *n* (p. 463) Multiply integers as usual as real numbers, then reduce mod *n*.	**Multiplicación mod *n*** (pág. 463) Multiplicar enteros al igual que se multiplican números reales, luego reducir mod *n*.
Mutually exclusive (disjoint) events (p. 344) Two events that cannot happen on the same trial.	**Sucesos (disjuntos) mutuamente exclusivos** (pág. 344) Dos sucesos que no pueden suceder en el mismo ensayo.

N

English	Español
Negative predictive value (NPV) (p. 72) Of the people who get a negative test result, the proportion who do not have the condition: $NPV = P(\text{condition absent} \mid \text{negative test})$	**Valor negativo predecible (VNP)** (pág. 72) De las personas que obtienen un resultado negativo en una prueba, la proporción de los que no padecen la condición: $VNP = P(\text{condición ausente} \mid \text{prueba negativa})$
Nonresponse bias (p. 384) A type of response bias where the people who have been selected for the sample but cannot be contacted or refuse to answer are different in a crucial way from those who give responses.	**Sesgo por falta de respuesta** (pág. 384) Tipo de respuesta sesgada donde las personas que han sido electas para la muestra pero que no se han podido contactar o rehúsan contestar, son significativamente diferentes de aquellas personas que sí responden.
Normal distribution A theoretical probability distribution that is bell-shaped, symmetric, continuous, and defined for all real numbers.	**Distribución normal** Distribución de probabilidad teórica con forma de campana, simétrica, continua y definida para todos los números reales.

O

English	Español
Objective function (p. 577) In a linear programming problem, an algebraic expression like $7x - 3y$ whose value is to be optimized within the constraints of the problem.	**Función objetiva** (pág. 577) En un problema de programación lineal, es una expresión algebraica como $7x - 3y$ cuyo valor debe ser optimizado dentro de las restricciones del problema.
Observed frequency (or, **observed count**) *O* (p. 43) The number of individuals in the sample that fall into a given category.	**Frecuencia observada** (o, **conteo observado**) *O* (pág. 43) Número de individuos en la muestra que pertenecen a una categoría dada.
One-to-one function (p. 132) A function with the property that for any value *r* in the range, there is exactly one *x* in the domain such that $f(x) = r$.	**Función inyectiva** (pág. 132) Función con la propiedad que para cualquier valor *r* en el rango, hay exactamente una *x* en el dominio, tal que, $f(x) = r$.

P

English	Español
Parameter (p. 377) A numerical characteristic of a population, such as the mean or the proportion of successes.	**Parámetro** (pág. 377) Característica numérica de una población, como la media o la proporción de éxitos.
Percentile A value *x* in a distribution lies at the *p*th percentile if *p*% of the values in the distribution are less than or equal to *x*.	**Percentil** Un valor *x* de una distribución se encuentra en el percentil *p*-ésimo si *p*% de los valores de la distribución son menores que o iguales a *x*.
Period (p. 117) The length of the smallest interval (in the domain) of a periodic function that contains one cycle of its graph.	**Período** (pág. 117) Longitud del menor intervalo (en el dominio) de una función periódica que contiene un ciclo de su gráfica.

689

English

Period of a circular function (p. 117) The length of a smallest interval (in the domain) that corresponds to a portion of the graph from one point to the point at which the graph starts repeating itself.

Periodic graph (p. 117) A graph of a pattern of change that repeats itself over and over again.

Permutation (p. 223) An arrangement of objects or numbers in which order matters, that is, different orderings are counted as different possibilities, and repetitions are not allowed.

Piecewise-defined function (p. 131) A function that is written using two or more expressions.

Placebo (p. 15) In an experiment, a fake treatment that has no medical value, but looks like a real treatment to the person receiving it.

Placebo effect (p. 15) The tendency for people to do better when they are given special attention or when they believe they are getting competent medical care even if they are receiving a treatment of no medical value.

Plane, equation of (p. 558) The three-dimensional analogue of a line. The standard form equation of a plane is $Ax + By + Cz = D$, where A, B, C, and D are real numbers, not all zero.

Plausible (p. 40) A hypothesis or guess about a characteristic of a population (or populations) is called plausible if it is reasonably consistent with a known random sample taken from that population (or, with known random samples taken from the populations).

Political apportionment (p. 647) Used in this text with specific reference to the apportionment of seats in the U.S. House of Representatives.

Population (pp. 39, 376) The entire group of people or things of interest.

Positive predictive value (PPV) (p. 72) Of the people who get a positive test result, the proportion who have the condition:
$PPV = P(condition\ present \mid positive\ test)$

Power function (p. 124) A function with rule of correspondence that can be expressed in the form $f(x) = ax^r$, $r \neq 0$.

Español

Período de una función circular (pág. 117) Longitud del menor intervalo (en el dominio) que corresponde a una porción de la gráfica desde un punto al punto en el cual la gráfica comienza a repetirse.

Gráfica periódica (pág. 117) Gráfica de un patrón de cambio que se repite una y otra vez.

Permutación (pág. 223) Distribución de objetos o números en la cual el orden sí importa. Es decir, los órdenes diferentes se cuentan como posibilidades diferentes y no se permiten las repeticiones.

Función definida por tramos (pág. 131) Una función que está escrita usando dos o más expresiones.

Placebo (pág. 15) En un experimento, tratamiento falso que no tiene valor médico pero que le parece un tratamiento real a la persona que lo recibe.

Efecto placebo (pág. 15) Tendencia de las personas a sentirse mejor cuando reciben atención especial o cuando creen que reciben cuidados médicos competentes aunque reciban tratamiento sin ningún valor médico.

Ecuación del plano (pág. 558) Analogía tridimensional de una recta. La forma estándar de la ecuación del plano es $Ax + By + Cz = D$, donde A, B, C y D son números reales, no todos cero.

Verosímil (pág. 40) Una hipótesis o conjetura sobre una característica de una población (o poblaciones) es verosímil si es razonablemente consistente con una muestra aleatoria conocida tomada de esa población (o con muestras aleatorias conocidas tomadas de las poblaciones).

Repartición política (pág. 647) Se usa en este texto haciendo referencia específica a la distribución de escaños en la Cámara de Representantes de los Estados Unidos.

Población (pág. 39, 376) Todo el grupo de personas o cosas de interés.

Valor positivo predecible (VPP) (pág. 72) De las personas que obtienen un resultado positivo en una prueba, la proporción que padece la condición:
$VPP = P(condición\ presente \mid prueba\ positiva)$

Función de potencia (pág. 124) Función con regla de correspondencia que se puede expresar de la forma $f(x) = ax^r$, $r \neq 0$.

English

Prefix-free code (p. 512) A code in which no code word is the prefix of any other code word, with the properties that it is made up of code words of different lengths (it is a variable-length code), every encoded message can be decoded to just one unique original message (it is uniquely decodable), and decoding requires no false starts or backtracking.

Present value of an annuity (p. 326) The balance of funds in an annuity at the time of retirement (when contributions stop and payout begins).

Principal (p. 268) In finance, the amount of money borrowed or invested.

Prismoidal formula (p. 570) The volume of any three-dimensional figure bounded by a quadratic surface and two parallel plans can be calculated using the prismoidal formula $V = \frac{B + 4M + T}{6}h$, where B is the area of the cross section at the base, M is the area of the cross section at the middle, T is the area of the cross section at the top, and h is the height of the figure.

Probability distribution (p. 342) A description of all possible quantitative (numerical) outcomes of a chance situation, along with the probability of each outcome.

Probability with equally likely outcomes (p. 242) If there are a finite number of possible outcomes, all of which are equally likely, then the probability of event A is given by

$$P(A) = \frac{number\ of\ outcomes\ corresponding\ to\ event\ A}{total\ number\ of\ possible\ outcomes}$$

Public-key cryptosystem (p. 467) A system used to securely send confidential information by encrypting and decrypting the information using keys, in which an individual creates a pair of related keys—a public key, which can be shared with everyone, and a private key, which is kept secret and is known only to the individual.

Español

Código sin prefijo (pág. 512) Código en el cual ninguna palabra clave es el prefijo de cualquier otra palabra clave, con las propiedades de que está formada por palabras clave de distintas longitudes (es un código de longitud variable), cada mensaje cifrado se puede descifrar en un único mensaje original (unívocamente descifrable) y el descifrado no requiere intentos fallidos o algoritmos de marcha atrás.

Valor presente de una anualidad (pág. 326) Saldo de fondos en una anualidad en el momento de la jubilación (cuando se dejan de hacer contribuciones y se comienzan a recibir los pagos).

Capital (pág. 268) En finanzas, la cantidad de dinero que se pide prestado o se invierte.

Fórmula del volumen de un prismatoide (pág. 570) El volumen de cualquier figura tridimensional delimitada por una superficie cuadrática y dos planos paralelos se puede calcular usando la fórmula del volumen de un prismatoide $V = \frac{B + 4M + T}{6}h$, donde B es el área de la sección transversal en la base, M es el área de la sección transversal en el medio, T es el área de la sección transversal en la parte superior y h es la altura de la figura.

Distribución de probabilidad (pág. 342) Descripción de todos los resultados cuantitativos posibles (numéricos) de una situación aleatoria, conjuntamente con la probabilidad de cada resultado.

Probabilidad con resultados equiprobables (pág. 242) Si hay un número finito de resultados posibles, todos equiprobables, entonces la probabilidad del suceso A viene dada por

$$P(A) = \frac{número\ de\ resultados\ correspondientes\ al\ suceso\ A}{número\ total\ de\ resultados\ posibles}$$

Sistema encriptado de clave pública (pág. 467) Sistema que se usa para enviar información confidencial de manera segura cifrando y descifrando la información con el uso de claves, para las cuales un individuo crea un par de claves relacionadas: una clave pública, que se puede compartir con todos y una clave privada, que se mantiene en secreto y solo la conoce el individuo.

English

P-value for a chi-square test (p. 94) In a test of homogeneity, the probability of getting a value of χ^2 as large as or larger than the one computed from your samples if they had been taken at random from two homogeneous populations. In a test of independence, the probability of getting a value of χ^2 as large as or larger than the one computed from your sample if it had been taken at random from a population where the two variables are independent. The larger the value of χ^2, the smaller the P-value and the more evidence against the hypothesis of homogeneous populations or independent variables.

P-value for a test of a proportion (p. 356) The probability of getting, just by chance, the number of successes or even more (or, in some scenarios, even fewer) than the number in the actual sample if you, instead, had conducted n binomial trials using the assumed or hypothesized probability of a success p. When the P-value is very small, then it is not plausible that p is the correct probability of a success in the population from which the actual sample was taken.

Pythagorean Theorem If the lengths of the sides of a right triangle are a, b, and c with the side of length c opposite the right angle, then $a^2 + b^2 = c^2$.

Q

Quadratic formula The formula $x = \dfrac{-b \pm \sqrt{b^2 - 4ac}}{2a}$ that gives the solutions of any quadratic equation in the form $ax^2 + bx + c = 0$, where a, b, and c are constants and $a \neq 0$.

Quadratic function (p. 124) A function with rule of correspondence that can be expressed in the form $f(x) = ax^2 + bx + c$, $(a \neq 0)$.

Questionnaire bias (p. 384) A type of response bias where the question is worded so as to lead people to one particular answer or worded so that people in the sample misunderstand the question.

R

Radian (p. 66) The measure of a central angle of a circle that intercepts an arc equal in length to the radius of the circle. One radian equals $\dfrac{180}{\pi}$ degrees, which is approximately 57.2958°.

Español

Valor de P para una prueba chi-cuadrado (pág. 94) En una prueba de homogeneidad, la probabilidad de obtener un valor de χ^2 igual o mayor que el calculado a partir de las muestras, si se han tomado al azar, de dos poblaciones homogéneas. En una prueba de independencia, la probabilidad de obtener un valor de χ^2 igual o mayor que el calculado a partir de la muestra tomada al azar de una población donde las dos variables son independientes. Mientras mayor sea el valor de χ^2, menor el valor de P y más evidencia contra la hipótesis de poblaciones homogéneas o variables independientes.

Valor de P para una prueba de una proporción (pág. 356) Probabilidad de obtener, solo por azar, el número de éxitos o hasta más (en algunos casos hasta menos) que el número en la muestra real si, en cambio, se hubiesen realizado n ensayos binomiales usando la probabilidad asumida o hipotética de un éxito p. Cuando el valor P es muy pequeño, no es verosímil que p sea la probabilidad correcta de un éxito en la población de la cual se tomó la muestra en sí.

Teorema de Pitágoras Si las longitudes de los lados de un triángulo rectángulo son a, b y c con el lado de longitud c opuesto al ángulo recto, entonces $a^2 + b^2 = c^2$.

Fórmula cuadrática La fórmula $x = \dfrac{-b \pm \sqrt{b^2 - 4ac}}{2a}$ que da las soluciones de cualquier ecuación cuadrática en la forma $ax^2 + bx + c = 0$, donde a, b y c son constantes y $a \neq 0$.

Función cuadrática (pág. 124) Función con regla de correspondencia que se puede expresar de la forma $f(x) = ax^2 + bx + c$, $(a \neq 0)$.

Sesgo del cuestionario (pág. 384) Tipo de sesgo de respuesta donde la pregunta se redacta para llevar a las personas hacia una respuesta en particular o para que las personas en la muestra interpreten mal la pregunta.

Radián (pág. 66) Medida del ángulo central de un círculo que intercepta un arco de igual longitud que el radio del círculo. Un radián equivale a $\dfrac{180}{\pi}$ grados, lo cual es aproximadamente 57.2958°.

English

Random assignment (p. 14) In an experiment, using chance to assign the treatments to the available subjects.

Random sampling (or, **simple random sampling**) (p. 378) A process equivalent to writing the names of all individuals in the population on identical slips of paper, mixing them up, and then selecting slips of paper at random. In a random sample of size n, each possible sample of n individuals has an equal chance of being the sample.

Ranked-choice voting (p. 608) A method of voting in which voters rank the candidates in order of preference.

Recursion A method, or description of a method, whereby a given step in a process is described in terms of the previous step(s).

Recursive formula A formula involving *recursion*. (Also called a recurrence relation or a difference equation.)

Recursive formula for a sequence (p. 136) A formula that expresses a given term in a sequence as a function of previous terms and possibly a constant.

Reduce mod n (p. 463) To reduce an integer mod n means to replace the integer by its remainder upon division by n.

Reflection across a line A transformation in which each point of the original figure (preimage) has an image that is the same perpendicular distance from the line of reflection as the original point but is on the opposite side of the line.

Relative risk (p. 9) The ratio of the absolute risk for two groups.

Response bias (p. 384) A method of getting a response from the sample that makes the estimate from repeated sampling larger or smaller than the population parameter, on average.

Response rate (p. 378) Percentage of individuals contacted to be in the sample who gave a response.

Response variable (p. 14) The outcome being measured in an experiment or observational study.

Español

Asignación aleatoria (pág. 14) En un experimento, usar el azar para asignar los tratamientos a los sujetos disponibles.

Muestreo aleatorio (o, **muestreo aleatorio simple**) (pág. 378) Proceso equivalente a escribir los nombres de todos los individuos en la población en tiras de papel idénticas, mezclarlas y elegir tiras al azar. En la selección de una muestra aleatoria de tamaño n, cada muestra posible de n individuos tiene igual oportunidad de ser la muestra.

Votación de selección clasificada (pág. 608) Método de votación en el cual los votantes clasifican a los candidatos en orden de preferencia.

Recursión Método o descripción de un método mediante el cual un paso dado en un proceso se describe en términos del paso o pasos previos.

Fórmula recurrente Fórmula que involucra *recursión*. (También llamada relación recurrente o ecuación de diferencia).

Fórmula recurrente de una sucesión (pág. 136) Fórmula que expresa un término dado en una sucesión como una función de términos previos y posiblemente una constante.

Reducir mod n (pág. 463) Reducir un entero mod n significa remplazar el entero por su residuo cuando se divide entre n.

Reflexión a través de una recta Transformación en la cual cada punto de la figura original (pre-imagen) tiene una imagen que está a la misma distancia perpendicular de la línea de reflexión que el punto original pero en el lado opuesto de la recta.

Riesgo relativo (pág. 9) La proporción del riesgo absoluto para dos grupos.

Sesgo de respuestas (pág. 384) Método para obtener una respuesta de la muestra que hace que la estimación de muestreos repetitivos sea, en promedio, más grande o más pequeña que el parámetro de la población.

Tasa de respuestas (pág. 378) Porcentaje de individuos contactados para estar en la muestra que dieron una respuesta.

Variable de respuestas (pág. 14) Resultado que se mide en un experimento o en un estudio por observación.

English

Rigid transformation A transformation of points in the plane that preserves all distances. Such a transformation repositions a figure in a plane without changing its shape or size.

RSA public-key cryptosystem (p. 468) A particular public-key cryptosystem, named after the developers Ronald L. Rivest, Adi Shamir, and Leonard Adleman, in which the keys are constructed using large prime numbers and modular arithmetic.

S

Sample (p. 39) A group of people or things taken from a specified population.

Sample selection bias (p. 383) A method of selecting a sample that makes the estimate from repeated sampling larger or smaller than the population parameter, on average.

Sample survey (p. 376) A poll or survey that collects information from only some of the individuals in a population.

Sampling (p. 345) Selecting a sample from a population.

Sampling distribution (p. 401) A display of the estimates of the population parameter from repeated random sampling. Can be exact if all possible random samples are included or approximate if constructed using a sample of random samples.

Sampling error (or, **chance error** or **error attributable to sampling**) (p. 401) The absolute value of the difference between the estimate from the sample and the population parameter. A larger sample size tends to result in a smaller sampling error.

Sampling variability (or, **variability in sampling**) (p. 402) The fact that the estimate of the population parameter varies from random sample to random sample.

Sampling with replacement (p. 345) Each sampled individual is replaced in the population before the next is drawn. Thus, probabilities do not change from draw to draw.

Español

Transformación rígida Transformación de los puntos en el plano que conserva todas las distancias. Tal transformación reposiciona una figura en un plano sin cambiar el tamaño o la forma de la figura.

Sistema encriptado RSA de clave pública (pág. 468) Sistema encriptado de clave pública particular nombrado en honor a sus desarrolladores Ronald L. Rivest, Adi Shamir y Leonard Adleman, en el cual las claves se construyen con números primos grandes y aritmética modular.

Muestra (pág. 39) Grupo de personas o cosas tomadas de una población en particular.

Sesgo en selección de muestras (pág. 383) Método de selección de una muestra que hace que la estimación de muestreos repetitivos sea, en promedio, mayor o menor que el parámetro de la población.

Encuesta por muestreo (pág. 376) Sondeo o cuestionario que reúne información solo de algunos individuos en una población.

Muestreo (pág. 345) Elegir una muestra de una población.

Distribución muestral (pág. 401) Representación de las estimaciones del parámetro de la población a partir del muestreo aleatorio repetido. Puede ser exacta si todas las muestras aleatorias posibles se incluyen o aproximada cuando se construye usando una muestra de muestras aleatorias.

Error muestral (o, **error aleatorio** o **error atribuible al muestreo**) (pág. 401) Valor absoluto de la diferencia entre la estimación de la muestra y el parámetro de la población. Un mayor tamaño de la muestra tiende a dar como resultado un error de muestreo menor.

Variabilidad muestral (o **variabilidad en muestreo**) (pág. 402) El hecho de que la estimación de los parámetros de la población varíe de muestra aleatoria a muestra aleatoria.

Muestreo con reemplazo (pág. 345) Cada individuo de la muestra se reemplaza en la población antes de sacar el próximo. De esta manera las probabilidades no cambian de selección en selección.

English

Sampling without replacement (p. 345) Each sampled is not replaced in the population before the next is drawn. Thus, probabilities can change from draw to draw.

Sensitivity of a screening test (p. 72) Of the people who have the condition, the proportion who get a positive test result: $P(positive\ test\ |\ condition\ present)$.

Set (p. 230) A well-defined unordered collection of objects.

Set difference (NOT) (p. 438) The Boolean operator NOT corresponds to the set operation set difference. The set difference of sets A and B is the set containing all elements that are in A and not in B, which is denoted $A - B$ or $A\backslash B$.

Similarity transformation Composite of a size transformation and a rigid transformation. Such a transformation resizes a figure in a plane without changing its shape.

Simple interest (pp. 35, 268) The amount of simple interest earned on a principal amount P can be calculated by using the formula $I = Prt$, where r is the annual interest rate expressed as a decimal and t is the time of the investment. The time units for r and t must agree.

Single blind (p. 15) An experiment where the subject does not know which treatment he or she is getting. That is, subjects in all treatment groups appear to be treated exactly the same way.

Single-digit substitution error (p. 488) An error in an ID number in which one digit is incorrect while all others are correct.

Size bias (p. 383) A type of sample selection bias where the method of sampling causes larger individuals to be overrepresented in the sample.

Specificity of a screening test (p. 72) Of the people who do not have the condition, the proportion who get a negative test result: $P(negative\ test\ |\ condition\ absent)$.

Español

Muestreo sin reemplazo (pág. 345) Cada individuo de la muestra no se reemplaza en la población antes de sacar el próximo. De esta manera las probabilidades pueden cambiar de selección en selección.

Sensibilidad de una prueba de detección (pág. 72) De las personas que padecen la condición, la proporción que obtiene un resultado positivo de la prueba: $P(prueba\ positiva\ |\ condición\ presente)$.

Conjunto (pág. 230) Colección de objetos, sin orden y bien definida.

Diferencia de conjuntos (NO) (pág. 438) El operador lógico NO corresponde a la operación de conjuntos diferencia de conjuntos. La diferencia de los conjuntos A y B es el conjunto que contiene todos los elementos que pertenecen a A y no a B; se denota como $A - B$ o $A\backslash B$.

Transformación de semejanza Combinación de una transformación de tamaño y una transformación rígida. Tal transformación cambia las medidas de una figura en el plano sin cambiar su forma.

Interés simple (pág. 35, 268) La cantidad de interés simple ganado sobre un capital P se puede calcular con la fórmula $I = Prt$, donde r es la tasa de interés anual expresada como decimal y t es el tiempo de la inversión. Las unidades de tiempo de r y t deben concordar.

Experimento a ciegas (pág. 15) Experimento donde el sujeto no sabe qué tratamiento recibe. Es decir, los sujetos en todos los grupos de tratamiento parecen recibir exactamente el mismo tratamiento.

Error de sustitución de un dígito (pág. 488) Error en un número de identificación en el cual un dígito es incorrecto mientras que todos los demás son correctos.

Sesgo por tamaño (pág. 383) Tipo de sesgo en selección de muestras donde el método de muestreo hace que individuos más grandes estén sobrerepresentados en la muestra.

Especificidad de una prueba de detección (pág. 72) De las personas que no padecen la condición, la proporción que recibe un resultado negativo de la prueba: $P(prueba\ negativa\ |\ condición\ ausente)$.

English

Sphere (p. 555) The set of points in three dimensions that are a fixed distance r, called the *radius*, from a given point O, called the *center*. A sphere with center at the origin and radius r can be represented by an equation of the form $x^2 + y^2 + z^2 = r^2$, where r is a positive number.

Stacked bar graph (or, **segmented bar graph**) (p. 6) A type of bar graph where each group is represented by a single bar, divided proportionally according to the frequency (or percentage) of the group that falls into each category.

Standard deviation, s (p. 59) A measurement of how much the values in a distribution vary from their mean, based on the sum of the squared deviations from the mean: $s = \sqrt{\dfrac{\Sigma (x - \bar{x})^2}{n - 1}}$

Statistical significance for a chi-square test (p. 48) Getting a value of χ^2 that is large enough to reasonably conclude, in a test of homogeneity, that the populations from which the random samples were taken are not homogeneous or, in a test of independence, that the variables are not independent in the population from which the random sample was taken. In an experiment, the conclusion that it is not reasonable to attribute the difference in the response between treatment groups solely to the particular random assignment of treatments to subjects.

Statistical significance for a test of a proportion (p. 356) A result from a sample that gives a small P-value. That is, the number of successes from the sample or a number even more extreme has a low probability of happening just by chance if the assumed or hypothesized proportion p of successes in the population is correct. Thus it is reasonable to conclude that the assumed or hypothesized value of p is not correct.

Strata (p. 378) Division of a population into non-overlapping groups of similar individuals, such as men and women or freshmen, sophomores, juniors, and seniors.

Español

Esfera (pág. 555) Conjunto de puntos en tres dimensiones que están a una distancia fija r llamada *radio*, de un punto dado O, llamado *centro*. Una esfera con centro en el origen y radio r se puede representar por una ecuación de la forma $x^2 + y^2 + z^2 = r^2$, donde r es un número positivo.

Gráfica de barras agrupadas (o **gráfica de barras segmentadas**) (pág. 6) Tipo de gráfica de barras donde cada grupo está representado por una sola barra, dividida proporcionalmente según la frecuencia (o porcentaje) de los grupos pertenecientes a cada categoría.

Desviación estándar, s (pág. 59) Medida que muestra cuánto varían los valores en una distribución respecto a su media, en base a la suma de las desviaciones elevadas al cuadrado respecto a la media: $s = \sqrt{\dfrac{\Sigma (x - \bar{x})^2}{n - 1}}$

Significación estadística para una prueba chi-cuadrado (pág. 48) Obtener un valor de χ^2 suficientemente grande para concluir razonablemente, en una prueba de homogeneidad, que las poblaciones de las cuales se tomaron las muestras aleatorias no son homogéneas o, en una prueba de independencia, que las variables no son independientes en la población de la cual se tomó la muestra aleatoria. En un experimento, la conclusión de que no es razonable atribuirle la diferencia en la respuesta entre grupos de tratamiento únicamente a la asignación aleatoria específica de los tratamientos a los sujetos.

Significación estadística de una prueba de proporciones (pág. 356) Resultado de una muestra que produce un valor P pequeño. Es decir, el número de éxitos de la muestra o un número aún más extremo tiene una probabilidad baja de que suceda solo por azar si la proporción p asumida o hipotética de éxitos en la población es correcta. Por lo tanto, es razonable concluir que el valor de p asumido o hipotético, no es correcto.

Estratos (pág. 378) División de una población en grupos no solapados de individuos semejantes, como hombres y mujeres o estudiantes de primer año, segundo año, penúltimo año y último año.

English

Stratified random sampling (p. 378) Selecting a sample by first dividing the population into strata. Then a random sample is taken from each stratum. The sizes of the random samples usually are proportional to the sizes of the strata.

Subjects (p. 14) In an experiment, the available group of people (or animals, plants, or things) to whom treatments are randomly assigned.

Subset of a set (pp. 230, 442) A set A is a subset of set B, denoted $A \subseteq B$, if and only if every element of A is also an element of B.

Symmetric-key cryptosystem (p. 457) A system used to securely send confidential information by encrypting and decrypting the information using keys, in which the same key is used to encrypt and decrypt (the exact same key might not be used, but at least it is easy to calculate the keys from each other).

T

Topographic profile (p. 536) A vertical cross-section view along a line drawn across a portion of a map.

Traces (p. 558) Cross sections of a surface formed by the intersection of the coordinate planes with the surface.

Transformation A correspondence between all points in a plane such that (1) each point in the plane has a unique image point in the plane, and (2) for each point in the plane, there is a point that is the preimage of the given point.

Transposition error (p. 490) An error in an ID number in which two digits are transposed.

Treatments (p. 14) In an experiment, the conditions to be randomly assigned to the subjects so their responses can be compared.

Trials, binomial (p. 341) A set of independent trials where the result of each trial is a "success" or a "failure" and the probability of a success is the same on each trial.

Español

Muestreo aleatorio estratificado (pág. 378) Elegir una muestra dividiendo primero la población en estratos. Luego, se toma una muestra aleatoria de cada estrato. Los tamaños de las muestras aleatorias normalmente son proporcionales a los tamaños de los estratos.

Sujetos (pág. 14) En un experimento, el grupo disponible de personas (o animales, plantas o cosas) a quienes se les asignan tratamientos aleatoriamente.

Subconjunto de un conjunto (pág. 230, 442) Un conjunto A es un subconjunto del conjunto B, denotado $A \subseteq B$, si y solo si cada elemento de A también es un elemento de B.

Sistema encriptado de clave simétrica (pág. 457) Sistema que se usa para enviar información confidencial de manera segura cifrando y descifrando la información con el uso de claves, en el cual se usa la misma clave para cifrar y descifrar. Quizás no se use exactamente la misma clave, pero por lo menos es fácil diferenciar las claves entre sí mediante el cálculo.

Perfil topográfico (pág. 536) Vista vertical de una sección transversal a lo largo de una recta dibujada a través de una parte de un mapa.

Trazas (pág. 558) Secciones transversales de una superficie formada por la intersección de planos de coordenadas con la superficie.

Transformación Correspondencia entre todos los puntos en un plano tal que (1) cada punto en el plano tiene un único punto imagen en el plano y (2) para cada punto en el plano, existe un punto que es preimagen del punto dado.

Error de transposición (pág. 490) Error en un número de identificación en el cual se transponen dos dígitos.

Tratamientos (pág. 14) En un experimento, las condiciones a ser aleatoriamente asignadas a los sujetos de manera que las respuestas se puedan comparar.

Ensayos binomiales (pág. 341) Conjunto de ensayos independientes donde el resultado de cada uno es "éxito" o "fracaso" y la probabilidad de éxito es la misma en cada ensayo.

English	Español

Trials, independent (p. 341) Identical chance situations where the result of one does not change the probability of any outcome on another.

Ensayos independientes (pág. 341) Situaciones de posibilidad idénticas donde el resultado de una no cambia la probabilidad de cualquier resultado en la otra.

Two-way frequency table (p. 6) A table that summarizes how many individuals fall into each pair of categories for two categorical variables. For example, groups (explanatory variable) may define the different columns and the levels of the response variable may define the different rows. For a single group, two different categorical variables define the rows and columns.

Tabla de frecuencias de doble entrada (pág. 6) Tabla que resume cómo se distribuyen varios individuos en cada par de categorias para dos variables categóricas. Por ejemplo, los grupos (variable explicativa) pueden definir las diferentes columnas y los niveles de la variable de respuesta pueden definir las diferentes filas. Para un grupo único, dos variables categóricas diferentes definen las filas y columnas.

U

Union (OR) (p. 436) The Boolean operator OR corresponds to the set operation union. The union of sets A and B is the set containing all elements that are in either A or B or both, denoted $A \cup B$.

Unión (O) (pág. 436) El operador lógico O corresponde a la operación de conjuntos unión. La unión de los conjuntos A y B es el conjunto que contiene todos los elementos que están o en A o en B o en ambos, denotado como $A \cup B$.

Universal set (p. 441) The set of all objects (elements) that are under consideration.

Conjunto universal (pág. 441) Conjunto de todos los objetos (elementos) en consideración.

V

Variability in sampling (p. 402) See *Sampling variability*.

Variabilidad en el muestreo (pág. 402) Ver *Variabilidad de muestreo*.

Venn diagram (p. 435) A diagram, typically intersecting circles, used to represent sets and their relationships to each other.

Diagrama de Venn (pág. 435) Diagrama, típicamente de círculos intersecados, que se usa para representar conjuntos y sus relaciones entre sí.

Verhoeff code (p. 523) A code used for ID numbers, named for its developer Jacobus Verhoeff, in which the check digit for the ID number is computed using the group of symmetries of a regular pentagon, the dihedral group D_{10}.

Código de Verhoeff (pág. 523) Código que se usa para números de identificación, llamado por su desarrollador Jacobus Verhoeff, en el cual el dígito de marca para el número de identificación se calcula a partir del grupo de simetrías de un pentágono regular, el grupo diedral D_{10}.

Vertical asymptote (p. 125) A function f has a vertical asymptote at $x = a$ if and only if $|f(x)|$ increases without bound as x approaches a.

Asíntota vertical (pág. 125) Una función f tiene una asíntota vertical en $x = a$ si y solo si $|f(x)|$ crece sin límite cuando x tiende a a.

Vertical stretch or compression (p. 171) If f is any function and k is a positive constant, then the graph of $y = kf(x)$ is a vertically stretched version of the graph of $y = f(x)$ when $k > 1$, or a vertically compressed version of the graph of $y = f(x)$ when $0 < k < 1$.

Compresión o extensión vertical (pág. 171) Si f es una función y k una constante positiva, entonces la gráfica $y = kf(x)$ es una versión estirada verticalmente de la gráfica $y = f(x)$ cuando $k > 1$, o una versión comprimida verticalmente de la gráfica de $y = f(x)$ cuando $0 < k < 1$.

Vertical translation (p. 145) A transformation that "slides" all points in the plane the same vertical distance.

Traslación vertical (pág. 145) Transformación que "desliza" todos los puntos en el plano a la misma distancia vertical.

English

Español

Voluntary response bias (p. 383) A type of sample selection bias that can result from allowing people to decide on their own whether to be in the sample.

Sesgo de respuesta voluntaria (pág. 383) Tipo de sesgo de selección de una muestra que resulta de permitir que las personas decidan si pertenecerán o no a la muestra.

W

Weighted voting (p. 621) A voting system in which each voter can have more than one vote. For example, a shareholder of a corporation has a number of votes equal to the number of shares owned. The **weight** of a person's vote is the number of votes the person has.

Votación ponderada (pág. 621) Sistema de votación en el cual cada votante puede tener más de un voto. Por ejemplo, un accionista de una corporación tiene un número de votos igual al número de acciones que posee. El **peso** del voto de una persona es el número de votos que posee la persona.